Calculus:

Pure and Applied

**A J Sherlock
E M Roebuck
M G Godfrey**

Millfield

Edward Arnold

© A J Sherlock, E M Roebuck, and M G Godfrey, 1982

First published in 1982
by Edward Arnold (Publishers) Ltd
41 Bedford Square, London WC1B 3DQ

British Library Cataloguing in Publication Data

Sherlock, A. J.
  Calculus.
  1. Calculus
  I. Title   II. Roebuck, E. M.
  III. Godfrey, M. G.
  515   QA303

  ISBN 0–7131–3446–1

Typeset by Macmillan India Ltd, Bangalore

Printed in Great Britain by
Thomson Litho Ltd, East Kilbride, Scotland

Preface

Any situation involving a rate of change is described mathematically in the language of calculus. Thus a knowledge of calculus is a basic necessity for a scientific education and is a prerequisite for a serious study of the physical, biological, environmental, and social sciences, and of economics. It is not surprising, therefore, that all 'A' level mathematics courses (and 'O' level additional mathematics courses) contain a large amount of calculus; nor that it features in many 'O' level syllabuses as an optional topic.

This book has been written by three very experienced teachers for the following main reasons.

(1) With the advent of the electronic calculator, the numerical approach has added a new dimension to the understanding of calculus.

(2) Most calculus texts were written for the very able pupils studying 'A' level mathematics. These texts tended towards brief, abstract explanations of the theory without paying sufficient attention to the basic understanding and techniques. This book is an attempt to rectify this situation.

The ideas of calculus have been developed logically and thoroughly from the start, at a pace which increases as the students deepen and strengthen their basic technique. The calculator is used to demonstrate numerically many of the fundamental processes such as gradient finding, area finding, and evaluating limits. Not only are the principles explained fully, but they are also shown to be plausible. Great attention has been given to detail and the material has been thoroughly tested over several years with students from a wide range of ability. Worked examples cover not only the standard problems but also those in which difficulties are often encountered. There are many exercises of graded questions, revision exercises and sets of questions from most examination boards.

The book covers all 'A' and 'S' level single-subject mathematics syllabuses besides the 'A' level double-subject syllabuses. However, an introduction to analysis has not been included since this, rightly, belongs to a second course.

The text will be found to be of great value to both experienced and inexperienced teachers, but it will be invaluable to the student since it may be used with the minimum of instruction. It will therefore appeal to those studying on their own, at evening classes or under day-release schemes. The effectiveness of the material is attested by the success of the students who have used it.

We should like to thank our colleagues at Millfield for their help and advice: in particular David Elton and Don Gillis who have worked through many of the exercises, and our retired colleague John Carr who has helped with the proof reading. We should also like to thank the following examination boards for permission to use some of their questions: Associated Examining Board (AEB); University of Cambridge Local Examinations Syndicate (C); Joint Matriculation Board (JMB); University of London University Entrance and Schools Examination Council (L); Oxford Delegacy of Local Examinations (O); Oxford and Cambridge Schools Examination Board (O and C, MEI, SMP); Southern Universities' Joint Board (S); and Welsh Joint Education Committee (W). The worked examples on pp. 465, 489 and 491 are entirely the responsibility of the authors and not of the University of London University Entrance and Schools Examination Council. All other worked examples and all answers to exercises are entirely the responsibility of the authors and not of the examination boards.

Millfield   Alan Sherlock
1980   Elizabeth Roebuck
                                                                                   Maurice Godfrey

Contents

Preface   iii

1 Gradients   1
Introduction   1
1.1   Curves and functions   1
1.2   Gradient of a line   4
1.3   Gradient of a curve by drawing   6
1.4   Calculating the gradient of a curve at a point   7
1.5   Formula for the gradient of $y = x^n$ 10
1.6   Formula for the gradient of $y = kx^n$ 12

2 Differentiation   14
2.1   The derived function: differentiation of x^n 14
2.2   Fractional and negative powers   16
2.3   Differentiation of powers   19
2.4   Other notations   21
2.5   Finding the gradient at a point   21
2.6   The gradient function   22
2.7   Working from first principles   25
2.8   When the gradient does not exist   31

3 Applications of differentiation   34
3.1   Repeated differentiation   34
3.2   Curve sketching using the derived function   36
3.3   Stationary points   38
3.4   The derived function as a rate of change   42
3.5   Problems involving maxima and minima   43
3.6   The equation of the tangent to a curve   45
3.7   The equation of the normal to a curve   46

Revision questions A   49
Revision paper A1   49
Revision paper A2   49

4 Areas   51
4.1   Area under a curve, by drawing   51
4.2   Area under a curve, by calculation   52
4.3   Formula for the area under $y = x^n$ 54
4.4   Area under other curves   56

5 The fundamental theorem of calculus   57
  5.1   The area function   57
  5.2   The gradient of the area function   59
  5.3   The fundamental theorem   59
  5.4   The definite integral   63

6 Integration I   66
        Introduction   66
  6.1   The reverse of differentiation   66
  6.2   Simple applications of integration   69
  6.3   Evaluation of definite integrals   72
  6.4   Problems encountered with definite integrals   74

7 Applications of definite integrals   76
  7.1   Area between a curve and the x-axis   76
  7.2   Area between two curves   78
  7.3   Area between a curve and the y-axis   80
  7.4   Average values   82
  7.5   Odd and even functions   83
  7.6   Volumes of revolution about the x-axis   85
  7.7   Rotation about other lines   88

8 Functions given as a table of values   90
        Introduction   90
  8.1   Gradient finding   90
  8.2   Area finding   94
  8.3   Volumes of revolution   100
  8.4   The length of a curve   101
  8.5   Drawing a curve, given a table of values for $\dfrac{dy}{dx}$ 102

Revision questions B   105
  Revision paper B1   105
  Revision paper B2   105
  Miscellaneous revision questions: paper B3   106

9 Algebraic functions   108
  9.1   Sketching graphs of products and quotients   108
  9.2   Differentiation of products   111
  9.3   Differentiation of quotients   113
  9.4   Composite functions   114
  9.5   Differentiation of algebraic functions   117
  9.6   Integration of algebraic functions   119
  9.7   Inverse functions   123
  9.8   Differentiation of x^n 127

10 Circular functions 129

       Introduction 129

   10.1 Sketching graphs and gradient functions 129

   10.2 Differentiation of circular functions 132

   10.3 Integration of circular functions 137

   10.4 Differential equations of the type $\dfrac{d^2 y}{dx^2} = -n^2 y$ 139

   10.5 Inverse circular functions 142

   10.6 Integration using the inverse circular functions 146

11 Motion in a straight line 149

       Introduction 149

   11.1 Displacement, velocity and acceleration 149

   11.2 Application of integration 155

   11.3 Motion in stages 161

   11.4 Oscillations 165

12 Rates of change 170

   12.1 Dependent and independent variables 170

   12.2 Maxima and minima 173

   12.3 Rates of change with time 176

   12.4 Differential equations 181

   12.5 Formation of differential equations 184

   12.6 Problems involving differential equations 188

   12.7 Small changes 191

   12.8 Proportionate and percentage changes 193

   12.9 Application to errors 197

Revision questions C 200

   Revision paper C1 200

   Revision paper C2 201

   Miscellaneous revision questions: paper C3 202

13 Curves 204

   13.1 Stationary points 204

   13.2 Points of inflexion 206

   13.3 Curve sketching 208

   13.4 Implicit relations 218

   13.5 The conics 222

   13.6 Limits by calculator 230

   13.7 Evaluating limits 232

   13.8 Continuity 236

14 Parametric equations 239

   14.1 Sketching curves from parametric equations 239

   14.2 Differentiation 242

   14.3 Tangents and normals 245

   14.4 Areas 247

   14.5 Length of a curve 249

15 Integration II   253
     Introduction   253
  15.1   Integration by substitution   253
  15.2   Definite integrals by substitution   259
  15.3   Integration by inspection   264
  15.4   Integration by parts   265
  15.5   Methods of integration   269

16 Numerical calculus   277
     Introduction   277
  16.1   The numerical solution of equations, using Newton's method   277
  16.2   Numerical integration: Simpson's rule   281
  16.3   Numerical differentiation, and its applications   284
  16.4   The Lagrange interpolating formula   289

Revision questions D   293
  Revision paper D1   293
  Revision paper D2   294
  Miscellaneous revision questions: paper D3   295

17 Infinite series and applications   297
     Introduction   297
  17.1   Polynomials   298
  17.2   The best fitting polynomial at a general point   301
  17.3   Maclaurin's series   303
  17.4   Taylor's series   307
  17.5   Other methods for finding series   309
  17.6   Power series for integrals   314
  17.7   Solution of differential equations   315
  17.8   L'Hôpital's rule for limits   318
  17.9   The accuracy of numerical methods   319

18   The exponential function   326
  18.1   The exponential function   326
  18.2   Sketching exponential functions   329
  18.3   Differentiation of exponential functions   330
  18.4   Integration of exponential functions   331
  18.5   Differential equations of type $\dfrac{dy}{dx} = ky$ 335
  18.6   Maclaurin series for e^x 336
  18.7   Limits involving exponential functions   339

19 Logarithmic functions   343
  19.1   The logarithmic function   343
  19.2   Differentiation of logarithmic functions   345
  19.3   Integration using logarithmic functions   349
  19.4   Maclaurin series for $\ln(1 + x)$ 356
  19.5   Limits involving $\ln x$ 358

20 Differential equations   362

20.1   Introduction   362

20.2   Differential equations of the form $a\dfrac{dy}{dx}+by=f(x)$ (where a and b are constants)   366

20.3   Differential equations with separable variables   372

20.4   The integrating factor method   375

20.5   Use of a substitution   380

20.6   Second order differential equations of the form $a\dfrac{d^2y}{dx^2}+b\dfrac{dy}{dx}+cy=0$ (where a, b and c are constants)   384

20.7   Second order differential equations of the form $a\dfrac{d^2y}{dx^2}+b\dfrac{dy}{dx}+cy=f(x)$ (where a, b and c are constants)   391

20.8   Miscellaneous differential equations   400

21 Applications of differential equations   403

Introduction   403

21.1   Applications in mechanics   403

21.2   Applications in physics   409

21.3   Applications in chemistry and biology   413

21.4   Numerical methods   417

Revision questions E   422

Revision paper E1   422

Revision paper E2   423

Miscellaneous revision questions: paper E3   424

22 Integration III   428

22.1   Systematic integration: standard integrals, and algebraic functions   428

22.2   Systematic integration: circular functions and parts   435

22.3   Systematic integration: further algebraic functions   440

22.4   Miscellaneous integration   442

22.5   Properties of definite integrals   443

22.6   Evaluating definite integrals   448

22.7   Improper integrals   451

22.8   Reduction formulae   453

22.9   Definite integrals with variable limits   456

22.10 Integrals, series and inequalities   459

23 Hyperbolic functions   462

23.1   Definitions and graphs   462

23.2   Formulae and equations   463

23.3   Differentiation and integration   467

23.4   Inverse hyperbolic functions   470

23.5   Maclaurin series   475

24 Curves, areas and surfaces   477

24.1   Length of a curve   477
24.2   The area of a sector   479
24.3   Curvature   481
24.4   Curved surface area of revolution   485
24.5   Centroids   488
24.6   Polar coordinates   491

25 Partial differentiation   494

25.1   Surfaces   494
25.2   Stationary points   498
25.3   Application to inaccuracies in measurement   500

Revision questions F   503

Revision paper F1   503
Revision paper F2   504
Miscellaneous revision questions: paper F3   505

Answers to exercises   509

Index   533

1

Gradients

Introduction

Calculus is a very useful branch of mathematics, with applications in physics, chemistry, biology and every other scientific subject.

There are two basic ideas. The first is the gradient (or slope) of a curve. This enables us to deal with problems involving change, and rates of change. This is useful because many laws of nature refer to rates of change. In Fig. 1.1, the gradient of the curve at A is

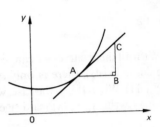

Fig. 1.1

$$\frac{BC}{AB}$$

The second idea is that of the area under a curve. As well as finding areas, this also enables us to find volumes, lengths, centres of mass, and so on. In Fig. 1.2, the area under the curve is the shaded area.

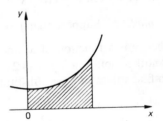

Fig. 1.2

1.1 Curves and functions

The two basic ideas are concerned with curves, so first of all we shall indicate the type of curves we are interested in. We show a curve on a graph, and some typical shapes of curves are illustrated in Figs 1.3 to 1.7.

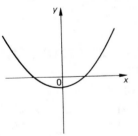

Fig. 1.3

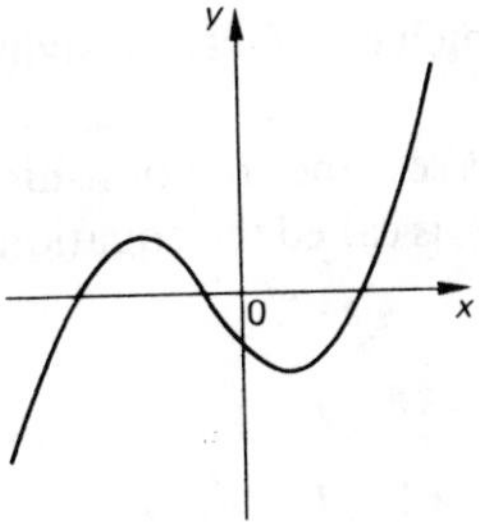

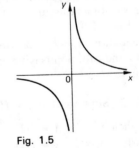

Fig. 1.5

Fig. 1.4

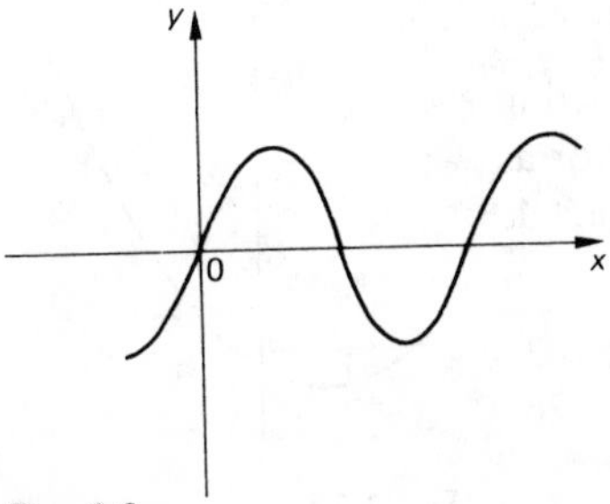

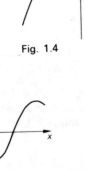

Fig. 1.6

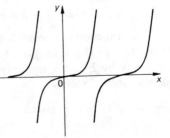

Fig. 1.7

1

We shall also meet graphs which consist of straight lines, like those in Figs 1.8 to 1.10.

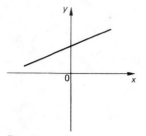

Fig. 1.8

Fig. 1.9

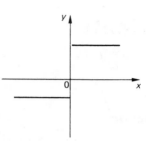

Fig. 1.10

With all these curves we can see that, given a value of x (for which the curve exists), there is only *one* corresponding value of y (see Fig. 1.11). When this is the case, we say that y **is a function of x**, and we may write $y = f(x)$ (said 'y equals f of x').

There may be a simple formula telling us how to calculate the value of y from any given value of x.

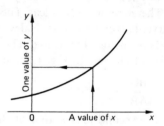

Fig. 1.11

Example 1   Suppose that $y = 2x + 3$.

Then y is a function of x, and we can write $y = f(x) = 2x + 3$. [Another notation is $f: x \to 2x + 3$.] $f(x)$ now stands for $2x + 3$. To find values of y we substitute values for x. For example,

when $x = 1$, $y = f(1) = 2 \times 1 + 3 = 5$
when $x = 2$, $y = f(2) = 2 \times 2 + 3 = 7$
when $x = 3$, $y = f(3) = 2 \times 3 + 3 = 9$

and so on. When we put these points on a graph, and join them together, we obtain the curve $y = 2x + 3$, which is in fact a straight line (see Fig. 1.12).

The equation $y = 2x + 3$, which expresses the relationship between x and y for points lying on the curve, is called the equation of the curve.

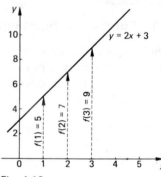

Fig. 1.12

Example 2   Suppose that $y = f(x) = 2x^2 + 3x - 1$.

Then   when $x = 1$, $y = f(1) = 2 \times 1^2 + 3 \times 1 - 1 = 4$
when $x = 2$, $y = f(2) = 2 \times 2^2 + 3 \times 2 - 1 = 13$
when $x = 3$, $y = f(3) = 2 \times 3^2 + 3 \times 3 - 1 = 26$

Calculating values of this function can be made easier by rewriting $2x^2 + 3x - 1$ as $(2x + 3)x - 1$. This is evaluated on a calculator as follows: enter x; multiply by 2; add 3; multiply by x; subtract 1.
We now complete a table of values.

When $x =$	-2	-1.5	-1	-0.5	0	0.5	1	1.5	2	2.5	3
$y =$	1	-1	-2	-2	-1	1	4	8	13	19	26

Fig. 1.13 shows the curve $y = 2x^2 + 3x - 1$.

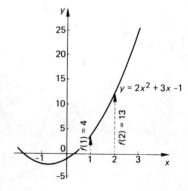

Fig. 1.13

The technique we used to evaluate this function is called **nested multiplication**, or **nesting**. It can be used for any polynomial function $(a + bx + cx^2 + dx^3 + \ldots)$. For example, the function $3x^3 - 5x^2 + 4x - 2$ would be rewritten as $[(3x - 5)x + 4]x - 2$ and calculated as follows: enter x; multiply by 3; subtract 5; multiply by x; add 4; multiply by x; subtract 2.

Nesting is particularly useful when the function has to be evaluated when x is a number like 2.5037. The value of x is stored in the memory, and only one key has to be pressed to recall it.

Example 3 Suppose that $y = f(x) = \dfrac{1}{x}$.

We can easily obtain a table of values.

When
$x =$	-2	-1.5	-1	-0.5	0	0.5	1	1.5	2
$y =$	-0.5	-0.67	-1	-2	?	2	1	0.67	0.5

What happens when $x = 0$? Substituting in the equation gives $y = \dfrac{1}{0}$, but this has no meaning since we cannot divide by zero. Therefore the graph should not show a value of y when $x = 0$. We now examine values of x close to 0.

When
$x =$	-0.4	-0.3	-0.2	-0.1	0.1	0.2	0.3	0.4
$y =$	-2.5	-3.33	-5	-10	10	5	3.33	2.5

It is clear that the curve $y = \dfrac{1}{x}$ consists of two parts, separated at $x = 0$ (see Fig. 1.14).

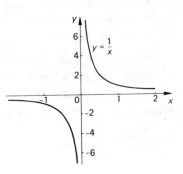

Fig. 1.14

If we wish to speak of several different functions, we may call them $g(x)$, $h(x)$, and so on.

Exercise 1.1 (Drawing graphs of functions)

Keep the graphs which you draw in questions 1, 5, 6 and 7. You will need them in exercises later in this chapter.

In questions 1 to 3, calculate the values of y when $x = -2, -1.5, -1, \ldots, +1.5, +2$, and draw a graph of the curve. Use a scale of 2 cm = 1 unit on the x-axis. The scale for the y-axis is given after each question.

1 $y = x^2$ [2 cm = 1 unit] **2** $y = x^3$ [1 cm = 1 unit] **3** $y = x^4$ [1 cm = 2 units]
4 For the function $y = x^2 - 3x + 8$, use the nested form $y = (x - 3)x + 8$ to calculate values of y when $x = -2, -1.5, -1, \ldots, +2.5, +3$, and draw a graph of the curve. Use scales of 2 cm = 1 unit on the x-axis and 1 cm = 1 unit on the y-axis.
5 Draw a graph of the curve $y = 3 - 2x - x^2$ for values of x between -3 and $+2$. [The nested form is $y = (-x - 2)x + 3$.] Use a scale of 2 cm = 1 unit on both axes.
6 Draw a graph of the curve $y = -x^3 + 2x^2 - x + 2$ for values of x between -0.5 and $+1.5$. Use a scale of 4 cm = 1 unit on both axes.

4   Gradients

In questions 7 and 8, explain why there is no value of y when $x = 0$. Draw a graph of the curve for values of x between -4 and $+4$.

7 $y = x + \dfrac{9}{x}$ [Use scales of 2 cm = 1 unit on the x-axis and 1 cm = 2 units on the y-axis.]

8 $y = \dfrac{1}{x^2}$ [Use a scale of 2 cm = 1 unit on both axes.]

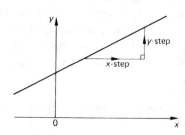

Fig. 1.15

1.2   Gradient of a line

The **gradient** (or slope) of a line is defined as

$$\frac{y\text{-step}}{x\text{-step}}$$

(see Fig. 1.15) and this shows how steep the line is.

The values of the x-step and the y-step must be read from the scales on the x-axis and y-axis respectively. Care must be taken if the scales are different.

x-steps to the right are positive, and to the left negative
y-steps upwards are positive, and downwards negative

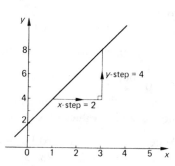

Fig. 1.16

Example 1   In Fig. 1.16, the gradient of the line is

$$\frac{y\text{-step}}{x\text{-step}} = \frac{+4}{+2} = 2$$

Example 2   In Fig. 1.17, the gradient of the line is

$$\frac{y\text{-step}}{x\text{-step}} = \frac{-1}{+3} = -\frac{1}{3}$$

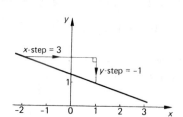

Fig. 1.17

Example 3   Find the gradient of the line joining the points $(4, -1)$ and $(2, 6)$.

With reference to Fig. 1.18, from $(4, -1)$ to $(2, 6)$,

the x-step $= 2 - 4 = -2$
the y-step $= 6 - (-1) = 7$

So the gradient of the line is

$$\frac{y\text{-step}}{x\text{-step}} = \frac{+7}{-2} = -3.5$$

Notice that a line such as  has a positive gradient;

while a line such as   has a negative gradient;
and a horizontal line ——— has zero gradient.

The gradient of a line is the same no matter which part of the line we look at. This enables us to find the equation of a line, as follows.

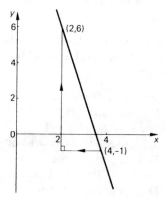

Fig. 1.18

(a)   Suppose the line has gradient m, and cuts the y-axis at $(0, c)$. In Fig. 1.19, if (x, y) is any point on the line, the gradient of the line is

$$\frac{y\text{-step}}{x\text{-step}} = \frac{y-c}{x}$$

We have

$$\frac{y-c}{x} = m$$

so $y = mx + c$. The equation of the line is

$$\boxed{y = mx + c}$$

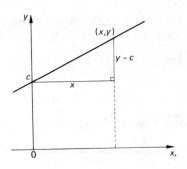

Fig. 1.19

(b)   Suppose the line has gradient m, and passes through a given point (x_1, y_1). In Fig. 1.20, if (x, y) is any other point on the line, the gradient is

$$\frac{y-y_1}{x-x_1}$$

We have

$$\frac{y-y_1}{x-x_1} = m$$

so $y - y_1 = m(x - x_1)$. The equation of the line is

$$\boxed{y - y_1 = m(x - x_1)}$$

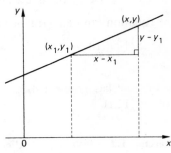

Fig. 1.20

Example 4   Find the equation of the line with gradient -3, cutting the y-axis at $(0, 4)$.

Here $m = -3$ and $c = 4$. The equation of the line is $y = -3x + 4$.

Example 5   Find the equation of the line passing through the point $(2, -4)$ with gradient $5/3$.

Here $m = 5/3$, $x_1 = 2$ and $y_1 = -4$. The equation of the line is

$$y - (-4) = \frac{5}{3}(x - 2)$$

i.e. $y + 4 = \frac{5}{3}(x - 2)$

which may be rearranged as

$$y = \frac{5}{3}x - \frac{22}{3}$$

or as

$$5x - 3y = 22$$

Example 6 Find the equation of the line passing through the points (2, 3) and (4, − 2).

The gradient of the line is

$$\frac{y\text{-step}}{x\text{-step}} = \frac{-2-3}{4-2} = -\frac{5}{2}$$

and it passes through the point (2, 3), so its equation is

$$y - 3 = -\frac{5}{2}(x - 2)$$

or $5x + 2y = 16$

Example 7 Sketch the line whose equation is $4x - 3y = 10$.

The equation can be rearranged as

$$y = \frac{4}{3}x - \frac{10}{3}$$

So the line has gradient 4/3, and cuts the y-axis at (0, − 10/3) (see Fig. 1.21).

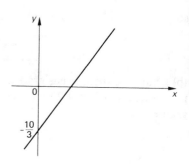

Fig. 1.21

Exercise 1.2 (Gradients and equations of lines)

1 Find the gradient of the line joining the points:
   (i) (1, 1) and (2, 4)   (ii) (2, 5) and (− 1, 6)
 (iii) (2, 4) and (− 2, − 4)   (iv) (− 9, 2) and (− 4, − 1)
2 Find the equation of the line:
   (i) with gradient 2, cutting the y-axis at (0, − 1);
  (ii) with gradient $-\frac{1}{5}$, cutting the y-axis at (0, 3);
 (iii) passing through the point (2, 5), with gradient 3;
 (iv) passing through the point (− 1, 3), with gradient $-\frac{1}{2}$;
  (v) passing through the points (4, 1) and (6, 5);
 (vi) passing through the points (− 2, 7) and (4, 2).
3 Sketch the line whose equation is:
   (i) $y = 2x - 3$ (ii) $y = 5 - \frac{1}{2}x$
 (iii) $3x - 5y = 8$ (iv) $10x + 4y = 7$

1.3 Gradient of a curve by drawing

Consider the curve shown in Fig. 1.22. To find the gradient of the curve at the point A, we draw a straight line passing through A, in the same direction as the curve at A. This line is the **tangent** to the curve at A (it touches the curve at A). The gradient of the curve at A is the gradient of this tangent. This can be found by taking two points on the line, measuring the y-step and the x-step and working out $\frac{y\text{-step}}{x\text{-step}}$ (see Fig. 1.23). In this case, $\frac{y\text{-step}}{x\text{-step}} = \frac{2.9}{2} = 1.45$, so the gradient of the curve at A is approximately 1.45.

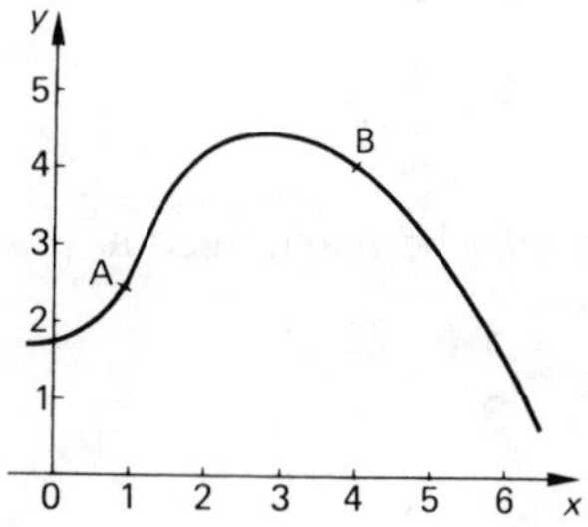

Fig. 1.22

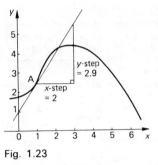

Fig. 1.23

The tangent at the point B will have a negative gradient (about −0.75), so it is clear that the gradient of a curve depends on the point we are considering. We cannot find a single value for the gradient of the curve; any value we find is the gradient of the curve at a particular point.

Exercise 1.3   (Gradients by drawing)

By drawing tangents on your graphs (from Exercise 1.1), find the gradients of the following curves at the given points.

1 $y = x^2$ at $(1.5, 2.25)$

2 $y = x^2$ at $(-1, 1)$

3 $y = -x^3 + 2x^2 - x + 2$ at $(0, 2)$

4 $y = x + \dfrac{9}{x}$ at $(2, 6.5)$

If you compare your answers to these questions with those obtained by others in your class, you will probably notice some considerable variations. It is not easy to draw a tangent accurately, and the value of a gradient found in this way can only be approximate.

We shall now show how the gradient of the curve at a point may be found without accurate drawing, provided that the equation of the curve is known.

1.4 Calculating the gradient of a curve at a point

Consider the point A(2, 4) on the curve $y = x^2$. Let P be another point on the curve, and draw in the line AP (see Fig. 1.24). If P is close to A, the gradient of the line AP will be approximately the same as the gradient of the curve at A (and the line AP will be close to the tangent at A). The closer P is taken to A, the more accurate these approximations will become.

In Table 1.1, the gradient of the line AP is calculated for various values of the x-step. For example, when the x-step from A to P is 0.1, the x-coordinate of P is 2.1, and so the y-coordinate is $(2.1)^2 = 4.41$. The y-step is therefore $4.41 - 4 = 0.41$, and the gradient of AP is

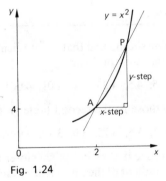

Fig. 1.24

$$\frac{y\text{-step}}{x\text{-step}} = \frac{0.41}{0.1} = 4.1$$

The x-steps are taken smaller and smaller, so that P is becoming closer and closer to A.

Use a calculator to complete the table. You should work across each row without clearing the calculator, since each value is immediately used to calculate the next one (as indicated).

We could equally well take P on the left of A (see Fig. 1.25). The x-step and the y-step from A to P are now both negative. For example, when the x-step from A to P is −0.1, the x-coordinate of

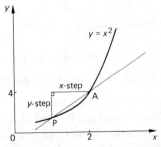

Fig. 1.25

Table 1.1

x-step from A to P	x-coord of P	y-coord of P	y-step from A to P	Gradient of AP
Enter on calculator →	add 2 →	square →	subtract 4 →	divide by x-step
1	3	9	5	5
0.5	2.5	6.25	2.25	4.5
0.25	2.25	5.0625	1.0625	4.25
0.1	2.1	4.41	0.41	4.1
0.01	2.01	4.0401		
0.001	2.001			
0.0001				

P is 1.9, so the y-coordinate is $(1.9)^2 = 3.61$, and the y-step from A to P is $3.61 - 4 = -0.39$.

Complete Table 1.2.

Table 1.2

x-step from A to P	x-coord of P	y-coord of P	y-step from A to P	Gradient of AP
Enter →	add 2 →	square →	subtract 4 →	divide by x-step
−1	1	1	−3	3
−0.5	1.5	2.25	−1.75	3.5
−0.25	1.75	3.0625	−0.9375	3.75
−0.1	1.9	3.61	−0.39	
−0.01	1.99	3.9601		
−0.001	1.999			
−0.0001				

You should find that the gradients of AP in the first table (with P on the right of A) are

5, 4.5, 4.25, 4.1, 4.01, 4.001, 4.0001

and those in the second table (with P on the left of A) are

3, 3.5, 3.75, 3.9, 3.99, 3.999, 3.9999

We see that as P is taken closer and closer to A (on either side), the gradient of the line AP approaches the value 4, which is called a **limiting value**. It seems reasonable to say that the gradient of the curve at A is 4. (In fact, we *define* the gradient of the curve at A to be this limiting value; see p. 26.)

We cannot find the gradient of the curve at A by considering an x-step of 0, for then the y-step would also be 0, and the gradient of AP would be $\frac{0}{0}$, which has no meaning. This is why the limiting process above is necessary.

This method can be used to find the gradient at any given point on any curve. Usually we can estimate the limiting value of the gradient of AP after trying just a few x-steps. It is sensible to include one step to the left. This limiting value is the gradient of the curve at A.

Example 1 Find the gradient of the curve $y = x^3$ at the point A(3, 27).

Table 1.3

x-step from A to P	x-coord of P	y-coord of P	y-step from A to P	Gradient of AP
Enter $\longrightarrow$	add 3 $\longrightarrow$	cube $\longrightarrow$	subtract 27 $\longrightarrow$	divide by x-step
0.01	3.01	27.271	0.271	27.090
0.001	3.001	27.027	0.027	27.009
− 0.001	2.999	26.973	−0.027	26.991

The limiting value of the gradient is 27.0, so the gradient of the curve at A is 27.0.

Notes

(1)   All the numbers in Table 1.3 have been rounded to three decimal places, but the gradient is calculated from the unrounded values (still on the calculator). Serious errors may result if the rounded values of the y-step are used. This is because the y-step is obtained by subtracting two numbers which are almost equal, and significant figures are lost in this type of calculation.

(2)   If we use a very small x-step, then (for the same reason as above) the errors due to the calculator itself will noticeably affect the value obtained for the gradient. The following values were obtained using a typical eight figure calculator. (These values will vary from calculator to calculator. Try it on yours.)

x-step from A to P	Gradient of AP
0.1	27.91
0.01	27.090104
0.001	27.009 002
0.0001	27.000 87
0.00001	26.998 2
0.000001	26.976
0.0000001	26.7

For x-steps of less than 0.000 01, the calculator errors have a marked effect, and the values of the gradient no longer get closer and closer to the limiting value.

Example 2 Find the gradient of the curve $y = x + \dfrac{9}{x}$ at the point

A(2, 6.5).

Table 1.4

x-step from A to P	x-coord of P	y-coord of P	y-step from A to P	Gradient of AP
		Calculate		
Enter $\longrightarrow$	add 2 $\longrightarrow$	$x + \dfrac{9}{x}$ $\longrightarrow$	subtract 6.5 $\longrightarrow$	divide by x- step
0.01	2.01	6.488	− 0.012	− 1.239
0.001	2.001	6.499	− 0.001	− 1.249
− 0.001	1.999	6.501	0.001	− 1.251

The gradient of the curve at A is − 1.25.

Exercise 1.4   (Gradients by calculation)

In each case, take x-steps of 0.01 and 0.001 to the right, and then one of 0.001 to the left. Estimate the limiting value.

1   [Keep your answers to this question. You will need them in Exercise 1.5.]
Find the gradients of the following curves, all at the point (1, 1).

(i) $y = x^2$ (ii) $y = x^3$ (iii) $y = x^4$ (iv) $y = x^5$

(v) $y = \sqrt{x}$ (vi) $y = \dfrac{1}{x}$ (vii) $y = \dfrac{1}{x^2}$

Your results should show that the gradient of the curve $y = x^n$ at the point (1, 1) is n.

[Note that $\sqrt{x}$ may be written as $x^{\frac{1}{2}}$, $\dfrac{1}{x}$ as x^{-1} and $\dfrac{1}{x^2}$ as x^{-2}.]

In questions 2 to 5 find the gradients of the following curves at the given points.

2 $y = x^4$ at (2, 16)   **3** $y = \sqrt{x}$ at (4, 2)

4 $y = 3 - 2x - x^2$ at (1, 0)   **5** $y = \dfrac{1}{x - 2}$ at (3, 1)

6   On your graph of $y = 3 - 2x - x^2$ (from Exercise 1.1), draw a line through the point (1, 0) with the gradient found in question 4. Does this line look like a tangent to the curve?

1.5   Formula for the gradient of $y = x^n$

We shall now try to find a formula for the gradient of the curve $y = x^2$ at a general point.
   We first find the gradient at the point (1, 1).

Table 1.5

x-step from A to P	x-coord of P	y-coord of P	y-step from A to P	Gradient of AP
Enter $\longrightarrow$	add 1 $\longrightarrow$	square $\longrightarrow$	subtract 1 $\longrightarrow$	divide by x-step
0.01	1.01	1.020	0.020	2.01
0.001	1.001	1.002	0.002	2.001
− 0.001	0.999	0.998	− 0.002	1.999

The gradient at (1, 1) is 2.

Now consider the point (2, 4) on the curve. Using values of the
x-step twice what they were above, we obtain Table 1.6.

Table 1.6

x-step from A to P	x-coord of P	y-coord of P	y-step from A to P	Gradient of AP
Enter $\longrightarrow$	add 2 $\longrightarrow$	square $\longrightarrow$	subtract 4 $\longrightarrow$	divide by x-step
0.02	2.02	4.080	0.080	4.02
0.002	2.002	4.008	0.008	4.002
− 0.002	1.998	3.992	− 0.008	3.998

These values of the gradient are twice those in Table 1.5, so the
gradient at (2, 4) is twice 2, or 4.

Now consider the point (3, 9) on the curve. Using values of the
x-step 3 times those in the Table 1.5, we obtain Table 1.7.

Table 1.7

x-step from A to P	x-coord of P	y-coord of P	y-step from A to P	Gradient of AP
Enter $\longrightarrow$	add 3 $\longrightarrow$	.square $\longrightarrow$	subtract 9 $\longrightarrow$	divide by x-step
0.03	3.03	9.181	0.181	6.03

Without going any further, we can see that the values of the
gradient will be 3 times those in Table 1.5, so the gradient at (3, 9) is
$3 \times 2 = 6$.

Similarly, the gradient at the point (4, 16) is $4 \times 2 = 8$, and so on.

At a general value of x, the gradient appears to be $x \times 2 = 2x$, so
the gradient at a general point on the curve $y = x^2$ seems to be
given by the formula $2x$.

Exercise 1.5   (Investigating the gradients of $y = x^n$)

1   Look at the table you used to find the gradient of the curve $y = x^3$ at the point (1, 1) [Exercise 1.4,
question 1(ii)]. Using values for the x-step twice those in this table, i.e. x-steps of 0.02, 0.002, and
− 0.002, show that the gradient at (2, 8) is 4 times the gradient at (1, 1). Find the gradients (i) at (3, 27)
by taking an x-step of 0.03 and (ii) at (4, 64) by taking an x-step of 0.04. (iii) What is the gradient at a
general value of x?

In questions 2 to 5, find a formula for the gradient at a general point on the given curve by taking an
x-step of 0.02 at $x = 2$, of 0.03 at $x = 3$, of 0.04 at $x = 4$, and comparing with the corresponding results
at $x = 1$ [obtained in Exercise 1.4, question 1].

2 $y = x^4$ **3** $y = x^5$ **4** $y = \dfrac{1}{x}$ **5** $y = \dfrac{1}{x^2}$

6 Complete Table 1.8.

Table 1.8

Curve	Formula for the gradient
$y = x$	1
$y = x^2$	$2x$
$y = x^3$	
$y = x^4$	
$y = x^5$	
$y = 1/x$	
$y = 1/x^2$	

Your results should demonstrate that the gradient at a general point on the curve $y = x^n$ is given by the formula

$$\boxed{nx^{n-1}}$$

1.6   Formula for the gradient of $y = kx^n$

We shall now consider the curve $y = 2x^2$.
   At the point (1, 2):

x-step from A to P	x-coord of P	y-coord of P	y-step from A to P	Gradient of AP
Enter $\longrightarrow$	add 1 $\rightarrow$	square $\rightarrow \times 2 \rightarrow$	subtract 2 $\longrightarrow$	divide by x-step
0.01	1.01	2.040	0.040	4.02

At the point (2, 8):

x-step from A to P	x-coord of P	y-coord of P	y-step from A to P	Gradient of AP
Enter $\longrightarrow$	add 2 $\rightarrow$	square $\rightarrow \times 2 \rightarrow$	subtract 8 $\longrightarrow$	divide by x-step
0.02	2.02	8.161	0.161	8.04

At the point (3, 18):

x-step from A to P	x-coord of P	y-coord of P	y-step from A to P	Gradient of AP
Enter $\longrightarrow$	add 3 $\rightarrow$	square $\rightarrow \times 2 \rightarrow$	subtract 18 $\longrightarrow$	divide by x-step
0.03	3.03	18.362	0.362	12.06

These values of the gradients are twice the values found for the curve $y = x^2$ at the points (1, 1), (2, 4) and (3, 9), so the gradient of the curve $y = 2x^2$ at (1, 2) is 4, at (2, 8) is 8, and at (3, 18) is 12. Similarly, for any value of x the gradient of the curve $y = 2x^2$ will be twice the gradient of the curve $y = x^2$. So at a general value of x, the gradient of the curve $y = 2x^2$ seems to be given by the formula $2 \times (2x) = 4x$.

Exercise 1.6   (Investigating the gradients of $y = kx^n$)

1   Consider the curve $y = 3x^2$. Show that the gradients at $(1, 3)$, $(2, 12)$ and $(3, 27)$ are three times the gradients of the curve $y = x^2$ at $(1, 1)$, $(2, 4)$ and $(3, 9)$. What is the formula for the gradient of the curve $y = 3x^2$ at a general value of x?

2   Write down the formula for the gradient at a general point on the following curves:
(i) $y = 5x^2$ (ii) $y = 9x^2$ (iii) $y = \frac{1}{2}x^2$ (iv) $y = kx^2$

3   Consider the curve $y = 2x^3$. Compare the gradients at $(1, 2)$, $(2, 16)$, $(3, 54)$ with the gradients of the curve $y = x^3$ at $(1, 1)$, $(2, 8)$ and $(3, 27)$ [found in Exercise 1.5, question 1]. What is the formula for the gradient of the curve $y = 2x^3$ at a general value of x?

4   Write down the formula for the gradient at a general point on the following curves:
(i) $y = 3x^3$ (ii) $y = 4x^3$ (iii) $y = \frac{1}{4}x^3$ (iv) $y = kx^3$

5   Write down the formula for the gradient at a general point on the following curves.
(i) $y = 2x^4$ (ii) $y = 3x^4$ (iii) $y = kx^4$ (iv) $y = 2x^5$ (v) $y = 3x^5$

(vi) $y = kx^5$ (vii) $y = \dfrac{2}{x}$ (viii) $y = \dfrac{3}{x}$ (ix) $y = \dfrac{k}{x}$ (x) $y = kx^n$

Your results should demonstrate that the gradient of the curve $y = kx^n$ is given by the formula

$$\boxed{knx^{n-1}}$$

2

Differentiation

2.1   The derived function: differentiation of x^n

If y is a function of x which is represented by the curve $y = f(x)$, then the gradient of the curve varies from point to point. The function which gives the gradient of the curve in terms of x is called the **derived function**.

For example, if $y = x^2$, we have seen (p. 11) that the gradient at a general point is given by the formula $2x$, and so the derived function is $2x$.

We use the notation

$$\frac{dy}{dx}$$

(said 'dee y by dee x', or simply 'dee y dee x') for the derived function. (Notice that in $\frac{dy}{dx}$, x and y are in italic since they are variables.) This is just a notation, and it must not be thought of as dy divided by dx.

If $y = x^2$, then $\frac{dy}{dx} = 2x$.

If $y = x^3$, the gradient of the curve is given by $3x^2$, and so $\frac{dy}{dx} = 3x^2$.

If $y = x^4$, then $\frac{dy}{dx} = 4x^3$, and so on.

If y is a constant, for example $y = 3$, the curve is a straight line parallel to the x-axis, and so the gradient is always zero; thus $\frac{dy}{dx} = 0$.

We therefore have

if $y = c$ (constant), then $\frac{dy}{dx} = 0$

if $y = x$,   then $\frac{dy}{dx} = 1$

if $y = x^2$,   then $\frac{dy}{dx} = 2x$

if $y = x^3$,   then $\frac{dy}{dx} = 3x^2$

if $y = x^4$,   then $\dfrac{dy}{dx} = 4x^3$

$\vdots$

if $y = x^n$,   then $\dfrac{dy}{dx} = nx^{n-1}$

Multiples and sums

If $y = 2x^2$, we have seen (p. 13) that the gradient at any value of x is twice the gradient of the curve $y = x^2$; so $\dfrac{dy}{dx} = 2(2x) = 4x$. Similarly,

if $y = 3x^2$, then $\dfrac{dy}{dx} = 3(2x) = 6x$

if $y = 7x^5$, then $\dfrac{dy}{dx} = 7(5x^4) = 35x^4$

and so on.

If $y = x^2 + x^4$, then $\dfrac{dy}{dx}$ is the sum of the derived functions of x^2 and of x^4 (see p. 30 for a justification of this). So $\dfrac{dy}{dx} = 2x + 4x^3$. For any sum or difference of functions, we can find the derived function term by term. For example, if

$$y = 3x^6 - 2x^4 + 5x - 7$$

then

$$\frac{dy}{dx} = 3(6x^5) - 2(4x^3) + 5(1) - 0$$
$$= 18x^5 - 8x^3 + 5$$

The process of finding the derived function is called **differentiation**. If the functions are written in terms of x, then we say we are **differentiating with respect to x**.

Example 1   Differentiate $\frac{1}{3}x^8 - 7x$.

If $y = \frac{1}{3}x^8 - 7x$, then

$$\frac{dy}{dx} = \frac{1}{3}(8x^7) - 7(1) = \frac{8}{3}x^7 - 7$$

Example 2   Differentiate $(2 + x)(3 + 4x - x^2)$.

We cannot do this by finding the derived functions of $(2 + x)$ and $(3 + 4x - x^2)$, and simply multiplying them together. We must first multiply out the brackets. (In Chapter 9 we shall show how to differentiate such a product without multiplying out the brackets.) If

$$y = (2 + x)(3 + 4x - x^2) = 6 + 11x + 2x^2 - x^3$$

then

$$\frac{dy}{dx} = 0 + 11(1) + 2(2x) - 3x^2 = 11 + 4x - 3x^2$$

Exercise 2.1   (Differentiation of polynomials)

Differentiate the following functions.

1 x^5	**2** x^9	**3** x^{99}	**4** $8x^7$
5 $\frac{2}{5}x^4$	**6** $x^3 + x$	**7** $x^6 - x^4$	**8** $x^2 - 6x$
9 $16 - 4x$	**10** $3x^2 - 7x - 8$	**11** $8x^3 + 6x^2 - 37x + 4$	**12** $3 - 2x^2 + x^4 - 5x^6$
13 $x(x + 2)$	**14** $(x - 4)(2x + 3)$	**15** $(2 - x)(x^3 + 3x - 4)$	**16** $(2x + 1)^2$
17 $x(x - 1)(x - 2)$	**18** kx^n	**19** x^{n+1}	**20** $ax^2 + bx + c$

2.2 Fractional and negative powers

Our basic result, that if $y = x^n$ then $\dfrac{dy}{dx} = nx^{n-1}$, is in fact true for all values of n, including fractional and negative values (we shall justify this in Chapter 9). We shall first explain what we mean by fractional and negative powers of x.

If n is a positive integer, then $x^n = (x)(x)(x) \ldots (x)$ (where there are n factors). So, for example, $x^1 = x$, $x^2 = (x)(x)$ and so on.

We have the following laws which we know:

(i) $(x^m)(x^n) = x^{m+n}$ (ii) $\dfrac{x^m}{x^n} = x^{m-n}$

(iii) $(x^m)^n = x^{mn}$ (iv) $(ab)^n = a^n b^n$ (v) $\left(\dfrac{a}{b}\right)^n = \dfrac{a^n}{b^n}$

The definitions are extended in such a way that these laws will remain true even if m and n are fractions (or negative numbers). In the following, m and n are positive integers and r is a positive fraction.

Zero power

$$x^0 = x^{n-n} = \frac{x^n}{x^n} = 1$$

This is true for any x (except $x = 0$).

For example, $5^0 = 1$; $(-17)^0 = 1$.

Roots and powers

$$(x^{\frac{1}{n}})^n = x^{\frac{1}{n} \times n} = x^1 = x$$

So $x^{\frac{1}{n}}$ is the nth root of x, i.e.

$$x^{\frac{1}{n}} = \sqrt[n]{x}$$

(The square root $x^{\frac{1}{2}} = \sqrt[2]{x}$ is usually written simply as $\sqrt{x}$.) For example,

$$16^{\frac{1}{2}} = \sqrt{16} = 4; \qquad 8^{\frac{1}{3}} = \sqrt[3]{8} = 2$$

$$\left(\frac{1}{64}\right)^{\frac{1}{6}} = \frac{1^{\frac{1}{6}}}{64^{\frac{1}{6}}} = \frac{1}{\sqrt[6]{64}} = \frac{1}{2}$$

$$\left(\frac{-8}{27}\right)^{\frac{1}{3}} = \frac{(-8)^{\frac{1}{3}}}{(27)^{\frac{1}{3}}} = \frac{\sqrt[3]{-8}}{\sqrt[3]{27}} = \frac{-2}{3}$$

Positive fractional powers

$$x^{\frac{m}{n}} = x^{m \times \frac{1}{n}} = (x^m)^{\frac{1}{n}} = \sqrt[n]{x^m}$$

Also $\quad x^{\frac{m}{n}} = x^{\frac{1}{n} \times m} = (x^{\frac{1}{n}})^m = (\sqrt[n]{x})^m$

The first form is normally used when we write a power in root form (for example, $x^{\frac{3}{4}} = \sqrt[4]{x^3}$). When evaluating a power, the second form is usually more convenient. For example,

$$8^{\frac{2}{3}} = (\sqrt[3]{8})^2 = 2^2 = 4$$

$$81^{\frac{5}{4}} = (\sqrt[4]{81})^5 = 3^5 = 243$$

$$\left(\frac{32}{243}\right)^{\frac{3}{5}} = \frac{32^{\frac{3}{5}}}{243^{\frac{3}{5}}} = \frac{(\sqrt[5]{32})^3}{(\sqrt[5]{243})^3} = \frac{2^3}{3^3} = \frac{8}{27}$$

$$(-64)^{\frac{4}{3}} = (\sqrt[3]{-64})^4 = (-4)^4 = 256$$

Negative powers

$$x^r x^{-r} = x^{r+(-r)} = x^0 = 1$$

and so $\quad x^{-r} = \dfrac{1}{x^r} = \left(\dfrac{1}{x}\right)^r$

For example,

$$3^{-2} = \frac{1}{3^2} = \frac{1}{9}; \qquad 4^{-\frac{1}{2}} = \frac{1}{4^{\frac{1}{2}}} = \frac{1}{\sqrt{4}} = \frac{1}{2}$$

$$\left(\frac{2}{5}\right)^{-3} = \left(\frac{5}{2}\right)^3 = \frac{125}{8}$$

$$(-8)^{-\frac{5}{3}} = \frac{1}{(-8)^{\frac{5}{3}}} = \frac{1}{(\sqrt[3]{-8})^5} = \frac{1}{(-2)^5} = -\frac{1}{32}$$

Summary

We have

$$x^n = (x)(x)(x) \ldots (x)$$
$$x^1 = x$$
$$x^0 = 1$$
$$x^{\frac{1}{n}} = \sqrt[n]{x}$$
$$x^{\frac{m}{n}} = \sqrt[n]{x^m} = (\sqrt[n]{x})^m$$
$$x^{-r} = \frac{1}{x^r} = \left(\frac{1}{x}\right)^r$$

Example 1 Simplify $\dfrac{x^{-\frac{1}{2}}}{x^{-\frac{3}{2}}}$.

$$\frac{x^{-\frac{1}{2}}}{x^{-\frac{3}{2}}} = x^{-\frac{1}{2}-(-\frac{3}{2})} = x^1 = x$$

Example 2 Simplify $\dfrac{(x^{\frac{3}{2}}+x)(2-x^{\frac{1}{2}})}{x^{\frac{1}{2}}}$.

$$\frac{(x^{\frac{3}{2}}+x)(2-x^{\frac{1}{2}})}{x^{\frac{1}{2}}} = \frac{2x^{\frac{3}{2}} - x^{\frac{3}{2}}x^{\frac{1}{2}} + 2x - xx^{\frac{1}{2}}}{x^{\frac{1}{2}}}$$

$$= \frac{2x^{\frac{3}{2}} - x^2 + 2x - x^{\frac{3}{2}}}{x^{\frac{1}{2}}}$$

$$= \frac{x^{\frac{3}{2}} - x^2 + 2x}{x^{\frac{1}{2}}} = \frac{x^{\frac{3}{2}}}{x^{\frac{1}{2}}} - \frac{x^2}{x^{\frac{1}{2}}} + \frac{2x}{x^{\frac{1}{2}}}$$

$$= x - x^{\frac{3}{2}} + 2x^{\frac{1}{2}}$$

This can be written in root form as

$$x - \sqrt{x^3} + 2\sqrt{x}$$

Exercise 2.2 (Powers)

Evaluate the following (questions 1 to 20).

1 $4^{\frac{1}{2}}$

2 $64^{\frac{1}{2}}$

3 $27^{\frac{1}{3}}$

4 $\left(\dfrac{4}{9}\right)^{\frac{1}{2}}$

5 $\left(\dfrac{1}{16}\right)^{\frac{1}{4}}$

6 $\left(\dfrac{1}{8}\right)^{0}$

7 $4^{\frac{5}{2}}$

8 $64^{\frac{2}{3}}$

9 $81^{\frac{3}{4}}$

10 $\left(\dfrac{1}{32}\right)^{\frac{2}{5}}$

11 $\left(\dfrac{16}{25}\right)^{\frac{3}{2}}$

12 $\left(\dfrac{125}{8}\right)^{\frac{4}{3}}$

13 2^{-3}

14 20^{-1}

15 $9^{-\frac{1}{2}}$

16 $\left(\dfrac{4}{7}\right)^{-2}$

17 $\left(\dfrac{1}{8}\right)^{-\frac{1}{3}}$

18 $\left(\dfrac{27}{8}\right)^{-\frac{5}{3}}$

19 $(-32)^{\frac{3}{5}}$

20 $(-2)^{-4}$

If you have a calculator with a y^x key, work through questions 1 to 20 again to check your solutions. You may find that your method of calculation has to be adjusted if you have to find the power of a negative number.

Express the following as powers of x (questions 21 to 25).

21 $\sqrt{x}$ 22 $\dfrac{1}{x}$ 23 $\dfrac{1}{x^4}$ 24 $\dfrac{1}{\sqrt[3]{x^2}}$ 25 $\sqrt[4]{\dfrac{1}{x^3}}$

Write the following powers of x in root form (questions 26 to 30).

26 $x^{\frac{1}{2}}$ 27 $x^{\frac{2}{3}}$ 28 $x^{-\frac{4}{5}}$ 29 $\dfrac{1}{x^{\frac{3}{2}}}$ 30 $x^{-\frac{3}{4}}$

Simplify the following (questions 31 to 35). Give your answers in both power and root form.

31 $\dfrac{x^{\frac{5}{2}}}{x^{\frac{1}{2}}}$ 32 $\dfrac{x^{-\frac{2}{3}}}{x^{-\frac{5}{3}}}$ 33 $x^{\frac{2}{3}}(x^{\frac{1}{3}}+x^{-\frac{1}{3}})$ 34 $\dfrac{x^2-3x^{\frac{3}{2}}}{x^{\frac{1}{2}}}$ 35 $\dfrac{(1+x^{\frac{1}{3}})(x^{\frac{5}{3}}-x^2)}{x^{\frac{1}{3}}}$

2.3   Differentiation of powers

We can differentiate any power of x by applying the result that if $y = x^n$, then $\dfrac{dy}{dx} = nx^{n-1}$. Note that we usually give the result in the form of the question, so that if the question is in root form, then we give the answer in root form.

Example 1   Differentiate   (i) $\sqrt{x}$ (ii) $x^{\frac{5}{3}}$ (iii) $\dfrac{1}{x^2}$ (iv) $\dfrac{1}{\sqrt[3]{x}}$.

(i)   If $y = \sqrt{x} = x^{\frac{1}{2}}$, then

$$\frac{dy}{dx} = \frac{1}{2}x^{-\frac{1}{2}} = \frac{1}{2\sqrt{x}}$$

(ii)   If $y = x^{\frac{5}{3}}$, then

$$\frac{dy}{dx} = \frac{5}{3}x^{\frac{2}{3}}$$

(iii)   If $y = \dfrac{1}{x^2} = x^{-2}$, then

$$\frac{dy}{dx} = (-2)x^{-3} = \frac{-2}{x^3}$$

(iv)   If $y = \dfrac{1}{\sqrt[3]{x}} = x^{-\frac{1}{3}}$, then

$$\frac{dy}{dx} = \left(-\frac{1}{3}\right)x^{-\frac{4}{3}} = -\frac{1}{3x^{\frac{4}{3}}} = \frac{-1}{3\sqrt[3]{x^4}}$$

We can now differentiate multiples and sums of powers of x. Sometimes it is necessary to multiply or divide out first.

Example 2   Differentiate

(i) $2\sqrt{x} + \dfrac{3}{x}$ (ii) $(\sqrt{x} - 1)(x + 2)$

(iii) $\dfrac{3x^6 - 5}{x^4}$ (iv) $\dfrac{(1 + x)(1 - x)}{\sqrt[3]{x}}$

(i) If $y = 2\sqrt{x} + \dfrac{3}{x} = 2x^{\frac{1}{2}} + 3x^{-1}$, then

$$\frac{dy}{dx} = 2\left(\frac{1}{2}x^{-\frac{1}{2}}\right) + 3(-1)x^{-2} = \frac{1}{\sqrt{x}} - \frac{3}{x^2}$$

(ii) If $y = (\sqrt{x} - 1)(x + 2) = (x^{\frac{1}{2}} - 1)(x + 2) = x^{\frac{3}{2}} + 2x^{\frac{1}{2}} - x - 2$, then

$$\frac{dy}{dx} = \frac{3}{2}x^{\frac{1}{2}} + 2\left(\frac{1}{2}x^{-\frac{1}{2}}\right) - 1 - 0 = \frac{3}{2}\sqrt{x} + \frac{1}{\sqrt{x}} - 1$$

(iii) If $y = \dfrac{3x^6 - 5}{x^4} = \dfrac{3x^6}{x^4} - \dfrac{5}{x^4} = 3x^2 - 5x^{-4}$, then

$$\frac{dy}{dx} = 3(2x) - 5(-4x^{-5}) = 6x + \frac{20}{x^5}$$

(iv) If $y = \dfrac{(1 + x)(1 - x)}{\sqrt[3]{x}} = \dfrac{1 - x^2}{x^{\frac{1}{3}}} = \dfrac{1}{x^{\frac{1}{3}}} - \dfrac{x^2}{x^{\frac{1}{3}}} = x^{-\frac{1}{3}} - x^{\frac{5}{3}}$, then

$$\frac{dy}{dx} = -\frac{1}{3}x^{-\frac{4}{3}} - \frac{5}{3}x^{\frac{2}{3}} = -\frac{1}{3x^{\frac{4}{3}}} - \frac{5}{3}x^{\frac{2}{3}} = -\frac{1}{3\sqrt[3]{x^4}} - \frac{5}{3}\sqrt[3]{x^2}$$

Exercise 2.3   (Differentiation of powers)

Differentiate the following functions.

1 $\sqrt[3]{x}$ 2 $x^{\frac{2}{5}}$ 3 $\dfrac{1}{x}$ 4 $\dfrac{1}{x^3}$

5 $\dfrac{1}{\sqrt[5]{x}}$ 6 $5x^{-\frac{3}{2}}$ 7 $2x^{\frac{4}{3}} - x^{\frac{2}{3}}$ 8 $5x^3 + \dfrac{4}{x} - \dfrac{7}{x^3}$

9 $\left(3 - \dfrac{1}{x}\right)\left(5 - \dfrac{2}{x}\right)$ 10 $(x^{\frac{1}{4}} + x^{-\frac{1}{4}})^2$ 11 $\dfrac{2x^5 + 3x^4}{x^2}$ 12 $\dfrac{2x + 5}{x^3}$

13 $\dfrac{4 - 2x^4}{3x^2}$ 14 $\dfrac{9 - x^2}{5x^4}$ 15 $\dfrac{1 + 3x}{\sqrt{x}}$ 16 $\dfrac{x^{\frac{1}{3}} + x^{-\frac{1}{3}}}{x^3}$

17 $\dfrac{3x^2 - 7x + 5}{x^4}$ 18 $\sqrt{x}\left(\sqrt{x} + \dfrac{1}{\sqrt{x}}\right)^2$ 19 $\sqrt[n]{x}$ 20 $\dfrac{1}{x^n}$

2.4 Other notations

When differentiating a function, for example $x^2 + 3x$, we say: 'if $y = x^2 + 3x$, then $\dfrac{dy}{dx} = 2x + 3$'. We may express the fact that the derived function of $(x^2 + 3x)$ is $(2x + 3)$ by writing

$$\frac{d}{dx}(x^2 + 3x) = 2x + 3$$

$\dfrac{d}{dx}$ (said 'dee by dee x of') is a notation meaning 'the derived function of'. If we use $f(x)$ to represent a function of x, its derived function is written as $f'(x)$ (said 'f dashed of x'). For example, if $f(x) = x^2 + 3x$, then $f'(x) = 2x + 3$.

The differentiation of $x^2 + 3x$ may be written out in any of the following ways:

(a) if $y = x^2 + 3x$, then $\dfrac{dy}{dx} = 2x + 3$

(b) $\dfrac{d}{dx}(x^2 + 3x) = 2x + 3$

(c) if $f(x) = x^2 + 3x$, then $f'(x) = 2x + 3$

The derived function may also be called the **gradient function** or the **derivative**, or the **differential coefficient**.

Exercise 2.4 (Alternative notations)

1 Find (i) $\dfrac{d}{dx}(x^3)$ (ii) $\dfrac{d}{dx}(3 + 4x - 5x^2)$ (iii) $\dfrac{d}{dx}\left(\dfrac{3}{x^2}\right)$ (iv) $\dfrac{d}{dx}(x^n)$

For the following functions, find $f'(x)$.

2 $f(x) = x^2$ 3 $f(x) = 2x^4$ 4 $f(x) = x^2 - 7x$

5 $f(x) = 2x^3 - x^2 + 5x - 8$ 6 $f(x) = 3x^{\frac{1}{2}}$ 7 $f(x) = x + \dfrac{9}{x}$

8 $f(x) = \dfrac{x^2 - 4}{\sqrt{x}}$ 9 $f(x) = (x + 3)(2x - 5)$ 10 $f(x) = x^{\frac{1}{3}}(x^2 + 1)$

2.5 Finding the gradient at a point

The derived function $\dfrac{dy}{dx}$ gives the gradient of a curve in terms of x. If we wish to find the gradient at a particular point, we first differentiate $\left(\text{to find } \dfrac{dy}{dx}\right)$ and then substitute in the appropriate value for x.

Example 1 Find the gradient of the curve
$y = 2x^3 - 4x^2 + 5x - 3$ at the point (2, 7).

We have $\dfrac{dy}{dx} = 2(3x^2) - 4(2x) + 5(1) - 0 = 6x^2 - 8x + 5$

When $x = 2$, $\dfrac{dy}{dx} = 6 \times 4 - 8 \times 2 + 5 = 13$, so the gradient at the
point (2, 7) is 13.

If the equation of the curve is $y = f(x)$, the derived function is
$\dfrac{dy}{dx} = f'(x)$. The gradient at $x = 1$ is $f'(1)$ and the gradient at
$x = 2$ is $f'(2)$.

Example 2 If $f(x) = x^2 + \dfrac{4}{x}$ find $f'(2)$ and $f'(-\tfrac{1}{3})$.

We have $f(x) = x^2 + 4x^{-1}$

and so $f'(x) = 2x + 4(-x^{-2}) = 2x - 4x^{-2}$

So $f'(2) = 2 \times 2 - 4 \times \tfrac{1}{4} = 3$

and $f'(-\tfrac{1}{3}) = 2 \times (-\tfrac{1}{3}) - 4 \times (-\tfrac{1}{3})^{-2} = -\tfrac{2}{3} - 4 \times 9 = -36\tfrac{2}{3}$

Exercise 2.5 (Calculation of gradients)

In questions 1 to 5 find the gradients of the curves at the given points.

1 $y = 3x^2$ (i) at (2, 12); (ii) at $(-1, 3)$.   2 $y = 2x^3 - 3x^2$ (i) at $(1, -1)$; (ii) at $(-2, -28)$.

3 $y = (4 + x)(2x - 1)$ at $(\tfrac{1}{2}, 0)$.   4 $y = \sqrt{x}$ (i) at (4, 2); (ii) at $(\tfrac{1}{9}, \tfrac{1}{3})$.

5 $y = x - \dfrac{4}{x}$ at $(-\tfrac{1}{2}, 7\tfrac{1}{2})$.

6 If $f(x) = 3 + 5x - 2x^2$ find $f'(0)$ and $f'(3)$.

7 If $f(x) = \dfrac{1}{x^2}$ find $f'(-1)$ and $f'(\tfrac{1}{2})$.

8 If $f(x) = \dfrac{\sqrt{x} - 1}{x^2}$ find $f'(1)$ and $f'(4)$.

2.6 The gradient function

The derived function $\dfrac{dy}{dx}$ gives the gradient of a curve in terms of x.

When we draw the graph of the derived function, we shall usually
call it the gradient function. It is important to be able to visualise
the relationship between the graph of a function and the graph of
its gradient function.

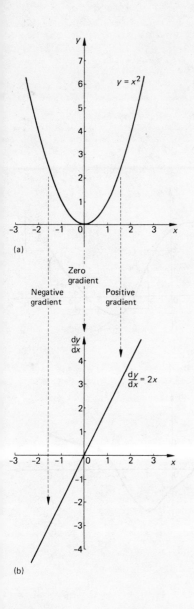

Fig. 2.1

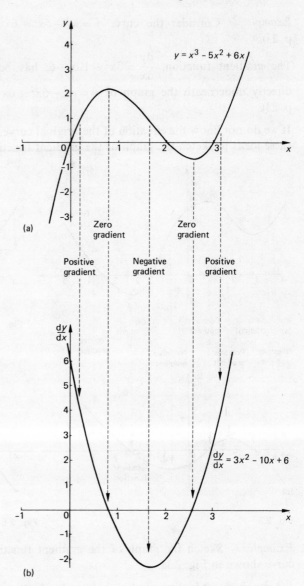

Fig. 2.2

Example 1   Consider the curve $y = x^2$ (Fig. 2.1a).

If we plot the graph of the gradient function, $\dfrac{dy}{dx} = 2x$, directly

underneath that of the curve $y = x^2$, we can see how the two graphs
are related (Fig. 2.1b).

   When the gradient of $y = x^2$ is positive, the graph of $\dfrac{dy}{dx}$ is above

the x-axis. When the gradient of $y = x^2$ is negative, the graph of $\dfrac{dy}{dx}$

is below the x-axis.

Example 2 Consider the curve $y = x^3 - 5x^2 + 6x$ (Fig. 2.2a, p. 23).

The gradient function, $\dfrac{dy}{dx} = 3x^2 - 10x + 6$, has been plotted directly underneath the graph of $y = x^3 - 5x^2 + 6x$ (Fig. 2.2b, p. 23).

If we do not know the equation of the original curve, we can use these ideas to sketch the graph of the gradient function.

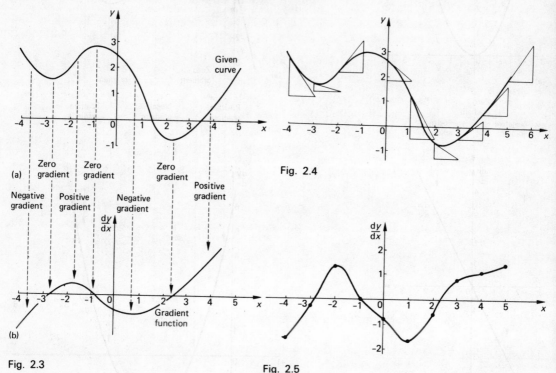

Fig. 2.4

Fig. 2.3

Fig. 2.5

Example 3 Sketch the graph of the gradient function for the curve shown in Fig. 2.3a.

A rough sketch is shown in Fig. 2.3b: we have not put a scale on the $\dfrac{dy}{dx}$-axis. We have considered where the gradient $\dfrac{dy}{dx}$ is positive or negative, but we have not found any numerical values of the gradient.

If we require a more accurate graph of the gradient function, we must find the gradient of the curve at several values of x (for example, by drawing tangents), and then plot these values.

Example 4 For the curve given in Example 3, draw a graph of the gradient function.

Tangents to the curve have been drawn at $x = -4, -3, \ldots, 4, 5$ (Fig. 2.4). For example, the tangent at $x = -2$ is drawn between

$x = -2$ and $x = -1$, and its gradient is $\dfrac{3.4 - 2.1}{1} = 1.3$. The

gradients of all these tangents are given below.

x	-4	-3	-2	-1	0	1	2	3	4	5
Gradient of tangent	-1.6	-0.3	1.3	0	-0.8	-1.7	-0.6	0.8	1.1	1.4

We use these values to plot the gradient function, as shown in Fig. 2.5.

Exercise 2.6   (The gradient function)

1   Draw a graph of the function $y = 1 + x^2 - \frac{1}{3}x^3$ for values of x between -1.5 and $+3.5$ (using scales of $2\,\text{cm} = 1$ unit on the x-axis and $1\,\text{cm} = 1$ unit on the y-axis). Differentiate to find $\dfrac{dy}{dx}$, and draw a graph of the gradient function underneath the first graph. (Use a scale of $1\,\text{cm} = 1$ unit on the $\dfrac{dy}{dx}$-axis). Indicate how the two graphs are related.

In questions 2 and 3, plot the given points on a graph, and join them by a smooth curve (suitable scales are given after each question). Sketch roughly the graph of the gradient function.

2

x	-2	-1	0	1	2	3	4	5
y	4.2	2.4	1.2	0.6	0.4	0.7	0.9	2.8

($2\,\text{cm} = 1$ unit on both axes.)

3

x	0	1	2	3	4	5	6	7	8	9	10	11	12	13
y	4.2	5.4	6.3	6.1	4.3	1.3	0.7	1.6	2.5	3.0	2.8	2.1	1.0	-0.8

($1\,\text{cm} = 1$ unit on both axes.)

4   For the curve given in question 3, draw tangents at $x = 0, 2, 4, \ldots, 12$, and measure their gradients. Draw a graph of the gradient function.

2.7   Working from first principles

In Chapter 1, we demonstrated numerically that the gradient of the curve $y = x^2$ is given by the formula $2x$. However, showing that the formula appears to work for a few values of x does not prove the result. We shall now see how this (and similar results) can be established. First we must decide exactly what we mean by the gradient of a curve.

Consider again the point A(2, 4) on the curve $y = x^2$ (see Fig. 2.6). Let P be any other point on the curve. The gradient of the line AP is $\dfrac{y\text{-step}}{x\text{-step}}$.

We shall use δx (said 'delta x') for the x-step from A to P. The notation δx means the change in x; it should be regarded as a single symbol representing a number (it must *not* be thought of as delta times x).

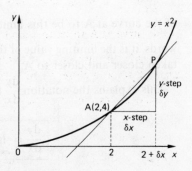

Fig. 2.6

Similarly, let δy (the change in y) be the y-step from A to P.

The x-coordinate of P is $(2 + \delta x)$, and so the y-coordinate of P is $(2 + \delta x)^2$; hence the y-step from A to P is

$$\delta y = (2 + \delta x)^2 - 4 = 4 + 4\,\delta x + (\delta x)^2 - 4$$
$$= 4\,\delta x + (\delta x)^2$$

and the gradient of AP is

$$\frac{y\text{-step}}{x\text{-step}} = \frac{\delta y}{\delta x} = \frac{4\,\delta x + (\delta x)^2}{\delta x} = 4 + \delta x$$

This is a general formula applying to any point P on the curve. It is still valid if P is to the left of A: then δx is negative, and the gradient of AP, $\dfrac{\delta y}{\delta x} = 4 + \delta x$, is less than 4.

If P is close to A, then δx is close to zero (though it may be positive or negative), and the gradient of AP, $\dfrac{\delta y}{\delta x} = 4 + \delta x$, is close to 4. As δx gets closer and closer to zero, either through positive or negative values, $\dfrac{\delta y}{\delta x}$ approaches a **limiting value** of 4.

We write $\dfrac{\delta y}{\delta x} \to 4$ as $\delta x \to 0$ (and say $\dfrac{\delta y}{\delta x}$ tends to 4 as δx tends to zero) or

$$\lim_{\delta x \to 0} \frac{\delta y}{\delta x} = 4$$

(and say the limit, as δx tends to zero, of $\dfrac{\delta y}{\delta x}$ is 4).

We now define the gradient of the curve at A to be this limiting value. The gradient at A is 4.

Definition of the gradient

Suppose A is a point on a curve (Fig. 2.7), and let δx and δy be the x-step and y-step measured from A to another point P on the curve.

If $\dfrac{\delta y}{\delta x}$ has a limiting value as $\delta x \to 0$, then we define the gradient of the curve at A to be this limiting value, i.e. $\lim\limits_{\delta x \to 0} \dfrac{\delta y}{\delta x}$. In other words, it is the limiting value of the gradient of the line AP, as P is taken closer and closer to A.

This explains the notation $\dfrac{\mathrm{d}y}{\mathrm{d}x}$ for the derived function. In fact,

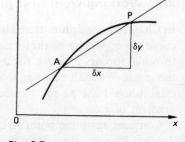

Fig. 2.7

$$\boxed{\frac{\mathrm{d}y}{\mathrm{d}x} = \lim_{\delta x \to 0} \frac{\delta y}{\delta x}}$$

Finding the derived function

The method used above for the point (2, 4) on the curve $y = x^2$ could be applied to find the gradient at any particular point on a curve, but it is also possible to work with a general point on the curve.

Let $A(x, y)$ be a general point on the curve $y = x^2$ (Fig. 2.8). If P is another point on the curve, and the x-step and y-step from A to P are δx and δy, then the coordinates of P are $(x + \delta x, y + \delta y)$.

Since A and P lie on the curve $y = x^2$, we have

$$y = x^2$$
$$\text{and} \quad y + \delta y = (x + \delta x)^2$$

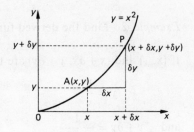

Fig. 2.8

We must remember that x and δx are quite separate quantities. Subtracting,

$$\delta y = (x + \delta x)^2 - x^2 = x^2 + 2x(\delta x) + (\delta x)^2 - x^2$$
$$= 2x(\delta x) + (\delta x)^2$$

Dividing by δx,

$$\frac{\delta y}{\delta x} = 2x + \delta x$$

As $\delta x \to 0$, $2x + \delta x \to 2x$, and so the derived function is

$$\frac{dy}{dx} = \lim_{\delta x \to 0} \frac{\delta y}{\delta x} = 2x$$

We have now shown that the gradient of the curve $y = x^2$ is indeed given by the formula $2x$.

Example 1   Find the derived function $\dfrac{dy}{dx}$ if $y = 2x^2 - 3x + 7$.

If (x, y) and $(x + \delta x, y + \delta y)$ are two points on the curve, then

$$y = 2x^2 - 3x + 7$$
$$\text{and} \quad y + \delta y = 2(x + \delta x)^2 - 3(x + \delta x) + 7$$

Subtracting,

$$\delta y = 2(x + \delta x)^2 - 3(x + \delta x) + 7 - (2x^2 - 3x + 7)$$
$$= 2x^2 + 4x(\delta x) + 2(\delta x)^2 - 3x - 3\delta x + 7 - 2x^2 + 3x - 7$$
$$= 4x(\delta x) + 2(\delta x)^2 - 3\delta x$$

So $\dfrac{\delta y}{\delta x} = 4x + 2(\delta x) - 3$

The derived function is

$$\frac{dy}{dx} = \lim_{\delta x \to 0} \frac{\delta y}{\delta x} = 4x - 3$$

Example 2   Find the derived function, $\dfrac{dy}{dx}$, if $y = \dfrac{1}{x}$.

If (x, y) and $(x + \delta x, y + \delta y)$ are two points on the curve, then

$$y = \frac{1}{x}$$

and $$y + \delta y = \frac{1}{x + \delta x}$$

Subtracting,

$$\delta y = \frac{1}{x + \delta x} - \frac{1}{x} = \frac{x - (x + \delta x)}{x(x + \delta x)}$$

$$= \frac{-\delta x}{x(x + \delta x)}$$

So $$\frac{\delta y}{\delta x} = \frac{-1}{x(x + \delta x)}$$

The derived function is

$$\frac{dy}{dx} = \lim_{\delta x \to 0} \frac{\delta y}{\delta x} = \frac{-1}{x(x)} = -\frac{1}{x^2}$$

Example 3   Find the derived function $\dfrac{dy}{dx}$ if $y = \sqrt{x}$.

If δx is a change in x, causing a change δy in y, then

$$y = \sqrt{x}$$

and $$y + \delta y = \sqrt{x + \delta x}$$

Subtracting,

$$\delta y = \sqrt{x + \delta x} - \sqrt{x} = \left(\sqrt{x + \delta x} - \sqrt{x} \right) \times \frac{(\sqrt{x + \delta x} + \sqrt{x})}{\sqrt{x + \delta x} + \sqrt{x}}$$

$$= \frac{(x + \delta x) - x}{\sqrt{x + \delta x} + \sqrt{x}} = \frac{\delta x}{\sqrt{x + \delta x} + \sqrt{x}}$$

So $$\frac{\delta y}{\delta x} = \frac{1}{\sqrt{x + \delta x} + \sqrt{x}}$$

The derived function is

$$\frac{dy}{dx} = \lim_{\delta x \to 0} \frac{\delta y}{\delta x} = \frac{1}{\sqrt{x} + \sqrt{x}} = \frac{1}{2\sqrt{x}}$$

In these examples we have found the derived function by direct evaluation of the limit $\lim\limits_{\delta x \to 0} \dfrac{\delta y}{\delta x}$. This is called differentiating from first principles.

Alternative notation

If we use $y = f(x)$ for the equation of a curve, and if (x, y) and $(x + \delta x, y + \delta y)$ are two points on the curve, then

$$y = f(x)$$

and

$$y + \delta y = f(x + \delta x)$$

Subtracting,

$$\delta y = f(x + \delta x) - f(x)$$

and so

$$\frac{\delta y}{\delta x} = \frac{f(x + \delta x) - f(x)}{\delta x}$$

The derived function is

$$f'(x) = \lim_{\delta x \to 0} \frac{\delta y}{\delta x} = \lim_{\delta x \to 0} \left(\frac{f(x + \delta x) - f(x)}{\delta x} \right)$$

We can of course use a single letter (instead of δx) for the x-step. In practice, h is often used, and we have

$$f'(x) = \lim_{h \to 0} \left(\frac{f(x + h) - f(x)}{h} \right)$$

Multiples

The function $2x^2$ may be regarded as twice the function x^2; similarly $-\dfrac{3}{x} = (-3)\left(\dfrac{1}{x}\right)$ is -3 times the function $\dfrac{1}{x}$.

In general, suppose that $y = ku$, where k is a constant and u is a function of x.

If δx is a change in x, this causes a change δu in u which then causes a change δy in y. The new values of y and u are $y + \delta y$ and $u + \delta u$, so we have

$$y + \delta y = k(u + \delta u)$$

and

$$y = ku$$

Subtracting,

$$\delta y = k(u + \delta u) - ku = k\delta u$$

Dividing by δx,

$$\frac{\delta y}{\delta x} = k\frac{\delta u}{\delta x}$$

If, as $\delta x \to 0$, $\dfrac{\delta u}{\delta x} \to \dfrac{du}{dx}$, then

$$\frac{dy}{dx} = \lim_{\delta x \to 0} \frac{\delta y}{\delta x} = k\frac{du}{dx}$$

Thus if $y = ku$, then

$$\frac{dy}{dx} = k\frac{du}{dx} \quad \text{(where } k \text{ is a constant)}$$

Hence we are justified in saying that

if $y = 2x^2$, then $\dfrac{dy}{dx} = 2(2x) = 4x$

if $y = -\dfrac{3}{x}$, then $\dfrac{dy}{dx} = (-3)\left(\dfrac{-1}{x^2}\right) = \dfrac{3}{x^2}$

and so on.

Sums and differences

The function $x^2 + \dfrac{4}{x}$ is the sum of the two functions x^2 and $\dfrac{4}{x}$. In general, suppose that $y = u + v$, where u and v are functions of x. If δx is a change in x and this causes changes of δu in u and δv in v, which together cause a change δy in y, then

$$y + \delta y = (u + \delta u) + (v + \delta v)$$
and $\qquad y = u + v$

Subtracting,

$$\delta y = \delta u + \delta v$$

Dividing by δx,

$$\frac{\delta y}{\delta x} = \frac{\delta u}{\delta x} + \frac{\delta v}{\delta x}$$

If, as $\delta x \to 0$,

$$\frac{\delta u}{\delta x} \to \frac{du}{dx} \quad \text{and} \quad \frac{\delta v}{\delta x} \to \frac{dv}{dx}$$

then $\qquad \dfrac{dy}{dx} = \lim\limits_{\delta x \to 0} \dfrac{\delta y}{\delta x} = \dfrac{du}{dx} + \dfrac{dv}{dx}$

Thus if $y = u + v$, then

$$\frac{dy}{dx} = \frac{du}{dx} + \frac{dv}{dx}$$

Similarly, if $y = u - v$, then

$$\frac{dy}{dx} = \frac{du}{dx} - \frac{dv}{dx}$$

For example, if $y = x^2 + \dfrac{4}{x}$, then

$$\frac{dy}{dx} = 2x + \left(-\frac{4}{x^2}\right) = 2x - \frac{4}{x^2}$$

and so on.

Exercise 2.7   (First principles)

1   A is the point $(1, 2)$ on the curve $y = 2x^2$. If P is another point on the curve, and the x-step from A to P is δx, find an expression for the y-step δy. Hence find $\dfrac{\delta y}{\delta x}$, and deduce the value of the gradient of the curve at A.

2   Repeat question 1 for the point $(2, -1)$ on the curve $y = x^2 - 5x + 5$.

For the following functions (questions 3 to 14), find the derived function $\dfrac{dy}{dx}$ from first principles.

3   $y = 3x^2$ 4 $y = x^2 + 3x$ 5 $y = 9 - 5x^2$
6   $y = 4x^2 - 9x + 5$ 7 $y = 3x - 2$ 8 $y = 2 - 5x$
9   $y = x^3$ [first verify that $(x + \delta x)^3 = x^3 + 3x^2(\delta x) + 3x(\delta x)^2 + (\delta x)^3$.]

10   $y = x^4$ 11 $y = x + \dfrac{1}{x}$ 12 $y = \dfrac{1}{x+1}$ 13 $y = \dfrac{1}{2x-3}$ 14 $y = \dfrac{1}{\sqrt{x}}$

15   If $y = u - kv$, where u and v are functions of x, and k is a constant, show from first principles that

$$\frac{dy}{dx} = \frac{du}{dx} - k\frac{dv}{dx}$$

2.8   When the gradient does not exist

We now have a precise mathematical definition of the gradient of a curve at a point A as the limit of $\dfrac{\delta y}{\delta x}$ as $\delta x \to 0$, provided that this limit exists. If $\dfrac{\delta y}{\delta x}$ does not approach a limiting value as $\delta x \to 0$ (taking both positive and negative values) then we say that the curve does not have a gradient at A. At such a point, $\dfrac{dy}{dx}$ does not exist, and we say that the function y is not differentiable.

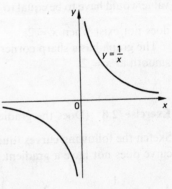

Example 1 $y = \dfrac{1}{x}$

With reference to Fig. 2.9, if $x \neq 0$, the gradient is given by $\dfrac{dy}{dx} = -\dfrac{1}{x^2}$. If $x = 0$, then $y = \dfrac{1}{0}$ is not defined, and there is no point on the curve corresponding to $x = 0$. So the curve does not exist, and therefore does not have a gradient, when $x = 0$.

Fig. 2.9

Example 2 $y = \begin{cases} 1 & \text{if } x \leqslant 1 \\ 2 & \text{if } x > 1 \end{cases}$

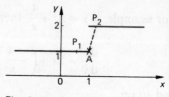

This curve (Fig. 2.10) consists of two straight lines: $y = 1$ up to and including the point $(1, 1)$, and $y = 2$ starting from, but *not* including, the point $(1, 2)$. If $x \neq 1$, the gradient is clearly zero.

Now consider the point $A(1, 1)$ on the curve. If P_1 is a point on the curve to the left of A, then the gradient of AP_1 is zero, i.e. if δx is negative, then $\dfrac{\delta y}{\delta x} = 0$. If P_2 is a point on the curve to the right of A, then the gradient of AP_2 becomes arbitrarily large as P_2 approaches $x = 1$; in fact if δx is positive, then $\delta y = 1$, so $\dfrac{\delta y}{\delta x} = \dfrac{1}{\delta x}$.

$\dfrac{\delta y}{\delta x}$ does not approach a limiting value as $\delta x \to 0$, so this curve does not have a gradient at the point $(1, 1)$.

Fig. 2.10

The curve has a break in it at $x = 1$. We say that the curve is not continuous, and that $x = 1$ is a point of discontinuity (see p. 236 for a fuller discussion of continuity). A curve can never have a gradient at such a point of discontinuity.

Example 3 $y = \begin{cases} x & \text{if } x \leqslant 2 \\ 4 - x & \text{if } x \geqslant 2 \end{cases}$

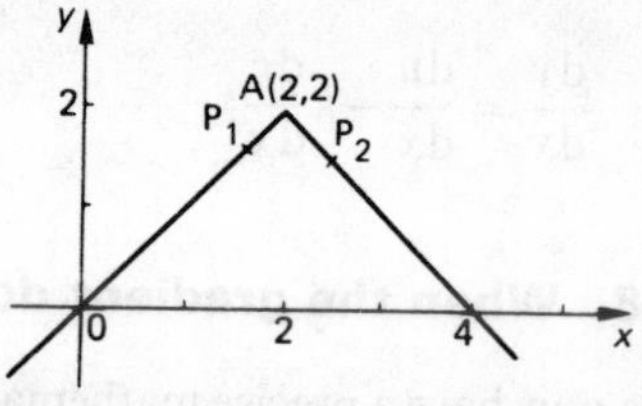

This graph (Fig. 2.11) consists of two straight lines which meet at the point $(2, 2)$.

If $x < 2$, the gradient is $+1$. If $x > 2$, the gradient is -1.

Now consider the point $A(2, 2)$. If P_1 is a point on the line to the left of A, then the gradient of AP_1 is $+1$. If P_2 is a point on the line to the right of A, then the gradient of AP_2 is -1. So $\dfrac{\delta y}{\delta x} = +1$ if δx is negative, and $\dfrac{\delta y}{\delta x} = -1$ if δx is positive. Thus $\dfrac{\delta y}{\delta x}$ does not approach a limiting value as $\delta x \to 0$ (because if it did, the limiting value would have to be equal to $+1$ and also equal to -1), and $\dfrac{dy}{dx}$ does not exist when $x = 2$.

Fig. 2.11

The graph has a sharp corner at A; we say that the graph is not smooth at $x = 2$.

Exercise 2.8   (Does the gradient exist?)

Sketch the following curves (questions 1 to 6) and state, with reasons, any values of x for which the curve does not have a gradient.

1 $y = \dfrac{1}{x - 2}$

2 $y = \begin{cases} 3 & \text{if } x \leqslant 4 \\ 2 & \text{if } x > 4 \end{cases}$

3 $y = \begin{cases} 2 + x & \text{if } x \leqslant 0 \\ 2 - x & \text{if } x \geqslant 0 \end{cases}$

4 $y = \dfrac{1}{x} - \dfrac{1}{x+1}$ **5** $y = \begin{cases} x+3 & \text{if } x \leqslant 1 \\ 3 & \text{if } x > 1 \end{cases}$ **6** $y = \begin{cases} 1+x & \text{if } x \leqslant 0 \\ 1-x & \text{if } 0 \leqslant x \leqslant 1 \\ x-1 & \text{if } x \geqslant 1 \end{cases}$

7 A is the point $(1, \frac{1}{2})$ on the curve

$$y = \begin{cases} \frac{1}{2} & \text{if } x \leqslant 1 \\ \frac{1}{2}x^2 & \text{if } x \geqslant 1 \end{cases}$$

If δx and δy are the changes in x and y measured from A, find expressions for $\dfrac{\delta y}{\delta x}$ (i) when δx is

negative, and (ii) when δx is positive. [Hint: when δx is positive, $\delta y = \frac{1}{2}(1+\delta x)^2 - \frac{1}{2}$.]

Does the curve have a gradient at A?

8 Repeat question 7 for the point $A(1, \frac{1}{2})$ on the curve

$$y = \begin{cases} x - \frac{1}{2} & \text{if } x \leqslant 1 \\ \frac{1}{2}x^2 & \text{if } x \geqslant 1 \end{cases}$$

3

Applications of differentiation

3.1 Repeated differentiation

If $y = f(x)$, the derived function (or derivative) $\dfrac{dy}{dx} = f'(x)$ is also a function of x, and is sometimes called the first derivative. Usually we can differentiate again, and we then obtain the second derivative, written $\dfrac{d^2y}{dx^2} = f''(x)$ (said 'dee two y by dx squared' and 'f double dashed of x'): $\dfrac{d^2y}{dx^2}$ is short for $\dfrac{d}{dx}\left(\dfrac{dy}{dx}\right)$. If we differentiate yet again we obtain the third derivative $\dfrac{d^3y}{dx^3} = f'''(x)$ (said 'dee three y by dx cubed' and 'f treble dashed of x'), and so on.

These higher derivatives are useful for investigating stationary points on curves (see p. 39) and for expressing a given function as a power series (see Chapter 17).

Example 1   If $y = 3x^4 - 7x^2 + 6x - 5$, find $\dfrac{dy}{dx}$ and $\dfrac{d^2y}{dx^2}$.

$$\frac{dy}{dx} = 3(4x^3) - 7(2x) + 6(1) - 0 = 12x^3 - 14x + 6$$

Differentiating again,

$$\frac{d^2y}{dx^2} = 12(3x^2) - 14(1) + 0 = 36x^2 - 14$$

Example 2   If $f(x) = 3\sqrt{x} - \dfrac{1}{x}$, find $f'(4)$ and $f''(4)$.

We have

$$f(x) = 3x^{\frac{1}{2}} - x^{-1}$$

so $$f'(x) = 3\left(\frac{1}{2}x^{-\frac{1}{2}}\right) - (-x^{-2})$$

$$= \frac{3}{2}x^{-\frac{1}{2}} + x^{-2}$$

so $$f'(4) = \frac{3}{2} \times \frac{1}{2} + \frac{1}{16} = \frac{13}{16}$$

Differentiating again,

$$f''(x) = \frac{3}{2}\left(-\frac{1}{2}x^{-\frac{3}{2}}\right) + (-2x^{-3})$$

$$= -\frac{3}{4}x^{-\frac{3}{2}} - 2x^{-3}$$

so $f''(4) = -\frac{3}{4} \times \frac{1}{8} - 2 \times \frac{1}{64} = -\frac{1}{8}$

Example 3 If $y = 3x^2 + 4x + \frac{2}{3}$, find $\frac{dy}{dx}$ and $\frac{d^2y}{dx^2}$, and verify that y

satisfies the equation

$$\frac{d^2y}{dx^2} - 2\frac{dy}{dx} + 3y = 9x^2$$

We have $\frac{dy}{dx} = 6x + 4$

and, differentiating again, $\frac{d^2y}{dx^2} = 6$. Thus

$$\frac{d^2y}{dx^2} - 2\frac{dy}{dx} + 3y = 6 - 2(6x+4) + 3\left(3x^2 + 4x + \frac{2}{3}\right)$$

$$= 9x^2$$

An equation involving derivatives, for example the equation

$$\frac{d^2y}{dx^2} - 2\frac{dy}{dx} + 3y = 9x^2$$

in Example 3 is called a **differential equation**.
Then

$$y = 3x^2 + 4x + \frac{2}{3}$$

is said to be a solution of the differential equation (since we have
shown that it satisfies the equation).

Exercise 3.1 (Repeated differentiation)

1 For the following functions, find $\frac{dy}{dx}$ and $\frac{d^2y}{dx^2}$.

(i) x^4 (ii) $7x^3$ (iii) $\sqrt{x}$ (iv) $\frac{1}{x}$ (v) $\frac{1}{x^4}$

(vi) $3x - 8$ (vii) $x^3 + 3x$ (viii) $(x-3)(x-4)$ (ix) $x + \frac{1}{x}$ (x) $2x^2 + \frac{3}{x^2}$

2 (i) If $f(x) = 3 - x + 4x^2 - 2x^3$, find $f(-2), f'(-2)$ and $f''(-2)$.

(ii) If $f(x) = \sqrt[3]{x}$, find $f(8)$, $f'(8)$ and $f''(8)$.

(iii) If $f(x) = 1 - \dfrac{4}{x} + \dfrac{2}{x^2}$, find $f(\frac{1}{2})$, $f'(\frac{1}{2})$ and $f''(\frac{1}{2})$.

3   If $y = 2x^2 + 3x + \dfrac{5}{4}$, find $\dfrac{dy}{dx}$ and $\dfrac{d^2y}{dx^2}$, and verify that y satisfies the equation

$$\frac{d^2y}{dx^2} - 3\frac{dy}{dx} + 4y = 8x^2$$

4   Verify that $y = x^3 - 3x^2 + 6$ satisfies $\dfrac{d^2y}{dx^2} + \dfrac{dy}{dx} + y = x^3$.

5   Verify that $y = \sqrt{x}$ satisfies $4x^2\dfrac{d^2y}{dx^2} + y = 0$.

6   Verify that $y = 2x + \dfrac{3}{x}$ satisfies $x^2\dfrac{d^2y}{dx^2} + x\dfrac{dy}{dx} - y = 0$.

7   Find the value of a if $y = x^2 + ax$ satisfies $x\dfrac{d^2y}{dx^2} - \dfrac{dy}{dx} = 3$.

8   Find the values of a and b if $y = x^2 + ax + b$ satisfies $\dfrac{d^2y}{dx^2} - 2\dfrac{dy}{dx} + y = x^2$.

3.2   Curve sketching using the derived function

It is often useful to be able to sketch the general shape of a graph, without plotting points accurately, and this is a technique which we shall develop in later chapters. The derived function $\dfrac{dy}{dx}$ can help us to do this, because it tells us whether the gradient of the curve is positive or negative.

When $\dfrac{dy}{dx}$ is positive, the graph has a positive gradient. As we move along the x-axis in the positive direction, the value of y increases. We say that y is an **increasing function** of x.

When $\dfrac{dy}{dx}$ is negative, the graph has a negative gradient, and as we move along the x-axis, the value of y decreases. We say that y is a **decreasing function** of x.

It is certainly possible for a function to be increasing in some regions and decreasing in others. For example, the function whose graph is shown in Fig. 3.1 is increasing when $x < 1$, decreasing when $1 < x < 3$, and increasing again when $x > 3$.

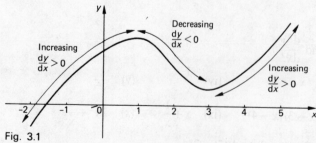

Fig. 3.1

Example 1 Sketch the curve $y = x^2 - 4x + 5$.

The derived function is $\dfrac{dy}{dx} = 2x - 4$. The graph of this is a straight

line which we can sketch easily (Fig. 3.2). We see that $\dfrac{dy}{dx} < 0$ when

$x < 2$, $\dfrac{dy}{dx} = 0$ when $x = 2$, and $\dfrac{dy}{dx} > 0$ when $x > 2$.

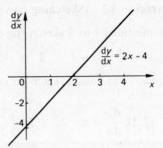

Fig. 3.2

When $x = 2$, $y = 4 - 8 + 5 = 1$ so we begin our sketch of the curve by marking in the point $(2, 1)$. When $x < 2$, y is a decreasing function of x, so we draw the graph sloping downwards towards the point $(2, 1)$. When $x > 2$, y is an increasing function of x, so we draw the graph sloping upwards away from the point $(2, 1)$. Note that when $x = 0$, $y = 5$, so the graph (Fig. 3.3) crosses the y-axis at the point $(0, 5)$.

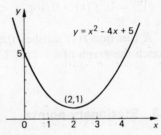

Fig. 3.3

Example 2 Sketch the curve which passes through the point $(3, 2)$ and which has the gradient function $\dfrac{dy}{dx} = 3 - x$.

We first sketch the graph of $\dfrac{dy}{dx} = 3 - x$, which is a straight line

(Fig. 3.4). We see that $\dfrac{dy}{dx} > 0$ when $x < 3$, $\dfrac{dy}{dx} = 0$ when $x = 3$ and

$\dfrac{dy}{dx} < 0$ when $x > 3$.

We begin our sketch with the given point $(3, 2)$. When $x < 3$, the graph slopes upwards towards this point, and when $x > 3$, the graph slopes downwards away from it. We can now sketch the curve (Fig. 3.5).

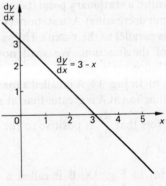

Fig. 3.4

Example 3 Sketch the graph of a function $y = f(x)$ which satisfies all the following conditions: $f(0) = 3$; $f'(x) < 0$ for $x < 0$; $f'(0) = 0$; $f'(x) > 0$ for $0 < x < 2$; $f'(2) = 0$; $f'(x) < 0$ for $x > 2$.

The condition $f(0) = 3$ tells us that the point $(0, 3)$ is on the graph. The remaining conditions give us information about the gradient $f'(x)$. The function is decreasing for $x < 0$ (since $f'(x) < 0$), increasing for $0 < x < 2$ (since $f'(x) > 0$), and decreasing for $x > 2$ (since $f'(x) < 0$). The sketch is shown in Fig. 3.6.

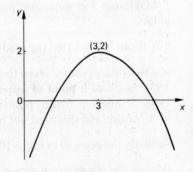

Fig. 3.5

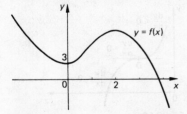

Fig. 3.6

Exercise 3.2   (Sketching curves)

In questions 1 to 3 sketch the curves, by first finding the derived functions.

1 $y = x^2 - 2x + 4$ **2** $y = 3 + 6x - x^2$ **3** $y = 2x^2 - 6x + 3$

In questions 4 and 5 sketch the curve which passes through the given point, and has the given gradient function.

4 $(2, 1)$; $\dfrac{dy}{dx} = x - 2$ **5** $\left(\dfrac{3}{2}, -1\right)$; $\dfrac{dy}{dx} = 3 - 2x$

In questions 6 and 7 sketch the graph of a function $y = f(x)$ which satisfies all the given conditions.

6 $f(0) = 0$; $f'(x) > 0$ for $x < 0$; $f'(0) = 0$; $f'(x) < 0$ for $x > 0$.
7 $f(5) = 1$; $f'(x) > 0$ for $x < -1$; $f'(-1) = 0$; $f'(x) < 0$ for $-1 < x < 5$; $f'(5) = 0$; $f'(x) > 0$ for $x > 5$.
8 A function $f(x)$ satisfies the conditions $f(0) = 0$; $f'(x) = -1$ for $x < 0$; $f'(x) = +1$ for $x > 0$. Sketch the graph of $y = f(x)$. Can you say anything about $f'(0)$?

3.3   Stationary points

A point on a graph at which the gradient is zero (that is, $\dfrac{dy}{dx} = 0$) is called a **stationary point** (because the function is neither increasing nor decreasing). At a stationary point, the curve is 'flat' (the tangent is parallel to the x-axis). The value of y is called a **stationary value** of the function. We shall now examine the different types of stationary points.

(i) In Fig. 3.7, A is called a **maximum point** (since the value of the function at A is greater than at any other point in the locality of A). Note that $\dfrac{dy}{dx}$ is positive to the left of A and negative to the right of A.

(ii) In Fig. 3.8, B is called a **minimum point**. Note that $\dfrac{dy}{dx}$ is negative to the left of B and positive to the right of B.

Maximum and minimum points are sometimes called **turning points**.

(iii) It can happen that the gradient $\left(\dfrac{dy}{dx}\right)$ is positive on both sides of a stationary point C (even though the gradient at C is zero): Fig. 3.9. C is called a **point of inflexion**. (Generally we use the term 'point of inflexion' for any point where a curve changes its direction of curvature, and this need not be at a stationary point; see p. 206.)

Similarly the point D in Fig. 3.10 (for which $\dfrac{dy}{dx}$ is negative on both sides of the stationary point) is also a point of inflexion.

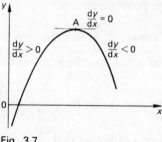

Fig. 3.7

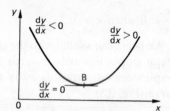

Fig. 3.8

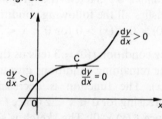

Fig. 3.9

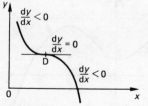

Fig. 3.10

(iv) If the gradient $\dfrac{dy}{dx}$ is zero for all points on the curve, we then have y as a constant (for example, $y = 2$ in Fig. 3.11).

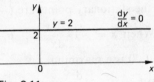

Fig. 3.11

We can use the second derivative $\dfrac{d^2y}{dx^2}$ to help classify stationary points. As $\dfrac{d^2y}{dx^2} = \dfrac{d}{dx}\left(\dfrac{dy}{dx}\right)$, this tells us whether $\dfrac{dy}{dx}$ is increasing or decreasing.

Suppose P is a stationary point. At P, $\dfrac{dy}{dx} = 0$.

If $\dfrac{d^2y}{dx^2}$ is positive, $\dfrac{dy}{dx}$ is increasing; so, since its value at P is zero, it must be negative to the left and positive to the right. We therefore have a **minimum point**:

If $\dfrac{d^2y}{dx^2}$ is negative, $\dfrac{dy}{dx}$ is decreasing; so, since its value at P is zero, it must be positive to the left and negative to the right. We therefore have a **maximum point**.

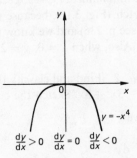

Maximum point

Fig. 3.12

However, if $\dfrac{d^2y}{dx^2}$ is zero, this in itself gives us no information about the nature of the stationary point. In this case it is usually best to investigate the gradient $\dfrac{dy}{dx}$ on both sides of P.

Figs 3.12, 3.13 and 3.14 show three curves with stationary points when $x = 0$ and with $\dfrac{d^2y}{dx^2} = 0$ at that point.

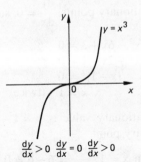

Point of inflexion

Fig. 3.13

Example 1   Find and classify the stationary points on the curve $y = x^3 - 3x^2 - 9x + 2$. Sketch the curve.

We have

$$\frac{dy}{dx} = 3x^2 - 6x - 9$$

and $$\frac{d^2y}{dx^2} = 6x - 6$$

For stationary points, $\dfrac{dy}{dx} = 0$, so

$$3x^2 - 6x - 9 = 0$$
i.e. $$x^2 - 2x - 3 = 0$$
i.e. $$(x + 1)(x - 3) = 0$$
i.e. $$x = -1 \text{ or } 3$$

Substituting in the equation of the curve to find the stationary values:

when $x = -1$, $y = -1 - 3 + 9 + 2 = 7$;
when $x = 3$, $y = 27 - 27 - 27 + 2 = -25$;

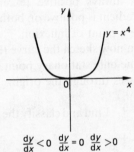

Minimum point

Fig. 3.14

so the stationary points are $(-1, 7)$ and $(3, -25)$.

When $x = -1$, $\dfrac{d^2y}{dx^2} = -6-6 = -12$, which is negative; so $(-1, 7)$ is a maximum.

When $x = 3$, $\dfrac{d^2y}{dx^2} = 18-6 = 12$, which is positive; so $(3, -25)$ is a minimum.

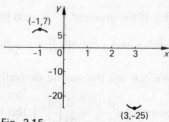

Fig. 3.15

To sketch the curve, we first mark in the maximum at $(-1, 7)$ and the minimum at $(3, -25)$; see Fig. 3.15. We can easily complete the sketch (Fig. 3.16), because the graph must be a continuous curve (see p. 236) and we know that there are no other stationary points. Also, when $x = 0$, $y = 2$.

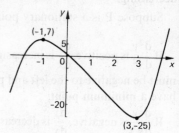

Fig. 3.16

Example 2   Find and classify the stationary points on the curve $y = x^3 - 3x^2 + 3x$. Sketch the curve.

We have

$$\frac{dy}{dx} = 3x^2 - 6x + 3$$

and

$$\frac{d^2y}{dx^2} = 6x - 6$$

For stationary points, $\dfrac{dy}{dx} = 0$, so

$$3x^2 - 6x + 3 = 0$$
i.e. $$x^2 - 2x + 1 = 0$$
i.e. $$(x-1)^2 = 0$$
i.e. $$x = 1 \quad \text{(twice)}$$

The stationary value is $y = 1 - 3 + 3 = 1$, so $(1, 1)$ is the only stationary point.

When $x = 1$, $\dfrac{d^2y}{dx^2} = 6-6 = 0$. This gives no further informa-

tion, so we return to $\dfrac{dy}{dx}$. Now $\dfrac{dy}{dx} = 3x^2 - 6x + 3 = 3(x-1)^2$, which is always positive (except at $x = 1$ where it is zero). So the gradient is positive on both sides of the point $(1, 1)$, which is therefore a point of inflexion.

We can now sketch the curve (Fig. 3.17), because we know that $(1, 1)$ is the only stationary point. Also, when $x = 0$, $y = 0$, so the curve passes through the origin.

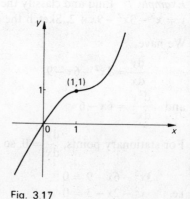

Fig. 3.17

Example 3   Find and classify the stationary values of the function $f(x) = 4x + \dfrac{1}{x}$.

We have

$$f(x) = 4x + \frac{1}{x} = 4x + x^{-1}$$

so $f'(x) = 4 - x^{-2} = 4 - \dfrac{1}{x^2}$

and $f''(x) = 2x^{-3} = \dfrac{2}{x^3}$

For stationary points, $f'(x) = 0$, so

$$4 - \frac{1}{x^2} = 0$$

i.e. $x^2 = \frac{1}{4}$

i.e. $x = \pm \frac{1}{2}$

The stationary values are thus $f(-\frac{1}{2}) = -2 - 2 = -4$ and $f(\frac{1}{2}) = 2 + 2 = 4$. But

$f''(-\frac{1}{2}) = \dfrac{2}{-\frac{1}{8}} = 16$; so -4 is a maximum value (when $x = -\frac{1}{2}$)

$f''(\frac{1}{2}) = \dfrac{2}{\frac{1}{8}} = 16$; so 4 is a minimum value (when $x = \frac{1}{2}$).

It may seem odd that the maximum value is less than the minimum value. $f(x)$ is not defined when $x = 0$ (since $\frac{1}{0}$ has no meaning). When the graph $y = 4x + \dfrac{1}{x}$ is plotted, it is found to consist of two parts, separated by $x = 0$ (Fig. 3.18).

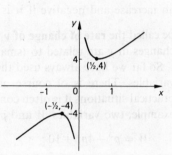

Fig. 3.18

Remember that maxima and minima are local terms: a maximum point gives a maximum value of the function *in that locality*, but this is not necessarily the maximum value of the function when viewed as a whole. For example, consider a function defined on the domain $-2 \leqslant x \leqslant 3$ whose graph is shown in Fig. 3.19. (The set of values taken by x is called the **domain** whilst the corresponding set of values taken by $f(x)$ is called the **range**.)

In Fig. 3.19, A is a local maximum, but in the given domain, the maximum value of the function occurs at C. Similarly, B is a local minimum, but the minimum value in the domain occurs at D.

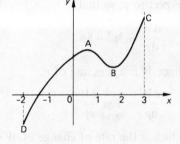

Fig. 3.19

Exercise 3.3   (Stationary points)

In questions 1 to 8 find and classify the stationary points on the following curves. Sketch the curves.

1   $y = x^2 - 2x + 4$ 2 $y = 1 - 4x - 2x^2$ 3 $y = x^3 + 3x^2 - 9x - 4$

4   $y = 8 + 24x + 3x^2 - x^3$ 5 $y = x^3 + 6x^2 + 12x + 9$ 6 $y = 1 - x^3$

7   $y = x^4 + 3$ 8 $y = 2 - x^4$

Find and classify the stationary values of the following functions (questions 9 to 13).

9   $f(x) = x^3 - 12x$ 10 $f(x) = 3 + 4x^3 - x^4$ 11 $f(x) = x + \dfrac{9}{x}$

12   $f(x) = x - \sqrt{x}$ 13 $f(x) = \dfrac{1}{x} + \dfrac{1}{x^2} - \dfrac{1}{x^3}$

Use your sketches from questions 3 and 5 to find the maximum and minimum values taken by the following functions in the given domain.

14 $y = x^3 + 3x^2 - 9x - 4;$ $-4 \leqslant x \leqslant 4.$ **15** $y = x^3 + 6x^2 + 12x + 9;$ $-3 \leqslant x \leqslant 0.$

3.4   The derived function as a rate of change

The derived function $\dfrac{dy}{dx}$ is defined as the limit as $\delta x \to 0$ of $\dfrac{\delta y}{\delta x}$.

Now $\dfrac{\delta y}{\delta x} = \dfrac{y\text{-step}}{x\text{-step}} = \dfrac{\text{change in } y}{\text{change in } x}$, where a change is positive if it is

an increase and negative if it is a decrease. $\dfrac{dy}{dx}$ can therefore be

be called the **rate of change of y with respect to x** as it shows how changes in y are related to (small) changes in x.

So far we have always used the letters x and y to represent the variables. There is of course no magic in these letters, and in practical situations it is often convenient to use different ones. For example, two variables W and p may be related by

$$W = p^3 - 4p^2 + 10$$

If a graph is drawn of W against p, its gradient is given by the derived function $\dfrac{dW}{dp}$ which is found by differentiating W with respect to p, so that

$$\frac{dW}{dp} = 3p^2 - 8p$$

Since W is a function of p,

$$\frac{dW}{dp} = \lim_{\delta p \to 0} \frac{\delta W}{\delta p}$$

which is the rate of change of W with respect to p.

We can differentiate again to obtain $\dfrac{d^2W}{dp^2} = 6p - 8$.

Many quantities vary with time, and rates of change with respect to time are particularly important. The letter t is normally used to represent time, so $\dfrac{dW}{dt}$ is the rate of change of W with respect to time, which may be called simply the rate of change of W.

Exercise 3.4   (Variables other than x and y)

1   If $u = 2q - 4q^2 + 3q^3$, find $\dfrac{du}{dq}$ and $\dfrac{d^2u}{dq^2}$.

2   If $x = t^4 - 2t + \dfrac{1}{t}$, find $\dfrac{dx}{dt}$ and $\dfrac{d^2x}{dt^2}$.

3 If $A = 4\pi r^2$, find $\dfrac{\mathrm{d}A}{\mathrm{d}r}$ and $\dfrac{\mathrm{d}^2 A}{\mathrm{d}r^2}$ (π is a constant).

4 If $V = \dfrac{4}{3}\pi r^3$, find $\dfrac{\mathrm{d}V}{\mathrm{d}r}$ and $\dfrac{\mathrm{d}^2 V}{\mathrm{d}r^2}$.

5 If $B = \pi r^2 + \pi r l$, find: (i) $\dfrac{\mathrm{d}B}{\mathrm{d}r}$ if l is constant; (ii) $\dfrac{\mathrm{d}B}{\mathrm{d}l}$ if r is constant.

6 If $s = ut + \dfrac{1}{2}gt^2$, find $\dfrac{\mathrm{d}s}{\mathrm{d}t}$ and $\dfrac{\mathrm{d}^2 s}{\mathrm{d}t^2}$ (u and g are constants).

3.5 Problems involving maxima and minima

The general method for tackling problems on maxima and minima is as follows.

i) Express the quantity to be maximised (or minimised), say W, in terms of letters related to the variables in the problem.

ii) Use the given conditions to form relationships between the variables, and express W in terms of one variable only, say p.

iii) Differentiate with respect to p to obtain $\dfrac{\mathrm{d}W}{\mathrm{d}p}$.

iv) Put $\dfrac{\mathrm{d}W}{\mathrm{d}p} = 0$ (since maximum and minimum values of W occur when $\dfrac{\mathrm{d}W}{\mathrm{d}p} = 0$) and find the value of p.

v) Check $\left(\text{usually by finding } \dfrac{\mathrm{d}^2 W}{\mathrm{d}p^2}\right)$ that this value of p does give a maximum (or minimum).

vi) Calculate the corresponding value of W (if required).

Example 1   120 m of fencing is to be used to form three sides of a rectangular enclosure, the fourth side being an existing wall. Find the maximum possible area of the enclosure.

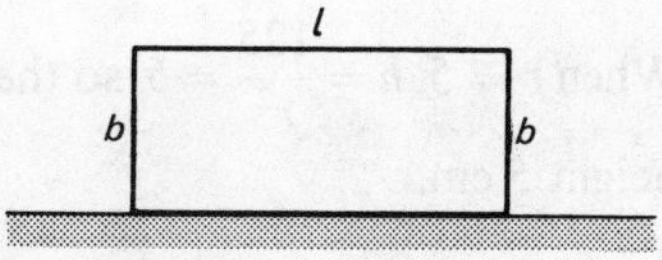

If the length of the enclosure is l m and the breadth b m, then the area of the enclosure is $A = lb$ (see Fig. 3.20). Since the total length of fencing is 120 m, we have $l + 2b = 120$. Thus

$$A = (120 - 2b)b = 120b - 2b^2$$

So $\dfrac{\mathrm{d}A}{\mathrm{d}b} = 120 - 4b$

For a maximum, $\dfrac{\mathrm{d}A}{\mathrm{d}b} = 0$, so $120 - 4b = 0$, i.e. $b = 30$. Since

$\dfrac{\mathrm{d}^2 A}{\mathrm{d}b^2} = -4$, this does give a maximum value. Substituting $b = 30$

into the equation for A, we obtain the maximum area as $A = (120 - 2b)b = 60 \times 30 = 1800 \text{ m}^2$ (in which case the enclosure measures 60 m by 30 m).

Fig. 3.20

Example 2   A metal can in the shape of a cylinder, open at one end, is to have a volume of 125π cm³. What should the dimensions be if the area of metal used is to be a minimum?

If the radius is r cm and the height h cm (see Fig. 3.21), then the area of metal used is $A = \pi r^2 + 2\pi r h$. This is in terms of two variables r and h (π of course is a constant).

Fig. 3.21

Since the volume is 125π cm³, we have

$$\pi r^2 h = 125\pi$$

so $$h = \frac{125}{r^2}$$

Thus

$$A = \pi r^2 + 2\pi r \left(\frac{125}{r^2}\right) = \pi r^2 + \frac{250\pi}{r}$$

This is now in terms of one variable only, i.e. r. So

$$\frac{dA}{dr} = 2\pi r - \frac{250\pi}{r^2}$$

For a minimum, $\dfrac{dA}{dr} = 0$, so

$$2\pi r = \frac{250\pi}{r^2}$$

i.e. $r^3 = 125$

i.e. $r = 5$

We have

$$\frac{d^2 A}{dr^2} = 2\pi + \frac{500\pi}{r^3}$$

which is positive when $r = 5$, so we do have a minimum value. When $r = 5$, $h = \dfrac{125}{r^2} = 5$, so the can should have radius 5 cm and height 5 cm.

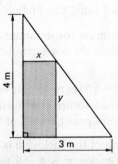

Fig. 3.22

Exercise 3.5   (Maxima and minima)

1   80 m of fencing is to be used for the 4 sides of a rectangular enclosure. Find the maximum area of the enclosure.

2   96 m of rope is to be used to mark out 3 sides of a rectangular enclosure, the other side being a hedge. Find the dimensions of the enclosure when its area is a maximum.

3   The sides p and q of a rectangle satisfy the condition $q = p + \dfrac{8}{p} - 2$. Find (i) the minimum area of the rectangle (ii) the minimum perimeter of the rectangle.

4   A rectangle with sides x and y fits inside a right-angled triangle with sides 3 m, 4 m, 5 m as shown in Fig. 3.22. Show that $4x + 3y = 12$, and find the maximum area of the rectangle.

5   A rectangular box has a square base, and no top. If the total area of the material used to make the box is 1200 cm², find the maximum volume of the box.

6   A metal rod has a square cross-section. If the volume of the rod is 512 cm³, find the dimensions of the rod when its total surface area (including the ends) is a minimum.

7   From a card measuring 24 cm by 9 cm, squares of side x cm are cut from each corner (see Fig. 3.23), and the resulting flaps are folded up to form an open box of height x cm. Find the maximum volume of this box.

8   A hollow cylinder is open at one end and closed at the other. Find the maximum volume if the total outside surface area is 300π cm².

9   A solid cylinder has volume 432π cm³. Find the dimensions of the cylinder when its total surface area is a minimum.

10   A cone is such that the diameter of its base added to its perpendicular height is 3 m. Find the maximum volume of the cone.

3.6   The equation of the tangent to a curve

The **tangent** to a curve at a point A is the straight line which passes through A and has the same gradient as the curve at A (see Fig. 3.24). It is the straight line which 'best fits' the curve in the neighbourhood of the point A.

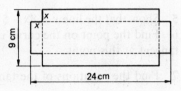

Fig. 3.23

If A is the point (x_1, y_1) we can find the gradient m_1 of the curve at A by differentiating and substituting $x = x_1$ in the gradient function. The tangent is then the line passing through (x_1, y_1) with gradient m_1, and so its equation is

$$y - y_1 = m_1(x - x_1)$$

(see p. 5).

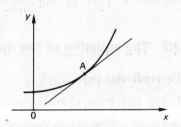

Fig. 3.24

Example 1   Find the equation of the tangent to the curve $y = 2x^3 - 7x^2 + 9$ at the point $(2, -3)$.

The gradient of the curve is given by $\dfrac{dy}{dx} = 6x^2 - 14x$. Substituting $x = 2$, the gradient at $(2, -3)$ is $6 \times 4 - 14 \times 2 = -4$. The tangent is the line through $(2, -3)$ with gradient -4, and so its equation is

$$y + 3 = -4(x - 2) \quad \text{or} \quad 4x + y = 5$$

Example 2   Show that the line $y = 4x - 3$ is a tangent to the curve $y = x^2 + 1$, and find the point of contact.

To find where the line meets the curve, we solve $y = 4x - 3$ and $y = x^2 + 1$ as simultaneous equations. Thus

$$x^2 + 1 = 4x - 3$$
i.e. $$x^2 - 4x + 4 = 0$$
i.e. $$(x - 2)^2 = 0$$

i.e. $x = 2$ and so $y = 5$ (using either equation). The gradient of the curve $y = x^2 + 1$ is given by $\dfrac{dy}{dx} = 2x$. When $x = 2$ the gradient is 4.

The tangent to the curve at $(2, 5)$ has a gradient of 4. Also, the line $y = 4x - 3$ passes through $(2, 5)$ and has a gradient of 4. Hence the line *is* the tangent to the curve at (2, 5). (When we solved $y = 4x - 3$ and $y = x^2 + 1$, we found that our equation had equal roots i.e. $x = 2$ (twice). In other words the line cut the curve at two points which were identical, i.e. the line was a tangent to the curve.)

Exercise 3.6   (Tangents)

In questions 1 to 4 find the equation of the tangent to the curve at the given point.

1   $y = x^2$ at (3, 9).

2   $y = x^3 - 4x$ at (−2, 0).

3   $y = 2x^3 + 7x^2 + 8x$ at (−1, −3).

4   $y = 2x - \dfrac{1}{x^2}$ at $(\frac{1}{2}, -3)$.

5   Show that the line $y = 8x - 5$ is a tangent to the curve $y = 2x^2 + 3$, and find the point of contact.

6   Find the point on the curve $y = x^2 - 3x$ at which the gradient is 5, and find the equation of the tangent at this point.

7   Find the equations of the tangents to the curve $y = x - \dfrac{8}{x}$ which have gradient 3.

8   A(4, 1) and B(−16, 16) are two points on the curve $y = \frac{1}{16}x^2$. The tangents at A and B intersect at the point C. Find the coordinates of C, and calculate the ratio $\dfrac{AC}{BC}$.

3.7   The equation of the normal to a curve

Perpendicular lines

In Fig. 3.25, P and Q are two points on a line, and the gradient of PQ is $m_1 = \dfrac{b}{a}$. Rotate the triangle QPR about Q through a right angle (anticlockwise) to obtain the triangle QST. Then QS is a line perpendicular to PQ, and the gradient of QS is $m_2 = \dfrac{TS}{QT} = \dfrac{-a}{b}$.

So we have

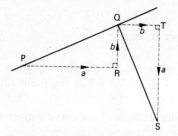

Fig. 3.25

$$m_1 m_2 = \left(\frac{b}{a}\right) \times \left(\frac{-a}{b}\right) = -1$$

If two lines are perpendicular, the product of their gradients is −1. If a line has gradient m_1, a perpendicular line has gradient

$$-\frac{1}{m_1}$$

Example   Find the equation of the perpendicular from the point (1, −3) to the line $3x - 5y + 7 = 0$.

The line $3x - 5y + 7 = 0$, or $y = \frac{3}{5}x + \frac{7}{5}$, has gradient $\frac{3}{5}$; so the perpendicular has gradient $-\frac{5}{3}$. The required line passes through (1, −3) and has gradient $-\frac{5}{3}$, and so its equation is

$$y + 3 = -\tfrac{5}{3}(x - 1) \quad \text{or} \quad 5x + 3y + 4 = 0$$

The normal to a curve

The **normal** to a curve at a point A is the line through A perpendicular to the tangent there (see Fig. 3.26).

If the gradient of the curve at $A(x_1, y_1)$ is m_1, then the gradient of the normal is $-\dfrac{1}{m_1}$, and so the equation of the normal is

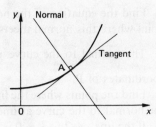

Fig. 3.26

$$y - y_1 = -\frac{1}{m_1}(x - x_1)$$

Example 1 Find the equation of the normal to the curve $y = 3x - 2x^2$ at the point $(3, -9)$.

We have $\dfrac{dy}{dx} = 3 - 4x$, so when $x = 3$, the gradient of the curve is $3 - 12 = -9$. The gradient of the normal is therefore $\frac{1}{9}$, and its equation is

$$y + 9 = \tfrac{1}{9}(x - 3) \quad \text{or} \quad x - 9y = 84$$

Example 2 The normal to the curve $y = 2x^2 - 1$ at the point $(1, 1)$, meets the curve again at P. Find the coordinates of P.

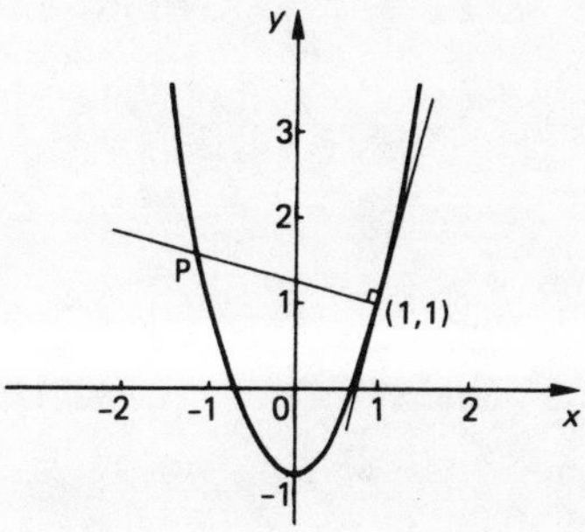

A sketch of the curve is shown in Fig. 3.27. We have $\dfrac{dy}{dx} = 4x$, so when $x = 1$ the gradient of the curve is 4. Hence the gradient of the normal is $-\frac{1}{4}$, and its equation is

$$y - 1 = -\tfrac{1}{4}(x - 1) \quad \text{or} \quad x + 4y = 5$$

This meets the curve $y = 2x^2 - 1$ where

$$x + 4(2x^2 - 1) = 5$$
i.e. $\qquad 8x^2 + x - 9 = 0$

Fig. 3.27

We know that $x = 1$ is one solution, and this helps us to factorise:

$$(x - 1)(8x + 9) = 0$$

Hence the normal meets the curve again when $x = -\frac{9}{8}$, and $y = 2(-\frac{9}{8})^2 - 1 = 49/32$. So P is $(-9/8, 49/32)$.

Exercise 3.7 (Normals)

1 Which of the following pairs of lines are perpendicular, which are parallel, and which are neither?
(i) $y = 4 - 3x$ (ii) $y = 4x - 1$ (iii) $3x - 5y = 10$ (iv) $y = x + 4$
 $3y = x + 7$ $\qquad$ $y = \tfrac{1}{4}x + 8$ $\qquad$ $10y = 6x - 13$ $\qquad$ $x + y = 5$
2 Find the equation of the perpendicular from the point $(-2, 5)$ to the line $4x + 7y = 12$.

In questions 3 to 5 find the equation of the normal to the curve at the given point.

3 $y = 2x^2$ at $(2, 8)$. $\qquad$ **4** $y = 1 + \dfrac{1}{x^2}$ at $(2, \tfrac{5}{4})$. $\qquad$ **5** $y = 4x - x^2$ at $(2, 4)$.

6   Find the equation of the normal to the curve $y = 2x^2 - 9x + 10$ at the point $(3, 1)$, and find the point where this normal meets the curve again.

7   The normal to the curve $y = \dfrac{1}{x}$ at the point $(-2, -\frac{1}{2})$ meets the curve again at P. Find the coordinates of P.

8   Find the points where the line $x + 2y = 7$ intersects the curve $y = 5 - x^2$, and show that the line is the normal to the curve at one of these points.

9   If the line $x - 2y + k = 0$ is a normal to the curve $y = x^3 - 3x^2 - 2x$, find the two possible values of k.

10   Show that the normal to the curve $y = \dfrac{1}{x^2}$ at $(1, 1)$ is also a tangent to the curve $y = 1 + \frac{1}{8}x^2$ and find the point of contact.

Revision questions A

Revision paper A1

1 If $f(x) = 3x^2 - \dfrac{1}{x} + 2$, find the values of

(a) $f(1)$ (b) $f(-2)$ (c) $f'(x)$ (d) $f'(2)$ (e) $f'(a)$ (f) $f''(2)$ (g) $f''(a)$

2 Calculate the coordinates of the points of zero slope on the curve $y = x^2(x-2)$ and sketch the curve from $x = -2$ to $x = +4$.

3 For the curve $y = \sqrt{x^3}$ with x and y both positive, find the coordinates of the point at which the tangent has a gradient of 3.

Find the equation of the normal to the curve at this point.

4 (i) Find the gradient of the curve $y = 3x^2 - x^3$ at the points $(3, 0)$ and $(-1, 4)$ on the curve. Show that the tangents to the curve at these points are parallel.

(ii) Find the coordinates of the 'turning points' on the curve $y = 3x^2 - x^3$, distinguishing between the 'maximum' and 'minimum' turning points.

(iii) Find the coordinates of the point on the curve where the gradient is 3 and show that at this point $\dfrac{d^2y}{dx^2} = 0$.

5 A farmer wishes to enclose an area of 5000 m^2 of a field. He wishes to use as little fencing as possible and to make use of a straight wall along one side of the field.

(i) If he encloses a rectangle with the wall as one side, calculate the minimum length of fencing required, and the length of wall used.

(ii) If he encloses a semicircle with the wall as diameter, calculate the length of fencing required.

Revision paper A2

1 (i) Differentiate $f(x) = (2x^2 + 1)(x - 1)$ to find the derived function $f'(x)$. Evaluate $f(1)$ and $f'(1)$.

(ii) Find the derivative with respect to x of the function

$$y = 3x - x^{-1}$$

and evaluate this derivative when $x = \frac{1}{2}$. For what value of x other than $\frac{1}{2}$ is the tangent to the curve $y = 3x - x^{-1}$ parallel to the tangent at the point where $x = \frac{1}{2}$?

2 If $y = x^3 + 3x^2 - 9x$ find the values of x which make $\dfrac{dy}{dx} = 0$ and the associated values of y.

Calculate y also for $x = -4$, 0 and $+4$ and hence sketch the graph of y against x in the interval $-4 \leqslant x \leqslant 4$. (It is not necessary to use graph paper.)

3 Find the equation of the tangent to the curve $4y = x^2 - 6x + 13$ at the point P on the curve where $x = 7$. Calculate the coordinates of the point A where this tangent meets the x-axis.

Find the equation of the line through A parallel to the normal at P. Prove that this line is also a tangent to the curve and find the coordinates of the point of contact Q.

Prove that the length PQ is equal to the sum of the distances from P and Q to the x-axis.

4 (i)   Find the maximum and minimum values of the function $x^3 + 3x^2 - 24x + 20$, stating, with a reason, the nature of each.

(ii)   Find the possible values of n if $y = x^n$ satisfies the differential equation $9x^2 \dfrac{d^2y}{dx^2} + 2y = 0$.

5   An ice cream cone has the form of a cone of base radius 5 cm and height 15 cm. It is required to make a cylindrical ice cream which will sit inside the cone and stick out by 3 cm. Fig. A.1 shows a vertical cross-section through the centre of the cone and ice cream cylinder.

If the radius of the cylinder is r cm and the height of the ice cream cylinder within the cone is h cm, prove that

$$h + 3r = 15$$

Hence prove that the volume, V cubic centimetres, of ice cream is given by the formula $V = 3\pi r^2 (6 - r)$, and find the value of r which gives the greatest volume of ice cream. (Do not neglect to show that this value of r does give a maximum value of V.)   (O)

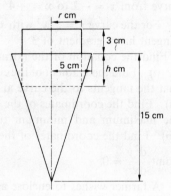

Fig. A.1

4
Areas

4.1 Area under a curve, by drawing

The first basic idea of calculus was that of finding the gradient of a curve. We now look at the second idea, that of finding the area 'under a curve.' The **area under a curve** is the area of the region bounded by the curve, by the x-axis and by two lines parallel to the y-axis.

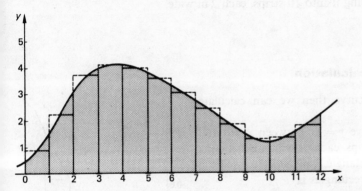

Fig. 4.1

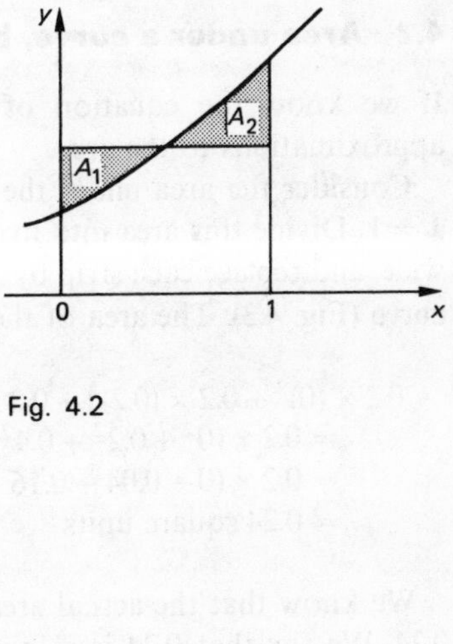

Fig. 4.2

In Fig. 4.1, the area under the curve, between $x = 0$ and $x = 12$, is shown shaded. We can find this area by dividing it into strips (we have chosen 12 strips, each 1 unit wide). To estimate the area of the first strip (between $x = 0$ and $x = 1$) we replace it by a rectangle drawn so that the areas A_1 and A_2 seem to 'balance' (Fig. 4.2). The height of this rectangle is 0.85. Similarly we replace the other 11 strips by rectangles.

The total area of these rectangles is

$$(1 \times 0.85) + (1 \times 2.3) + (1 \times 3.9) + (1 \times 4.25) + (1 \times 4.2)$$
$$+ (1 \times 3.7) + (1 \times 3.15) + (1 \times 2.5) + (1 \times 1.9) + (1 \times 1.3)$$
$$+ (1 \times 1.4) + (1 \times 1.85)$$
$$= 31.3 \text{ square units}$$

and this is an approximation to the area under the curve.

Exercise 4.1 (Areas by drawing)

1 Plot the following points on a graph (using a scale of 2 cm = 1 unit on both axes), and join them by a smooth curve.

x	-1	0	1	2	3	4	5
y	5	3	2	2	3	5	8

Estimate the area under the curve between $x = 0$ and $x = 4$, by dividing it into 4 strips one unit wide.

2 The following table gives the depth, at intervals of 2 m, across a river 20 m wide.

Distance (in metres) from one bank	0	2	4	6	8	10	12	14	16	18	20
Depth (in metres)	0	1.2	2.5	4.0	5.6	6.3	6.7	6.5	5.3	3.1	0

Draw a graph showing the depth as a function of the distance from one bank. Estimate the cross-sectional area of the river, by dividing it into 10 strips, each 2 m wide.

4.2 Area under a curve, by calculation

If we know the equation of a curve then we can calculate approximations to the area.

Consider the area under the curve $y = x^2$ between $x = 0$ and $x = 1$. Divide this area into five strips, each of which is 0.2 units wide, and replace each strip by a rectangle lying *entirely below* the curve (Fig. 4.3). The area of these rectangles is

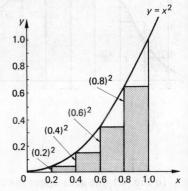

Fig. 4.3

$$0.2 \times (0)^2 + 0.2 \times (0.2)^2 + 0.2 \times (0.4)^2 + 0.2 \times (0.6)^2 + 0.2 \times (0.8)^2$$
$$= 0.2 \times (0^2 + 0.2^2 + 0.4^2 + 0.6^2 + 0.8^2)$$
$$= 0.2 \times (0 + 0.04 + 0.16 + 0.36 + 0.64)$$
$$= 0.24 \text{ square units}$$

We know that the actual area under the curve is greater than 0.24. We say that 0.24 is a **lower bound** for the area under the curve.

We can obtain a better approximation by using 10 strips, each 0.1 units wide (Fig. 4.4). The area of these rectangles is

$$0.1 \times (0^2 + 0.1^2 + 0.2^2 + 0.3^2 + \ldots + 0.8^2 + 0.9^2)$$
$$= 0.285 \text{ square units}$$

We can continue to divide the area into more and more narrower strips.

Table 4.1 gives lower bounds for the area under the curve $y = x^2$ between $x = 0$ and $x = 1$. It looks as if the lower bounds are approaching a limiting value of 0.333 (to 3 decimal places).

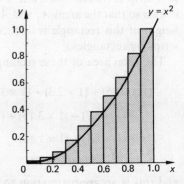

Fig. 4.4

Table 4.1

Number of strips	Width of strip	Area of rectangles
5	0.2	0.24
10	0.1	0.285
100	0.01	0.328
1000	0.001	0.333
10 000	0.0001	0.333

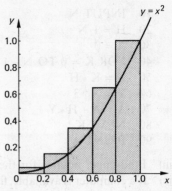

Fig. 4.5

Alternatively, we can replace the strips by rectangles 'above' the curve. Using five strips (Fig. 4.5), the area of these rectangles is

$$0.2 \times (0.2)^2 + 0.2 \times (0.4)^2 + 0.2 \times (0.6)^2 + 0.2 \times (0.8)^2 + 0.2 \times (1.0)^2$$
$$= 0.2 \times (0.2^2 + 0.4^2 + 0.6^2 + 0.8^2 + 1.0^2)$$
$$= 0.44 \text{ square units}$$

The actual area under the curve is less than this. We say that 0.44 is an **upper bound** for the area under the curve.

If we use 10 strips, each 0.1 units wide (Fig. 4.6), then the area of these rectangles is

$$0.1 \times (0.1^2 + 0.2^2 + 0.3^2 + \ldots + 0.9^2 + 1.0^2)$$
$$= 0.385 \text{ square units}$$

We can continue to divide the area into more and more narrower strips.

Table 4.2 shows upper bounds for the area under the curve $y = x^2$ between $x = 0$ and $x = 1$.

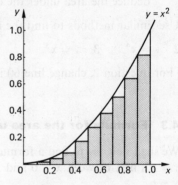

Fig. 4.6

Table 4.2

Number of strips	Width of strip	Area of rectangles
5	0.2	0.44
10	0.1	0.385
100	0.01	0.338
1000	0.001	0.334
10 000	0.0001	0.333

The upper bounds seem to be approaching a limiting value of 0.333 (to 3 decimal places).

The true area under the curve lies between any lower bound and any upper bound. Since the lower bounds and the upper bounds both have the same limiting value of 0.333, the area under the curve $y = x^2$, between $x = 0$ and $x = 1$, is 0.333 (to 3 decimal places).

Exercise 4.2 (Areas by calculation)

1 Consider the area under the curve $y = x^3$ between $x = 0$ and $x = 1$.
 (i) Using 5 strips, and then 10 strips, find lower bounds for the area.
 (ii) If a computer is available, find lower bounds using 100, 1000 and 10 000 strips, and hence find the limiting value of the lower bounds. The following BASIC program may be used (N is the number of strips).

```
10    INPUT N
20    H = 1/N
30    A = 0
40    FOR K = 0 TO N − 1
50    X = K * H
60    Y = X ↑ 3
70    A = A + H * Y
80    NEXT K
90    PRINT A
```

(iii)   Find upper bounds for the area using 5 strips, and then 10 strips.

(iv)   If a computer is available, find upper bounds using 100, 1000 and 10 000 strips. (Change line 40 in the program to 40 FOR K = 1 TO N.) Hence find the limiting value of the upper bounds, and deduce the area under the curve (to 3 decimal places).

Use similar methods to find the areas under the following curves, between $x = 0$ and $x = 1$.

2 $y = x^4$ **3** $y = x^5$

[For question 2, change line 60 in the program to 60 $Y = X ↑ 4$, and so on.]

4.3   Formula for the area under $y = x^n$

We shall now try to find a formula for the area under the curves $y = x^n$, first between $x = 0$ and $x = 1$, then between $x = 0$ and $x = b$.

If $n = 0$, the curve is $y = x^0 = 1$, which is a straight line (Fig. 4.7). The area under the curve between $x = 0$ and $x = 1$ is clearly 1.

If $n = 1$, the curve is $y = x^1 = x$, which is also a straight line (Fig. 4.8). Since the area of a triangle is $\frac{1}{2}$ × base × height, the area under this curve between $x = 0$ and $x = 1$ is $\frac{1}{2} \times 1 \times 1 = \frac{1}{2}$.

Putting these results, together with your answers to the previous Exercise, into a table, you should obtain the values given in Table 4.3.

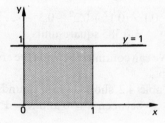

Fig. 4.7

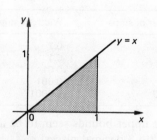

Fig. 4.8

Table 4.3

Curve	Area under curve between $x = 0$ and $x = 1$
$y = 1$	1.000
$y = x$	0.500
$y = x^2$	0.333
$y = x^3$	0.250
$y = x^4$	0.200
$y = x^5$	0.167

It looks as if these areas are $1, \frac{1}{2}, \frac{1}{3}, \frac{1}{4}, \frac{1}{5}, \frac{1}{6}$; i.e. for $n = 0, 1, 2, 3, 4, 5$, the area under the curve $y = x^n$, between $x = 0$ and $x = 1$, is $\dfrac{1}{n+1}$.

Now consider the area under the curve $y = x^2$ between $x = 0$ and $x = 2$. If we use 10 strips, the width of each strip is 0.2 units (Fig. 4.9). The area of the rectangles below the curve is

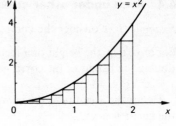

Fig. 4.9

$$0.2 \times [(0)^2 + (0.2)^2 + (0.4)^2 + (0.6)^2 + \ldots + (1.8)^2]$$
$$= 0.2 \times [4(0)^2 + 4(0.1)^2 + 4(0.2)^2 + 4(0.3)^2 + \ldots + 4(0.9)^2]$$
$$= 8 \times 0.1 \times [(0)^2 + (0.1)^2 + (0.2)^2 + (0.3)^2 + \ldots + (0.9)^2]$$
$$= 8 \times (\text{the area of the 10 rectangles below}$$
$$\text{the curve between } x = 0 \text{ and } x = 1)$$
$$= 8 \times 0.285$$

Similarly, if we use 20 strips the area of the rectangles below the curve will be

$$8 \times (\text{the area of the 20 rectangles below}$$
$$\text{the curve between } x = 0 \text{ and } x = 1)$$

and so on. So the area between $x = 0$ and $x = 2$ is 8 times the area between $x = 0$ and $x = 1$. Hence the true area between $x = 0$ and $x = 2$ is $8 \times \frac{1}{3}$.

Exercise 4.3   (Investigation of the area under $y = x^n$)

1   For the area under the curve $y = x^2$ between $x = 0$ and $x = 3$, using 10 strips, show that the area of the rectangles below the curve is 27 times the area of 10 rectangles below the curve between $x = 0$ and $x = 1$.

Deduce that the required area is $27 \times \frac{1}{3}$.

2   Show that the area under the curve $y = x^2$ between $x = 0$ and $x = 4$ is $64 \times \frac{1}{3}$.

It now appears that the area under the curve $y = x^2$ between $x = 0$ and $x = b$ is $b^3 \times \frac{1}{3}$.

3   Using the value $\frac{1}{4}$ for the area under the curve $y = x^3$ between $x = 0$ and $x = 1$, find the area under the curve $y = x^3$:
   (i)   between $x = 0$ and $x = 2$;
   (ii)   between $x = 0$ and $x = 3$;
   (iii)   between $x = 0$ and $x = b$.

4   Using the value $\frac{1}{5}$ for the area under the curve $y = x^4$ between $x = 0$ and $x = 1$, find the area under the curve $y = x^4$:
   (i)   between $x = 0$ and $x = 2$;
   (ii)   between $x = 0$ and $x = 3$;
   (iii)   between $x = 0$ and $x = b$.

5   Find the area under the curve $y = 1$ between $x = 0$ and $x = b$.

6   Find the area under the curve $y = x$ between $x = 0$ and $x = b$.

Your results should demonstrate that for $n = 0, 1, 2, \ldots$ the area under the curve $y = x^n$ between $x = 0$ and $x = b$ is

$$\boxed{\dfrac{b^{n+1}}{n+1}}$$

(see Fig. 4.10).

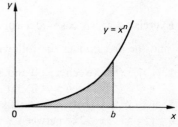

Fig. 4.10

4.4   Area under other curves

Example 1   Consider the curve $y = 2x^2$.

For any strip, the height of a rectangle below the curve $y = 2x^2$ is double the height of the corresponding rectangle below the curve $y = x^2$ (Fig. 4.11).

Hence the area under the curve $y = 2x^2$ is twice the area under the curve $y = x^2$.

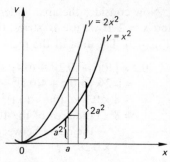

Fig. 4.11

Example 2   Consider the curve $y = 1 + x + x^2$.

Figs 4.11 to 4.15 illustrate that the area under the curve $y = 1 + x + x^2$ is equal to

  (area under $y = 1$) + (area under $y = x$) + (area under $y = x^2$)

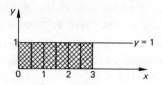

Fig. 4.12

Example 3   Find the area under the curve $y = 3 + 2x + x^2$ between $x = 0$ and $x = 4$.

The required area is the sum of the areas under $y = 3$, $y = 2x$ and $y = x^2$ (all between $x = 0$ and $x = 4$). Hence the area is

$$3 \times 4 + 2 \times \frac{4^2}{2} + \frac{4^3}{3} = 12 + 16 + 21\tfrac{1}{3} = 49\tfrac{1}{3} \text{ square units}$$

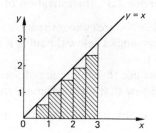

Fig. 4.13

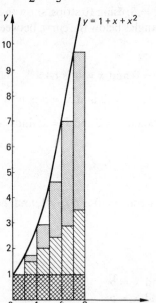

Fig. 4.15

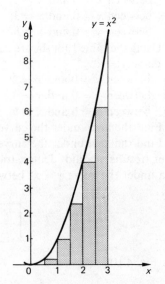

Fig. 4.14

Exercise 4.4   (Areas under curves)

Find the areas under the following curves.

1   $y = 3x^2$ between $x = 0$ and $x = 2$.
2   $y = 6x^4$ between $x = 0$ and $x = 1$.
3   $y = x^2 + x^3$ between $x = 0$ and $x = 3$.
4   $y = 5 + 3x + 2x^2$ between $x = 0$ and $x = 4$.
5   $y = x + 6x^2 + 4x^3$ between $x = 0$ and $x = 2$.
6   $y = 1 + x^2 + x^4 + x^6$ between $x = 0$ and $x = 1$.

5

The fundamental theorem of calculus

We have now investigated the two basic ideas of calculus—the ideas of 'gradient finding' and 'area finding'. The connection between these two fundamental concepts eluded mathematicians for hundreds of years. Before we can fully understand it we have to consider area functions.

5.1 The area function

We look again at the curve we met at the beginning of the last chapter. To find the areas of the strips, we have used the method of 'balancing' the areas by eye (see p. 51). We can see that in Fig. 5.1 (overleaf), the area under the curve between $x = 0$ and $x = 3$ is approximately $0.85 + 2.3 + 3.9 = 7.05$. We can easily obtain a table of values of the areas from $x = 0$ to any particular value of x, and this is shown below.

x	0	1	2	3	4	5	6	7	8	9	10	11	12
A	0	0.85	3.15	7.05	11.3	15.5	19.2	22.35	24.85	26.75	28.05	29.45	31.3

The function A specified by this table is called the **area function**. We can draw a graph of A against x as shown in Fig. 5.2.

The area function for the curves $y = x^n$

If we start with one of the curves $y = x^n$, we can use the results we discovered in Chapter 4 to find the area function.

For example, if the original curve is $y = x^2$, the area under the curve between 0 and b is $\frac{1}{3}b^3$ (see p. 55), and so the area A between 0 and x is given by the formula $A = \frac{1}{3}x^3$.

Exercise 5.1 (The area function)

Keep your answers to this Exercise. You will need them in Exercise 5.2.

1   Plot the following points on a graph, and join them by a smooth curve.

x	0	2	4	6	8	10	12	14	16	18	20
y	4.0	4.6	4.5	3.3	1.8	1.9	2.9	3.3	3.2	2.7	1.6

Divide the area under the curve into strips, and tabulate approximate values of the area A between 0 and x, using the method of 'balancing areas'. Draw a graph of A against x. (For the original curve, use scales 1 cm = 2 units on x-axis, 1 cm = 1 unit on y-axis. For the area curve, use 1 cm = 2 units on x-axis, 1 cm = 5 units on A-axis.)

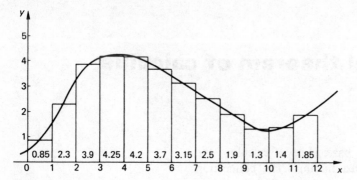

Fig. 5.1

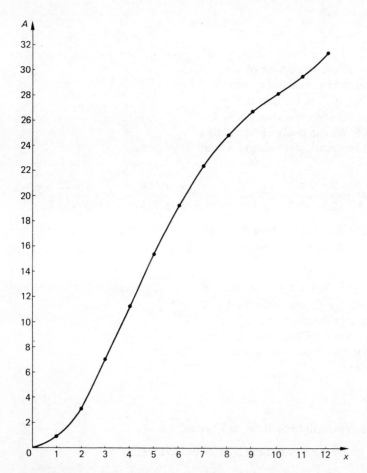

Fig. 5.2

2   With x-axis marked from 0 to 10, and y-axis from 0 to 5 (using a scale of 1 cm = 1 unit on both axes) draw any curve. Divide the area under the curve into strips one unit wide, and tabulate values of the area A between 0 and x. Draw a graph of A against x (using scales of 1 cm = 1 unit on the x-axis and 1 cm = 2 units on the A-axis).

3   For the following curves, use the results discovered in Exercise 4.3 to find a formula for the area A in terms of x.

(i) $y = x^3$ (ii) $y = x^4$ (iii) $y = 1$ (iv) $y = x$ (v) $y = x^n$

5.2   The gradient of the area function

We shall now find the gradient function for the area curve drawn in the previous section. In Fig. 5.3 (p. 60) tangents have been drawn (as accurately as we can) at $x = 0$, $x = 1$, $x = 2, \ldots, x = 12$, and their gradients measured, as follows:

x	0	1	2	3	4	5	6	7	8	9	10	11	12
Gradient	0.6	1.3	2.9	4.5	4.5	4.1	3.4	2.8	2.2	1.6	1.2	1.4	2.0

We draw a graph (Fig. 5.4) of these gradients, which are of course (approximate) values of $\dfrac{dA}{dx}$. This is remarkably similar to the original curve from which the areas were obtained.

This example demonstrates that there is a connection between the two basic ideas of gradient finding and area finding. If we start with any curve, draw a graph of its area function, then find the gradient function for this area curve, we obtain the original curve.

Since finding the area function and then the gradient function, takes us back where we started, we may regard gradient finding and area finding as 'opposites' of each other.

Confirmation for the curve $y = x^2$

If we start with the curve $y = x^2$, we have seen that the area function is $A = \frac{1}{3}x^3$. To find the gradient of A we now differentiate, obtaining $\dfrac{dA}{dx} = \frac{1}{3}(3x^2) = x^2$, which is the same as the original function.

Exercise 5.2   (Gradient of the area curve)

1 and 2   For each of the area curves drawn in Exercise 5.1, questions 1 and 2, draw tangents and tabulate values of the gradient. Draw a graph of the gradient function (using the same scales as the original curve), and compare this with your original curve.

3   (i)–(v)   Differentiate each of the area functions found in Exercise 5.1, question 3, and verify that $\dfrac{dA}{dx}$ is the same as the original function.

5.3   The fundamental theorem

We have demonstrated that, if A is the area under a curve $y = f(x)$, then the gradient function of A is the same as the original function,

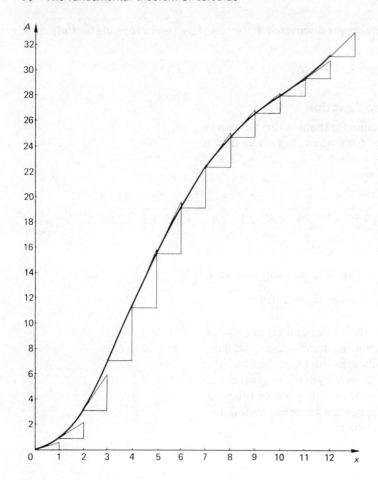

Fig. 5.3

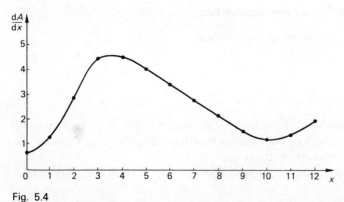

Fig. 5.4

i.e. $\dfrac{\mathrm{d}A}{\mathrm{d}x} = f(x)$. This result is true for any (continuous) curve, and it is of great importance because it links together the two basic ideas of gradients and areas. It is called the **fundamental theorem of calculus**.

So far, for simplicity, A has always been the area under the curve between 0 and x, but in fact the area could be measured from any fixed point (instead of O). In this section we shall refer to A as the area up to x.

Statement of the theorem. If $y = f(x)$ is a continuous curve, and A is the area under the curve up to x, then $\dfrac{dA}{dx} = f(x)$.

Proof of the theorem. Let δx be a change in x, and let δA be the corresponding change in A. δx may be positive or negative. We shall consider these two cases separately.

If δx is positive, then

$$A + \delta A = \text{area up to } x + \delta x$$
$$A = \text{area up to } x$$

Subtracting,

$$\delta A = \text{area between } x \text{ and } (x + \delta x)$$

The area δA is shown shaded in Fig. 5.5; δA is greater than the area of the rectangle below the curve. If the height of this rectangle is p, then its area is $p\delta x$ (p is, in fact, the minimum value taken by the function between x and $x + \delta x$). Similarly, δA is less than the area of the rectangle above the curve, of height q; the area of this rectangle is $q\delta x$. So we have

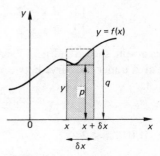

Fig. 5.5

$$p\delta x \leqslant \delta A \leqslant q\delta x$$

Dividing by δx,

$$p \leqslant \frac{\delta A}{\delta x} \leqslant q$$

If δx is negative, then $(x + \delta x)$ is to the left of x, and

$$A = \text{area up to } x$$
$$A + \delta A = \text{area up to } x + \delta x$$

Subtracting,

$$-\delta A = \text{area between } (x + \delta x) \text{ and } x$$

Now the shaded area is $(-\delta A)$. If p and q are the heights of rectangles below and above the curve (Fig. 5.6), then

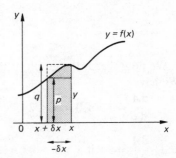

Fig. 5.6

$$p(-\delta x) \leqslant -\delta A \leqslant q(-\delta x)$$

Dividing by $-\delta x$ (which is positive),

$$p \leqslant \frac{\delta A}{\delta x} \leqslant q$$

So, in both cases, $\dfrac{\delta A}{\delta x}$ lies between p and q.

As $\delta x \to 0$, since the curve is a continuous one, $p \to y$ and $q \to y$, and hence $\dfrac{\delta A}{\delta x} \to y$. Now, since $\dfrac{dA}{dx} = \lim\limits_{\delta x \to 0} \dfrac{\delta A}{\delta x}$, we have

$$\frac{dA}{dx} = y \quad \text{or} \quad \frac{dA}{dx} = f(x)$$

In order to prove the fundamental theorem, it was necessary to assume that the curve $y = f(x)$ was continuous. If we start with a discontinuous curve, the result $\dfrac{dA}{dx} = f(x)$ may not be true.

Example   Consider the function

$$y = f(x) = \begin{cases} 1 & \text{if } x \leqslant 2 \\ 3 & \text{if } x > 2 \end{cases}$$

This graph, shown in Fig. 5.7, has a discontinuity at $x = 2$. If A is the area under the curve between 0 and x, we can find expressions for A.

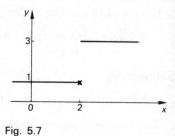

Fig. 5.7

If $0 \leqslant x \leqslant 2$

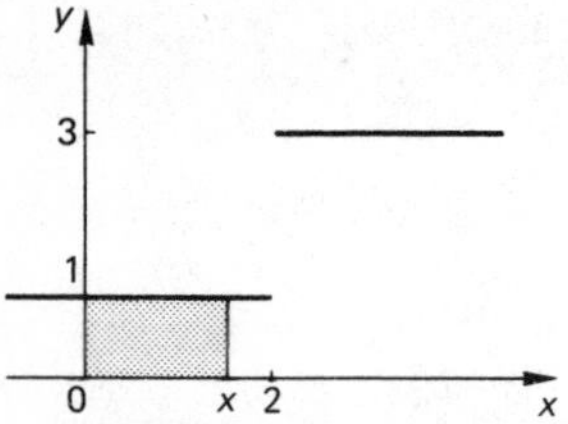

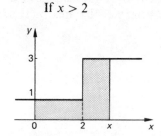

$A = x$

Fig. 5.8

If $x > 2$

$A = 2 + 3(x - 2) = 3x - 4$

Fig. 5.9

We now sketch the graph of the area function A (Fig. 5.10). We have

$$\frac{dA}{dx} = 1 \quad \text{if } 0 < x < 2$$

and $\dfrac{dA}{dx} = 3 \quad \text{if } x > 2$

But $\dfrac{dA}{dx}$ does not exist when $x = 2$ (since the graph of A is not smooth at this point). So the result $\dfrac{dA}{dx} = f(x)$ is true for all (positive) values of x except $x = 2$.

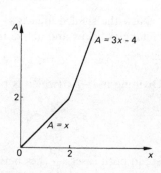

Fig. 5.10

Exercise 5.3   (The fundamental theorem)

1 Sketch the graph of the function $y = f(x) = \begin{cases} 2 & \text{if } x \leqslant 1 \\ 3 & \text{if } x > 1 \end{cases}$

Find expressions for the area A under the curve between 0 and x, and sketch the graph of A (for positive values of x). When is the result $\dfrac{\mathrm{d}A}{\mathrm{d}x} = f(x)$ true, and when is it false?

2 Repeat question 1 for the function $y = f(x) = \begin{cases} 3 & \text{if } x \leqslant 1 \\ 1 & \text{if } 1 < x \leqslant 3 \\ 2 & \text{if } x > 3 \end{cases}$

3 Sketch the graph of the function $y = \begin{cases} x & \text{if } x \leqslant 2 \\ 2 & \text{if } x > 2 \end{cases}$

Find expressions for the area A under the curve between 0 and x, and sketch the graph of A. Show that $\dfrac{\mathrm{d}A}{\mathrm{d}x} = f(x)$ for all (positive) values of x, including $x = 2$. Why is this to be expected?

5.4 The definite integral

We now consider the area under a curve between two given values of x, say between $x = a$ and $x = b$ (Fig. 5.11). This can be found by a limiting process (similar to that used for $y = x^2$ on p. 52), taking either rectangles below the curve (Fig. 5.12) or rectangles above the curve (Fig. 5.13).

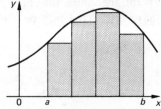

Fig. 5.11

As the strips are taken narrower and narrower, we have found that the limiting value is the same in the two cases; this limiting value is the required area under the curve. It follows that this area is also the limiting value of any area which is between that of the rectangles below the curve and that of the rectangles above the curve. In this section it will be convenient to take rectangles with height equal to that of the curve at the left-hand edge of each strip.

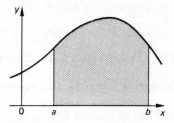

Fig. 5.12

Suppose the values of x at the left-hand edges of the strips are x_1, x_2, x_3, x_4 and the corresponding values of y are y_1, y_2, y_3, y_4. Let the width of each strip be δx.

The first rectangle has height y_1 and width δx, so its area is $y_1 \delta x$. The second rectangle has area $y_2 \delta x$, and so on. In Fig. 5.14, the total area of the shaded rectangles is

$$y_1 \delta x + y_2 \delta x + y_3 \delta x + y_4 \delta x = \sum_{r=1}^{r=4} y_r \delta x$$

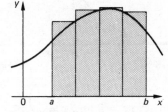

Fig. 5.13

If the area is between $x = a$ and $x = b$, we may write such a sum as

$$\sum_{x=a}^{x=b} y \, \delta x$$

where a typical rectangle, between x and $x + \delta x$, has height y and width δx. If the equation of the curve is $y = f(x)$, the sum is

$$\sum_{x=a}^{x=b} f(x) \delta x$$

This is only an approximation to the area under the curve, but we can make the approximation as good as we please by using narrower strips (and hence increasing the number of strips).

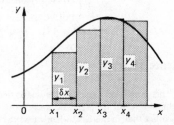

Fig. 5.14

We define the **definite integral**, $\int_a^b f(x)\,dx$ (said 'the integral from a to b of $f(x)$ dee x'), to be the limiting value of the sum $\sum_{x=a}^{x=b} f(x)\delta x$ as the widths (δx) of the strips tend to zero*, i.e.

$$\int_a^b f(x)\,dx = \lim_{\delta x \to 0}\left(\sum_{x=a}^{x=b} f(x)\delta x\right)$$

Fig. 5.15

Thus, provided that the curve $y = f(x)$ is above the x-axis (so far we have always assumed this), $\int_a^b f(x)\,dx$ is the area under the curve $y = f(x)$ between $x = a$ and $x = b$. However, our definition can still be applied when the curve is below the x-axis (i.e. when $f(x)$ is negative), and we shall see later how the definite integral may be interpreted in terms of areas in this case (see p. 76).

Also, the definite integral, which is defined as the limit of a sum, can represent quantities other than area (for example, distance, volume, mass, and so on).

It is clear how the notation $\int_a^b f(x)dx$ arises from $\sum_{x=a}^{x=b} f(x)\delta x$ ($\int$ is the old English S, standing for sum).

Evaluation using the area function

Example   Consider $\int_2^5 x^2\,dx$.

This is the area under the curve $y = x^2$ between $x = 2$ and $x = 5$. We have discovered (see p. 57) that the area under this curve between 0 and x is given by $A = \frac{1}{3}x^3$. With reference to Figs 5.16 to 5.18, we can see that

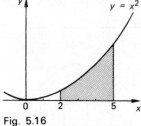

Fig. 5.16

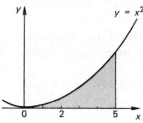

Fig. 5.17

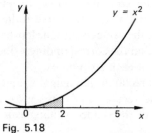

Fig. 5.18

$$\begin{pmatrix}\text{Area under the curve}\\ \text{between } x = 2 \text{ and } x = 5\end{pmatrix}$$
$$= \begin{pmatrix}\text{Area under the curve}\\ \text{between } x = 0 \text{ and } x = 5\end{pmatrix} - \begin{pmatrix}\text{Area under the curve}\\ \text{between } x = 0 \text{ and } x = 2\end{pmatrix}$$
$$= (\tfrac{1}{3} \times 5^3) - (\tfrac{1}{3} \times 2^3)$$
$$= 39$$

Hence $\int_2^5 x^2\,dx = 39$.

---

* This expression only holds provided that there is a limiting value; otherwise the definite integral $\int_a^b f(x)\,dx$ does not exist.

Notice that, in this case, when we write $\delta x \to 0$ we are assuming that δx is always positive (as it is the width of a strip).

We can use this method to evaluate any definite integral $\int_a^b f(x)\,dx$, provided that we know the area function A (i.e. the area under the curve $y = f(x)$ between 0 and x). Then $\int_a^b f(x)\,dx$, which is the area under the curve between $x = a$ and $x = b$, is equal to

$$\left(\begin{array}{c}\text{area under the curve}\\ \text{up to } b\end{array}\right) - \left(\begin{array}{c}\text{area under the curve}\\ \text{up to } a\end{array}\right)$$

(see Figs 5.19 to 5.21) and we can find these two areas by substituting $x = b$ and $x = a$ into our formula for A.

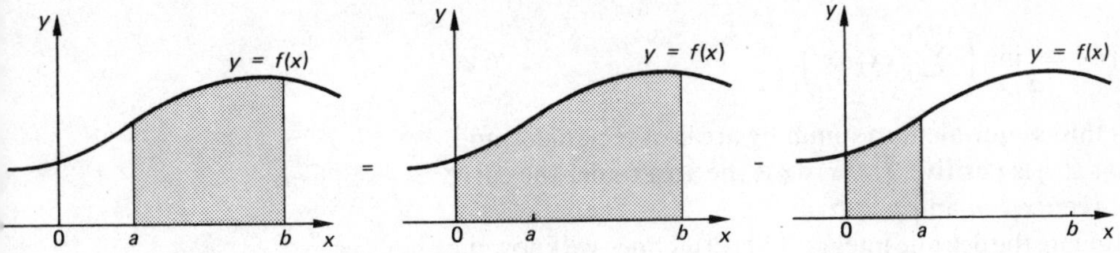

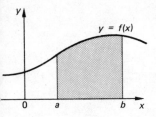

Fig. 5.19

Fig. 5.20

Fig. 5.21

The quantity $\int_a^b f(x)\,dx$ is the difference between the values of A when $x = b$ and when $x = a$, as shown in the sketch of the area function in Fig. 5.22.

Now suppose that we wish to evaluate $\int_a^b f(x)\,dx$, but we do not know the area function A for the curve $y = f(x)$. By the fundamental theorem, $\dfrac{dA}{dx} = f(x)$, and so we do know $\dfrac{dA}{dx}$. The problem is, knowing $\dfrac{dA}{dx}$ (it is equal to $f(x)$), to find A. We shall study this problem in the next chapter.

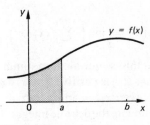

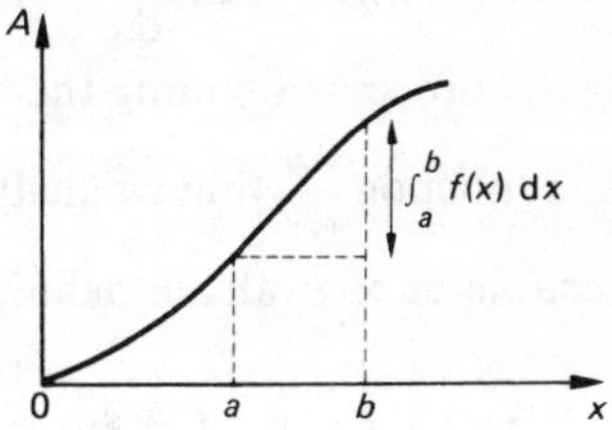

Fig. 5.22

Exercise 5.4   (Definite integrals from the area function)

1   Use the formula $A = \frac{1}{3}x^3$ for the area under the curve $y = x^2$ between 0 and x to evaluate

(i) $\displaystyle\int_1^4 x^2\,dx$ (ii) $\displaystyle\int_2^3 x^2\,dx$ (iii) $\displaystyle\int_0^6 x^2\,dx$

2   Use the area functions [see Exercise 4.3 and Exercise 5.1, question 3] to evaluate

(i) $\displaystyle\int_1^2 x^3\,dx$ (ii) $\displaystyle\int_1^3 x^3\,dx$ (iii) $\displaystyle\int_2^3 x^4\,dx$ (iv) $\displaystyle\int_3^7 x\,dx$ (v) $\displaystyle\int_{1/2}^4 x\,dx$

3   The area A under a certain curve $y = f(x)$ between 0 and x is given by the formula $A = 2x^2 + 5x$. Evaluate

(i) $\displaystyle\int_0^2 f(x)\,dx$ (ii) $\displaystyle\int_1^3 f(x)\,dx$

Use the fundamental theorem $\dfrac{dA}{dx} = f(x)$ to find the equation of the curve.

4   The area A under a certain curve $y = f(x)$ between 0 and x is given by the formula $A = \sqrt{x^3}$. Evaluate $\int_4^9 f(x)\,dx$. Use the fundamental theorem to find the function $f(x)$.

6

Integration I

Introduction

In the previous chapter, we defined the definite integral as the limit of a sum,

$$\int_a^b f(x)\,dx = \lim_{\delta x \to 0} \left(\sum_{x=a}^{x=b} f(x)\,\delta x \right)$$

We saw that this is equivalent to summing areas of rectangles, and, provided that $f(x)$ is positive, $\int_a^b f(x)\,dx$ is the area under the curve $y = f(x)$ between $x = a$ and $x = b$.

We can evaluate the definite integral $\int_a^b f(x)\,dx$ once we know the area function A for the curve $y = f(x)$. By the fundamental theorem, we know that $\dfrac{dA}{dx} = f(x)$. So, in this chapter, we shall study the process of finding the function A when we are given the derived function $\dfrac{dA}{dx}$ (but we shall use the letter y instead of A). This will enable us to evaluate definite integrals.

6.1 The reverse of differentiation

The process of differentiation is, given a function y, to find the derived function $\dfrac{dy}{dx}$. We now look at this process in reverse: given the derived function $\dfrac{dy}{dx}$, to find y. This is called **integration**.

For example, suppose we know that the derived function is

$$\frac{dy}{dx} = 2x + 3$$

To find y, we require a function which, when differentiated, gives $2x + 3$. One such function is

$$y = x^2 + 3x$$

(check this by differentiating it), but this is not the only one. Other possibilities are

$$y = x^2 + 3x + 2$$
$$y = x^2 + 3x - 5$$

and generally

$$y = x^2 + 3x + C$$

where C is any constant.

We say that the **integral** of $2x + 3$, with respect to x, is $x^2 + 3x + C$, and we write

$$\int (2x + 3)\, dx = x^2 + 3x + C$$

This is often called an **indefinite integral**; C is the **arbitrary constant of integration.**

The notation $\int f(x)\, dx$ means the integral of $f(x)$.

Integration of powers of x

For the integral of x, i.e. $\int x\, dx$, we require a function which, when differentiated, gives x. Now

$$\frac{d}{dx}(x^2) = 2x \quad \text{and} \quad \frac{d}{dx}(\tfrac{1}{2}x^2) = \tfrac{1}{2}(2x) = x$$

Thus
$$\int x\, dx = \tfrac{1}{2}x^2 + C$$

For $\int x^2\, dx$, we have

$$\frac{d}{dx}(x^3) = 3x^2 \quad \text{and} \quad \frac{d}{dx}(\tfrac{1}{3}x^3) = \tfrac{1}{3}(3x^2) = x^2$$

Thus
$$\int x^2\, dx = \tfrac{1}{3}x^3 + C$$

For $\int k\, dx$, where k is a constant, we have $\dfrac{d}{dx}(kx) = k$. Thus

$$\int k\, dx = kx + C$$

We therefore have

$$\int k\, dx = kx + C$$

$$\int x\, dx = \tfrac{1}{2}x^2 + C$$

$$\int x^2\, dx = \tfrac{1}{3}x^3 + C$$

Generally, for $\int x^n\, dx$, we have

$$\frac{d}{dx}(x^{n+1}) = (n+1)x^n$$

and $\quad \dfrac{d}{dx}\left(\dfrac{x^{n+1}}{n+1}\right) = \dfrac{(n+1)x^n}{n+1} = x^n$

Thus

$$\int x^n \, dx = \frac{x^{n+1}}{n+1} + C$$

This result tells us how to integrate powers of x. It is valid for all values of n (including fractional and negative values) except $n = -1$. It fails when $n = -1$ because we have to divide by $(n+1)$, which would then be zero. The integral of x^{-1}, i.e. $\int \frac{1}{x} \, dx$, cannot be found by considering powers of x.

Example 1 Integrate (i) x^9 (ii) $x^{\frac{5}{3}}$ (iii) $\frac{1}{\sqrt{x}}$ (iv) $\frac{1}{x^2}$.

(i) $\int x^9 \, dx = \frac{1}{10} x^{10} + C$

(ii) $\int x^{\frac{5}{3}} \, dx = \frac{x^{\frac{8}{3}}}{\frac{8}{3}} + C = \frac{3}{8} x^{\frac{8}{3}} + C$

(iii) $\int \frac{1}{\sqrt{x}} \, dx = \int x^{-\frac{1}{2}} \, dx = \frac{x^{\frac{1}{2}}}{\frac{1}{2}} + C = 2x^{\frac{1}{2}} + C = 2\sqrt{x} + C$

(iv) $\int \frac{1}{x^2} \, dx = \int x^{-2} \, dx = \frac{x^{-1}}{-1} + C = -\frac{1}{x} + C$

An expression consisting of sums and multiples of powers of x may be integrated term by term, provided that none of the terms is a multiple of $\frac{1}{x}$. Sometimes it is necessary to multiply or divide out first.

Example 2 Integrate (i) $3x^2 - 5x + 7$ (ii) $\frac{3}{x^3} - \frac{1}{4x^2}$ (iii) $(x^2+1)(3x-8)$ (iv) $\frac{x-3}{2\sqrt{x}}$

(i) $\int (3x^2 - 5x + 7) \, dx = 3(\frac{1}{3}x^3) - 5(\frac{1}{2}x^2) + 7x + C$

$$= x^3 - \tfrac{5}{2}x^2 + 7x + C$$

(ii) $\int \left(\frac{3}{x^3} - \frac{1}{4x^2} \right) dx = \int \left(3x^{-3} - \frac{1}{4} x^{-2} \right) dx$

$$= 3\left(\frac{x^{-2}}{-2} \right) - \frac{1}{4} \left(\frac{x^{-1}}{-1} \right) + C = -\frac{3}{2x^2} + \frac{1}{4x} + C$$

(iii) $\int (x^2+1)(3x-8) \, dx = \int (3x^3 - 8x^2 + 3x - 8) \, dx$

$$= 3(\tfrac{1}{4}x^4) - 8(\tfrac{1}{3}x^3) + 3(\tfrac{1}{2}x^2) - 8x + C$$

$$= \tfrac{3}{4}x^4 - \tfrac{8}{3}x^3 + \tfrac{3}{2}x^2 - 8x + C$$

(iv) $\displaystyle\int \frac{x-3}{2\sqrt{x}}\,dx = \int \frac{x-3}{2x^{\frac{1}{2}}}\,dx = \int \left(\frac{1}{2}x^{\frac{1}{2}} - \frac{3}{2}x^{-\frac{1}{2}}\right)dx = \frac{1}{2}\left(\frac{x^{\frac{3}{2}}}{\frac{3}{2}}\right) - \frac{3}{2}\left(\frac{x^{\frac{1}{2}}}{\frac{1}{2}}\right) + C$

$\qquad\qquad = \frac{1}{3}x^{\frac{3}{2}} - 3x^{\frac{1}{2}} + C = \frac{1}{3}\sqrt{x^3} - 3\sqrt{x} + C$

Exercise 6.1   (Integration of powers of x)

1 Find (i) $\displaystyle\int x^3\,dx$ (ii) $\displaystyle\int 7\,dx$ (iii) $\displaystyle\int (2x^3 + x^2 - 6x + 5)\,dx$ (iv) $\displaystyle\int x^{\frac{1}{4}}\,dx$ (v) $\displaystyle\int \frac{2}{x^3}\,dx$

In questions 2 to 20 integrate the given functions.

2 x^5	**3** x^{10}	**4** $2x^2 - 4x + 3$	**5** $5 + 3x - 4x^2$
6 $2x^3 - 7x$	**7.** $(2x-1)(4-x)$	**8** $(2x-5)^2$	**9** $(x^2+2)^2$
10 $x^{\frac{1}{3}}$	**11** $2x^{\frac{3}{2}} - 3x^{\frac{1}{2}}$	**12** $\frac{1}{2}x^{\frac{4}{3}}$	**13** $\sqrt{x}(x+3)$
14 $\dfrac{1}{x^4}$	**15** $\dfrac{3}{x^2} - \dfrac{5}{x^3}$	**16** $2 - \dfrac{1}{x^2}$	**17** $\dfrac{x^3 - 1}{x^2}$
18 $\dfrac{(2x+1)^2}{x^4}$	**19** $\sqrt{x} - \dfrac{3}{\sqrt{x}}$	**20** $\dfrac{2x^2 + 3x - 4}{\sqrt{x}}$	

6.2 Simple applications of integration

Solving differential equations

An equation such as $\dfrac{dy}{dx} = 2x + 3$, which gives the derived function, may be regarded as a differential equation. Integrating, we know that $y = x^2 + 3x + C$ satisfies the equation, and is therefore a solution of the differential equation, for any value of the constant C. (For example, $y = x^2 + 3x$, $y = x^2 + 3x + 4$, and so on, are different solutions of the equation.) We say that the general solution of the differential equation $\dfrac{dy}{dx} = 2x + 3$ is $y = x^2 + 3x + C$.

Thus the general solution contains an arbitrary constant. If we know the value of y for one particular value of x, then we can find the constant, and there is a unique function y which is a solution of the differential equation.

Example 1   Find the general solution of the differential equation $\dfrac{dy}{dx} = 4x - 3$.

The general solution is

$$y = \int (4x - 3)\,dx = 4(\tfrac{1}{2}x^2) - 3x + C$$

i.e. $y = 2x^2 - 3x + C$

Example 2 Solve the differential equation $\dfrac{dy}{dx} = 2x^2 - 5$, given that $y = 3$ when $x = 2$.

The general solution is

$$y = \int (2x^2 - 5)\, dx = 2(\tfrac{1}{3}x^3) - 5x + C$$

i.e. $y = \tfrac{2}{3}x^3 - 5x + C$

Substituting the values $y = 3$ when $x = 2$, we have

$$3 = \frac{16}{3} - 10 + C$$

so $C = \dfrac{23}{3}$. The solution is thus

$$y = \tfrac{2}{3}x^3 - 5x + \frac{23}{3}$$

Finding the equation of a curve

If we know the gradient function of a curve, then we know $\dfrac{dy}{dx}$ and we can find the equation of the curve by integrating. This will give us a family of curves.

Example 1 Find the family of curves which have gradient function $\dfrac{dy}{dx} = 2x + 3$.

We have $\dfrac{dy}{dx} = 2x + 3$, and so, integrating,

$$y = \int (2x + 3)\, dx = x^2 + 3x + C$$

Each value of C gives a different member of the family of curves. For example, the curves

$$y = x^2 + 3x$$
$$y = x^2 + 3x + 4$$
$$y = x^2 + 3x - 5$$

all have the same gradient function $\dfrac{dy}{dx} = 2x + 3$ (see Fig. 6.1).

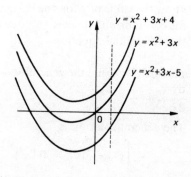

Fig. 6.1

However, if we also know one point on the curve, then we can determine the equation of the curve.

Example 2   Find the equation of the curve which passes through the point $(4, -1)$, and has gradient function $\dfrac{dy}{dx} = x - 4$.

Integrating,

$$y = \tfrac{1}{2}x^2 - 4x + C$$

Since the curve passes through $(4, -1)$, we substitute $x = 4$, $y = -1$; thus $-1 = 8 - 16 + C$, and so $C = 7$. The equation of the curve is $y = \tfrac{1}{2}x^2 - 4x + 7$.

Alternative notation   Using the $f(x)$ notation, the problem is: given the derived function $f'(x)$, to find $f(x)$. For example, if $f'(x) = 2x + 3$, then $f(x) = x^2 + 3x + C$.

Example   Find the function $f(x)$, given that $f'(x) = x^2 + 4x$ and $f(3) = 5$.

We have

$$f(x) = \int (x^2 + 4x)dx = \tfrac{1}{3}x^3 + 4(\tfrac{1}{2}x^2) + C$$

$$= \tfrac{1}{3}x^3 + 2x^2 + C$$

Since $f(3) = 5, 9 + 18 + C = 5$, so $C = -22$. Thus
$$f(x) = \tfrac{1}{3}x^3 + 2x^2 - 22$$

Exercise 6.2   (Applications of integration)

In questions 1 to 6 find the general solutions of the differential equations.

1 $\dfrac{dy}{dx} = 4x$ 　　　 2 $\dfrac{dy}{dx} = 4$ 　　　 3 $\dfrac{dy}{dx} = 3x^2 - 1$

4 $\dfrac{dy}{dx} + 4x = 2$ 　　 5 $x^2\dfrac{dy}{dx} = 3$ 　　 6 $x^2\dfrac{dy}{dx} = 2x^4 - 1$

In questions 7 to 10 solve the differential equations.

7 $\dfrac{dy}{dx} = 6x$ given that $y = 4$ when $x = 0$. 　　 8 $\dfrac{dy}{dx} = x^3 + x$ given that $y = 2$ when $x = 1$.

9 $\dfrac{dy}{dx} = 1 - \dfrac{1}{x^2}$ given that $y = -3$ when $x = \tfrac{1}{2}$. 　　 10 $\dfrac{dy}{dx} = 6\sqrt{x}$ given that $y = 7$ when $x = 4$.

11　Find the equation of the curve which passes through the point $(2, 1)$ and has gradient function

$$\dfrac{dy}{dx} = x - 2$$

12   Find the equation of the curve which passes through the point (1, 3) and has gradient function

$$\frac{dy}{dx} = 2 + \frac{1}{x^2}$$

13   Find the equation of the curve which passes through the point $(\frac{1}{4}, -2)$ and has gradient function

$$\frac{dy}{dx} = \frac{2x+1}{3\sqrt{x}}$$

In questions 14 and 15 use the given information to find $f(x)$.

14 $f'(x) = 2x, f(0) = 2.$ **15** $f'(x) = 7 - 4x, f(-3) = -1.$

6.3   Evaluation of definite integrals

We now return to the problem of evaluating a definite integral $\int_a^b f(x)\,dx$. If A is the area under the curve $y = f(x)$ up to x, the fundamental theorem states that $\dfrac{dA}{dx} = f(x)$. We can therefore find a formula for A by integrating $f(x)$, i.e. $A = \int f(x)\,dx$. Then the definite integral $\int_a^b f(x)\,dx$ is the difference between the values of A when $x = b$ and when $x = a$ (see p. 65).

Example 1   Consider $\int_1^3 (x^3 + 2x)dx$.

Now $A = \displaystyle\int (x^3 + 2x)dx = \frac{1}{4}x^4 + x^2 + C$

[Note that the area A may be measured from any fixed starting point (so far, this has usually been $x = 0$). The value of the constant C in the expression for A depends on which starting point is used.]
   When $x = 3$,

$$A = \frac{81}{4} + 9 + C$$

   When $x = 1$,

$$A = \frac{1}{4} + 1 + C$$

Thus $\displaystyle\int_1^3 (x^3 + 2x)dx = \left(\frac{81}{4} + 9 + C\right) - \left(\frac{1}{4} + 1 + C\right) = 28$

Notice that the arbitrary constant C cancels out in the subtraction. This will always happen, and in practice C may be omitted. In the previous example the evaluation is usually set out as follows:

$$\int_1^3 (x^3 + 2x)dx = \left[\frac{1}{4}x^4 + x^2\right]_1^3$$

$$= \left(\frac{81}{4} + 9\right) - \left(\frac{1}{4} + 1\right) = 28$$

The procedure is as follows.

To evaluate $\int_a^b f(x)\,dx$, we integrate $f(x)$ and write the result (omitting the arbitrary constant) inside square brackets, thus: $\left[\quad\right]_a^b$. We substitute first $x = b$ and then $x = a$ into the expression inside the brackets and subtract the two resulting values.

We have, so far, only considered the definite integral $\int_a^b f(x)\,dx$ in the cases when $f(x)$ is positive (i.e. the curve is above the x-axis) and $b > a$. However it can be shown that the method we have used remains valid if $f(x)$ takes negative values: in this case the definite integral may be a negative number. Also, we use the same method of evaluation when $b < a$.

Example 2

$$\int_2^4 (3 + 5x - x^2)\,dx = \left[\, 3x + \frac{5}{2}x^2 - \frac{1}{3}x^3 \,\right]_2^4$$

$$= \left(12 + \frac{80}{2} - \frac{64}{3}\right) - \left(6 + \frac{20}{2} - \frac{8}{3}\right) = 17\tfrac{1}{3}$$

Example 3

$$\int_{-3}^1 x(2x + 3)\,dx = \int_{-3}^1 (2x^2 + 3x)\,dx = \left[\, \frac{2}{3}x^3 + \frac{3}{2}x^2 \,\right]_{-3}^1$$

$$= \left(\frac{2}{3} + \frac{3}{2}\right) - \left(-\frac{54}{3} + \frac{27}{2}\right) = 6\tfrac{2}{3}$$

Example 4

$$\int_4^9 \left(\frac{1}{x^2} - \sqrt{x}\right)dx = \int_4^9 \left(x^{-2} - x^{\frac{1}{2}}\right)dx$$

$$= \left[\, -x^{-1} - \frac{2}{3}x^{\frac{3}{2}} \,\right]_4^9$$

$$= \left(-\frac{1}{9} - \frac{54}{3}\right) - \left(-\frac{1}{4} - \frac{16}{3}\right) = -12\tfrac{19}{36}$$

Remember,

an *indefinite integral* $\displaystyle\int f(x)\,dx$ is a *function of x*

a *definite integral* $\displaystyle\int_a^b f(x)\,dx$ is a *number*

a and b are called the **limits of integration**. In this context, the word limit means an end-point. This is quite distinct from its other meaning as the value approached by a limiting process.

Exercise 6.3   (Definite integrals)

Evaluate the following integrals.

1   $\displaystyle\int_{1}^{3} x^2\,dx$

2   $\displaystyle\int_{2}^{4} (x+3)\,dx$

3   $\displaystyle\int_{1}^{2} (x^2+5x-3)\,dx$

4   $\displaystyle\int_{-1}^{2} (4-x-3x^2)\,dx$

5   $\displaystyle\int_{-2}^{1} x^2(3-x)\,dx$

6   $\displaystyle\int_{-1}^{0} (3x^4-5x^5)\,dx$

7   $\displaystyle\int_{1}^{9} 3\sqrt{x}\,dx$

8   $\displaystyle\int_{1}^{8} \sqrt[3]{x}\,dx$

9   $\displaystyle\int_{0}^{1} x^{\frac{1}{4}}(2x-5)\,dx$

10   $\displaystyle\int_{-27}^{-1} x^{-\frac{1}{3}}\,dx$

11   $\displaystyle\int_{-3}^{-2} \frac{1}{x^2}\,dx$

12   $\displaystyle\int_{-3}^{-2} \frac{1}{x^3}\,dx$

13   $\displaystyle\int_{1}^{2} \left(\frac{2}{x^2}-\frac{5}{x^4}\right)dx$

14   $\displaystyle\int_{\frac{1}{4}}^{1} \left(\sqrt{x}-\frac{1}{\sqrt{x}}\right)dx$

15   $\displaystyle\int_{1}^{4} \frac{2x^2+3}{\sqrt{x}}\,dx$

6.4   Problems encountered with definite integrals

We have seen that there are times when $\dfrac{dy}{dx}$ does not exist (see p. 31) and we might expect to encounter similar problems with integration. For example, suppose we are given that the derived function is $\dfrac{dy}{dx}=\dfrac{1}{x^2}=x^{-2}$. The integral is $y=\dfrac{x^{-1}}{-1}+C$, i.e. $y=-\dfrac{1}{x}+C$.

However, when $x=0$, y is not defined, and hence $\dfrac{dy}{dx}$ does not exist.

So the required result $\dfrac{dy}{dx}=\dfrac{1}{x^2}$ is not true when $x=0$. The integration $\displaystyle\int\frac{1}{x^2}\,dx=-\frac{1}{x}+C$ is not valid when $x=0$.

Now consider the usual evaluation of a definite integral $\int_{a}^{b} f(x)\,dx$. If we write A for the integral of $f(x)$, then the evaluation starts $\int_{a}^{b} f(x)\,dx=[A]_{a}^{b}$. The values of A at $x=b$ and $x=a$ are then calculated and subtracted. This method will work provided that the result $\dfrac{dA}{dx}=f(x)$ holds true throughout the range of integration, i.e. between $x=a$ and $x=b$. So we should always check that $\dfrac{dA}{dx}$ does exist (and is equal to $f(x)$) for all values of x between a and b. If there are any values of x between a and b for which A is not defined, or A is not continuous, then $\dfrac{dA}{dx}$ will not exist at these values, and the evaluation of $\int_{a}^{b} f(x)\,dx$ will not be valid.

Example   Comment on the following evaluations:

(i) $\displaystyle\int_{-1}^{2}\frac{1}{x^2}dx = \left[-\frac{1}{x}\right]_{-1}^{2} = \left(-\frac{1}{2}\right)-\left(+1\right) = -\frac{3}{2}$

(ii) $\displaystyle\int_{-3}^{-1}\frac{1}{x^2}dx = \left[-\frac{1}{x}\right]_{-3}^{-1} = \left(+1\right)-\left(+\frac{1}{3}\right) = \frac{2}{3}$

(i)   The integral $A = -\dfrac{1}{x}$ is not defnied when $x = 0$, and so the

result $\dfrac{\mathrm{d}A}{\mathrm{d}x} = \dfrac{1}{x^2}$ is not true when $x = 0$. Since 0 lies between -1 and

2, the evaluation is not valid.

(ii)   Although the integral $A = -\dfrac{1}{x}$ is not defined when $x = 0$, the

result $\dfrac{\mathrm{d}A}{\mathrm{d}x} = \dfrac{1}{x^2}$ is true for all values of x between -3 and -1. So

the evaluation is perfectly correct.

Exercise 6.4   (Problems with definite integrals)

1   Criticise $\displaystyle\int_{-2}^{1}\frac{1}{x^3}dx = \left[-\frac{1}{2x^2}\right]_{-2}^{1} = \left(-\frac{1}{2}\right)-\left(-\frac{1}{8}\right) = -\frac{3}{8}.$

2   Criticise $\displaystyle\int_{-1}^{1}\left(1-\frac{1}{x^2}\right)dx = \left[x+\frac{1}{x}\right]_{-1}^{1} = (2)-(-2) = 4.$

In questions 3 to 8 state, giving reasons, whether the evaluation is valid or not.

3   $\displaystyle\int_{-3}^{1}\frac{1}{x^2}dx = \left[-\frac{1}{x}\right]_{-3}^{1} = -\frac{4}{3}$ 4 $\displaystyle\int_{1}^{2}\frac{1}{x^3}dx = \left[-\frac{1}{2x^2}\right]_{1}^{2} = \frac{3}{8}$

5   $\displaystyle\int_{-1}^{3}\left(2x+\frac{1}{x^2}\right)dx = \left[x^2-\frac{1}{x}\right]_{-1}^{3} = 6\frac{2}{3}$ 6 $\displaystyle\int_{-4}^{-2}\left(\frac{3}{x^2}-\frac{2}{x^3}\right)dx = \left[-\frac{3}{x}+\frac{1}{x^2}\right]_{-4}^{-2} = \frac{15}{16}$

7   $\displaystyle\int_{-8}^{1}x^{\frac{1}{3}}dx = \left[\frac{3}{4}x^{\frac{4}{3}}\right]_{-8}^{1} = -11\frac{1}{4}$ 8 $\displaystyle\int_{-1}^{27}x^{-\frac{4}{3}}dx = \left[-3x^{-\frac{1}{3}}\right]_{-1}^{27} = -4$

7

Applications of definite integrals

7.1 Area between a curve and the x-axis

Curves above the x-axis

If a curve $y = f(x)$ is above the x-axis, the area of a typical rectangle is $y\,\delta x$, and so the area under the curve between $x = a$ and $x = b$ is approximately $\sum_{x=a}^{x=b} y\,\delta x$ (Fig. 7.1). This area is $\lim_{\delta x \to 0} \sum_{x=a}^{x=b} y\,\delta x$, which is given by the definite integral $\int_a^b y\,dx$.

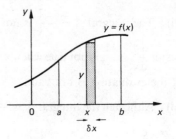

Fig. 7.1

Example Find the area under the curve $y = x^2 - 4x + 5$ between $x = -1$ and $x = 2$.

A sketch of the curve is shown in Fig. 7.2.
 The area is

$$\int_{-1}^{2} y\,dx = \int_{-1}^{2} (x^2 - 4x + 5)\,dx$$

$$= \left[\frac{1}{3}x^3 - 2x^2 + 5x \right]_{-1}^{2}$$

$$= \left(\frac{8}{3} - 8 + 10 \right) - \left(-\frac{1}{3} - 2 - 5 \right) = 12$$

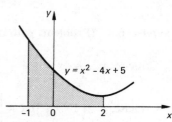

Fig. 7.2

Curves below the x-axis

If a curve $y = f(x)$ is below the x-axis, y is negative (Fig. 7.3). Thus $y\,\delta x$ is negative, but its numerical value is still the area of a rectangle. So $\int_a^b y\,dx$, which is $\lim_{\delta x \to 0} \sum_{x=a}^{x=b} y\,\delta x$, still gives the area between the curve and the x-axis, from $x = a$ to $x = b$, but it is negative.

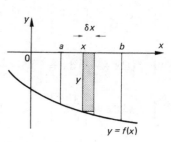

Fig. 7.3

Example Find the area between the curve $y = 1 - x^2$ and the x-axis, from $x = 2$ to $x = 5$.

Between $x = 2$ and $x = 5$, this curve is below the x-axis (Fig. 7.4).

$$\int_{2}^{5} y\,dx = \int_{2}^{5} (1 - x^2)\,dx = \left[x - \frac{1}{3}x^3 \right]_{2}^{5}$$

$$= \left(5 - \frac{125}{3} \right) - \left(2 - \frac{8}{3} \right) = -36$$

The area is therefore 36.

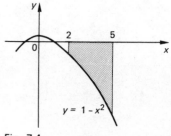

Fig. 7.4

Curves partly above and partly below the x-axis

If a curve is partly above and partly below the x-axis, we first find where the curve crosses the x-axis. Areas above and below the x-axis are then calculated separately.

Example   Find the total area between the curve $y = x^2 - 3x + 2$ and the x-axis, from $x = 0$ to $x = 3$.

The curve, shown in Fig. 7.5, crosses the x-axis when $y = 0$,

i.e. $x^2 - 3x + 2 = 0$
i.e. $(x - 1)(x - 2) = 0$
i.e. $x = 1$ or $x = 2$

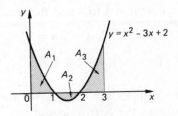

Fig. 7.5

The required area (shown shaded) is partly above and partly below the x-axis. We find the areas A_1, A_2, and A_3 separately.

$$\int_0^1 y\,dx = \int_0^1 (x^2 - 3x + 2)\,dx = \left[\frac{1}{3}x^3 - \frac{3}{2}x^2 + 2x\right]_0^1$$

$$= \left(\frac{1}{3} - \frac{3}{2} + 2\right) - (0) = \frac{5}{6}$$

So the area $A_1 = \frac{5}{6}$.

$$\int_1^2 y\,dx = \int_1^2 (x^2 - 3x + 2)\,dx = \left[\frac{1}{3}x^3 - \frac{3}{2}x^2 + 2x\right]_1^2$$

$$= \left(\frac{8}{3} - 6 + 4\right) - \left(\frac{1}{3} - \frac{3}{2} + 2\right) = -\frac{1}{6}$$

This is negative because the area is below the x-axis. The area $A_2 = \frac{1}{6}$.

$$\int_2^3 y\,dx = \int_2^3 (x^2 - 3x + 2)\,dx = \left[\frac{1}{3}x^3 - \frac{3}{2}x^2 + 2x\right]_2^3$$

$$= \left(9 - \frac{27}{2} + 6\right) - \left(\frac{8}{3} - 6 + 4\right) = \frac{5}{6}$$

So the area $A_3 = \frac{5}{6}$.

The total area between the curve and the x-axis is therefore

$$\tfrac{5}{6} + \tfrac{1}{6} + \tfrac{5}{6} = 1\tfrac{5}{6}$$

Notice that the definite integral from $x = 0$ to $x = 3$ is

$$\int_0^3 y\,dx = \int_0^3 (x^2 - 3x + 2)\,dx = \left[\frac{1}{3}x^3 - \frac{3}{2}x^2 + 2x\right]_0^3$$

$$= \left(9 - \frac{27}{2} + 6\right) - (0) = 1\tfrac{1}{2}$$

This is, in fact, the sum of the separate areas, but counting areas below the x-axis as negative, i.e. $\frac{5}{6} - \frac{1}{6} + \frac{5}{6} = 1\frac{1}{2}$. If we wish to find the actual area, it is essential to calculate the areas separately, as above.

Exercise 7.1   (Area between a curve and the x-axis)

In questions 1 to 4 find the area under the curve between the given values of x.

1   $y = 3x^2 + 2$; $x = 0, x = 2$.   2 $y = 5 - 4x$; $x = -1, x = 1$.

3   $y = \dfrac{1}{\sqrt{x}}$; $x = \frac{1}{4}, x = 9$.   4 $y = \dfrac{3}{x^2}$; $x = -2, x = -1$.

In questions 5 to 7 the curves are below the x-axis. Find the area between the curve and the x-axis between the given values of x.

5   $y = 1 - x^2$; $x = 1, x = 3$.   6 $y = x^2 + 7x + 6$; $x = -3, x = -2$.
7   $y = 2 - \sqrt{x}$; $x = 4, x = 9$.
8   Show that the curve $y = x^2 - 4x + 3$ cuts the x-axis when $x = 1$ and $x = 3$. Calculate the area between the curve and the x-axis: (i) from $x = 0$ to $x = 1$; (ii) from $x = 1$ to $x = 3$; (iii) from $x = 3$ to $x = 4$. Hence find the total area between the curve and the x-axis between $x = 0$ and $x = 4$.
   Find $\int_0^4 (x^2 - 4x + 3)dx$, and interpret this value.

9   Find the total area between the curve $y = \dfrac{1}{x^2} - 1$ and the x-axis, between $x = \dfrac{1}{2}$ and $x = 2$.

   Find $\displaystyle\int_{\frac{1}{2}}^{2} \left(\dfrac{1}{x^2} - 1\right)dx$ and interpret this value.

In questions 10 to 12 find the area enclosed by the curve and the x-axis.

10   $y = 6 + x - x^2$ 11 $y = x(x - 2)(x - 3)$ 12 $y = x^3 - 4x$

7.2   Area between two curves

Consider two curves $y = f(x)$ and $y = g(x)$ as shown in Fig. 7.6. If the area between the two curves, from $x = a$ to $x = b$, is divided into strips, a typical rectangle has height $f(x) - g(x)$ and area $\{f(x) - g(x)\}\delta x$. So the total area is approximately $\displaystyle\sum_{x=a}^{x=b} \{f(x) - g(x)\}\delta x$. Taking the limit as $\delta x \to 0$, the area between the two curves is

$$\int_a^b \left\{f(x) - g(x)\right\}dx$$

So long as the curve $y = f(x)$ is above the curve $y = g(x)$, the position of the x-axis in relation to the area does not matter. The height of a typical rectangle is always $f(x) - g(x)$, and the area is given by the above definite integral.

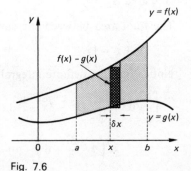

Fig. 7.6

Example 1   Find the area between the curves $y = x^2$ and $y = x^2 + 2x + 3$, from $x = 1$ to $x = 2\frac{1}{2}$.

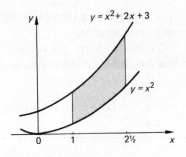

Between $x = 1$ and $x = 2\frac{1}{2}$, the curve $y = x^2 + 2x + 3$ is above the curve $y = x^2$ (see Fig. 7.7). So the required area is

$$\int_1^{2\frac{1}{2}} \left\{ (x^2 + 2x + 3) - x^2 \right\} dx = \int_1^{2\frac{1}{2}} (2x + 3)\, dx$$

$$= \left[x^2 + 3x \right]_1^{2\frac{1}{2}}$$

$$= \left(6\frac{1}{4} + 7\frac{1}{2} \right) - \left(1 + 3 \right) = 9\frac{3}{4}$$

Fig. 7.7

We often have to find the area enclosed between two curves. We first find the values of x where the curves intersect, and then we integrate between these values.

Example 2   Find the area enclosed between the curve $y = 4 + 5x - x^2$ and the line $y = 4 - x$.

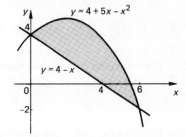

A sketch is shown in Fig. 7.8. The curve meets the line when

$$4 + 5x - x^2 = 4 - x$$
i.e. $\quad x^2 - 6x = 0$
i.e. $\quad x(x - 6) = 0$
i.e. $\quad x = 0 \quad \text{or} \quad 6$

We therefore integrate between $x = 0$ and $x = 6$. The curve $y = 4 + 5x - x^2$ is above $y = 4 - x$ between these values and so the required area (shown shaded) is

Fig. 7.8

$$\int_0^6 \left\{ (4 + 5x - x^2) - (4 - x) \right\} dx = \int_0^6 (6x - x^2)\, dx$$

$$= \left[3x^2 - \frac{1}{3}x^3 \right]_0^6$$

$$= (108 - 72) - (0) = 36$$

Note that although some of the area is beneath the x-axis, we are evaluating $\int_0^6 (6x - x^2)\, dx$, and $6x - x^2$ is greater than or equal to zero throughout the range of integration.

Example 3   Find the total area enclosed between the curves $y = x^3 + x^2 - 5x$ and $y = x^2 - x$.

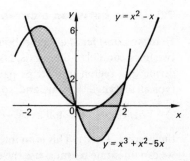

A sketch is shown in Fig. 7.9. The curves meet when

$$x^3 + x^2 - 5x = x^2 - x$$
i.e. $\quad x^3 - 4x = 0$
i.e. $\quad x(x^2 - 4) = 0$
i.e. $\quad x(x + 2)(x - 2) = 0$
i.e. $\quad x = -2, 0 \quad \text{or} \quad 2$

Fig. 7.9

The required area (shown shaded) consists of two parts, which we must find separately.

From $x = -2$ to $x = 0$, $y = x^3 + x^2 - 5x$ is above $y = x^2 - x$, so the area between the curves is

$$\int_{-2}^{0} \left\{ (x^3 + x^2 - 5x) - (x^2 - x) \right\} dx = \int_{-2}^{0} (x^3 - 4x) dx$$

$$= \left[\tfrac{1}{4}x^4 - 2x^2 \right]_{-2}^{0}$$

$$= (0) - (4 - 8) = 4$$

From $x = 0$ to $x = 2$, $y = x^2 - x$ is above $y = x^3 + x^2 - 5x$, so the area between the curves is

$$\int_{0}^{2} \left\{ (x^2 - x) - (x^3 + x^2 - 5x) \right\} dx = \int_{0}^{2} (4x - x^3) dx$$

$$= \left[2x^2 - \tfrac{1}{4}x^4 \right]_{0}^{2}$$

$$= (8 - 4) - (0) = 4$$

The total area enclosed by the two curves is therefore $4 + 4 = 8$.

Exercise 7.2   (Areas between curves)

In questions 1 to 4 find the area bounded by the two given curves and the given values of x.

1   $y = x^2$, $y = x^2 + 4x$; $x = 1$, $x = 2$.   2 $y = x^2$, $y = x^3$; $x = 1$, $x = 3$.
3   $y = x^2 - 3x$, $y = x + 2$; $x = 0$, $x = 4$.   4 $y = 3 - x^2$, $y = x - x^2$; $x = -2$, $x = 1$.

In questions 5 to 10 find the area enclosed between the two given curves

5   $y = x^2$; $y = 3x$.   6 $y = x^2 - 1$; $y = 3 - 3x$.   7 $y = x^2 - 3$; $y = 5 - x^2$.
8   $y = x^2 - 5x$; $y = 6 - x - x^2$.   9 $y = x^3$; $y = 9x$.   10 $y = x^3 + x^2$; $y = x^2 + 4x$.

7.3   Area between a curve and the y-axis

Here we consider the region bounded by a curve, the y-axis, and two lines parallel to the x-axis, say $y = p$ and $y = q$ (Fig. 7.10). If we divide this region into strips parallel to the x-axis, the area of a typical rectangle is $x\,\delta y$, and so the total area is approximately $\sum_{y=p}^{y=q} x\,\delta y$. Taking the limit as $\delta y \to 0$, the area between the curve and the y-axis is $\int_{p}^{q} x\,dy$. This is an integral with respect to y, and before we can integrate we must use the equation of the curve to express x as a function of y.

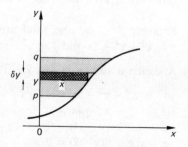

Fig. 7.10

Example 1 Find the area between the curve $y^2 = 1 + 2x$ and the y-axis, from $y = 2$ to $y = 3$.

Rearranging the equation, we have $x = \frac{1}{2}(y^2 - 1)$ and so the required area (shown shaded in Fig. 7.11) is

$$\int_2^3 x \, dy = \int_2^3 \frac{1}{2}(y^2 - 1) dy = \left[\frac{1}{6} y^3 - \frac{1}{2} y \right]_2^3$$

$$= \left(\frac{9}{2} - \frac{3}{2} \right) - \left(\frac{4}{3} - 1 \right)$$

$$= 2\frac{2}{3}$$

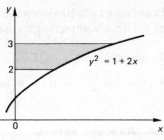

Fig. 7.11

Another method is illustrated by the following example.

Example 2 Find the area between the curve $y = (x + 1)^2$ and the y-axis, from $y = 4$ to $y = 9$, for $x \geqslant 0$.

We find the required area as (Fig. 7.12)

[Area of rectangle OPQR] – Area A – Area B

$$= 18 - 4 - \int_1^2 (x + 1)^2 dx$$

$$= 18 - 4 - \int_1^2 (x^2 + 2x + 1) dx$$

$$= 14 - \left[\frac{1}{3} x^3 + x^2 + x \right]_1^2$$

$$= 14 - \left\{ \left(\frac{8}{3} + 4 + 2 \right) - \left(\frac{1}{3} + 1 + 1 \right) \right\} = 14 - 6\frac{1}{3}$$

$$= 7\frac{2}{3}$$

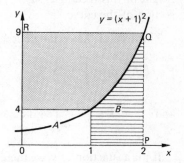

Fig. 7.12

Alternatively we can rearrange the equation as

$$y^{\frac{1}{2}} = x + 1 \quad \text{or} \quad x = y^{\frac{1}{2}} - 1$$

The area is

$$\int_4^9 x \, dy = \int_4^9 (y^{\frac{1}{2}} - 1) dy = \left[\frac{2}{3} y^{\frac{3}{2}} - y \right]_4^9$$

$$= \left(18 - 9 \right) - \left(\frac{16}{3} - 4 \right) = 7\frac{2}{3} \quad \text{as before}$$

Exercise 7.3 (Area between a curve and the y-axis)

Find the area bounded by the given curve, the y-axis, and two given values of y.

1 $y^2 = x$; $y = 1, y = 3$.   2 $y^3 = \frac{1}{4}x$; $y = 0, y = 2$.   3 $y^2 = 1 - x$; $y = 0, y = 1$.

4 $y^4 = 2x$; $y = 0, y = 2$.   5 $y = x^3$; $y = 1, y = 8$.   6 $y = \frac{1}{x^3}$; $y = 8, y = 27$.

In questions 7 to 10, sketch the region bounded by the given curve, the y-axis, and the two given values of y. Use a 'subtraction method' to find the area.

7 $y = x^3$; $y = 1$, $y = 27$.

8 $y = x^2 + x$; $y = 0$, $y = 6$, for $x \geqslant 0$.

9 $y = 5 - \dfrac{1}{x^2}$; $y = 1$, $y = 4$, for $x \geqslant 0$.

10 $y = x^2 - 9$; $y = -5$, $y = 0$, for $x \geqslant 0$.

7.4 Average values

Consider a curve $y = f(x)$ between $x = a$ and $x = b$. The area under this curve is $\int_a^b y\,dx$ (see Fig. 7.13).

We can draw a rectangle whose area is equal to the area under the curve between $x = a$ and $x = b$. It is reasonable to say that the height of this rectangle, say k, is the average value of y over this section of the curve.

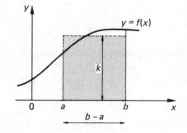

Fig. 7.13

The area of the rectangle is $k(b - a)$, so we have

$$k(b - a) = \int_a^b y\,dx$$

and

$$k = \frac{1}{b - a} \int_a^b y\,dx$$

If y is a function of x, we therefore define the **average value** of y over the interval $a \leqslant x \leqslant b$ (i.e. for values of x between a and b) to be

$$\boxed{\frac{1}{b - a} \int_a^b y\,dx}$$

If y takes negative values (i.e. the curve is wholly or partly below the x-axis), then the interpretation in terms of areas will be more complicated, but the average value is still defined as $\dfrac{1}{b-a} \int_a^b y\,dx$.

Example 1 If $y = (x - 1)^2$, find the average value of y over the interval $0 \leqslant x \leqslant 3$.

The curve is shown in Fig. 7.14. The average value is

$$\frac{1}{3 - 0} \int_0^3 (x - 1)^2 \, dx = \frac{1}{3} \int_0^3 (x^2 - 2x + 1)\,dx$$

$$= \frac{1}{3}\left[\frac{1}{3}x^3 - x^2 + x \right]_0^3 = \frac{1}{3}\left\{ (9 - 9 + 3) - (0) \right\}$$

$$= 1$$

The value of y is always positive, and takes values between 0 and 4 in the given interval, so this seems a reasonable average value.

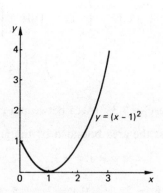

Fig. 7.14

Example 2 Find the average value of $y = x^2(2-x)$ over the interval $-1 \leqslant x \leqslant 3$.

The curve is shown in Fig. 7.15. The average value is

$$\frac{1}{3-(-1)} \int_{-1}^{3} x^2(2-x)\,dx = \frac{1}{4} \int_{-1}^{3} (2x^2 - x^3)\,dx$$

$$= \frac{1}{4}\left[\frac{2}{3}x^3 - \frac{1}{4}x^4 \right]_{-1}^{3}$$

$$= \frac{1}{4}\left\{ \left(18 - \frac{81}{4} \right) - \left(-\frac{2}{3} - \frac{1}{4} \right) \right\} = -\frac{1}{3}$$

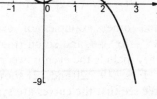

Fig. 7.15

Note that y takes both positive and negative values in the given interval.

Exercise 7.4   (Average values)

For the following functions, find the average value of y over the given interval.

1   $y = x^2$, $0 \leqslant x \leqslant 4$.   2 $y = 2x - 1$, $-1 \leqslant x \leqslant 4$.   3 $y = x(x-3)$, $0 \leqslant x \leqslant 4$.

4   $y = x(x^2 - 4)$, $-1 \leqslant x \leqslant 3$.   5 $y = \frac{1}{x^2}$, $1 \leqslant x \leqslant 3$.   6 $y = \frac{1}{\sqrt{x}}$, $\frac{1}{9} \leqslant x \leqslant \frac{1}{4}$.

7.5   Odd and even functions

When we have to evaluate a definite integral whose limits are plus and minus the same number, for example the integral $\int_{-2}^{2} (2x^3 + 3x^2 - 9x + 6)\,dx$, we can often simplify the calculation by using symmetry.

An **odd** function $f(x)$ is one for which $f(-x) = -f(x)$, i.e. when x is replaced by $-x$, the numerical value of the function is unaltered, but its sign is changed. For example, $f(x) = x^5$ is an odd function, since

$$f(-x) = (-x)^5 = -x^5 = -f(x)$$

Also $f(x) = x^3 - 4x$ is odd, since

$$f(-x) = (-x)^3 - 4(-x) = -x^3 + 4x = -(x^3 - 4x) = -f(x)$$

and other examples of odd functions are x, x^3, x^7, $2x^9 + 3x^5 - 7x$, and so on (they are called odd functions because they include the odd powers of x).

Figs 7.16, 7.17 and 7.18 show the graphs of some odd functions. We see that the curves have rotational symmetry about the origin.

When we integrate an odd function $f(x)$ between $x = -a$ and $x = +a$, the areas above the x-axis between $x = 0$ and $x = a$ cancel with the equal areas below the x-axis between $x = -a$ and $x = 0$ (which count as negative), and *vice versa*. So the definite integral is zero, i.e.

$$\int_{-a}^{a} f(x)\,dx = 0$$

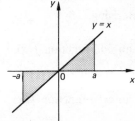

Fig. 7.16

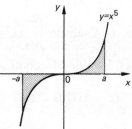

Fig. 7.17

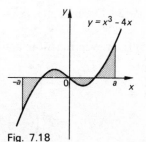

Fig. 7.18

An **even** function $f(x)$ is one for which $f(-x) = f(x)$, i.e. when x is replaced by $-x$, the value of the function is unchanged. For example $f(x) = x^4 - 5x^2 + 4$ is even, since

$$f(-x) = (-x)^4 - 5(-x)^2 + 4 = x^4 - 5x^2 + 4 = f(x)$$

and other examples of even functions are 1, x^2, x^4, x^6, $x^{10} - 8x^4 + 7$, and so on (they are called even functions because they include the even powers of x).

Figs 7.19, 7.20 and 7.21 show the graphs of some even functions. We see that the curves are symmetrical about the y-axis.

When we integrate an even function $f(x)$ between $x = -a$ and $x = +a$, the areas between $x = 0$ and $x = a$ are repeated between $x = -a$ and $x = 0$. So the definite integral is twice that between $x = 0$ than $x = -a$.

$$\int_{-a}^{a} f(x)\,dx = 2\int_{0}^{a} f(x)\,dx$$

This simplifies the calculation, because it is easier to substitute $x = 0$ than $x = -a$.

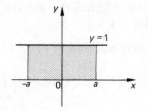

Fig. 7.19

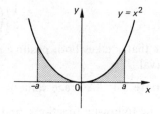

Fig. 7.20

Summarising,

for an odd function $f(x)$, $\qquad \int_{-a}^{a} f(x)\,dx = 0$

(i.e. $f(-x) = -f(x)$)

for an even function $f(x)$, $\qquad \int_{-a}^{a} f(x)\,dx = 2\int_{0}^{a} f(x)\,dx$

(i.e. $f(-x) = f(x)$)

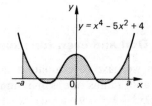

Fig. 7.21

Example 1 Evaluate $\displaystyle\int_{-3}^{3} (x^7 - 2x^3)\,dx$.

$(x^7 - 2x^3)$ is an odd function, and so

$$\int_{-3}^{3} (x^7 - 2x^3)\,dx = 0$$

Example 2 Evaluate $\displaystyle\int_{-4}^{4} (x^2 - 5)\,dx$.

$(x^2 - 5)$ is an even function, and so

$$\int_{-4}^{4} (x^2 - 5)\,dx = 2\int_{0}^{4} (x^2 - 5)\,dx = 2\left[\frac{1}{3}x^3 - 5x\right]_{0}^{4}$$

$$= 2\left\{\left(\frac{64}{3} - 20\right) - 0\right\} = 2\tfrac{2}{3}$$

Some functions are neither odd nor even. For example, if

$$f(x) = 2x^3 + 3x^2 - 9x + 6$$

then

$$f(-x) = 2(-x)^3 + 3(-x)^2 - 9(-x) + 6$$
$$= -2x^3 + 3x^2 + 9x + 6$$

and this is not equal to $-f(x)$ nor to $f(x)$. However, we can consider the odd and even parts separately.

Example 3 Evaluate $\displaystyle\int_{-2}^{2} (2x^3 + 3x^2 - 9x + 6)\,dx$.

$$\int_{-2}^{2} (2x^3 + 3x^2 - 9x + 6)\,dx = \int_{-2}^{2} (2x^3 - 9x)\,dx + \int_{-2}^{2} (3x^2 + 6)\,dx$$

$$= 0 + 2\int_{0}^{2} (3x^2 + 6)\,dx$$

$$= 2\Big[x^3 + 6x \Big]_{0}^{2}$$

$$= 2\{(8 + 12) - 0\} = 40$$

Exercise 7.5 (Odd and even functions)

In questions 1 to 8 state whether the given function is odd, even, or neither.

1 x^9 2 $3x^8$ 3 $x^4 - x^2 + 1$ 4 $x^4 + 2x$

5 $4x^5 - 9x$ 6 $x^7 - 3x^3 + 1$ 7 $x^{\frac{4}{3}}$ 8 $x^{\frac{5}{3}}$

In questions 9 to 16 evaluate the definite integrals.

9 $\displaystyle\int_{-8}^{8} x^5\,dx$ 10 $\displaystyle\int_{-3}^{3} (4x^3 + 5x)\,dx$ 11 $\displaystyle\int_{-2}^{2} (x^4 - x^2)\,dx$

12 $\displaystyle\int_{-3}^{3} (3x^2 - 4)\,dx$ 13 $\displaystyle\int_{-3}^{3} (x^5 - 6x^3 + 3x^2 + 12x - 4)\,dx$ 14 $\displaystyle\int_{-5}^{5} (2x^7 - 3x^5 + 8x^3 + 3)\,dx$

15 $\displaystyle\int_{-1}^{1} (7x^3 - x^2 + 6x - 2)\,dx$ 16 $\displaystyle\int_{-6}^{6} x^{\frac{5}{3}}\,dx$

7.6 Volumes of revolution about the x-axis

Consider the area between a curve and the x-axis, from $x = a$ to $x = b$.

  If this area is rotated about the x-axis, it will generate a solid, called a **solid of revolution**. The solid has the x-axis as an axis of symmetry; all cross-sections taken perpendicular to this axis are circular (see Fig. 7.22).

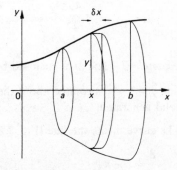

Fig. 7.22

To find the volume of this solid, we consider a typical rectangle of height y and width δx. When this is rotated about the x-axis it generates a circular disc, of radius y and thickness δx. The volume of this disc is $\pi y^2 \delta x$, and so the total volume of the solid is approximately $\sum_{x=a}^{x=b} \pi y^2 \, \delta x$.

We can make this approximation as close as we please to the true volume by taking narrower and narrower strips. Thus the volume of the solid is $\lim_{\delta x \to 0} \sum_{x=a}^{x=b} \pi y^2 \, \delta x$. This is the limit of a sum, which can be written as the definite integral $\int_a^b \pi y^2 \, dx$. (Remember that a definite integral $\int_a^b f(x) \, dx$ is defined as $\lim_{\delta x \to 0} \sum_{x=a}^{x=b} f(x) \, \delta x$; here $f(x)$ is replaced by πy^2.)

Hence, the volume of the solid of revolution is

$$\int_a^b \pi y^2 \, dx$$

We must use the equation of the curve to express y^2 in terms of x; then the definite integral may be evaluated in the usual way.

Example 1 The area under the curve $y = x^2 + 2x$, from $x = 1$ to $x = 2$, is rotated about the x-axis. Find the volume of the solid generated.

The volume (Fig. 7.23) is

$$\int_1^2 \pi y^2 \, dx = \int_1^2 \pi (x^2 + 2x)^2 \, dx = \pi \int_1^2 (x^4 + 4x^3 + 4x^2) \, dx$$

$$= \pi \left[\frac{1}{5}x^5 + x^4 + \frac{4}{3}x^3 \right]_1^2$$

$$= \pi \left\{ \left(\frac{32}{5} + 16 + \frac{32}{3} \right) - \left(\frac{1}{5} + 1 + \frac{4}{3} \right) \right\}$$

$$= \frac{458}{15} \pi$$

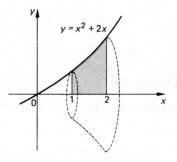

Fig. 7.23

Example 2 The area enclosed between the curve $y = \dfrac{8}{x}$ and the line $y = 6 - x$ is rotated about the x-axis. Find the volume of the solid generated.

The curve meets the line (Fig. 7.24) when

$$\frac{8}{x} = 6 - x$$

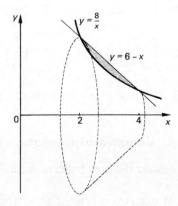

Fig. 7.24

i.e. $8 = 6x - x^2$

i.e. $x^2 - 6x + 8 = 0$

i.e. $(x - 2)(x - 4) = 0$

i.e. $x = 2$ or $x = 4$

The volume of the solid generated when the area under the line $y = 6 - x$, from $x = 2$ to $x = 4$, is rotated about the x-axis is

$$\int_2^4 \pi y^2 \, dx = \int_2^4 \pi (6 - x)^2 \, dx = \pi \int_2^4 (36 - 12x + x^2) \, dx$$

$$= \pi \left[36x - 6x^2 + \frac{1}{3} x^3 \right]_2^4$$

$$= \pi \left\{ \left(144 - 96 + \frac{64}{3} \right) - \left(72 - 24 + \frac{8}{3} \right) \right\} = \frac{56}{3} \pi$$

The volume of the solid generated when the area under the curve $y = \frac{8}{x}$, from $x = 2$ to $x = 4$, is rotated about the x-axis is

$$\int_2^4 \pi y^2 \, dx = \int_2^4 \pi \left(\frac{64}{x^2} \right) dx$$

$$= \pi \left[-\frac{64}{x} \right]_2^4 = \pi \left\{ (-16) - (-32) \right\} = 16\pi$$

The required volume is the difference between these two, i.e.

$$\frac{56}{3} \pi - 16\pi = \frac{8}{3} \pi$$

Deriving formulae

Some common solids, such as spheres and cones, may be regarded as solids of revolution. We can then use integration to derive the formulae for their volumes.

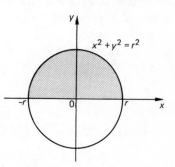

Fig. 7.25

Example   Derive the formula $\frac{4}{3}\pi r^3$ for the volume of a sphere of radius r.

A sphere may be generated by rotating a circle about a diameter. The circle of radius r, centred at the origin, has equation $x^2 + y^2 = r^2$ (see Fig. 7.25). If the area between this circle and the x-axis, from $x = -r$ to $x = r$, is rotated about the x-axis, it will generate a sphere of radius r.

The volume of the sphere is

$$\int_{-r}^r \pi y^2 \, dx = \int_{-r}^r \pi (r^2 - x^2) \, dx = \pi \left[r^2 x - \frac{1}{3} x^3 \right]_{-r}^r$$

$$= \pi \left\{ \left(r^3 - \frac{1}{3} r^3 \right) - \left(-r^3 + \frac{1}{3} r^3 \right) \right\} = \frac{4}{3} \pi r^3$$

Exercise 7.6   (Volumes of revolution)

In questions 1 to 6, find the volume of the solid generated when the area described is rotated about the x-axis. In each case sketch the solid.

1 The area between the curve $y = x^2$ and the x-axis, from $x = 1$ to $x = 3$.

2 The area between the curve $y = \dfrac{1}{x}$ and the x-axis, from $x = \frac{1}{2}$ to $x = 2$.

3 The area between the curve $4x^2 + y^2 = 16$ and the x-axis from $x = -2$ to $x = 2$.
4 The area enclosed between the curve $y = x(3 - x)$ and the x-axis.
5 The area enclosed between the curve $y = 3x - x^2$ and the line $y = x$.
6 The area enclosed between the curves $y = 5 - x^2$ and $y = x^2 + 3$.
7 Part of a sphere of radius r is obtained by rotating the area between the circle $x^2 + y^2 = r^2$ and the x-axis, from $x = \frac{2}{3}r$ to $x = r$, about the x-axis. We call this a cap of height $\frac{1}{3}r$. Find its volume. Find also the volumes of caps of height (i) $\frac{1}{2}r$ (ii) h.
8 The area between the line $y = mx$ and the x-axis, from $x = 0$ to $x = h$, is rotated about the x-axis. Find the volume of the cone generated. If the radius of the base of the cone is r, express m in terms of r and h, and hence derive the formula $\frac{1}{3}\pi r^2 h$ for the volume of the cone.

7.7   Rotation about other lines

When an area is rotated about a line other than the x-axis, we can determine which integral is required by considering a typical rectangle.

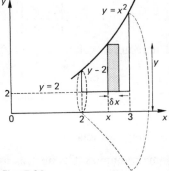

Example 1   The area bounded by the curve $y = x^2$ and the lines $y = 2$, $x = 2$ and $x = 3$, is rotated about the line $y = 2$. Find the volume of the solid generated.

The typical rectangle shown in Fig. 7.26 has height $(y - 2)$. When it is rotated it generates a disc of volume $\pi(y - 2)^2 \, \delta x$. Thus the required volume is

Fig. 7.26

$$\lim_{\delta x \to 0} \sum_{x=2}^{x=3} \pi(y - 2)^2 \, \delta x$$

which is

$$\int_2^3 \pi(y - 2)^2 \, dx = \int_2^3 \pi(x^2 - 2)^2 \, dx$$

$$= \pi \int_2^3 (x^4 - 4x^2 + 4) \, dx$$

$$= \pi \left[\frac{1}{5}x^5 - \frac{4}{3}x^3 + 4x \right]_2^3$$

$$= \pi \left\{ \left(\frac{243}{5} - 36 + 12 \right) - \left(\frac{32}{5} - \frac{32}{3} + 8 \right) \right\}$$

$$= \frac{313}{15} \pi$$

Example 2   The area between the curve $3x^2 + y^2 = 9$ and the
y-axis, from $y = -3$ to $y = 3$, is rotated about the y-axis. Find the
volume of the solid generated.

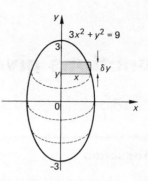

The typical rectangle shown in Fig. 7.27 has length x and width δy.
When it is rotated about the y-axis it generates a disc of radius x
and thickness δy, having volume $\pi x^2\, \delta y$. Thus the required volume
is

$$\lim_{\delta y \to 0} \sum_{y=-3}^{y=3} \pi x^2\, \delta y$$

which is
$$\int_{-3}^{3} \pi x^2\, dy = \pi \int_{-3}^{3} \frac{1}{3}(9 - y^2)\, dy = \pi \int_{-3}^{3} \left(3 - \frac{1}{3}y^2\right) dy$$

Fig. 7.27

$$= \pi \left[3y - \frac{1}{9}y^3\right]_{-3}^{3} = \pi\{(9 - 3) - (-9 + 3)\} = 12\pi$$

Example 3   The area enclosed between the two branches of the
curve $4x^2 - y^2 = 9$ and the lines $y = 4$ and $y = -4$ is rotated
about the y-axis. Calculate the volume of the solid of revolution
formed.

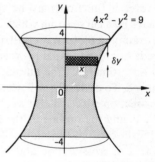

A rectangle of length x and width δy, when rotated about the y-
axis, generates a disc with volume $\pi x^2\, \delta y$ (Fig. 7.28). Hence the
volume of the solid of revolution is

Fig. 7.28

$$\lim_{\delta y \to 0} \sum_{y=-4}^{y=4} \pi x^2\, \delta y$$

which is
$$\int_{-4}^{4} \pi x^2\, dy = \pi \int_{-4}^{4} \frac{1}{4}(y^2 + 9)\, dy = \frac{1}{4}\pi \left[\frac{1}{3}y^3 + 9y\right]_{-4}^{4}$$

$$= \frac{1}{4}\pi\left\{\left(\frac{64}{3} + 36\right) - \left(-\frac{64}{3} - 36\right)\right\} = \frac{86}{3}\pi$$

Exercise 7.7   (Rotation about other lines)

Find the volume of the solid generated when the area described is rotated about the given axis.

1   The area between the curve $y = 2x^2$ and the line $y = 1$ from $x = 1$ to $x = 3$; rotated about the line
$y = 1$.

2   The area between the curve $y = \dfrac{1}{x^2}$ and the line $y = -2$, from $x = 1$ to $x = 2$; rotated about the
line $y = -2$.

3   The area between the curve $y = 3x^2$ and the y-axis, from $y = 1$ to $y = 4$; rotated about the y-axis.

4   The area between the curve $4x^2 + y^2 = 16$ and the y-axis, from $y = -4$ to $y = 4$; rotated about the
y-axis. [Compare your answer with question 3 of Exercise 7.6.]

5   The area between the curve $y^3 = x^2$ and the y-axis, from $y = 0$ to $y = 2$; rotated about the y-axis.

6   The area between the curve $y = x^2$ and the line $x = 1$, from $y = 1$ to $y = 4$; rotated about the line
$x = 1$.

8

Functions given as a table of values

Introduction

So far we have considered functions which have been represented by an algebraic equation or by a curve. We have looked at the problems of 'gradient finding' and 'area finding' for such functions. However a function may be specified by a table of values or perhaps by a curve (from which a table of values can be drawn up). For example a motorway constructor may have to remove part of a hillside in order to build his road. This means that he will need to find an estimate of the amount of earth which will have to be removed. To do this he will need to know the approximate area of cross-section of the hill. He will have no idea of the equation of the curve giving the outline of the hill, but by surveying (or by test-boring) he can build up a table of values from which he can estimate the area of cross-section and hence the volume of material to be removed. Fig. 8.1 shows a scale drawing of a cross-section of the hillside. It is then possible to make a table of values:

x	0	1	2	3	4	5	6	7	8	9
y	0	7	10	15	17	21	19	18	9	2

Here x might be in units of 10 metres, the height y in metres above the projected roadway.

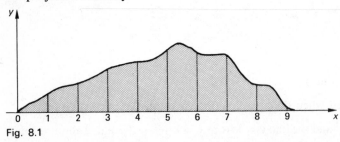

Fig. 8.1

In this chapter we shall be looking at functions given by a table like this. Usually the values of x will be equally spaced.

8.1 Gradient finding

We shall consider the problem of finding the gradient from a table of values. Suppose, for example, that the table above gives values of a function $f(x)$ and we wish to estimate the gradient of the curve $y = f(x)$ when $x = 3$, i.e. $f'(3)$.

90

From part of our table,

x	2	3	4
$y = f(x)$	10	15	17

we obtain three points P, Q and R on the curve, and we require the gradient at Q.

First method We find the gradients of the chords PQ and QR, one on each side of Q, and take the average of these two gradients (Fig. 8.2).

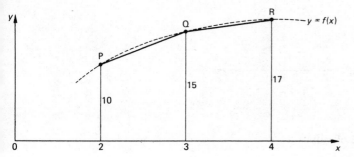

Fig. 8.2

The gradient of PQ is $\dfrac{15 - 10}{3 - 2} = 5$

The gradient of QR is $\dfrac{17 - 15}{4 - 3} = 2$

Thus $f'(3) \approx \frac{1}{2}(5 + 2) = 3.5$.

Second method We find the gradient of the chord PR (Fig. 8.3) (since this is approximately parallel to the tangent at Q). Thus

$$f'(3) \approx \frac{17 - 10}{4 - 2} = 3.5$$

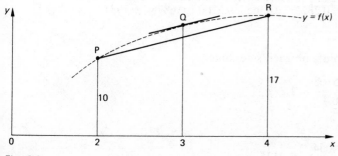

Fig. 8.3

We see that the answer is the same in both cases; the two methods are effectively the same, although the second method was quicker. We cannot say how accurate our answer of 3.5 is likely to be, but it is the best we can do with the given information. All we know is that the curve passes through the given points, and the true gradient at Q could be very different from 3.5 (Fig. 8.4).

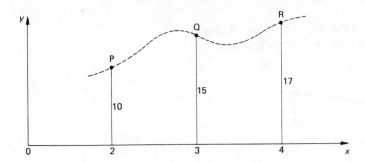

Fig. 8.4

In general, suppose that the table of values is

x	x_1	x_2	x_3	x_4	...
$f(x)$	y_1	y_2	y_3	y_4	...

To estimate $f'(x_2)$ we consider the points on each side of (x_2, y_2) (see Fig. 8.5). These are (x_1, y_1) and (x_3, y_3), and so

$$f'(x_2) \approx \frac{y_3 - y_1}{x_3 - x_1} = \frac{f(x_3) - f(x_1)}{x_3 - x_1}$$

Similarly,

$$f'(x_3) \approx \frac{y_4 - y_2}{x_4 - x_2} = \frac{f(x_4) - f(x_2)}{x_4 - x_2}$$

and so on.

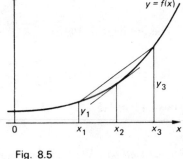

Fig. 8.5

Example 1 The following table gives values of a function $f(x)$.

x	0.6	0.8	1.0	1.2	1.4	1.6	1.8	2.0
$f(x)$	-0.4108	-0.1231	0.1000	0.2823	0.4365	0.5700	0.6878	0.7931

Tabulate approximate values of $f'(x)$.

To estimate $f'(0.8)$, we consider the points on each side; thus

$$f'(0.8) \approx \frac{0.1000 - (-0.4108)}{1.0 - 0.6} = \frac{0.5108}{0.4} = 1.277$$

Similarly,

$$f'(1.0) \approx \frac{0.2823 - (-0.1231)}{1.2 - 0.8} = \frac{0.4054}{0.4} = 1.0135$$

and so on. We obtain the following values:

x	0.6	0.8	1.0	1.2	1.4	1.6	1.8	2.0
$f(x)$	-0.4108	-0.1231	0.1000	0.2823	0.4365	0.5700	0.6878	0.7931
$f'(x)$			1.2770	1.0135	0.8413	0.7193	0.6283	0.5578

If we know the equation of the curve, we can use it to obtain a table of values. Then we can apply the same method as above. In this case we can improve the accuracy by taking the points closer together.

Example 2   If $f(x) = \dfrac{x-1}{2x-x^2}$, find approximate values for $f'(1.5)$ and $f'(3)$.

We cannot (as yet) differentiate $f(x)$, so we shall work from a table of values. To estimate $f'(1.5)$, we take a small x-step, say 0.01, to each side, and use the formula $\dfrac{x-1}{2x-x^2}$ to obtain a table of values.

When $x = 1.49$,

$$f(1.49) = \frac{1.49 - 1}{(2 \times 1.49) - (1.49)^2} = 0.644822$$

and so on. This gives the following set of values.

x	1.49	1.5	1.51
$f(x)$	0.644 822	0.666 667	0.689 282

Then

$$f'(1.5) \approx \frac{f(1.51) - f(1.49)}{1.51 - 1.49}$$

$$= \frac{0.689\,282 - 0.644\,822}{0.02}$$

$$= 2.223$$

(Using a calculator, we first calculate $f(1.49)$ and store it in the memory. We then calculate $f(1.51)$, subtract the memory, and divide by 0.02.) The actual value of $f'(1.5)$ is 2.2.

Similarly, for $f'(3)$, taking an x-step of 0.01 to each side, we obtain

x	2.99	3	3.01
$f(x)$	$-0.672\,275$	$-0.666\,667$	$-0.661\,162$

and $f'(3) \approx \dfrac{f(3.01) - f(2.99)}{3.01 - 2.99}$

$$= \frac{-0.661\,162 - (-0.672\,275)}{0.02}$$

$$= 0.5556$$

Exercise 8.1   (Gradients from a table of values)

1   The following table gives values of a function $f(x)$. Tabulate approximate values of $f'(x)$ for $x = 0.1, 0.2, \ldots, 0.9$.

x	0.0	0.1	0.2	0.3	0.4	0.5	0.6	0.7	0.8	0.9	1.0
$f(x)$	0.5040	0.5438	0.5832	0.6217	0.6591	0.6950	0.7291	0.7611	0.7910	0.8186	0.8438

2   The following table gives values of a function $f(x)$. Tabulate approximate values of $f'(x)$ for $x = -0.25, 0, \ldots, 1.75$.

x	-0.5	-0.25	0	0.25	0.5	0.75	1	1.25	1.5	1.75	2
$f(x)$	2.090	2.558	2.866	2.996	2.940	2.701	2.295	1.745	1.087	0.362	-0.387

3   If $f(x) = x^3$, form a table of values for $x = 0.9, 1, 1.1$, and hence find an approximate value for $f'(1)$.

Check your result by differentiation.

4   If $f(x) = \sqrt{x}$, find an approximate value for $f'(25)$, by taking an x-step of 0.5 on each side.

5   If $f(x) = \dfrac{x+1}{5-x}$, find an approximate value for $f'(0)$, by taking an x-step of 0.01 to each side.

8.2   Area finding

We return to the problem of finding the cross-sectional area of the hill given on p. 90. The table of values is

x	0	1	2	3	4	5	6	7	8	9
y	0	7	10	15	17	21	19	18	9	2

and these known points are shown in Fig. 8.6. We require the area under this curve between $x = 0$ and $x = 9$.

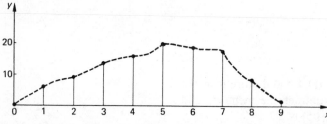

Fig. 8.6

First method: lower and upper bounds   Taking rectangles below the curve (we have assumed that the curve is increasing when $x < 5$ and decreasing when $x > 5$); the total area of the rectangles shown in Fig. 8.7 is

$$(0 \times 1) + (y_1 \times 1) + (y_2 \times 1) + (y_3 \times 1) + (y_4 \times 1) + (y_6 \times 1) + (y_7 \times 1) + (y_8 \times 1) + (y_9 \times 1)$$

(notice that y_5 is omitted)

$$= 0 + 7 + 10 + 15 + 17 + 19 + 18 + 9 + 2 = 97 \text{ square units}$$

This is a lower bound for the area.

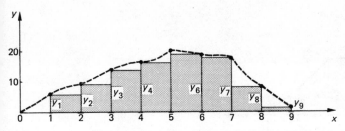

Fig. 8.7

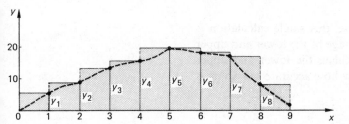

Fig. 8.8

Taking rectangles above the curve, the total area of the rectangles in Fig. 8.8 is

$$(y_1 \times 1) + (y_2 \times 1) + (y_3 \times 1) + (y_4 \times 1) + (y_5 \times 1) + (y_5 \times 1) + (y_6 \times 1) + (y_7 \times 1) + (y_8 \times 1)$$

(notice that y_5 occurs twice)

$$= 7 + 10 + 15 + 17 + 21 + 21 + 19 + 18 + 9$$
$$= 137 \text{ square units}$$

This is an upper bound for the area. For an estimate of the area, we take the average of the lower and upper bounds. Thus the area is approximately $\dfrac{97 + 137}{2} = 117$ square units.

As for the accuracy of this value, we can say that the true area certainly lies between 97 and 137 square units.

Second method: the trapezium rule   We join the given points by straight lines, and find the areas of the trapezia so formed (Fig. 8.9). The area of a trapezium is $\frac{1}{2}$(sum of parallel sides) × width, and so the total area of these trapezia is

$$\tfrac{1}{2}(0 + y_1) \times 1 + \tfrac{1}{2}(y_1 + y_2) \times 1 + \tfrac{1}{2}(y_2 + y_3) \times 1 + \ldots + \tfrac{1}{2}(y_7 + y_8) \times 1 + \tfrac{1}{2}(y_8 + y_9) \times 1$$
$$= y_1 + y_2 + y_3 + y_4 + y_5 + y_6 + y_7 + y_8 + \tfrac{1}{2}y_9$$
$$= 7 + 10 + 15 + 17 + 21 + 19 + 18 + 9 + 1$$
$$= 117 \text{ square units}$$

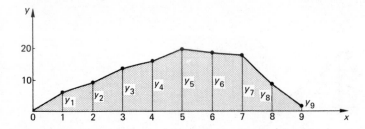

Fig. 8.9

This gives the same value as before; this single calculation is effectively the same as taking the average of the lower and upper bounds. However, if we do not calculate the lower and upper bounds separately, we shall not know how accurate our value is likely to be.

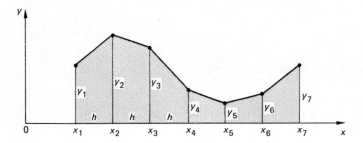

Fig. 8.10

In general, if the table of values is

x	x_1	x_2	x_3	$\ldots$	x_n
y	y_1	y_2	y_3	$\ldots$	y_n

and the differences between successive values of x are all equal to h (Fig. 8.10), then the area under the curve, between $x = x_1$ and $x = x_n$, is approximately

$$\tfrac{1}{2}(y_1 + y_2)h + \tfrac{1}{2}(y_2 + y_3)h + \tfrac{1}{2}(y_3 + y_4)h + \ldots + \tfrac{1}{2}(y_{n-1} + y_n)h$$
$$= h(\tfrac{1}{2}y_1 + \tfrac{1}{2}y_2 + \tfrac{1}{2}y_2 + \tfrac{1}{2}y_3 + \tfrac{1}{2}y_3 + \ldots + \tfrac{1}{2}y_{n-1} + \tfrac{1}{2}y_{n-1} + \tfrac{1}{2}y_n)$$

$$\boxed{= h(\tfrac{1}{2}y_1 + y_2 + y_3 + \ldots + y_{n-1} + \tfrac{1}{2}y_n)}$$

This is the **trapezium rule**.

The values of y are often called 'ordinates'. In Fig. 8.10 there are 7 ordinates (and 6 strips).

Example 1 The following table gives points on a curve.

x	3	3.2	3.4	3.6	3.8	4.0
y	6.44	8.91	9.97	9.46	7.46	4.27

Use the trapezium rule to find an approximation for the area under the curve between $x = 3$ and $x = 4$.

The width of the strips is $h = 0.2$, and the area is approximately

$$h(\tfrac{1}{2}y_1 + y_2 + y_3 + y_4 + y_5 + \tfrac{1}{2}y_6)$$

$$= 0.2(\tfrac{1}{2} \times 6.44 + 8.91 + 9.97 + 9.46 + 7.46 + \tfrac{1}{2} \times 4.27)$$

$$= 8.231$$

Example 2 Find an approximate value for the area under the curve $y = \dfrac{1}{x+4}$ between $x = 1$ and $x = 3$, using the trapezium rule with 5 ordinates.

We must first obtain a table of values. If there are to be 5 ordinates, there are 4 strips, so each strip has width $\tfrac{2}{4} = 0.5$, and the values of x are 1, 1.5, 2, 2.5, 3. We use the formula $y = \dfrac{1}{x+4}$ to calculate the corresponding values of y.

x	1	1.5	2	2.5	3
y	0.2	0.1818	0.1667	0.1538	0.1429

The area under the curve is approximately

$$0.5(\tfrac{1}{2} \times 0.2 + 0.1818 + 0.1667 + 0.1538 + \tfrac{1}{2} \times 0.1429) = 0.3369 \cdot$$

(The first value of y is calculated, divided by 2, and stored in the memory. The next value of y is calculated and added to the memory; similarly for the next two values. The last value of y is calculated, divided by 2, then added to the memory. Finally, the memory is recalled and multiplied by 0.5.)

The area under the curve is given by the definite integral $\displaystyle\int_1^3 \dfrac{1}{x+4}\, dx$, so we have shown that $\displaystyle\int_1^3 \dfrac{1}{x+4}\, dx \approx 0.3369$.

In this example, we could improve the accuracy by using more ordinates.

Even if the strips have different widths, we can still use trapezia to estimate the area under the curve.

Example 3   The following table gives points on a curve.

x	2.0	2.4	2.6	2.9	3.0
y	4.7	6.8	5.4	5.2	3.4

Find an approximation for the area under the curve between $x = 2$ and $x = 3$.

The area of the first trapezium in Fig. 8.11 is $\frac{1}{2}(4.7 + 6.8) \times 0.4$, and the total area is approximately

$$\frac{1}{2}(4.7 + 6.8) \times 0.4 + \frac{1}{2}(6.8 + 5.4) \times 0.2 + \frac{1}{2}(5.4 + 5.2) \times 0.3 + \frac{1}{2}(5.2 + 3.4) \times 0.1$$
$$= 5.54$$

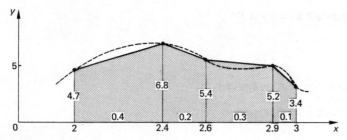

Fig. 8.11

The area of a circle

The evaluation of π has always interested mathematicians. We shall estimate it by considering lower and upper bounds for the area of a circle.

We take a circle of radius 2 units, centred at the origin, and we consider the area of the quadrant for which x and y are both positive.

The equation of the circle is $x^2 + y^2 = 2^2$, so $y = \sqrt{4 - x^2}$.

Suppose we divide the area into 10 strips, each of width 0.2. We make a table of values.

x	0	0.2	0.4	0.6	0.8	1.0	1.2	1.4	1.6	1.8	2.0
$y = \sqrt{4 - x^2}$	2.00	1.99	1.96	1.91	1.83	1.73	1.60	1.43	1.20	0.87	0

Considering rectangles below the curve (Fig. 8.12), the lower bound for the area is

$$(1.99 \times 0.2) + (1.96 \times 0.2) + \ldots + (0.87 \times 0.2) + 0$$
$$= 0.2\,(1.99 + 1.96 + \ldots + 0.87 + 0) = 2.904$$

Considering rectangles above the curve, the upper bound for the area is

$$(2.00 \times 0.2) + (1.99 \times 0.2) + \ldots + (1.20 \times 0.2) + (0.87 \times 0.2)$$
$$= 0.2\,(2.00 + 1.99 + \ldots + 1.20 + 0.87) = 3.304$$

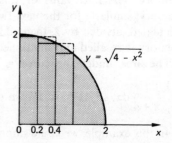

Fig. 8.12

The area is given by the definite integral $\int_0^2 \sqrt{4 - x^2}\,dx$, so we have shown that $\int_0^2 \sqrt{4 - x^2}\,dx$ lies between 2.904 and 3.304, i.e. $2.904 < \int_0^2 \sqrt{4 - x^2}\,dx < 3.304$.

Also, the area of the whole circle is $\pi \times 2^2 = 4\pi$; thus the area of the quadrant is π. We have shown that π lies between 2.904 and 3.304.

We can improve on this result by using more strips. With the help of a computer, the values shown in Table 8.1 were obtained.

Table 8.1

Number of strips	Width of strip	Lower bound	Upper bound
10	0.2	2.9045	3.3045
100	0.02	3.1204	3.1604
200	0.01	3.1312	3.1512
1000	0.002	3.1396	3.1436

(The true value of π is 3.14159 . . .)

Exercise 8.2 (Areas from a table of values)

In questions 1 to 3, find lower and upper bounds for the area under the given curve, by considering rectangles below and above the curve.

1 The curve given by the following table of values, between $x = 1$ and $x = 2$.

x	1.0	1.2	1.4	1.6	1.8	2.0
y	6.5	6.2	5.2	4.3	4.0	2.6

2 $y = \sqrt{1+x}$, between $x = 0$ and $x = 5$ (use strips 1 unit wide).

3 $y = \dfrac{1}{1+x^2}$, between $x = 0$ and $x = 1$ (use strips 0.2 units wide).

In questions 4 and 5, find lower and upper bounds for the definite integrals.

4 $\displaystyle\int_1^4 \frac{1}{x}\,dx$ (use strips 0.5 units wide).

5 $\displaystyle\int_{-20}^{-4} \sqrt{400 - x^2}\,dx$ (use strips 2 units wide).

In questions 6 to 8 use the trapezium rule to find an approximate value for the area under the given curve.

6 The curve given by the following table of values, between $x = 0$ and $x = 6$.

x	0	1	2	3	4	5	6
y	10.0	7.9	5.8	4.1	2.7	4.2	11.5

7 $y = \dfrac{1}{x-2}$, between $x = 3$ and $x = 8$ (use 6 ordinates).

8 $y = \dfrac{x^4}{x-1}$, between $x = 1.5$ and $x = 2$ (use 6 ordinates).

In questions 9 and 10, use the trapezium rule to find an approximate value for the definite integrals.

9 $\displaystyle\int_1^4 \sqrt{x^2 - 1}\,dx$ (use 7 ordinates).

10 $\displaystyle\int_2^3 \frac{1}{5x - x^2 - 4}\,dx$ (use 3 ordinates).

8.3 Volumes of revolution

We can use similar ideas to find lower and upper bounds for the volume of a solid of revolution.

Example The following table gives points on a curve.

x	1.0	1.1	1.2	1.3	1.4	1.5	1.6	1.7	1.8	1.9	2.0
y	1.26	1.30	1.35	1.41	1.48	1.56	1.65	1.75	1.86	1.98	2.11

Find lower and upper bounds for the volume of the solid formed when the area under this curve, between $x = 1$ and $x = 2$, is rotated about the x-axis.

When a rectangle below the curve is rotated about the x-axis, it generates a disc inside the solid of revolution (Fig. 8.13). The first such disc has volume $\pi \times 1.26^2 \times 0.1$, and the lower bound for the volume of revolution is

$$(\pi \times 1.26^2 \times 0.1) + (\pi \times 1.30^2 \times 0.1) + \ldots + (\pi \times 1.98^2 \times 0.1)$$
$$= \pi \times 0.1 \times (1.26^2 + 1.30^2 + \ldots + 1.98^2) = 7.815$$

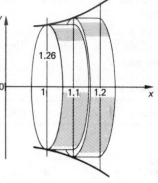

Fig. 8.13

Similarly, rectangles above the curve generate discs outside the solid of revolution (Fig. 8.14). The first such disc has volume $\pi \times 1.30^2 \times 0.1$, and the upper bound for the volume of revolution is

$$(\pi \times 1.30^2 \times 0.1) + (\pi \times 1.35^2 \times 0.1) + \ldots + (\pi \times 2.11^2 \times 0.1)$$
$$= \pi \times 0.1 \times (1.30^2 + 1.35^2 + \ldots + 2.11^2) = 8.715$$

Hence the volume lies between 7.815 and 8.715 cubic units. For an approximate value, we may take the average of these bounds; the volume is approximately $\frac{1}{2}(7.815 + 8.715) = 8.265$ cubic units.

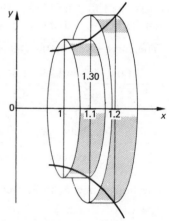

Fig. 8.14

Exercise 8.3 (Volumes from a table of values)

1 The following table gives points on a curve.

x	4	4.25	4.5	4.75	5
y	1.12	1.54	1.76	1.92	2.05

Find lower and upper bounds for the volume of the solid formed when the area under this curve, between $x = 4$ and $x = 5$, is rotated about the x-axis.

2 The following table gives points on a curve.

x	0	5	10	15	20	25	30	35	40
y	7.20	5.61	4.37	3.40	2.65	2.06	1.61	0.95	0

Find an approximate value for the volume of the solid formed when the area under this curve, between $x = 0$ and $x = 40$, is rotated about the x-axis, by taking the average of the lower and upper bounds.

3 Obtain a table of values for the function $y = \dfrac{x}{3-x}$ when $x = 1, 1.2, 1.4, 1.6, 1.8, 2$. Hence find lower and upper bounds for the volume of the solid formed when the area under this curve, between $x = 1$ and $x = 2$, is rotated about the x-axis.

8.4   The length of a curve

We can obtain an approximation for the length of a curve by joining points on it with straight lines (Fig. 8.15).

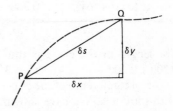

Fig. 8.15

Suppose P and Q are two points on the curve (Fig. 8.16). If the x-step is δx and the y-step δy, then, using Pythagoras' theorem, the length of the straight line PQ is

$$\delta s = \sqrt{(\delta x)^2 + (\delta y)^2}$$

We then add together the lengths of all such chords.

Since we have replaced each section of the curve by a straight line, this method will always give a lower bound for the length of the curve.

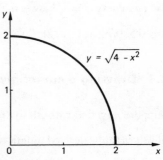

Fig. 8.16

Example 1   The following table gives points on a curve.

x	3	3.25	3.5	3.75	4
y	6.3	7.0	8.2	8.6	8.7

Find an approximation for the length of this curve between $x = 3$ and $x = 4$.

Taking the first two points (3, 6.3) and (3.25, 7.0), we have $\delta x = 0.25$ and $\delta y = 0.7$. The length of the straight line joining them is

$$\delta s = \sqrt{(\delta x)^2 + (\delta y)^2}$$
$$= \sqrt{(0.25)^2 + (0.7)^2} = 0.743$$

We may set out the calculation as follows.

x	3	3.25	3.5	3.75	4
y	6.3	7.0	8.2	8.6	8.7
δx		0.25	0.25	0.25	0.25
δy		0.7	1.2	0.4	0.1
$\delta s = \sqrt{(\delta x)^2 + (\delta y)^2}$		0.743	1.226	0.472	0.269

The length of the curve is thus calculated to be approximately $0.743 + 1.226 + 0.472 + 0.269 = 2.71$.

Example 2   Find an approximation for the length of the curve $y = \sqrt{4 - x^2}$ between $x = 0$ and $x = 2$.

This curve, sketched in Fig. 8.17, is the arc of a quadrant of a circle of radius 2 (see p. 98). We shall take 10 strips, as before. We use the formula $y = \sqrt{4 - x^2}$ to obtain a table of values, and then we proceed as above.

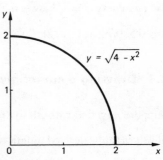

Fig. 8.17

x	0	0.2	0.4	0.6	0.8	1.0	1.2	1.4	1.6	1.8	2.0
$y = \sqrt{4 - x^2}$	2.00	1.99	1.96	1.91	1.83	1.73	1.60	1.43	1.20	0.87	0
δx		0.2	0.2	0.2	0.2	0.2	0.2	0.2	0.2	0.2	0.2
δy		-0.01	-0.03	-0.05	-0.08	-0.10	-0.13	-0.17	-0.23	-0.33	-0.87
$\delta s = \sqrt{(\delta x)^2 + (\delta y)^2}$		0.200	0.202	0.206	0.215	0.223	0.239	0.262	0.305	0.386	0.893

The length of the curve is thus calculated to be approximately $0.200 + 0.202 + \ldots + 0.893 = 3.13$.

The circumference of the circle is $2\pi \times 2 = 4\pi$, and so the length of this arc is π. We have shown that $\pi \approx 3.13$. We can obtain a better approximation by taking more strips. Using a computer, the values given in Table 8.2 were obtained.

Table 8.2

Number of strips	Width of strip	Approx. length of curve
10	0.2	3.132 264 62
100	0.02	3.141 298 57
1000	0.002	3.141 583 38

Exercise 8.4 (Lengths of curves)

Find approximations for the lengths of the following curves, between the values of x given.

1 A curve passing through the following points:

x	0	1	2	3	4	5
y	0.5	1.0	1.6	2.3	2.7	2.9

2 A curve passing through the following points:

x	3.0	3.1	3.2	3.3	3.4	3.5	3.6	3.7	3.8	3.9	4.0
y	5.58	5.52	5.51	5.56	5.55	5.46	5.36	5.36	5.42	5.44	5.45

3 The curve $y = x^2$ between $x = 0$ and $x = 1$ (first make a table of values for $x = 0, 0.2, 0.4, \ldots, 1$).

4 The curve $y = \dfrac{1}{x}$ between $x = 0.5$ and $x = 2$ (first make a table of values for $x = 0.5, 0.75, 1, \ldots, 2$).

8.5 Drawing a curve, given a table of values for $\dfrac{dy}{dx}$

Suppose now that we wish to draw a curve, but we are given a table of values for its derived function $\dfrac{dy}{dx}$. To do this, we must also know one point through which the curve passes.

For example, suppose that a curve passes through the point (1, 0), and values of its gradient $\dfrac{dy}{dx}$ are given in the following table.

x	1	2	3	4
$\dfrac{dy}{dx}$	1.00	0.50	0.33	0.25

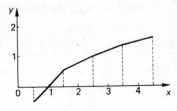

Fig. 8.18

We approximate to the curve by using straight line segments, with gradient 1 when $x = 1$, gradient 0.5 when $x = 2$ and so on. We start from (1, 0). As the gradient is given at integer points it would seem sensible to draw in the line segments for $0.5 \leqslant x \leqslant 1.5$, $1.5 \leqslant x \leqslant 2.5$, ... Hence we draw in a line of gradient 1 passing through (1, 0) for $0.5 \leqslant x \leqslant 1.5$. For $1.5 \leqslant x \leqslant 2.5$, draw in a line segment of gradient 0.5 joining onto the previous line at $x = 1.5$. Continuing in this way we obtain Fig. 8.18. This is the best we can do without further information. If however we knew the formula for $\dfrac{dy}{dx}$, it would be possible to consider narrower and narrower

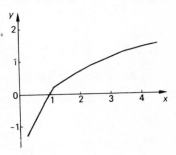

Fig. 8.19

strips. In this example, as you may have realised, $\dfrac{dy}{dx} = \dfrac{1}{x}$, so we can extend our table of values easily. Here we take strips of width 0.5:

x	0.5	1.0	1.5	2.0	2.5	3.0	3.5	4.0
$\dfrac{dy}{dx} = \dfrac{1}{x}$	2.00	1.00	0.67	0.50	0.40	0.33	0.29	0.25

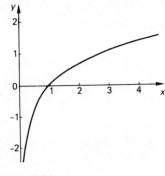

Fig. 8.20

We can also take a step 'backwards' drawing the line segment of gradient 2 for $0.25 \leqslant x \leqslant 0.75$ (Fig. 8.19).

If we 'smooth out' the corners we shall have an approximation to the curve which passes through (1, 0), and for which $\dfrac{dy}{dx} = \dfrac{1}{x}$ (see Fig. 8.20).

Exercise 8.5 $\left(\text{Constructing curves from values of } \dfrac{dy}{dx} \right)$

Draw straight line segments approximating to curves with the given properties.

1   Passing through the point (0, 1) and with gradient $\dfrac{dy}{dx}$ given as follows:

x	0	1	2	3	4	5
$\dfrac{dy}{dx}$	0	0.25	0.5	0.75	1.0	1.25

2 Passing through the point (1, 3) and with gradient $\dfrac{dy}{dx}$ given as follows:

x	0	0.5	1	1.5	2	2.5	3
$\dfrac{dy}{dx}$	0	4.5	3.7	1.7	0	-1.0	-1.5

3 Passing through the point (0, 0) and with gradient function $\dfrac{dy}{dx} = \sqrt{1-x^2}$. (First make a table of values of $\dfrac{dy}{dx}$ for $x = 0, 0.2, 0.4, \ldots, 1.0$.)

Revision questions B

Revision paper B1

1 (i) Integrate the following functions with respect to x:

(a) $x^2 - \dfrac{1}{x^2}$ (b) $x^{-\frac{1}{2}} - x^{-\frac{3}{2}}$

(ii) Evaluate $\int_2^8 \sqrt{2x}\, dx$

(iii) Evaluate $\int_2^5 (x^2 - 6x + 8)\, dx$ and interpret your result geometrically by means of a sketch.

2 (i) Find the area between the curve $y = x^2$, the y-axis and the line $y = 1$.

(ii) Find the coordinates of the two points where the curve $y = (5 - 3x)(x + 2)$ cuts the x-axis. Calculate the area enclosed between the curve and the x-axis.

3 (i) A curve is such that at any point (x, y) on the curve,

$$\frac{dy}{dx} = 24 - 12x^2$$

In addition, the curve passes through the point $(1, 12)$. Find its equation. Find, also, the coordinates of the points on the curve where the gradient is -24.

(ii) Evaluate $\int_{-1}^{2} x(1 - x)\, dx$.   (O)

4 Calculate the volume of the solid obtained by rotating the area bounded by the curve $y^2 = 4x$ and the ordinates $x = 0$ and $x = 16$ through two right angles about the x-axis.

A cylindrical hole of radius 1 unit which has the line $y = 0$ as its axis is now drilled through this solid. Calculate the volume of material remaining. [Leave π in your answer.]

5 The cost of running a cargo ship is £150 a day for wages and other crew expenses and £$\dfrac{8v^2}{3}$ a day

for fuel and maintenance, where v is the steady speed in km h^{-1}. Find in terms of v how many days a voyage of 2400 km takes and hence obtain a formula for the total cost of the whole journey.

Deduce the most economical speed and the least total cost of the whole voyage.

6 (i) By considering rectangles above and below the curve, find the upper and lower bounds of

$$\int_0^1 \frac{1}{x^2 + 1}\, dx$$

using strips 0.2 units wide.

(ii) If $f(x) = \dfrac{x - 1}{x^2 + 1}$, find an approximate value for $f'(1)$ by taking an x-step of 0.1 to each side.

Revision paper B2

1 (i) Evaluate the integrals

(a) $\displaystyle\int_1^2 \frac{x^2 - 2x - 3}{x^4}\, dx$ (b) $\displaystyle\int_1^2 (x^3 - 1)(x + 2)\, dx$

(ii)   The curve $x^2 + y^2 = 25$ is a circle, centre $(0, 0)$ and radius 5. Use this fact to evaluate the integral $\int_0^5 \sqrt{25 - x^2} \, dx$

2   Sketch the graphs of $y = x^2$ and $y = (x + 1)^2 - 4$ using the same axes for both graphs and find the coordinates of their point of intersection.

Calculate the areas

(a)   between $y = x^2$ and $y = 9$;   (b) bounded by $y = x^2$, $y = (x + 1)^2 - 4$ and the x-axis.   (O)

3   Find the area between the graph of $y = 2 + 3\sqrt{x}$ and the x-axis in the interval $1 \leqslant x \leqslant 4$. The positive value of the square root is to be taken and the unit is one centimetre.

Find also the volume created by rotating this area through $360°$ about the x-axis. Give the answer as a multiple of π.   (O and C)

4   (a)   ABCD is a rectangle in which $AB = 3a$ units and $AD = 4a$ units. E is on AB and F on BC so that $AE = BF = x$ units. Show that the area of the triangle DEF is

$$(6a^2 - 2ax + \tfrac{1}{2}x^2) \text{ square units}$$

Deduce that if a is fixed but x varies the least area of the triangle is one third of the area of the rectangle.

(b)   The equation of a curve is $y = a + bx - x^2$ where a and b are constants. The curve passes through the point $(3, 0)$ and its gradient there is -10. Find a and b and the area bounded by the axes and the arc of the curve between $x = 0$ and $x = 3$.   (O)

5   The depth y metres of a stream 10 metres wide at a point x metres from one particular side is given by the formula

$$y = \sqrt{x - \frac{1}{100}x^3}$$

Work out and record the depth at 2, 4, 6 and 8 metres from that side.

If the stream flows at 50 metres a minute write down a formula (including an integral) for the number of cubic metres of water passing per minute.

Use an approximation method of integration to estimate to two significant figures how much this is.

6   Find an approximation to the length of the curve given by $y = \sqrt{x}$ between $x = 0$ and $x = 2$. [Make out a table of values for $x = 0, 0.25, 0.50 \ldots, 2$.]

Miscellaneous revision questions: paper B3

1   (i)   Find a formula for y in terms of x if $\dfrac{dy}{dx} = \dfrac{x^2 - 1}{\sqrt{x}}$ and $y = 0$ when $x = 1$.

Calculate y when $x = 4$.

(ii)   Find the equation of the tangent at $(2, 1)$ to the curve $y = x^2 - x - 1$.

Find also the coordinates of the point on this curve at which the normal is parallel to the tangent at $(2, 1)$.

2   Find the gradients of the curve $y = (1 - x)(3 + x)$ at the points where it cuts the axes of coordinates.

Find also the area which lies within the curve and above the x-axis.

3   (i)   Find the equation of the tangent at the point $(4, 3)$ on the curve $xy = 12$.

(ii)   Find the equation of the curve which passes through the origin and satisfies the two conditions

(a) $\dfrac{d^2y}{dx^2} = 2$ at all points of the curve;   (b) $\dfrac{dy}{dx} = 1$ at the origin.

4   The arc of the parabola $x^2 = ky$ between $(0, 0)$ and another point on the curve is revolved about the y-axis through four right angles. A bowl which has the shape of this surface has a rim diameter 12 cm and a depth of 6 cm. Find the value of k, and find also the greatest volume of liquid the bowl will hold.

5   (a)   Find a positive value of x for which the gradient of the curve $y = 48x - x^3$ is zero. Calculate the corresponding value of y and show that it is a maximum. Denoting this point on the curve by P and the origin by O, write down the gradient of the line OP. Find the x coordinate of a point on the curve between O and P at which the tangent is parallel to OP.

(b)   Evaluate $\displaystyle\int_{1}^{2} 6(x+2)(x-1)\,dx.$ (O)

6   If $y = 2x^3 - 3x^2 - 12x + 20$ find the maximum and minimum values of y and the values of x for which these occur. Plot a graph of y against x for $-3 \leqslant x \leqslant 4$.

Indicate with arrows on your graph the 'maximum' and 'minimum' values found above. State also the largest and smallest values of y that occur for $-3 \leqslant x \leqslant 4$.

7   A metal plate is made in the shape formed by the part of the curve $y^2 = -x+1$ for which $x \geqslant 0$ and $y \geqslant 0$, and the part of the curve $y = 1 - \frac{1}{9}x^2$ between $x = 0$ and $x = 3$.

Sketch the shape of this metal plate and calculate its area, given that the units used are metres.

8   A doorknob is made in the shape formed by rotating the part of the curve $y^2 = 2(4-x)$ between $x = 2$ and $x = 4$ through two right angles about the x-axis and by rotating the part of the curve $y = \frac{1}{2} + \frac{3}{8}x^2$ between $x = 0$ and $x = 2$ through four right angles about the x-axis. Draw sketches of both curves in the appropriate regions and calculate the volume of the doorknob, giving your answer as a multiple of π.   (O)

9   An open tank, volume 500 m³, having a square base x metres in length and vertical height h metres was constructed from a sheet of metal. Show that $hx^2 = 500$ and that the surface area of the tank in square metres is

$$s = x^2 + \frac{2000}{x}$$

Hence find the values of x and h if the minimum area of sheet metal was used in the construction of the tank.

10   (i)   If the curve $ax^2 + bx + c$ passes through the points $(0, 4)$, $(1, 2)$ and $(3, 10)$ find the values of a, b and c. Find the equation of the line joining $(0, 4)$ and $(3, 10)$ and prove that the area between the line and the curve is $\int_{0}^{3}(6x - 2x^2)\,dx$. Evaluate this integral.

(ii)   Find the average value of $y = \dfrac{1}{72}(12 + 4x - x^2)$ over the range $0 \leqslant x \leqslant 6$.

(iii)   Show that $y = 2x - \dfrac{1}{x^2}$ satisfies the equation

$$x\frac{d^2y}{dx^2} + 3\frac{dy}{dx} = 6$$

11   Use the trapezium rule to find an approximate value of the integral $\int_{-1}^{1}\sqrt{1+x^3}\,dx$, using 5 ordinates.

12   By cutting the solid of revolution of $f(x) = (1+x^2)^{\frac{1}{3}}$ between $x = 1$ and $x = 2$ into 10 slices, find upper and lower bounds for this volume. Hence give an approximation for $\int_{1}^{2}(1+x^2)^{\frac{2}{3}}\,dx$.

9

Algebraic functions

9.1 Sketching graphs of products and quotients

It is often helpful to regard a function as the product or quotient of two or more simpler functions.

Example 1   Sketch the curve $y = (x-1)(2x-x^2)$.

To calculate a value of y corresponding to a given value of x (say $x = 4$), we first find the values of $(x-1)$ and $(2x-x^2)$ separately, starting in each case from the given value of x (when $x = 4$, $x-1 = 3$ and $2x-x^2 = -8$); then we multiply these values (when $x = 4$, $y = 3 \times (-8) = -24$). We say that y is the product of the two functions $(x-1)$ and $(2x-x^2)$.

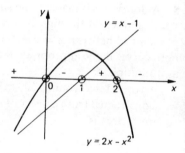

Fig. 9.1

We sketch the graphs of $(x-1)$ and $(2x-x^2)$ on the same diagram.

The product $y = (x-1)(2x-x^2)$ is zero when either $(x-1)$ or $(2x-x^2)$ is zero. Thus $y = 0$ when $x = 0$, 1 or 2. (In Fig. 9.1 these points are indicated by $\odot$.)

The sign of $y = (x-1)(2x-x^2)$ has been indicated on the x-axis. For example, when $x < 0$, $(x-1)$ and $(2x-x^2)$ are both negative (their graphs are below the x-axis); so y is positive.

If x is very large and positive (we say 'as x tends to plus infinity', and write 'as $x \to +\infty$'), then $(x-1)$ is large and positive $(x-1 \to +\infty)$ and $(2x-x^2)$ is large and negative (we say '$2x-x^2$ tends to minus infinity', and write $2x-x^2 \to -\infty$). So the product $y = (x-1)(2x-x^2)$ is large and negative $(y \to -\infty)$. We can check this by calculating a few values. For example, when $x = 4$, $y = -24$; when $x = 5$, $y = -60$, and so on.

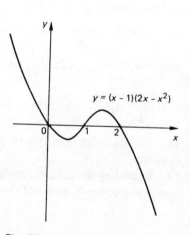

Similarly, as $x \to -\infty$, then $(x-1) \to -\infty$ and $(2x-x^2) \to -\infty$; so $y \to +\infty$. (Checking, when $x = -5$, $y = 210$; when $x = -10$, $y = 1320$.)

We know that the graph of $y = (x-1)(2x-x^2)$ is a smooth curve; we know where y is positive, negative and zero, and we know its behaviour as $x \to +\infty$ and as $x \to -\infty$. We can therefore sketch the curve (Fig. 9.2).

Fig. 9.2

Example 2 Sketch the curve $y = x(x-1)^2(3-x)$.

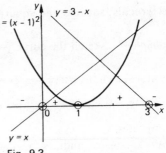

We first sketch the graphs of x, $(x-1)^2$ and $(3-x)$ on the same diagram (Fig. 9.3). $y = x(x-1)^2(3-x)$ is zero when $x = 0, 1,$ or 3. The sign of y is indicated on the x-axis (for example, when $x < 0$, two of the graphs are above the x-axis and one is below it; so the product y is negative).

As $\qquad x \to +\infty, y \to -\infty;$

and as $\quad x \to -\infty, y \to -\infty.$

Fig. 9.3

We can now sketch the curve (Fig. 9.4).

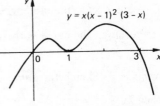

Fig. 9.4

Example 3 Sketch the curve $y = \dfrac{x-1}{2x-x^2}$.

We first sketch the graphs of $(x-1)$ and $(2x-x^2)$. In Fig. 9.5 the sign of the quotient $y = \dfrac{x-1}{2x-x^2}$ has been indicated on the x-axis.

y is zero when the numerator $(x-1)$ is zero, i.e. at $x = 1$ (the point marked $\odot$).

When the denominator $(2x-x^2)$ is zero, i.e. at $x = 0$ and $x = 2$, y is not defined (since we cannot divide by zero). If x is close to one of these values, y is numerically large. (For example, when $x = 1.9$, $y = 4.74$; when $x = 1.99$, $y = 49.75$; when $x = 2.01$, $y = -50.25$.) We have indicated this by drawing vertical dotted lines at $x = 0$ and $x = 2$.

As $x \to +\infty$ and as $x \to -\infty$, the highest powers of x in the numerator and denominator will dominate the other terms [see p. 210 for a justification of this method]; thus

$$y \approx \frac{x}{-x^2} = -\frac{1}{x}$$

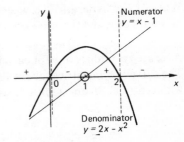

Fig. 9.5

As $x \to +\infty$, $y \to 0$; and as $x \to -\infty$, $y \to 0$. (Checking, when $x = 10$, $y = -0.11$; when $x = -10$, $y = 0.09$.)

We can now sketch the curve (Fig. 9.6). We know where y is positive, negative, and zero. As x approaches 0 or 2, the value of y becomes numerically large.

A line which a curve approaches arbitrarily closely is called an **asymptote** of the curve. Thus the lines $x = 0$ (the y-axis), $x = 2$, and $y = 0$ (the x-axis) are asymptotes of this curve.

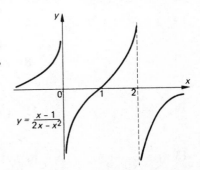

Fig. 9.6

110 Algebraic functions

Reciprocals, squares and square roots

Example 1 Sketch the curve $y = \dfrac{1}{2x - x^2}$.

We first sketch the curve $y = 2x - x^2$ (Fig. 9.7). When $2x - x^2 = 0$,
i.e. at $x = 0$ and $x = 2$, the curve $y = \dfrac{1}{2x - x^2}$ has vertical
asymptotes.

As $x \to +\infty$ and as $x \to -\infty$, $(2x - x^2) \to -\infty$, and so
$y = \dfrac{1}{2x - x^2} \to 0$. $\dfrac{1}{2x - x^2}$ is positive when $(2x - x^2)$ is positive, and
negative when $(2x - x^2)$ is negative. When $(2x - x^2)$ is increasing,
$\dfrac{1}{2x - x^2}$ is decreasing, and when $(2x - x^2)$ is decreasing, $\dfrac{1}{2x - x^2}$
is increasing. The curve is therefore as shown in Fig. 9.8.

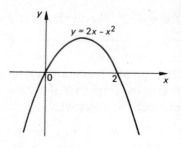

Fig. 9.7

Example 2 Sketch the curve $y = (2x - x^2)^2$.

When $2x - x^2 = 0$, $(2x - x^2)^2 = 0$. Otherwise, $(2x - x^2)^2$ is always
positive. The curve is shown in Fig. 9.9.

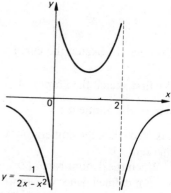

Example 3 Sketch the curve $y^2 = 2x - x^2$.

When $2x - x^2$ is negative, i.e. when $x < 0$ and when $x > 2$, there are
no values of y, and the curve does not exist in these regions. When
$2x - x^2 = 0$, $y = 0$. When $2x - x^2$ is positive, i.e. when $0 < x < 2$,
there are two values of y for each value of x, namely $\pm\sqrt{2x - x^2}$.
The curve is symmetrical about the x-axis (and is, in fact, a circle
with centre $(1, 0)$ and radius 1; see Fig. 9.10).

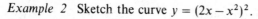
$y = \dfrac{1}{2x - x^2}$

Fig. 9.8

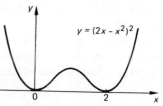

$y = (2x - x^2)^2$

Fig. 9.9

Exercise 9.1 (Sketching products and quotients)

Sketch the following curves.

1 $y = (x + 1)(2 - x)$ 2 $y = x(x^2 - 4)$ 3 $y = x(x - 2)(x - 3)$

4 $y = x^2(x + 2)$ 5 $y^2 = (x + 1)(2 - x)$ 6 $y = x^2 - 4$

7 $y = \dfrac{1}{x^2 - 4}$ 8 $y = (x^2 - 4)^2$ 9 $y^2 = x^2 - 4$

10 $y = \dfrac{1}{1 - x}$ 11 $y = 1 + x^2$ 12 $y = \dfrac{1}{1 + x^2}$

13 $y = \dfrac{x}{x^2 - 4}$ 14 $y^2 = \dfrac{x}{x^2 - 4}$ 15 $y = \dfrac{x^2 - 1}{x^2(x - 2)}$

9.2   Differentiation of products

We know that if $y = ku$, where k is a constant and u is a function of x, then

$$\frac{dy}{dx} = k\frac{du}{dx}$$

(for example, if $y = 3x^2$, then $\frac{dy}{dx} = 3(2x) = 6x$).

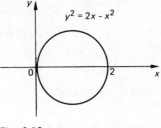

Fig. 9.10

Also, if $y = u + v$, where u and v are functions of x, then

$$\frac{dy}{dx} = \frac{du}{dx} + \frac{dv}{dx}$$

(for example, if $y = x^2 + x^4$, then $\frac{dy}{dx} = 2x + 4x^3$).

We now derive methods for differentiating

a product $y = uv$ (for example, $y = (x - 1)(2x - x^2)$)

and a quotient $y = \dfrac{u}{v}$ $\left(\text{for example, } y = \dfrac{x - 1}{2x - x^2}\right)$.

Products

Suppose $y = uv$, where u and v are functions of x.

Let δx be a change in x, causing changes δu and δv in u and v; these in turn cause a change δy in y.

The new values of y, u and v are $(y + \delta y)$, $(u + \delta u)$ and $(v + \delta v)$, so we have

$$y + \delta y = (u + \delta u)(v + \delta v)$$

Also, $y = uv$

Subtracting,

$$\delta y = (u + \delta u)(v + \delta v) - uv = u(\delta v) + (\delta u)v + (\delta u)(\delta v)$$

Dividing by δx,

$$\frac{\delta y}{\delta x} = u\frac{\delta v}{\delta x} + \frac{\delta u}{\delta x}v + (\delta u)\frac{\delta v}{\delta x}$$

As $\delta x \to 0$, assuming that $\dfrac{du}{dx}$ and $\dfrac{dv}{dx}$ exist, we have

$$\delta u \to 0, \quad \frac{\delta u}{\delta x} \to \frac{du}{dx} \quad \text{and} \quad \frac{\delta v}{\delta x} \to \frac{dv}{dx}$$

so $\dfrac{\delta y}{\delta x} \to u\dfrac{dv}{dx} + \dfrac{du}{dx}v + (0)\dfrac{dv}{dx} = u\dfrac{dv}{dx} + \dfrac{du}{dx}v$

Thus $\dfrac{dy}{dx} = \lim\limits_{\delta x \to 0} \dfrac{\delta y}{\delta x} = u\dfrac{dv}{dx} + \dfrac{du}{dx}v$

We therefore have

$$\text{if } y = uv \text{ then } \frac{dy}{dx} = u\frac{dv}{dx} + \frac{du}{dx}v$$

This is the **product rule**. It is valid for any value of x at which $\frac{du}{dx}$ and $\frac{dv}{dx}$ both exist.

Using the function notation, if

$$f(x) = u(x)v(x)$$

then $f'(x) = u(x)v'(x) + u'(x)v(x)$

Example Differentiate $(x-1)(2x-x^2)$.

Let $y = (x-1)(2x-x^2)$. We regard this as the product of the two functions $u = x - 1$ and $v = 2x - x^2$. Then

$$\frac{dy}{dx} = u\frac{dv}{dx} + \frac{du}{dx}v$$

$$= (x-1)(2-2x) + (1)(2x-x^2)$$

$$= 2x - 2x^2 - 2 + 2x + 2x - x^2 = -3x^2 + 6x - 2$$

Exercise 9.2 (Differentiation of products)

Use the product rule to differentiate the following (questions 1 to 8).

1   $(x-1)(x+3)$ 2 $x^2(4-x)$ 3 $(x-1)(x^2+1)$ 4 $2x^2(x^3+7)$
5   $(2x+1)^2$ [*consider as* $(2x+1)(2x+1).$]   6 $x^{\frac{1}{2}}(x^2+x)$ 7 $x^2(x^{\frac{1}{3}}-1)$ 8 $(2x-1)(3+x^{-2})$
9   Differentiate $(x+1)(3-2x)$: (i) by using the product rule; (ii) by first multiplying out the brackets.

10   Let $u = x^5, v = x^3$, and $y = uv = x^8$. Verify that the product rule gives the correct answer for $\frac{dy}{dx}$.

Also, verify the product rule when $u = x^m$ and $v = x^n$.

11   Suppose $y = uvw$, where u, v and w are functions of x. By first writing $y = u(vw)$, show that

$$\frac{dy}{dx} = \frac{du}{dx}vw + u\frac{dv}{dx}w + uv\frac{dw}{dx}$$

Use this result to differentiate

(i) $x(x-1)(x-2)$ (ii) $x^2(3x-1)(2-x^4)$

12   Criticise the following argument: 'Let $u = x^3$, $v = x^{-2}$ and $y = uv = x$. Then

$$\frac{dy}{dx} = u\frac{dv}{dx} + \frac{du}{dx}v = (x^3)\frac{dv}{dx} + (3x^2)v$$

When $x = 0$,

$$\frac{dy}{dx} = (0)\frac{dv}{dx} + (0)v = 0'$$

(but $y = x$, so $\frac{dy}{dx} = 1$ for all x).

9.3 Differentiation of quotients

Suppose that $y = \dfrac{u}{v}$, where u and v are functions of x.

Let δx be a change in x, causing changes δu, δv and δy in u, v and y. We have

$$y = \frac{u}{v}$$

and $\quad y + \delta y = \dfrac{u + \delta u}{v + \delta v}$

Subtracting,

$$\delta y = \frac{u + \delta u}{v + \delta v} - \frac{u}{v}$$

$$= \frac{v(u + \delta u) - u(v + \delta v)}{v(v + \delta v)} = \frac{v\delta u - u\delta v}{v(v + \delta v)}$$

Dividing by δx,

$$\frac{\delta y}{\delta x} = \frac{v\dfrac{\delta u}{\delta x} - u\dfrac{\delta v}{\delta x}}{v(v + \delta v)}$$

As $\delta x \to 0$, assuming that $\dfrac{du}{dx}$ and $\dfrac{dv}{dx}$ exist, we have

$$\delta v \to 0, \quad \frac{\delta u}{\delta x} \to \frac{du}{dx} \quad \text{and} \quad \frac{\delta v}{\delta x} \to \frac{dv}{dx}$$

so $\qquad \dfrac{dy}{dx} = \lim\limits_{\delta x \to 0} \dfrac{\delta y}{\delta x} = \dfrac{v\dfrac{du}{dx} - u\dfrac{du}{dx}}{v^2}$

Hence,

$$\boxed{\text{if } y = \frac{u}{v} \text{ then } \frac{dy}{dx} = \frac{v\dfrac{du}{dx} - u\dfrac{dv}{dx}}{v^2}}$$

This is the **quotient rule**. It is valid provided that $\dfrac{du}{dx}$ and $\dfrac{dv}{dx}$ both exist. We also need $v \neq 0$ (otherwise y is not defined).

Using the function notation, if

$$f(x) = \frac{u(x)}{v(x)}$$

then $\quad f'(x) = \dfrac{v(x)u'(x) - u(x)v'(x)}{[v(x)]^2}$

Example   Differentiate $\dfrac{x-1}{2x-x^2}$.

Let $y = \dfrac{x-1}{2x-x^2}$. We regard this as the quotient of the two

functions $u = x-1$ and $v = 2x-x^2$. Then

$$\frac{dy}{dx} = \frac{v\dfrac{du}{dx} - u\dfrac{dv}{dx}}{v^2}$$

$$= \frac{(2x-x^2)(1)-(x-1)(2-2x)}{(2x-x^2)^2} = \frac{2-2x+x^2}{(2x-x^2)^2}$$

This is not valid if $2x-x^2 = 0$, i.e. if $x = 0$ or $x = 2$.

Exercise 9.3   (Differentiation of quotients)

Use the quotient rule to differentiate the following (questions 1 to 8). State any values of x for which your answers are not valid.

1   $\dfrac{x+1}{5-x}$ 2 $\dfrac{x+3}{x^2-4}$ 3 $\dfrac{1}{2x+3}$ 4 $\dfrac{4}{1+x^2}$

5   $\dfrac{3x^2-1}{3x^2+2}$ 6 $\dfrac{x^2-3x+7}{x^2+x-2}$ 7 $\dfrac{2x-3}{x^3+1}$ 8 $\dfrac{x^2-1}{x^2(x-2)}$

9   Differentiate $\dfrac{5x^4-2}{x^3}$ (i) by using the quotient rule; (ii) by first dividing out.

10   Let $u = x^6, v = x^2$ and $y = \dfrac{u}{v} = x^4$. Verify that the quotient rule gives the correct answer for $\dfrac{dy}{dx}$.

Also, verify the quotient rule when $u = x^m$ and $v = x^n$.

9.4   Composite functions

Consider, for example, the function $y = (2x-x^2)^2$. To calculate a value of y (say when $x = 4$) we first find the value of $2x-x^2$ (when $x = 4$, $2x-x^2 = -8$); then we square this value (when $x = 4$, $y = (-8)^2 = 64$). The value of one function is used as the starting point for a second function: we say that y is the **composition** of the two functions. If we introduce a new variable u to represent the value of the first function, i.e. $u = 2x-x^2$, then $y = (2x-x^2)^2 = u^2$.

We have $y = u^2$ where $u = 2x-x^2$; y is a function of u, and u is a function of x. We say that y is a **function of a function** of x.

To sketch the graph of a composite function, we sketch the graph of the first function, i.e. $u = 2x-x^2$. Then we consider the effect of the second function on the values of u, i.e. we square the values of u. This is in fact what we did when we sketched the curve $y = (2x-x^2)^2$ on p. 110.

Example Express $y = \sqrt{4x^2 + 1}$ as a function of a function.

If we were calculating y for a given value of x, we would first find $4x^2 + 1$; so let $u = 4x^2 + 1$.

Then $y = \sqrt{u}$ where $u = 4x^2 + 1$.

Differentiation of composite functions

Suppose y is a function of u, and u is a function of x. Let δx be a change in x, causing a change δu in u. Since y is a function of u, this change δu in u will cause a change δy in y.

For example, suppose that $y = u^2$ where $u = 2x - x^2$, and consider a change in x from 4 to 4.1 (so that $\delta x = 0.1$):

$$x \qquad\qquad u = 2x - x^2 \qquad\qquad y = u^2$$

$$\left. 4 \atop 4.1 \right\} \begin{array}{c} \longrightarrow -8 \\ \delta x = 0.1 \\ \longrightarrow -8.61 \end{array} \left. \right\} \begin{array}{c} \longrightarrow 64 \\ \delta u = -0.61 \\ \longrightarrow 74.1321 \end{array} \left. \right\} \delta y = 10.1321$$

Here u changes from -8 to -8.61. Thus $\delta u = -0.61$, and $\dfrac{\delta u}{\delta x} = -6.1$. This change in u causes y to change from 64 to 74.1321. Thus $\delta y = 10.1321$, and $\dfrac{\delta y}{\delta u} = -16.61$.

Considering y as the composite function $y = (2x - x^2)^2$, the change $\delta x = 0.1$ in x causes a change $\delta y = 10.1321$ in y. Thus $\dfrac{\delta y}{\delta x} = 101.321$. Notice that $101.321 = (-16.61) \times (-6.1)$. This is because

$$\frac{\delta y}{\delta u} \times \frac{\delta u}{\delta x} = \frac{10.1321}{-0.61} \times \frac{-0.61}{0.1}$$

$$= \frac{10.1321}{0.1} = \frac{\delta y}{\delta x}$$

In general, we always have

$$\frac{\delta y}{\delta x} = \frac{\delta y}{\delta u} \times \frac{\delta u}{\delta x}$$

As $\delta x \to 0$ and assuming that $\dfrac{du}{dx}$ exists, we have $\dfrac{\delta u}{\delta x} \to \dfrac{du}{dx}$ and $\delta u \to 0$; assuming that $\dfrac{dy}{du}$ exists, we then have $\dfrac{\delta y}{\delta u} \to \dfrac{dy}{du}$, and so $\dfrac{\delta y}{\delta x} \to \dfrac{dy}{du} \times \dfrac{du}{dx}$. Hence

$$\boxed{\frac{dy}{dx} = \frac{dy}{du} \times \frac{du}{dx}}$$

This is called the **chain rule**, as it applies to a chain of variables y, u, x, in which each variable is a function of the one following it. It is valid for any value of x at which $\dfrac{du}{dx}$ exists and for which $\dfrac{dy}{du}$ exists at the corresponding value of u.

The chain rule can easily be extended to a chain of four variables y, u, v, x; then

$$\frac{dy}{dx} = \frac{dy}{du} \times \frac{du}{dv} \times \frac{dv}{dx}$$

Similarly it may be extended to a chain of any length.

Example 1   Differentiate $(2x - x^2)^2$.

Let $y = (2x - x^2)^2 = u^2$, where $u = 2x - x^2$. Then

$$\frac{dy}{dx} = \frac{dy}{du} \times \frac{du}{dx} = 2u \times (2 - 2x) = 2(2x - x^2)(2 - 2x)$$

Example 2   Differentiate $\sqrt{4x^2 + 1}$.

Let $y = \sqrt{4x^2 + 1} = \sqrt{u} = u^{\frac{1}{2}}$, where $u = 4x^2 + 1$. Then

$$\frac{dy}{dx} = \frac{dy}{du} \times \frac{du}{dx} = \tfrac{1}{2}u^{-\frac{1}{2}}(8x) = \frac{4x}{\sqrt{u}} = \frac{4x}{\sqrt{4x^2 + 1}}$$

Exercise 9.4   (Composite functions)

Express each of the following as a function of a function (questions 1 to 5).

1 $y = (x - 3)^5$ **2** $y = (2x + 1)^8$ **3** $y = (x^2 + 4x - 5)^3$ **4** $y = \sqrt{x^3 + 5}$ **5** $y = (1 + 2\sqrt{x})^4$

Differentiate the following functions (questions 6 to 15).

6 $(x - 3)^5$ **7** $(2x + 1)^8$ **8** $(x^2 + 4x - 5)^3$ **9** $\sqrt{x^3 + 5}$

10 $(1 + 2\sqrt{x})^4$ **11** $(1 - 4x)^3$ **12** $(2x^2 + x)^9$ **13** $\sqrt{3 + 2x - x^2}$

14 $(5x + 1)^{-2}$ **15** $\left(7 - \dfrac{1}{x}\right)^{\frac{4}{3}}$

16   If $y = u^3$, where $u = x^2 + x + 1$, express y and $\dfrac{dy}{dx}$ in terms of x.

If, instead, $y = u^2 + u + 1$ where $u = x^3$, express y and $\dfrac{dy}{dx}$ in terms of x.

17   Differentiate $(2x + 3)^2$ (i) by expressing it as a function of a function; (ii) by first multiplying out.

18   If $y = u^4$ where $u = x^2$, verify that the chain rule $\dfrac{dy}{dx} = \dfrac{dy}{du} \times \dfrac{du}{dx}$ gives the correct answer for $\dfrac{dy}{dx}$.

Also, verify the chain rule for $y = u^m$ where $u = x^n$.

19   (i)   If $y = u^3$, $u = 4 + \sqrt{v}$ and $v = 1 - 2x^2$, express y and $\dfrac{dy}{dx}$ in terms of x.

(ii)   Differentiate $\sqrt{(4x - 1)^3 + 1}$.

20   Let $y = u^{\frac{1}{3}}$ where $u = x^3$; so $y = (x^3)^{\frac{1}{3}} = x$.

Criticise the following calculation of $\dfrac{dy}{dx}$ when $x = 0$:

$$\text{`}\frac{dy}{dx} = \frac{dy}{du} \times \frac{du}{dx} = \frac{dy}{du} \times (3x^2)$$

When $x = 0$, $\dfrac{dy}{dx} = \dfrac{dy}{du} \times 0 = 0$.'

9.5   Differentiation of algebraic functions

It is sometimes necessary to use a combination of the product rule, the quotient rule and the chain rule.

Example 1   Differentiate $(x - 5)(4x^2 + 1)^5$.

Let $y = (x - 5)(4x^2 + 1)^5$. This is the product of the two functions $u = x - 5$ and $v = (4x^2 + 1)^5$. We can differentiate u easily but v is a function of a function.

Let $v = t^5$ where $t = 4x^2 + 1$; so

$$\frac{dv}{dx} = 5t^4(8x) = 40x(4x^2 + 1)^4$$

Using the product rule,

$$\frac{dy}{dx} = (x - 5)[40x(4x^2 + 1)^4] + (1)[(4x^2 + 1)^5]$$

$$= (4x^2 + 1)^4[40x(x - 5) + (4x^2 + 1)]$$

$$= (4x^2 + 1)^4[44x^2 - 200x + 1]$$

Example 2   Differentiate $\dfrac{x^2}{(x - 2)(x + 3)}$.

Let $y = \dfrac{x^2}{(x - 2)(x + 3)} = \dfrac{x^2}{x^2 + x - 6}$

This can be differentiated using the quotient rule:

$$\frac{dy}{dx} = \frac{(x^2 + x - 6)(2x) - (x^2)(2x + 1)}{(x^2 + x - 6)^2}$$

$$= \frac{x^2 - 12x}{(x^2 + x - 6)^2}$$

$$= \frac{x(x - 12)}{(x - 2)^2(x + 3)^2}$$

This is valid provided that $(x - 2)(x + 3)$ is not zero; so we need $x \neq 2$ and $x \neq -3$.

Example 3 Differentiate $\sqrt{\dfrac{2+x}{3-x^2}}$.

Let $y = \sqrt{\dfrac{2+x}{3-x^2}} = (2+x)^{\frac{1}{2}}(3-x^2)^{-\frac{1}{2}}$

This can be differentiated using the product rule, and the chain rule for $(2+x)^{\frac{1}{2}}$ and $(3-x^2)^{-\frac{1}{2}}$; i.e. if

$$u = (2+x)^{\frac{1}{2}}$$

$$\frac{du}{dx} = \frac{1}{2}(2+x)^{-\frac{1}{2}}(1)$$

and if $v = (3-x^2)^{-\frac{1}{2}}$

$$\frac{dv}{dx} = -\frac{1}{2}(3-x^2)^{-\frac{3}{2}}(-2x) = x(3-x^2)^{-\frac{3}{2}}$$

Hence

$$\frac{dy}{dx} = (2+x)^{\frac{1}{2}}[x(3-x^2)^{-\frac{3}{2}}] + \frac{1}{2}(2+x)^{-\frac{1}{2}}(3-x^2)^{-\frac{1}{2}}$$

$$= \frac{1}{2}(2+x)^{-\frac{1}{2}}(3-x^2)^{-\frac{3}{2}}[2(2+x)(x) + (3-x^2)]$$

(taking out the lowest power of each factor)

$$= \frac{4x+2x^2+3-x^2}{2(2+x)^{\frac{1}{2}}(3-x^2)^{\frac{3}{2}}}$$

$$= \frac{x^2+4x+3}{2\sqrt{(2+x)(3-x^2)^3}}$$

This is valid provided that $(2+x)$ and $(3-x^2)$ are not zero (this requires $x \neq -2$ and $x \neq \pm\sqrt{3}$), and that $\left(\dfrac{2+x}{3-x^2}\right)$ is positive (otherwise y is not a real number). So we must have $x < -2$ or $-\sqrt{3} < x < \sqrt{3}$.

Exercise 9.5 (Differentiation of algebraic functions)

Differentiate the following functions, stating any restrictions on the value of x.

1 $x(4x+9)^3$ 2 $x^4(5-2x)^5$ 3 $(3x-4)^5(2-x)^3$

4 $(x^2+1)^4(1-x)^7$ 5 $x^2\sqrt{2x-1}$ 6 $x\sqrt{1-x^2}$

7 $(4x-1)^{\frac{5}{3}}(x^2+2)^{\frac{2}{3}}$ 8 $\dfrac{2x^2}{(x-1)^3}$ 9 $\dfrac{(2x+5)^4}{(3-x)^7}$

10 $\dfrac{x^2}{\sqrt{2+x}}$ 11 $\dfrac{\sqrt{3x^2-2}}{x+1}$ 12 $\dfrac{x^2}{(x-1)(x+2)}$

13 $\dfrac{(x-1)(x-4)}{(x-2)(x-3)}$ 14 $\sqrt{\dfrac{3x+1}{3x-1}}$ 15 $\left(\dfrac{x}{x^2-1}\right)^{\frac{2}{3}}$ 16 $\dfrac{4x(3x+1)^5}{\sqrt{9-x^2}}$

9.6 Integration of algebraic functions

We first recall (see p. 68) that if we can express a function as a sum (or difference) of terms, where each term is a multiple of a power of x, then we may integrate term by term using $\int x^n dx = \dfrac{x^{n+1}}{n+1} + C$, provided that none of the terms is a multiple of $\dfrac{1}{x}$ (i.e. provided $n \neq -1$).

Example 1

$$\int \left(3x^3 - x^{\frac{4}{3}} - 2 + \frac{1}{x^2}\right) dx = \int (3x^3 - x^{\frac{4}{3}} - 2 + x^{-2}) dx$$

$$= 3(\tfrac{1}{4}x^4) - \tfrac{3}{7}x^{\frac{7}{3}} - 2x + (-1)x^{-1} + C$$

$$= \tfrac{3}{4}x^4 - \tfrac{3}{7}x^{\frac{7}{3}} - 2x - \frac{1}{x} + C$$

Example 2

$$\int (x-1)(2x+3) dx = \int (2x^2 + x - 3) dx$$

$$= \tfrac{2}{3}x^3 + \tfrac{1}{2}x^2 - 3x + C$$

Example 3

$$\int \frac{x^2 - 1}{x^4} dx = \int \left(\frac{x^2}{x^4} - \frac{1}{x^4}\right) dx = \int (x^{-2} - x^{-4}) dx$$

$$= -x^{-1} - \left(-\frac{1}{3}\right) x^{-3} + C = -\frac{1}{x} + \frac{1}{3x^3} + C$$

$$= \frac{1 - 3x^2}{3x^3} + C$$

The linear function rule

We shall now see how the chain rule of differentiation can help us to integrate certain functions.

Example 1 Integrate $(3x - 2)^6$.

We could do this by multiplying it out, but that would be very tedious. Our experience of differentiation suggests that the integral is something like $(3x - 2)^7$.

If $y = (3x - 2)^7$, then, using the chain rule,

$$\frac{dy}{dx} = 7(3x - 2)^6 (3) = 21(3x - 2)^6$$

Thus, if $y = \dfrac{1}{21}(3x-2)^7$, then $\dfrac{dy}{dx} = (3x-2)^6$. Hence

$$\int (3x-2)^6 \, dx = \frac{1}{21}(3x-2)^7 + C$$

It would be better if we could write down the answer without working backwards. We may regard $(3x-2)^6$ as a function of a function, $(3x-2)^6 = u^6$ where $u = 3x-2$. Integrating u^6 with respect to u gives $\frac{1}{7}u^7 + C = \frac{1}{7}(3x-2)^7 + C$, we need to divide this by $3\left(\text{which arises as } \dfrac{du}{dx}\right)$.

The general rule is:

> to integrate a function of u, where $u = ax+b$,
> integrate with respect to u,
> and then divide by a (the coefficient of x).

This is called the **linear function rule**, and it only applies when u is a *linear* function of x (i.e. $u = ax+b$, which gives a straight line graph).

We shall now show why this works. Suppose y is a function of u, where $u = ax+b$, and let

$$Q = \int y \, dx$$

be the required integral. Then

$$\frac{dQ}{dx} = y$$

By the chain rule,

$$\frac{dQ}{dx} = \frac{dQ}{du} \times \frac{du}{dx} = \frac{dQ}{du} \times a$$

and so $\dfrac{dQ}{du} \times a = y$

Thus $\dfrac{dQ}{du} = \dfrac{1}{a}y$

and $Q = \int \dfrac{1}{a}y \, du = \dfrac{1}{a}\int y \, du$

i.e. $\int y \, dx = \dfrac{1}{a}\int y \, du$

which is the linear function rule.

Example 1 above would be set out as

$$\int (3x-2)^6 \, dx = \frac{1}{3} \times \frac{1}{7}(3x-2)^7 + C$$

$$= \frac{1}{21}(3x-2)^7 + C$$

Example 2

$$\int \sqrt{8x+3}\, dx = \int (8x+3)^{\frac{1}{2}}\, dx \quad \text{(the coefficient of } x \text{ is 8)}$$

$$= \frac{1}{8} \times \frac{2}{3}(8x+3)^{\frac{3}{2}} + C = \frac{1}{12}\sqrt{(8x+3)^3} + C$$

Example 3

$$\int \frac{1}{\sqrt{x+5}}\, dx = \int (x+5)^{-\frac{1}{2}}\, dx \quad \text{(the coefficient of } x \text{ is 1)}$$

$$= 2(x+5)^{\frac{1}{2}} + C = 2\sqrt{x+5} + C$$

Example 4

$$\int \frac{1}{(5-2x)^2}\, dx = \int (5-2x)^{-2}\, dx \quad \text{(the coefficient of } x \text{ is } -2)$$

$$= \left(-\frac{1}{2}\right)(-1)(5-2x)^{-1} + C = \frac{1}{2(5-2x)} + C$$

Example 5

$$\int \left(1-\frac{1}{2}x\right)^4 dx = (-2)\frac{1}{5}\left(1-\frac{1}{2}x\right)^5 + C \left(\text{the coefficient of } x \text{ is } -\frac{1}{2}\right)$$

$$= -\frac{2}{5}\left(1-\frac{1}{2}x\right)^5 + C$$

In general, we have

$$\int (ax+b)^n\, dx = \frac{1}{a} \times \frac{(ax+b)^{n+1}}{n+1} + C$$

(provided $n \neq -1$).

It is important to realise that the rule can be used to integrate a function of a function only when the first function (u) is a linear function. There is no such rule for integrating a general function of a function.

For example, from

$$\frac{d}{dx}[(x^2+1)^3] = 6x(x^2+1)^2$$

it does *not* follow that

$$\int (x^2+1)^2\, dx = \frac{(x^2+1)^3}{6x} + C$$

because

$$\int (x^2 + 1)^2 \, dx = \int (x^4 + 2x^2 + 1) \, dx$$

$$= \frac{1}{5}x^5 + \frac{2}{3}x^3 + x + C$$

whereas

$$\frac{(x^2 + 1)^3}{6x} + C = \frac{x^6 + 3x^4 + 3x^2 + 1}{6x} + C$$

$$= \frac{1}{6}x^5 + \frac{1}{2}x^3 + \frac{1}{2}x + \frac{1}{6x} + C$$

There is little resemblance between these two expressions.

In general the situation is much worse. The functions $y = u^{\frac{1}{2}}$ and $u = x^3 + 1$ seem simple enough, but the composite function $y = (x^3 + 1)^{\frac{1}{2}}$ has an integral which cannot be expressed in terms of elementary functions.

Definite integrals

A definite integral can be evaluated by substituting the limits into the integral.

Example

$$\int_1^5 \frac{1}{\sqrt{3x + 1}} \, dx = \int_1^5 (3x + 1)^{-\frac{1}{2}} \, dx$$

$$= \left[\frac{1}{3} \times 2(3x + 1)^{\frac{1}{2}} \right]_1^5 = \left(\frac{2}{3} \times 4 \right) - \left(\frac{2}{3} \times 2 \right) = \frac{4}{3}$$

We should always check that the integral (the expression inside the square brackets) does indeed differentiate to give the original function for all values of x between the limits of integration. Otherwise the evaluation is not valid.

For example, consider the evaluation

$$\int_1^6 \frac{1}{(x - 2)^2} \, dx = \left[-\frac{1}{x - 2} \right]_1^6 = \left(-\frac{1}{4} \right) - (1)$$

$$= -\frac{5}{4}$$

When $x = 2$, the integral $-\dfrac{1}{x - 2}$ is not defined and so the result $\dfrac{d}{dx}\left(-\dfrac{1}{x - 2} \right) = \dfrac{1}{(x - 2)^2}$ fails to hold. Since $x = 2$ lies between $x = 1$ and $x = 6$, the evaluation is not valid, and the value $-\dfrac{5}{4}$ which we obtained is meaningless.

Exercise 9.6   (Integration of algebraic functions)

In questions 1 to 18, integrate the functions with respect to x.

1 $5x^2 - 2x^7 + 3$ **2** $\dfrac{2}{x^3} - \dfrac{1}{3x^2}$ **3** $x^{\frac{5}{2}} - 5x^{\frac{3}{2}} + 6x^{\frac{1}{2}}$

4 $x^2(2x - 3)$ **5** $\dfrac{x^3 - 4}{x^2}$ **6** $\dfrac{(x + 3)(x - 2)}{x^5}$

7 $(4x + 3)^5$ **8** $5(2x - 1)^4$ **9** $(x - 2)^7$

10 $(3 - 7x)^5$ **11** $(1 - 3x)^4$ **12** $\sqrt{2x + 5}$

13 $\dfrac{1}{\sqrt{5 - 4x}}$ **14** $(x + 3)^{-\frac{1}{4}}$ **15** $\dfrac{1}{(3x - 1)^3}$

16 $\left(3 + \dfrac{1}{3}x\right)^{-4}$ **17** $\left(1 - \dfrac{1}{2}x\right)^{-\frac{2}{3}}$ **18** $(3 - 4x)^{\frac{3}{5}}$

In questions 19 to 24, evaluate the definite integrals.

19 $\displaystyle\int_{\frac{1}{2}}^{1} \dfrac{1 - x}{x^3} \, dx$ **20** $\displaystyle\int_{-2}^{-1} (2x + 3)^4 \, dx$ **21** $\displaystyle\int_{1}^{5} \dfrac{1}{\sqrt{2x - 1}} \, dx$

22 $\displaystyle\int_{1}^{6} \sqrt{x + 3} \, dx$ **23** $\displaystyle\int_{0}^{1} \dfrac{1}{(4 + 5x)^2} \, dx$ **24** $\displaystyle\int_{3}^{42} (3x - 1)^{-\frac{2}{3}} dx$

25 Find $\int (2x + 1)^2 \, dx$ (i) by first multiplying out the brackets; (ii) by using the linear function rule. Are your two answers the same? Evaluate $\int_0^3 (2x + 1)^2 \, dx$ by both methods.

In questions 26 to 31, evaluate the definite integral if possible. If the definite integral cannot be evaluated, explain why.

26 $\displaystyle\int_{0}^{4} \dfrac{1}{(x - 3)^2} \, dx$ **27** $\displaystyle\int_{-3}^{0} \dfrac{1}{(2x - 1)^2} \, dx$ **28** $\displaystyle\int_{1}^{2} \dfrac{1}{(3 - 2x)^3} \, dx$

29 $\displaystyle\int_{1}^{6} \left(\dfrac{1}{x^2} + \dfrac{1}{(x - 5)^2}\right) dx$ **30** $\displaystyle\int_{3}^{12} \sqrt{x - 4} \, dx$ **31** $\displaystyle\int_{3}^{12} (x - 4)^{\frac{1}{3}} dx$

32 (i) Criticise

$$\int (x^3 + 2)^2 \, dx = \dfrac{\frac{1}{3}(x^3 + 2)^3}{3x^2} + C$$

(ii) Find $\int (x^3 + 2)^2 \, dx$.

9.7 Inverse functions

The inverse graph

The graph of a function $f(x)$ is the set of points (x, y) which satisfy $y = f(x)$. The **inverse graph** is the set of points (x, y) which satisfy $x = f(y)$, where f is the same function. Whenever (x, y) lies on the original graph $y = f(x)$, then (y, x) lies on the inverse graph. Since (y, x) is the reflection of (x, y) in the line $y = x$, the inverse graph is the reflection of the original graph in the line $y = x$ (see Fig. 9.11).

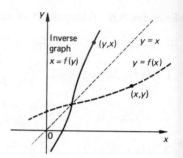

Fig. 9.11

Example 1   Sketch the inverse graphs for

(i) $f(x) = x^2$ (ii) $f(x) = x^3 + x$ (iii) $f(x) = \dfrac{1}{x} - 2$

See Figs 9.12, 9.13 and 9.14.

The inverse graph is not necessarily the graph of a function of x; for example, if $f(x) = x^2$, the inverse graph gives two values of y for each (positive) value of x.

When the inverse graph *does* represent a function, this is called the **inverse function** of $f(x)$. The equation of the inverse graph is $x = f(y)$, which then defines y as a function of x. We use the notation $f^{-1}(x)$ for the inverse function. Thus

$$\boxed{y = f^{-1}(x) \text{ is equivalent to } x = f(y)}$$

It may be possible to rearrange the equation $x = f(y)$ algebraically to obtain y in terms of x; we then obtain an expression for the inverse function $f^{-1}(x)$.

Example 2   Find the inverse function of $f(x) = \dfrac{1}{x} - 2$.

If $y = f^{-1}(x)$ is the inverse function, then $x = f(y)$, i.e.

$$x = \frac{1}{y} - 2$$

$$\frac{1}{y} = x + 2$$

$$y = \frac{1}{x + 2}$$

Hence the inverse function is $f^{-1}(x) = \dfrac{1}{x + 2}$.

The following example shows how we can find the gradient of the inverse graph without actually finding the inverse function.

Example 3   If $f(x) = x^3 + x$, show that the point $(10, 2)$ lies on the inverse graph, and find its gradient at this point.

First consider the curve $y = x^3 + x$. When $x = 2$, $y = 10$, so the point $(2, 10)$ lies on this curve. Hence, interchanging the coordinates, the point $(10, 2)$ lies on the inverse graph.

(i) $f(x) = x^2$

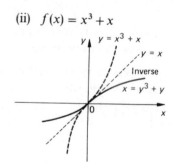

Fig. 9.12

(ii) $f(x) = x^3 + x$

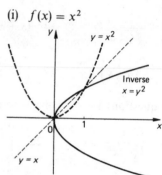

Fig. 9.13

(iii) $f(x) = \dfrac{1}{x} - 2$

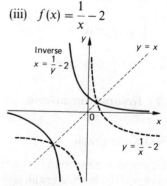

Fig. 9.14

If the inverse function is $y = f^{-1}(x)$, then

$$x = f(y)$$

i.e. $x = y^3 + y$

Differentiating with respect to y,

$$\frac{dx}{dy} = 3y^2 + 1$$

Hence $\dfrac{dy}{dx} = \dfrac{1}{\dfrac{dx}{dy}}^* = \dfrac{1}{3y^2 + 1}$

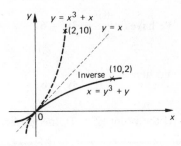

Fig. 9.15

This gives the gradient of the inverse graph in terms of y (see Fig. 9.15). At the point $(10, 2)$, $y = 2$, so the gradient is $\dfrac{dy}{dx} = \dfrac{1}{12 + 1} = \dfrac{1}{13}$.

When the inverse graph does not represent a function, we can still obtain an inverse function, by restricting our attention to a domain over which $f(x)$ is an increasing (or decreasing) function, as shown in the following example.

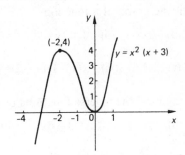

Fig. 9.16

Example 4   If $f(x) = x^2(x + 3)$, find a domain over which $f(x)$ is decreasing. Sketch the corresponding inverse function $f^{-1}(x)$, and find the gradient of the inverse graph at the point $(2, -1)$.

First we sketch the curve $y = f(x) = x^2(x + 3)$; see Fig. 9.16. Then $\dfrac{dy}{dx} = 3x^2 + 6x$, which is zero when $x = -2$ or 0. $(-2, 4)$ is a maximum point, and $(0, 0)$ is a minimum point.

$f(x)$ is decreasing when $-2 \leqslant x \leqslant 0$, and then $f(x)$ takes values between 0 and 4.

We now restrict our attention to that section of the curve $y = x^2(x + 3)$ lying between $x = -2$ and $x = 0$. We sketch the corresponding inverse function $y = f^{-1}(x)$ by reflecting this section of the curve in the line $y = x$ (see Fig. 9.17). $f^{-1}(x)$ is defined for $0 \leqslant x \leqslant 4$, and takes values between -2 and 0.

If $y = f^{-1}(x)$, then

$$x = f(y) = y^2(y + 3) = y^3 + 3y^2$$

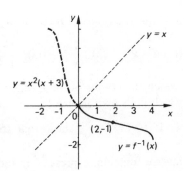

Fig. 9.17

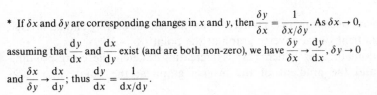

* If δx and δy are corresponding changes in x and y, then $\dfrac{\delta y}{\delta x} = \dfrac{1}{\delta x/\delta y}$. As $\delta x \to 0$, assuming that $\dfrac{dy}{dx}$ and $\dfrac{dx}{dy}$ exist (and are both non-zero), we have $\dfrac{\delta y}{\delta x} \to \dfrac{dy}{dx}$, $\delta y \to 0$ and $\dfrac{\delta x}{\delta y} \to \dfrac{dx}{dy}$; thus $\dfrac{dy}{dx} = \dfrac{1}{dx/dy}$.

We have $\quad \dfrac{dx}{dy} = 3y^2 + 6y$

and hence $\quad \dfrac{dy}{dx} = \dfrac{1}{3y^2 + 6y}$

At the point $(2, -1)$, the gradient is $\dfrac{dy}{dx} = \dfrac{1}{3-6} = -\dfrac{1}{3}$.

Exercise 9.7   (Inverse functions)

1   Sketch the inverse graphs for the following functions. In each case state whether the inverse graph is the graph of a function.

(i) $2x$ (ii) $3 - 2x$ (iii) $x + 2$ (iv) $x^2 + 1$

(v) x^3 (vi) $x(x-1)(x-2)$ (vii) $\dfrac{1}{x+1}$ (viii) $\dfrac{1}{1+x^2}$

2   Find the inverse functions of the following.

(i) $2x$ (ii) $3 - 4x$ (iii) $x + 3$ (iv) x^3

(v) $\dfrac{1}{x-4}$ (vi) $8 - \dfrac{3}{x}$ (vii) $\dfrac{2x+1}{3-4x}$ (viii) $\dfrac{1}{(x+4)^3}$

3   Find the inverse functions of the following.

(i) $4 - x$ (ii) $\dfrac{1}{x}$ (iii) $\dfrac{1-x}{1+x}$

Sketch the graphs of these functions. What condition must be satisfied by the graph of a function which is its own inverse?

4   If $f(x) = 1 + \dfrac{1}{x}$, show that the point $(\frac{3}{2}, 2)$ lies on the inverse graph. If $y = f^{-1}(x)$ is the inverse function, express $\dfrac{dy}{dx}$ in terms of y, and hence find the gradient of the inverse graph at the point $(\frac{3}{2}, 2)$.

Also, find the inverse function $f^{-1}(x)$, and check the value for the gradient by differentiating and putting $x = \frac{3}{2}$.

5   If $f(x) = 2x^3 + 3x$, find the gradient of the inverse graph at the points (i) $(5, 1)$ (ii) $(-22, -2)$.

6   If $f(x) = x^5 + x$, find the gradient of the inverse graph at the points (i) $(-2, -1)$ (ii) $(34, 2)$.

7   If $f(x) = x(x-2)$, find a domain over which $f(x)$ is decreasing. Sketch the corresponding inverse function $f^{-1}(x)$, and find the gradient of the inverse graph at the point $(8, -2)$.

8   If $f(x) = x^2(6-x)$, find a domain over which $f(x)$ is increasing. Sketch the corresponding inverse function $f^{-1}(x)$, and find the gradient of the inverse graph at the points (i) $(16, 2)$ (ii) $(27, 3)$.

9.8   Differentiation of x^n

We have assumed that $\dfrac{d}{dx}(x^n) = nx^{n-1}$. We demonstrated this
numerically in Chapter 1, and, working from first principles, we
have proved that it is true for $n = 2$, $n = -1$ and $n = \frac{1}{2}$ (see p. 27).
We shall now prove this result for all rational values of n.

Case I: n is a positive integer

Let $y = x^n$. If δx and δy are corresponding changes in x and y, then

$$y + \delta y = (x + \delta x)^n$$

Subtracting,

$$\delta y = (x + \delta x)^n - x^n$$

Using the binomial theorem,

$$(a + b)^n = a^n + na^{n-1}b + \frac{n(n-1)}{2!}a^{n-2}b^2 + \ldots + b^n$$

we have

$$\delta y = x^n + nx^{n-1}(\delta x) + \frac{n(n-1)}{2!}x^{n-2}(\delta x)^2 + \ldots + (\delta x)^n - x^n$$

$$= nx^{n-1}(\delta x) + \frac{n(n-1)}{2!}x^{n-2}(\delta x)^2 + \ldots + (\delta x)^n$$

Dividing by δx,

$$\frac{\delta y}{\delta x} = nx^{n-1} + \frac{n(n-1)}{2!}x^{n-2}(\delta x) + \ldots + (\delta x)^{n-1}$$

Hence $\dfrac{dy}{dx} = \lim\limits_{\delta x \to 0} \dfrac{\delta y}{\delta x} = nx^{n-1}$

Case II: $n = \dfrac{1}{m}$, where m is a positive integer

Let $y = x^n = x^{\frac{1}{m}}$. This is the inverse function of x^m. Then

$$x = y^m$$

Differentiating with respect to y, using Case I,

$$\frac{dx}{dy} = my^{m-1}$$

So $\dfrac{dy}{dx} = \dfrac{1}{my^{m-1}} = \dfrac{1}{m(x^{\frac{1}{m}})^{m-1}} = \dfrac{1}{mx^{1-\frac{1}{m}}} = \left(\dfrac{1}{m}\right)x^{\frac{1}{m}-1}$

$$= nx^{n-1}$$

Case III: n is a positive fraction

Suppose $n = \frac{p}{q}$, where p and q are positive integers, and let $y = x^n = x^{\frac{p}{q}} = (x^{\frac{1}{q}})^p$. This is a composite function: $y = u^p$ where $u = x^{\frac{1}{q}}$. Using Case I,

$$\frac{dy}{du} = pu^{p-1}$$

and, using Case II,

$$\frac{du}{dx} = \frac{1}{q}x^{\frac{1}{q}-1}$$

Using the chain rule,

$$\frac{dy}{dx} = \frac{dy}{du} \times \frac{du}{dx}$$

$$= pu^{p-1} \times \frac{1}{q}x^{\frac{1}{q}-1} = p(x^{\frac{1}{q}})^{p-1} \times \frac{1}{q}x^{\frac{1}{q}-1}$$

$$= \frac{p}{q}x^{(\frac{p-1}{q}+\frac{1}{q}-1)} = \frac{p}{q}x^{\frac{p}{q}-1}$$

$$= nx^{n-1}$$

Case IV: n is a negative rational number

Suppose $n = -r$, where r is positive, and let $y = x^n = x^{-r} = \frac{1}{x^r}$.

Using the quotient rule,

$$\frac{dy}{dx} = \frac{(x^r)(0) - (1)\dfrac{d}{dx}(x^r)}{(x^r)^2}$$

$$= \frac{-rx^{r-1}}{x^{2r}}$$

(using Case III)

$$= (-r)x^{-r-1}$$

$$= nx^{n-1}$$

10

Circular functions

Introduction

In this chapter, we assume that you are familiar with the 'circular functions', and have worked with radians.

The circular functions are $\sin x$, $\cos x$, $\tan x = \dfrac{\sin x}{\cos x}$, $\sec x = \dfrac{1}{\cos x}$, $\operatorname{cosec} x = \dfrac{1}{\sin x}$, and $\cot x = \dfrac{\cos x}{\sin x}$. ($\tan x$ and $\sec x$ are not defined when $\cos x = 0$; similarly, $\operatorname{cosec} x$ and $\cot x$ are not defined when $\sin x = 0$.)

We shall always assume that x is measured in radians (remember that π radians $= 180°$).

10.1 Sketching graphs and gradient functions

Figs 10.1 and 10.2 show the graphs of $\sin x$ and $\cos x$ respectively.

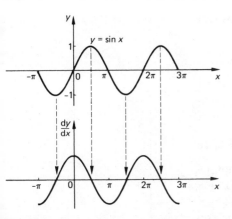

Fig. 10.2

Fig. 10.1

We now sketch the gradient function of $\sin x$. The gradient is zero when $x = -\frac{1}{2}\pi, \frac{1}{2}\pi, \frac{3}{2}\pi, \frac{5}{2}\pi, \ldots$. It is positive for the range $-\frac{1}{2}\pi < x < \frac{1}{2}\pi$, negative for $\frac{1}{2}\pi < x < \frac{3}{2}\pi$, and so on.

This is similar in shape to the graph of $\cos x$, but we do not know the height of the curve. We investigate this numerically.

We can find an approximate value for the gradient of $y = \sin x$ at a point A by calculating the gradient of the line joining points on

129

either side of A (see Fig. 10.3). For example, if $x = 0.4$, and we take an x-step of 0.01 to each side, the points on the curve are $Q(0.39, \sin(0.39))$ and $R(0.41, \sin(0.41))$, and the gradient of QR is

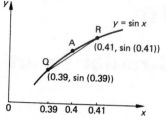

Fig. 10.3

$$\frac{\sin(0.41) - \sin(0.39)}{0.41 - 0.39} = \frac{0.398\,6093 - 0.380\,1884}{0.02}$$

$$= 0.92105$$

Using this method for $x = 0, 0.2, 0.4, \ldots, 2.0$, we obtain the following values.

x	0.0	0.2	0.4	0.6	0.8	1.0
Approximate gradient of $y = \sin x$	0.999 98	0.980 05	0.921 05	0.825 32	0.696 70	0.540 29

x	1.2	1.4	1.6	1.8	2.0
Approximate gradient of $y = \sin x$	0.362 35	0.169 96	−0.029 20	−0.227 20	−0.416 14

The corresponding values of $\cos x$ are as follows.

x	0.0	0.2	0.4	0.6	0.8	1.0
$\cos x$	1.000 00	0.980 07	0.921 06	0.825 34	0.696 71	0.540 30

x	1.2	1.4	1.6	1.8	2.0
$\cos x$	0.362 36	0.169 97	−0.029 20	−0.227 20	−0.416 15

and so it seems likely that the gradient function of $\sin x$ is $\cos x$, i.e.

$$\frac{d}{dx}(\sin x) = \cos x$$

Example   Sketch the graphs of (i) $\tan x$ (ii) $\sec x$ (iii) $\sec^2 x$. Show that it is plausible that $\dfrac{d}{dx}(\tan x) = \sec^2 x$.

(i)   Let $y = \tan x = \dfrac{\sin x}{\cos x}$. y is not defined when $\cos x = 0$, i.e. when $x = \pm\frac{1}{2}\pi, \pm\frac{3}{2}\pi, \ldots$, and the curve has vertical asymptotes at these values of x. $y = 0$ when $\sin x = 0$, i.e. when $x = 0, \pm\pi, \pm 2\pi, \ldots$

We first sketch the graphs of $\sin x$ and $\cos x$ using the same axes (see Fig. 10.4). The sign of the quotient $\tan x = \dfrac{\sin x}{\cos x}$ has been indicated on the x-axis.

Fig. 10.5 shows a sketch of the curve $y = \tan x$.

Next we sketch the gradient function of $\tan x$ (for use later in the question); see Fig. 10.6. The gradient is always positive, and it becomes arbitrarily large as x approaches the values $\pm\frac{1}{2}\pi, \pm\frac{3}{2}\pi, \ldots$

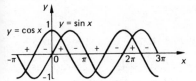
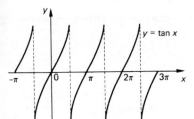

Fig. 10.4

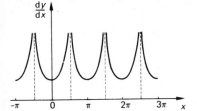

Fig. 10.5

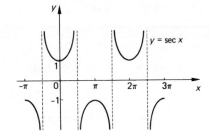

Fig. 10.7

Fig. 10.6

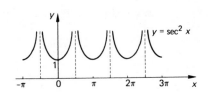

Fig. 10.8

(ii)   Let $y = \sec x = \dfrac{1}{\cos x}$. y is not defined when $\cos x = 0$; the curve has vertical asymptotes when $x = \pm\frac{1}{2}\pi, \pm\frac{3}{2}\pi, \ldots$ y is never zero; it is positive or negative in the same regions as $\cos x$. The curve is sketched in Fig. 10.7.

(iii)   For $y = \sec^2 x$, we square the values of $\sec x$ on the previous graph (see Fig. 10.8). $\sec^2 x$ is always positive. This curve is similar in shape to the gradient function of $\tan x$. So it is plausible that

$$\frac{d}{dx}(\tan x) = \sec^2 x.$$

We can also demonstrate $\dfrac{d}{dx}(\tan x) = \sec^2 x$ numerically, for any particular value of x; for example $x = 1.2$. We take points on either side of this, say $(1.19, \tan(1.19))$ and $(1.21, \tan(1.21))$. The gradient of $\tan x$ at $x = 1.2$ is approximately

$$\frac{\tan(1.21) - \tan(1.19)}{1.21 - 1.19} = \frac{2.65032 - 2.49790}{0.02}$$

$$= 7.621$$

Also $\sec^2(1.2) = (2.7597)^2 = 7.616$

Exercise 10.1   (Circular functions: graphs and numerical work)

1   Sketch the gradient function of $\cos x$, and show that it is plausible that $\dfrac{d}{dx}(\cos x) = -\sin x$.

2   Tabulate approximate values for the gradient of $y = \cos x$ when $x = 0, 0.2, 0.4, \ldots, 2.0$ radians, by taking an x-step of 0.01 to each side. Remember to set your calculator to radian mode.
   Compare these with the corresponding values of $-\sin x$.

3   Sketch the graphs of (i) $\cot x$ (ii) $\operatorname{cosec} x$ (iii) $-\operatorname{cosec}^2 x$.

   Show that it is plausible that $\dfrac{d}{dx}(\cot x) = -\operatorname{cosec}^2 x$.

4   Sketch the graphs of (i) $\sec x$ (ii) $\sec x \tan x$.

   Show that it is plausible that $\dfrac{d}{dx}(\sec x) = \sec x \tan x$.

5   Demonstrate the following results numerically. Remember to work in radians.

(i) $\dfrac{d}{dx}(\sin x) = \cos x$ when $x = 1.3$.   (ii) $\dfrac{d}{dx}(\cos x) = -\sin x$ when $x = 2.2$.

(iii) $\dfrac{d}{dx}(\tan x) = \sec^2 x$ when $x = 0.5$.   (iv) $\dfrac{d}{dx}(\cot x) = -\operatorname{cosec}^2 x$ when $x = 1.4$.

(v) $\dfrac{d}{dx}(\sec x) = \sec x \tan x$ when $x = 3.5$.

10.2   Differentiation of circular functions

We shall now prove the results which were suggested graphically and numerically in Section 10.1. First we need to show that, as $x \to 0, \dfrac{\sin x}{x} \to 1$.

   Since $\dfrac{\sin x}{x}$ is an even function $\left(\text{if } f(x) = \dfrac{\sin x}{x}, \text{ then} \right.$

$f(-x) = \dfrac{\sin(-x)}{-x} = \dfrac{-\sin x}{-x} = \dfrac{\sin x}{x} = f(x)\Big)$, we only need to consider the case where x is positive.

   In Fig. 10.9, A and P are points on a circle with centre O and radius r. The angle AOP is x radians. PT is the tangent to the circle at P, and thus OPT is a right-angled triangle. Hence $PT = r \tan x$.
   The area of the *triangle* OAP is

$$\tfrac{1}{2} \times OA \times OP \times \sin x = \tfrac{1}{2}r^2 \sin x$$

The area of the *sector* OAP is

$$\tfrac{1}{2}r^2 x$$

The area of the triangle OPT is

$$\tfrac{1}{2} \times OP \times PT = \tfrac{1}{2}r^2 \tan x$$

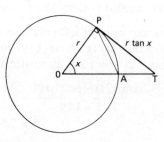

Fig. 10.9

So we have $\frac{1}{2}r^2 \sin x < \frac{1}{2}r^2 x < \frac{1}{2}r^2 \tan x$

or $\sin x < \ x \ < \tan x$

or $\sin x < \ x \ < \dfrac{\sin x}{\cos x}$

Dividing by $\sin x$,

$$1 < \frac{x}{\sin x} < \frac{1}{\cos x}$$

As $x \to 0$, $\cos x \to 1$, and $\dfrac{x}{\sin x}$ lies between two quantities

(1 and $\dfrac{1}{\cos x}$), which both tend to 1. Thus $\dfrac{x}{\sin x} \to 1$, and hence also

$\dfrac{\sin x}{x} \to 1$.

Now suppose $y = \sin x$. Let δx be a change in x, causing a change δy in y. Then

$$y + \delta y = \sin (x + \delta x)$$

and so $\delta y = \sin (x + \delta x) - \sin x$

$$= 2 \cos\left(\frac{(x + \delta x) + x}{2}\right) \sin\left(\frac{(x + \delta x) - x}{2}\right)$$

$$\left(\text{using } \sin P - \sin Q = 2 \cos\left(\frac{P + Q}{2}\right) \sin\left(\frac{P - Q}{2}\right)\right)$$

$$= 2 \cos\left(x + \frac{\delta x}{2}\right) \sin\left(\frac{\delta x}{2}\right)$$

Dividing by δx,

$$\frac{\delta y}{\delta x} = \frac{2 \cos\left(x + \dfrac{\delta x}{2}\right) \sin\left(\dfrac{\delta x}{2}\right)}{\delta x}$$

$$= \cos\left(x + \frac{\delta x}{2}\right) \times \frac{\sin\left(\dfrac{\delta x}{2}\right)}{\dfrac{\delta x}{2}}$$

Now $\dfrac{dy}{dx} = \lim\limits_{\delta x \to 0} \dfrac{\delta y}{\delta x}$, and, as $\delta x \to 0$, $\cos\left(x + \dfrac{\delta x}{2}\right) \to \cos x$, and

from our previous result, $\dfrac{\sin\left(\dfrac{\delta x}{2}\right)}{\dfrac{\delta x}{2}} \to 1$. Hence $\dfrac{dy}{dx} = \cos x$, i.e.

$$\boxed{\frac{d}{dx}(\sin x) = \cos x}$$

Now suppose $y = \cos x$. If dx and δy are corresponding changes, then

$$y + \delta y = \cos(x + \delta x)$$
and so $\quad \delta y = \cos(x + \delta x) - \cos x$

$$= -2 \sin\left(\frac{(x + \delta x) + x}{2}\right) \sin\left(\frac{(x + \delta x) - x}{2}\right)$$

$$\left(\text{using } \cos P - \cos Q = -2 \sin\left(\frac{P + Q}{2}\right) \sin\left(\frac{P - Q}{2}\right)\right)$$

$$= -2 \sin\left(x + \frac{\delta x}{2}\right) \sin\left(\frac{\delta x}{2}\right)$$

Dividing by δx,

$$\frac{\delta y}{\delta x} = \frac{-2 \sin\left(x + \frac{\delta x}{2}\right) \sin\left(\frac{\delta x}{2}\right)}{\delta x}$$

$$= -\sin\left(x + \frac{\delta x}{2}\right) \times \frac{\sin\left(\frac{\delta x}{2}\right)}{\frac{\delta x}{2}}$$

As $\delta x \to 0$, $\sin\left(x + \frac{\delta x}{2}\right) \to \sin x$, and $\dfrac{\sin\left(\dfrac{\delta x}{2}\right)}{\dfrac{\delta x}{2}} \to 1$, so

$$\frac{dy}{dx} = -\sin x$$

Hence

$$\boxed{\frac{d}{dx}(\cos x) = -\sin x}$$

Alternatively, we can find the derivative of $\cos x$ by using the results $\sin(\frac{1}{2}\pi - x) = \cos x$ and $\cos(\frac{1}{2}\pi - x) = \sin x$.

Let $y = \cos x = \sin(\frac{1}{2}\pi - x)$. This is a function of a function; $y = \sin u$ where $u = \frac{1}{2}\pi - x$. Now $\dfrac{dy}{du} = \cos u$ (from our first result),

and $\dfrac{du}{dx} = -1$. Hence

$$\frac{dy}{dx} = \frac{dy}{du} \times \frac{du}{dx}$$

$$= (\cos u) \times (-1) = -\cos(\tfrac{1}{2}\pi - x)$$

$$= -\sin x$$

The other circular functions can now be differentiated by first expressing them in terms of $\sin x$ and $\cos x$.

Example 1   Differentiate (i) $\tan x$; (ii) $\sec x$.

(i)   If $y = \tan x = \dfrac{\sin x}{\cos x}$, then, using the quotient rule,

$$\frac{dy}{dx} = \frac{(\cos x)(\cos x) - (\sin x)(-\sin x)}{\cos^2 x}$$

$$= \frac{\cos^2 x + \sin^2 x}{\cos^2 x} = \frac{1}{\cos^2 x} = \sec^2 x$$

(ii)   If $y = \sec x = \dfrac{1}{\cos x}$, then

$$\frac{dy}{dx} = \frac{(\cos x)(0) - (1)(-\sin x)}{\cos^2 x}$$

$$= \frac{\sin x}{\cos^2 x} = \frac{1}{\cos x} \times \frac{\sin x}{\cos x} = \sec x \tan x$$

Note that, when $x = \pm\frac{1}{2}\pi, \pm\frac{3}{2}\pi, \ldots$, $\tan x$ and $\sec x$ are not defined, and so their derivatives do not exist for these values of x.

The derivatives of the six circular functions are given in Table 10.1.

Table 10.1

y	$\dfrac{dy}{dx}$
$\sin x$	$\cos x$
$\cos x$	$-\sin x$
$\tan x$	$\sec^2 x$
$\sec x$	$\sec x \tan x$
$\operatorname{cosec} x$	$-\operatorname{cosec} x \cot x$
$\cot x$	$-\operatorname{cosec}^2 x$

Example 2   Differentiate the following.
(i) $\sin 3x$ (ii) $\operatorname{cosec}(\frac{1}{4}\pi - 2x)$ (iii) $\cos^3 x$
(iv) $\sec^2 x \tan x$ (v) $x \sin x$ (vi) $\sin(x^\circ)$
(Here x° means x degrees.)

(i)   Let $y = \sin 3x$. This is a function of a function; put $y = \sin u$, where $u = 3x$. Then

$$\frac{dy}{dx} = \frac{dy}{du} \times \frac{du}{dx} = (\cos u) \times 3 = 3 \cos 3x$$

(ii)   Let $y = \operatorname{cosec}(\frac{1}{4}\pi - 2x) = \operatorname{cosec} u$ where $u = \frac{1}{4}\pi - 2x$. Then

$$\frac{dy}{dx} = (-\operatorname{cosec} u \cot u) \times (-2)$$

$$= 2 \operatorname{cosec}(\tfrac{1}{4}\pi - 2x) \cot(\tfrac{1}{4}\pi - 2x)$$

(iii)   Let $y = \cos^3 x = u^3$ where $u = \cos x$. Then

$$\frac{dy}{dx} = 3u^2(-\sin x) = -3\cos^2 x \sin x$$

(iv)   Let $y = \sec^2 x \tan x$. Using the product rule,

$$\frac{dy}{dx} = \sec^2 x \frac{d}{dx}(\tan x) + \frac{d}{dx}(\sec^2 x)\tan x$$

$$= \sec^2 x \sec^2 x + (2\sec x \sec x \tan x)\tan x$$

$$= \sec^4 x + 2\sec^2 x \tan^2 x$$

(v)   Let $y = x \sin x$. Then

$$\frac{dy}{dx} = x\frac{d}{dx}(\sin x) + \frac{d}{dx}(x)\sin x$$

$$= x \cos x + \sin x$$

(vi)   We have x degrees $= \dfrac{\pi x}{180}$ radians, so if $y = \sin(x°)$

$= \sin\left(\dfrac{\pi x}{180}\right)$, then

$$\frac{dy}{dx} = \frac{\pi}{180}\cos\left(\frac{\pi x}{180}\right) = \frac{\pi}{180}\cos(x°)$$

Exercise 10.2   (Differentiation of circular functions)

1   Assuming the derivatives of $\sin x$ and $\cos x$, show that

(i) $\dfrac{d}{dx}(\cot x) = -\operatorname{cosec}^2 x$ (ii) $\dfrac{d}{dx}(\operatorname{cosec} x) = -\operatorname{cosec} x \cot x$

Differentiate the following functions (questions 2 to 29).

2 $\sin 6x$	3 $\cos 2x$	4 $3\sin 2x - 5\cos 3x$	5 $-4\sin\frac{3}{2}x$	
6 $\cos(\frac{1}{4}\pi - 3x)$	7 $\sin(x^2)$	8 $\sin^3 x$	9 $4\cos^2 x$	

10 $\sin^2 3x$	11 $-2\cos^3 4x$	12 $\tan 2x$	13 $3\sec 2x$
14 $\cot(2x+1)$	15 $-2\tan(\frac{1}{3}\pi + 3x)$	16 $\cot^3 x$	17 $-2\operatorname{cosec}^4 x$
18 $\sec^2 3x$	19 $\sin x \cos x$	20 $\sec x \tan x$	21 $\sin 3x \cos 5x$
22 $x \cos x$	23 $x^2 \cot x$	24 $2\cos x + 2x\sin x - x^2\cos x$	25 $\dfrac{\tan x}{x}$
26 $\dfrac{\cos 2x}{\sin 3x}$	27 $\sqrt{1+\sin x}$	28 $\dfrac{\sin x}{1-\cos x}$	29 $\sin^2(x^2)$

30   Differentiate $\sin^2 x$ and $\cos^2 x$. Hence find the derivative of $\sin^2 x + \cos^2 x$. Comment on your result.

31   By writing $x°$ (x degrees) $= \dfrac{\pi x}{180}$ radians, differentiate $\cos(x°)$.

32   Differentiate (i) $\tan(x°)$ (ii) $\cos(2x°)$ (iii) $\sin(60° - x°)$.

10.3 Integration of circular functions

From the derivatives of the circular functions, we have

$$\int \sin x \, dx = -\cos x + C$$
$$\int \cos x \, dx = \sin x + C$$
$$\int \sec^2 x \, dx = \tan x + C$$
$$\int \mathrm{cosec}^2 x \, dx = -\cot x + C$$
$$\int \sec x \tan x \, dx = \sec x + C$$
$$\int \mathrm{cosec}\, x \cot x \, dx = -\mathrm{cosec}\, x + C$$

There are some annoying gaps in this list. We can integrate $\sec^2 x$ and $\sec x \tan x$, but we cannot yet integrate $\sec x$ or $\tan x$, which seem 'simpler'.

When evaluating definite integrals, remember that x is in radians. The limits of integration are often given as multiples of π.

Example 1

$$\int_0^\pi \sin x \, dx = \left[-\cos x \right]_0^\pi$$
$$= (-\cos \pi) - (-\cos 0) = 1 - (-1) = 2$$

Example 2

$$\int_{\frac{1}{4}\pi}^{\frac{1}{3}\pi} \sec^2 x \, dx = \left[\tan x \right]_{\frac{1}{4}\pi}^{\frac{1}{3}\pi}$$
$$= (\tan \tfrac{1}{3}\pi) - (\tan \tfrac{1}{4}\pi) = \sqrt{3} - 1$$

Also, remember to check that the integral is valid for all values of x between the limits of integration. For example, the evaluation

$$\int_0^{\frac{3}{4}\pi} \sec^2 x \, dx = \left[\tan x \right]_0^{\frac{3}{4}\pi} = (-1) - (0) = -1$$

is nonsense, because when $x = \frac{1}{2}\pi$ (which is between the limits 0 and $\frac{3}{4}\pi$), $\tan x$ is not defined, and so the relationship $\frac{d}{dx}(\tan x) = \sec^2 x$ is not true throughout the range.

We next consider the integration of some composite functions.

Example 3 Find $\int \sin (2x) \, dx$.

Put $\sin (2x) = \sin u$, where $u = 2x$. $u = 2x$ is a linear function, and so we have

$$\int \sin (2x) \, dx = \tfrac{1}{2} \int \sin u \, du$$
$$= -\tfrac{1}{2} \cos u + C = -\tfrac{1}{2} \cos (2x) + C$$

Example 4   Find $\int \sec(4x) \tan(4x) \, dx$.

Put $\sec(4x) \tan(4x) = \sec u \tan u$, where $u = 4x$. Then

$$\int \sec(4x) \tan(4x) \, dx = \tfrac{1}{4} \int \sec u \tan u \, du$$

$$= \tfrac{1}{4} \sec u + C = \tfrac{1}{4} \sec(4x) + C$$

Example 5

$$\int \operatorname{cosec}^2(3x - 2) \, dx = -\tfrac{1}{3} \cot(3x - 2) + C$$

Example 6

$$\int_0^\pi \cos(\tfrac{1}{4}x) \, dx = \left[4 \sin(\tfrac{1}{4}x) \right]_0^\pi$$

$$= \left(4 \times \frac{1}{\sqrt{2}} \right) - (0) = 2\sqrt{2}$$

Trigonometrical identities can sometimes be used to express a function in terms of other functions whose integrals are known.

Example 7   Using $\sec^2 x = 1 + \tan^2 x$, we have

$$\int \tan^2 x \, dx = \int (\sec^2 x - 1) \, dx = \tan x - x + C$$

Example 8   Using $\cos 2x = 1 - 2\sin^2 x = 2\cos^2 x - 1$, we have

(i) $\int \sin^2 x \, dx = \int \tfrac{1}{2}(1 - \cos 2x) \, dx$

$$= \tfrac{1}{2}x - \tfrac{1}{4}\sin 2x + C$$

(ii) $\int \cos^2 x \, dx = \int \tfrac{1}{2}(1 + \cos 2x) \, dx$

$$= \tfrac{1}{2}x + \tfrac{1}{4}\sin 2x + C$$

The 'sum formulae', like

$$2 \sin A \cos B = \sin(A + B) + \sin(A - B)$$

can be used to integrate certain products.

Example 9

$$\int \sin 5x \cos 2x \, dx = \int \tfrac{1}{2}(\sin 7x + \sin 3x) \, dx$$

$$= -\tfrac{1}{14}\cos 7x - \tfrac{1}{6}\cos 3x + C$$

Example 10

$$\int_0^{\frac{1}{2}\pi} \cos x \cos 2x \, dx = \int_0^{\frac{1}{2}\pi} \frac{1}{2}(\cos 3x + \cos x) \, dx$$

$$= \left[\frac{1}{6}\sin 3x + \frac{1}{2}\sin x \right]_0^{\frac{1}{2}\pi}$$

$$= (-\tfrac{1}{6} + \tfrac{1}{2}) - (0) = \tfrac{1}{3}$$

Exercise 10.3   (Integration of circular functions)

In questions 1 to 15, integrate the functions with respect to x.

1 $\cos 3x$	**2** $\sin(2x - 3)$	**3** $3 \sin x + 2 \cos x$
4 $\cos(\tfrac{1}{4}\pi - \tfrac{1}{2}x)$	**5** $\sec 2x \tan 2x$	**6** $\operatorname{cosec} \tfrac{1}{3}x \cot \tfrac{1}{3}x$
7 $\sec^2(\tfrac{1}{5}\pi + \tfrac{2}{3}x)$	**8** $\operatorname{cosec}^2 5x$	**9** $\cot^2 x$
10 $\tan^2 2x$	**11** $\sin^2 3x$	**12** $\cos^2 2x$
13 $\sin 4x \cos 3x$	**14** $\cos 6x \cos 2x$	**15** $\sin x \sin 4x$

In questions 16 to 27, evaluate the definite integrals.

16 $\displaystyle\int_0^{\frac{1}{3}\pi} \cos x \, dx$	**17** $\displaystyle\int_0^{\frac{1}{2}\pi} \sin x \, dx$	**18** $\displaystyle\int_{\frac{1}{4}\pi}^{\frac{1}{2}\pi} \operatorname{cosec}^2 x \, dx$
19 $\displaystyle\int_0^{\frac{1}{3}\pi} \sec x \tan x \, dx$	**20** $\displaystyle\int_0^{\frac{1}{2}\pi} \sin 2x \, dx$	**21** $\displaystyle\int_{-\frac{1}{2}\pi}^{\frac{1}{2}\pi} \sec^2 \tfrac{1}{2}x \, dx$
22 $\displaystyle\int_0^{\frac{1}{4}\pi} \tan^2 x \, dx$	**23** $\displaystyle\int_0^{2\pi} \sin^2 x \, dx$	**24** $\displaystyle\int_0^{\frac{1}{4}\pi} \cos^2 2x \, dx$
25 $\displaystyle\int_0^{\pi} \sin^2 \tfrac{1}{2}x \, dx$	**26** $\displaystyle\int_0^{\frac{1}{2}\pi} \sin x \cos 3x \, dx$	**27** $\displaystyle\int_{-\frac{1}{4}\pi}^{\frac{1}{4}\pi} \cos 2x \cos 3x \, dx$

28   Use the formula $\sin 3x = 3 \sin x - 4 \sin^3 x$ to find $\int \sin^3 x \, dx$, and evaluate $\displaystyle\int_0^{\frac{1}{2}\pi} \sin^3 x \, dx$.

29   Show that $\dfrac{1}{1 + \cos x} = \dfrac{1}{2} \sec^2 \tfrac{1}{2}x$. Hence evaluate $\displaystyle\int_0^{\frac{1}{2}\pi} \dfrac{1}{1 + \cos x} \, dx$.

30   Criticise $\displaystyle\int_{-\frac{1}{4}\pi}^{\frac{1}{4}\pi} \operatorname{cosec}^2 x \, dx = \left[-\cot x \right]_{-\frac{1}{4}\pi}^{\frac{1}{4}\pi} = (-1) - (1) = -2.$

10.4   Differential equations of the type $\dfrac{d^2y}{dx^2} = -n^2y$

In Chapter 6 (see p. 69) we used integration to solve simple differential equations, when $\dfrac{dy}{dx}$ was given in terms of x. We shall now see how the circular functions are useful for solving a different type of differential equation $\dfrac{d^2y}{dx^2} = -n^2y$ (notice that $\dfrac{d^2y}{dx^2}$ is given in terms of y instead of x). This equation occurs frequently in mechanics.

If $y = \sin x$, then $\dfrac{dy}{dx} = \cos x$ and $\dfrac{d^2y}{dx^2} = -\sin x = -y.$

Also,

if $y = \cos x$, then $\dfrac{dy}{dx} = -\sin x$ and $\dfrac{d^2y}{dx^2} = -\cos x = -y.$

So $y = \sin x$ and $y = \cos x$ are solutions of the differential equation $\dfrac{d^2y}{dx^2} = -y.$

We can show that $y = A \sin x + B \cos x$, where A and B are any constants, also satisfies the equation, because

$$\frac{dy}{dx} = A \cos x - B \sin x$$

and $\dfrac{d^2 y}{dx^2} = -A \sin x - B \cos x = -y$

Thus $y = A \sin x + B \cos x$ is the general solution of the differential equation $\dfrac{d^2 y}{dx^2} = -y$.

Also if $y = A \sin nx + B \cos nx$

then $\dfrac{dy}{dx} = An \cos nx - Bn \sin nx$

and $\dfrac{d^2 y}{dx^2} = -An^2 \sin nx - Bn^2 \cos nx = -n^2 y$

So

$$\boxed{y = A \sin nx + B \cos nx}$$

is the general solution of the differential equation

$$\boxed{\frac{d^2 y}{dx^2} = -n^2 y}$$

$\dfrac{d^2 y}{dx^2} = -n^2 y$ is called a **second-order differential equation**

$\left(\text{because it involves the second derivative } \dfrac{d^2 y}{dx^2}\right)$, and its general solution contains two arbitrary constants (A and B). We shall need two pieces of information if we are to find these constants.

Example 1   Find the general solution of the differential equation $\dfrac{d^2 y}{dx^2} = -9y$.

This equation is of the form $\dfrac{d^2 y}{dx^2} = -n^2 y$ with $n = 3$, and so the general solution is

$$y = A \sin 3x + B \cos 3x$$

Example 2   Solve the differential equation $\dfrac{d^2 y}{dx^2} = -y$, given that $y = 2$ when $x = 0$, and $y = 5$ when $x = \frac{1}{2}\pi$.

The general solution is

$$y = A \sin x + B \cos x$$

Substituting $x = 0$, $y = 2$ gives $2 = B$; and $x = \frac{1}{2}\pi$, $y = 5$ gives $5 = A$. Hence the solution is

$$y = 5 \sin x + 2 \cos x$$

Example 3 Solve the differential equation $\dfrac{d^2y}{dx^2} + 16y = 0$, given

that $y = 3$ and $\dfrac{dy}{dx} = -8$ when $x = 0$.

The equation is $\dfrac{d^2y}{dx^2} = -16y$, and so the general solution is

$$y = A \sin 4x + B \cos 4x$$

Then $\dfrac{dy}{dx} = 4A \cos 4x - 4B \sin 4x$

Substituting $x = 0$, $y = 3$ gives $3 = B$; and $x = 0$, $\dfrac{dy}{dx} = -8$ gives

$-8 = 4A$, so $A = -2$. Hence the solution is

$$y = -2 \sin 4x + 3 \cos 4x$$

Many oscillating systems, such as springs, waves on water, sound waves, and electromagnetic waves, obey differential equations of

the form $\dfrac{d^2y}{dx^2} = -n^2y$. It follows that circular functions can be

used to describe these phenomena.

Exercise 10.4 $\left(\text{Differential equations: } \dfrac{d^2y}{dx^2} = -n^2y\right)$

1 Verify that $y = A \sin\left(\tfrac{3}{2}x\right) + B \cos\left(\tfrac{3}{2}x\right)$ is a solution of the differential equation $4\dfrac{d^2y}{dx^2} + 9y = 0$.

In questions 2 to 5, write down the general solution of the differential equation.

2 $\dfrac{d^2y}{dx^2} = -4y$ **3** $9\dfrac{d^2y}{dx^2} = -4y$ **4** $\dfrac{d^2y}{dx^2} = -3y$ **5** $16\dfrac{d^2y}{dx^2} + y = 0$

In questions 6 to 10, solve the differential equations.

6 $\dfrac{d^2y}{dx^2} = -y$, given that $y = 1$ when $x = 0$, and $y = 4$ when $x = \tfrac{1}{2}\pi$.

7 $\dfrac{d^2y}{dx^2} = -\dfrac{1}{4}y$, given that $y = 0$ when $x = 0$, and $y = 5$ when $x = 3\pi$.

8 $\dfrac{d^2y}{dx^2} = -9y$, given that $y = 4$ and $\dfrac{dy}{dx} = 2$ when $x = 0$.

9 $9\dfrac{d^2y}{dx^2} + y = 0$, given that $y = 0$ and $\dfrac{dy}{dx} = -1$ when $x = 0$.

10 $4\dfrac{d^2y}{dx^2} + 25y = 0$, given that $y = -6$ and $\dfrac{dy}{dx} = 4$ when $x = \tfrac{1}{5}\pi$.

11 Verify that $y = A \sin 2x + B \cos 2x + \tfrac{3}{4}$ is a solution of the differential equation $\dfrac{d^2y}{dx^2} + 4y = 3$.

12 Verify that $y = \tfrac{1}{2}x \sin x$ is a solution of the differential equation $\dfrac{d^2y}{dx^2} + y = \cos x$.

10.5   Inverse circular functions

In the same way that we considered inverse algebraic functions (see p. 123) we now investigate inverses of the circular functions.

By looking at Fig. 10.10, we can see that the inverse graph of sin x is certainly not the graph of a function.

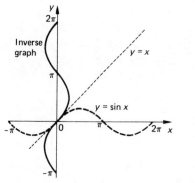

Fig. 10.10

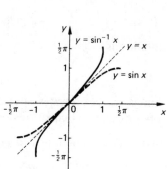

Fig. 10.11

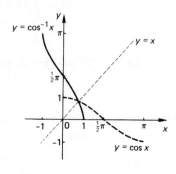

Fig. 10.12

Now $\sin x$ is increasing over the intervals
$$-\tfrac{1}{2}\pi \leqslant x \leqslant \tfrac{1}{2}\pi, \tfrac{3}{2}\pi \leqslant x \leqslant \tfrac{5}{2}\pi, \ldots$$
and decreasing over the intervals
$$-\tfrac{3}{2}\pi \leqslant x \leqslant -\tfrac{1}{2}\pi, \tfrac{1}{2}\pi \leqslant x \leqslant \tfrac{3}{2}\pi, \ldots$$
so we can obtain an inverse function by choosing any one of these intervals.

We choose $-\tfrac{1}{2}\pi \leqslant x \leqslant \tfrac{1}{2}\pi$, over which $\sin x$ is increasing. The corresponding inverse function is written $\sin^{-1}x$ (said 'sine minus one x' and sometimes written arc $\sin x$)*. $\sin^{-1}x$ is defined for $-1 \leqslant x \leqslant 1$; and $-\tfrac{1}{2}\pi \leqslant \sin^{-1}x \leqslant \tfrac{1}{2}\pi$ (see Fig. 10.11).

If $y = \sin^{-1}x$, then $\sin y = x$, so y is the angle (in radians, between $-\tfrac{1}{2}\pi$ and $\tfrac{1}{2}\pi$) whose sine is x.

$\cos x$ is a decreasing function for $0 \leqslant x \leqslant \pi$. The corresponding inverse function is written $\cos^{-1}x$. So $\cos^{-1}x$ is defined for $-1 \leqslant x \leqslant 1$; and $0 \leqslant \cos^{-1}x \leqslant \pi$ (see Fig. 10.12).

$\tan x$ is an increasing function for $-\tfrac{1}{2}\pi < x < \tfrac{1}{2}\pi$. The corresponding inverse function is written $\tan^{-1}x$. So $\tan^{-1}x$ is defined for all x; and $-\tfrac{1}{2}\pi < \tan^{-1}x < \tfrac{1}{2}\pi$ (see Fig. 10.13).

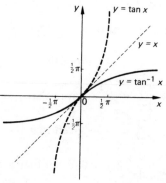

Fig. 10.13

---

* $\sin^{-1}x$ is not to be confused with $(\sin x)^{-1}$ which is $\dfrac{1}{\sin x} = \operatorname{cosec} x$.

We define $\operatorname{cosec}^{-1} x = \sin^{-1}\left(\dfrac{1}{x}\right)$, provided that $|x| \geqslant 1$ ($\operatorname{cosec}^{-1} x$ is not defined when $-1 < x < 1$). Thus $\operatorname{cosec}^{-1} x$ takes the same range of values as $\sin^{-1} x$, namely $-\frac{1}{2}\pi \leqslant \operatorname{cosec}^{-1} x \leqslant \frac{1}{2}\pi$ (see Fig. 10.14).

Similarly, we define $\sec^{-1} x = \cos^{-1}\left(\dfrac{1}{x}\right)$, provided that $|x| \geqslant 1$ (see Fig. 10.15); also $\cot^{-1} x = \tan^{-1}\left(\dfrac{1}{x}\right)$ when $x \neq 0$, and $\cot^{-1} 0 = \frac{1}{2}\pi$ (see Fig. 10.16).

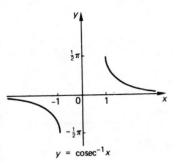

$y = \operatorname{cosec}^{-1} x$

Fig. 10.14

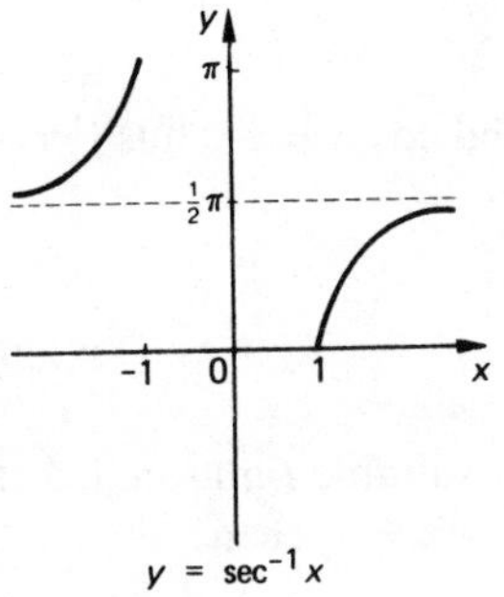

$y = \sec^{-1} x$

Fig. 10.15

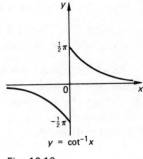

$y = \cot^{-1} x$

Fig. 10.16

Evaluation of inverse circular functions

Table 10.2

x	$-\frac{1}{2}\pi$	$-\frac{1}{3}\pi$	$-\frac{1}{4}\pi$	$-\frac{1}{6}\pi$	0	$\frac{1}{6}\pi$	$\frac{1}{4}\pi$	$\frac{1}{3}\pi$	$\frac{1}{2}\pi$	$\frac{2}{3}\pi$	$\frac{3}{4}\pi$	$\frac{5}{6}\pi$	π
$\sin x$	-1	$\dfrac{-\sqrt{3}}{2}$	$-\dfrac{1}{\sqrt{2}}$	$-\dfrac{1}{2}$	0	$\dfrac{1}{2}$	$\dfrac{1}{\sqrt{2}}$	$\dfrac{\sqrt{3}}{2}$	1				
$\cos x$					1	$\dfrac{\sqrt{3}}{2}$	$\dfrac{1}{\sqrt{2}}$	$\dfrac{1}{2}$	0	$-\dfrac{1}{2}$	$-\dfrac{1}{\sqrt{2}}$	$-\dfrac{\sqrt{3}}{2}$	-1
$\tan x$		$-\sqrt{3}$	-1	$-\dfrac{1}{\sqrt{3}}$	0	$\dfrac{1}{\sqrt{3}}$	1	$\sqrt{3}$					

Table 10.2 gives exact values of $\sin x$, $\cos x$ and $\tan x$ for 'common' angles in the domains relevant to the corresponding inverse functions.

We obtain values of $\sin^{-1} x$, $\cos^{-1} x$ and $\tan^{-1} x$ by interchanging the coordinates. For example,

$$\sin^{-1}\left(\frac{\sqrt{3}}{2}\right) = \frac{1}{3}\pi, \cos^{-1}\left(-\frac{1}{\sqrt{2}}\right) = \frac{3}{4}\pi, \tan^{-1}(-1) = -\frac{1}{4}\pi$$

Also, $\sec^{-1}\left(-\dfrac{2}{\sqrt{3}}\right) = \cos^{-1}\left(-\dfrac{\sqrt{3}}{2}\right) = \dfrac{5}{6}\pi$

Values of $\sin^{-1}x$ may be found directly on a calculator (using 'sin⁻¹' or 'inverse sin' or 'arc sin' keys, with the calculator set in radian mode). Similarly we can evaluate $\cos^{-1}x$ and $\tan^{-1}x$. Use your calculator to evaluate $\sin^{-1}(0.67)$ (the answer should be 0.7342) and $\tan^{-1}(-0.8)$ (the answer should be -0.6747). We would evaluate $\sec^{-1}(-2.7)$ as

$$\sec^{-1}(-2.7) = \cos^{-1}\left(\frac{1}{-2.7}\right) = \cos^{-1}(-0.3704) = 1.9502$$

Properties of inverse circular functions

Example 1 Show that $\sin^{-1}x = \frac{1}{2}\pi - \cos^{-1}x$.

Let $y = \cos^{-1}x$. Then $0 \leqslant y \leqslant \pi$ and $\cos y = x$. Thus $(\frac{1}{2}\pi - y)$ lies between $-\frac{1}{2}\pi$ and $\frac{1}{2}\pi$, and

$$\sin(\tfrac{1}{2}\pi - y) = \cos y = x$$
Hence $\quad \frac{1}{2}\pi - y = \sin^{-1}x$
i.e. $\quad \sin^{-1}x = \frac{1}{2}\pi - \cos^{-1}x$

Alternatively, we may consider a suitable right-angled triangle (Fig. 10.17). If $\sin^{-1}x = \theta$ and $\cos^{-1}x = \phi$ then

$$\theta = \frac{1}{2}\pi - \phi$$

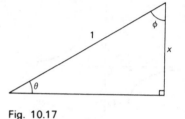

Fig. 10.17

Example 2 Show that $\tan^{-1}\frac{1}{2} + \tan^{-1}\frac{1}{3} = \frac{1}{4}\pi$.

Let $\alpha = \tan^{-1}\frac{1}{2}, \beta = \tan^{-1}\frac{1}{3}$. Then $\tan \alpha = \frac{1}{2}$ and $\tan \beta = \frac{1}{3}$, and so

$$\tan(\alpha + \beta) = \frac{\tan\alpha + \tan\beta}{1 - \tan\alpha\tan\beta} = \frac{\frac{1}{2} + \frac{1}{3}}{1 - \frac{1}{2} \times \frac{1}{3}} = \frac{\frac{5}{6}}{\frac{5}{6}} = 1$$

α and β are both between 0 and $\frac{1}{4}\pi$, and so $(\alpha + \beta)$ is between 0 and $\frac{1}{2}\pi$. Hence $\alpha + \beta = \tan^{-1}1 = \frac{1}{4}\pi$, i.e. $\tan^{-1}\frac{1}{2} + \tan^{-1}\frac{1}{3} = \frac{1}{4}\pi$.

Differentiation of inverse circular functions

Suppose $y = \sin^{-1}x$ (where $-1 < x < 1$). Then $\sin y = x$.

Differentiating with respect to y, $\dfrac{dx}{dy} = \cos y$. Thus

$$\frac{dy}{dx} = \frac{1}{\cos y} = \frac{1}{\sqrt{1 - \sin^2 y}} \overset{*}{=} \frac{1}{\sqrt{1 - x^2}}$$

Hence, provided $-1 < x < 1$,

$$\boxed{\frac{d}{dx}(\sin^{-1}x) = \frac{1}{\sqrt{1 - x^2}}}$$

*We can write $\cos y = +\sqrt{1 - \sin^2 y}$ because $-\frac{1}{2}\pi < y < \frac{1}{2}\pi$ and so $\cos y > 0$. If we had chosen a different range of values for $\sin^{-1}x$ then this might not be true.

Similarly, we find that

$$\frac{d}{dx}(\cos^{-1}x) = -\frac{1}{\sqrt{1-x^2}}$$

Now suppose $y = \tan^{-1}x$. Then $\tan y = x$, so $\dfrac{dx}{dy} = \sec^2 y$, and

$$\frac{dy}{dx} = \frac{1}{\sec^2 y} = \frac{1}{1+\tan^2 y} = \frac{1}{1+x^2}$$

Hence

$$\frac{d}{dx}(\tan^{-1}x) = \frac{1}{1+x^2}$$

It is interesting that the derivatives of the inverse circular functions are purely algebraic functions.

Example Differentiate the following functions.
(i) $\sin^{-1}(5x)$ (ii) $\cos^{-1}(x^3)$ (iii) $(\tan^{-1}x)^2$ (iv) $\sec^{-1}x$

(i) Let $y = \sin^{-1}5x = \sin^{-1}u$ where $u = 5x$; so

$$\frac{dy}{dx} = \frac{1}{\sqrt{1-u^2}} \times 5 = \frac{5}{\sqrt{1-(5x)^2}} = \frac{5}{\sqrt{1-25x^2}}$$

(ii) Let $y = \cos^{-1}(x^3) = \cos^{-1}u$ where $u = x^3$; so

$$\frac{dy}{dx} = -\frac{1}{\sqrt{1-u^2}} \times 3x^2 = \frac{-3x^2}{\sqrt{1-(x^3)^2}} = \frac{-3x^2}{\sqrt{1-x^6}}$$

(iii) Let $y = (\tan^{-1}x)^2 = u^2$ where $u = \tan^{-1}x$; so

$$\frac{dy}{dx} = 2u \times \frac{1}{1+x^2} = \frac{2\tan^{-1}x}{1+x^2}$$

(iv) Let $y = \sec^{-1}x = \cos^{-1}\left(\frac{1}{x}\right) = \cos^{-1}u$ where $u = \frac{1}{x}$; so

$$\frac{dy}{dx} = -\frac{1}{\sqrt{1-u^2}} \times \left(-\frac{1}{x^2}\right) = \frac{1}{x^2\sqrt{1-\frac{1}{x^2}}}$$

$$= \frac{1}{\sqrt{x^4\left(1-\frac{1}{x^2}\right)}} = \frac{1}{\sqrt{x^2(x^2-1)}}$$

Exercise 10.5   (Inverse circular functions)

1   Evaluate the following, leaving your answers as multiples of π.

(i) $\sin^{-1}\frac{1}{2}$, $\sin^{-1}1$, $\sin^{-1}0$, $\sin^{-1}\left(-\dfrac{\sqrt{3}}{2}\right)$, $\sin^{-1}(-1)$.

(ii) $\cos^{-1}\frac{1}{2}$, $\cos^{-1}\left(-\dfrac{\sqrt{3}}{2}\right)$, $\cos^{-1}(-1)$.

(iii) $\tan^{-1}1$, $\tan^{-1}0$, $\tan^{-1}(-\sqrt{3})$.

(iv) $\sec^{-1}2$, $\csc^{-1}(-1)$, $\cot^{-1}(\sqrt{3})$.

2   Using your calculator (remember to work in radians), obtain values of $\sin^{-1}x$ for $x = -1, -0.9, -0.8, \ldots, +1$, draw an accurate graph of $y = \sin^{-1}x$.
What happens if you try to evaluate $\sin^{-1}(1.1)$?

3   Obtain values of $\sec^{-1}x$ for $x = -5, -4.5, \ldots, -2, -1.8, -1.6, \ldots, -1, 1, 1.2, 1.4, \ldots, 2, 2.5, 3, \ldots, 5$, and draw a graph of $y = \sec^{-1}x$. What happens if you try to evaluate $\sec^{-1}(0.5)$?

4   Show that $\tan^{-1}\frac{2}{3} + \tan^{-1}\frac{4}{7} = \tan^{-1}2$.

5   Show that $\tan^{-1}3 - \tan^{-1}\frac{1}{2} = \frac{1}{4}\pi$.

Differentiate the following functions (questions 6 to 19).

6   $\sin^{-1}2x$　　　　7 $\cos^{-1}4x$　　　　8 $\tan^{-1}(\frac{1}{2}x)$　　　　9 $\sin^{-1}(x^2)$

10   $\tan^{-1}(2x+1)$　　11 $\tan^{-1}(x^4)$　　12 $(\sin^{-1}x)^2$　　13 $\cos^{-1}\left(\dfrac{x-1}{3}\right)$

14   $x^2\cos^{-1}x$　　　15 $\dfrac{\sin^{-1}x}{x}$　　　16 $\sin^{-1}(2\sin x)$　　17 $\tan^{-1}(\sqrt{x})$

18   $\sec^{-1}2x$　　　　19 $\csc^{-1}x$

20   Find $\dfrac{d}{dx}(\cot^{-1}x)$ (i)   by writing $x = \cot y$ and differentiating; (ii)   using the definition

$\cot^{-1}x = \tan^{-1}\left(\dfrac{1}{x}\right)$.

10.6   Integration using the inverse circular functions

The derivatives of the inverse circular functions are

$$\frac{d}{dx}(\sin^{-1}x) = \frac{1}{\sqrt{1-x^2}} \qquad\qquad \frac{d}{dx}(\cos^{-1}x) = -\frac{1}{\sqrt{1-x^2}}$$

$$\frac{d}{dx}(\tan^{-1}x) = \frac{1}{1+x^2} \qquad\qquad \frac{d}{dx}(\sec^{-1}x) = \frac{1}{\sqrt{x^2(x^2-1)}}$$

$$\frac{d}{dx}(\csc^{-1}x) = -\frac{1}{\sqrt{x^2(x^2-1)}} \qquad \frac{d}{dx}(\cot^{-1}x) = -\frac{1}{1+x^2}$$

We can therefore integrate the functions

$$\frac{1}{\sqrt{1-x^2}}, \frac{1}{1+x^2} \quad\text{and}\quad \frac{1}{\sqrt{x^2(x^2-1)}}$$

Both $\sin^{-1}x$ and $-\cos^{-1}x$ are integrals of $\dfrac{1}{\sqrt{1-x^2}}$. There is
no contradiction here, as $\sin^{-1}x = \frac{1}{2}\pi - \cos^{-1}x$ (see p. 144), so
these two integrals differ by a constant. We usually choose $\sin^{-1}x$
because it is simpler to work with (its value is always numerically
less than $\frac{1}{2}\pi$, and negative values of x give negative values of
$\sin^{-1}x$).

Both $\tan^{-1}x$ and $-\cot^{-1}x$ are integrals of $\dfrac{1}{1+x^2}$. There is a
very good reason for choosing $\tan^{-1}x$ ($\cot^{-1}x$ is discontinuous at
$x = 0$, and so cannot be used to evaluate a definite integral if 0 is
between the limits of integration).

So we have

$$\int \frac{1}{\sqrt{1-x^2}}\,dx = \sin^{-1}x + C$$

$$\int \frac{1}{1+x^2}\,dx = \tan^{-1}x + C$$

Example 1

$$\int_{-\frac{1}{2}}^{\frac{1}{2}} \frac{1}{\sqrt{1-x^2}}\,dx = \left[\sin^{-1}x\right]_{-\frac{1}{2}}^{\frac{1}{2}}$$

$$= \left(\frac{1}{6}\pi\right) - \left(-\frac{1}{6}\pi\right) = \frac{1}{3}\pi$$

Example 2

$$\int_{-1}^{\sqrt{3}} \frac{1}{1+x^2}\,dx = \left[\tan^{-1}x\right]_{-1}^{\sqrt{3}}$$

$$= \left(\frac{1}{3}\pi\right) - \left(-\frac{1}{4}\pi\right) = \frac{7}{12}\pi$$

Example 3

$$\int \frac{1}{\sqrt{1-9x^2}}\,dx = \int \frac{1}{\sqrt{1-(3x)^2}}\,dx$$

$$= \frac{1}{3}\sin^{-1}(3x) + C$$

Example 4

$$\int \frac{1}{9+25x^2}\,dx = \int \frac{1}{9\left(1+\frac{25}{9}x^2\right)}\,dx = \frac{1}{9}\int \frac{1}{1+\left(\frac{5}{3}x\right)^2}\,dx$$

$$= \frac{1}{9} \times \frac{3}{5} \tan^{-1}\left(\frac{5}{3}x\right) + C = \frac{1}{15}\tan^{-1}\left(\frac{5}{3}x\right) + C$$

Exercise 10.6 (Integration using inverse circular functions)

In questions 1 to 5, evaluate the definite integrals.

1 $\displaystyle\int_0^{1/\sqrt{2}} \frac{1}{\sqrt{1-x^2}}\,dx$ **2** $\displaystyle\int_{-1/2}^{\sqrt{3}/2} \frac{1}{\sqrt{1-x^2}}\,dx$ **3** $\displaystyle\int_0^1 \frac{1}{1+x^2}\,dx$

4 $\displaystyle\int_{-1/\sqrt{3}}^{1/\sqrt{3}} \frac{1}{1+x^2}\,dx$ **5** $\displaystyle\int_1^{\sqrt{3}} \frac{1}{1+x^2}\,dx$

In questions 6 to 11, integrate the functions with respect to x.

6 $\dfrac{1}{\sqrt{1-4x^2}}$ **7** $\dfrac{1}{1+16x^2}$ **8** $\dfrac{1}{9+4x^2}$

9 $\dfrac{1}{25+36x^2}$ **10** $\dfrac{1}{\sqrt{4-9x^2}}$ **11** $\dfrac{1}{\sqrt{25-16x^2}}$

12 Criticise the following.

(i) $\displaystyle\int_0^{\frac{1}{2}} \frac{1}{\sqrt{1-x^2}}\,dx = \left[\sin^{-1}x\right]_0^{\frac{1}{2}} = \left(\frac{5}{6}\pi\right) - (0) = \frac{5}{6}\pi$

(ii) $\displaystyle\int_0^1 \frac{1}{1+x^2}\,dx = \left[\tan^{-1}x\right]_0^1 = 45°$

(iii) $\displaystyle\int_{-1}^1 \frac{1}{1+x^2}\,dx = \left[-\cot^{-1}x\right]_{-1}^1 = \left(-\frac{1}{4}\pi\right) - \left(\frac{1}{4}\pi\right) = -\frac{1}{2}\pi$

11

Motion in a straight line

Introduction

In practice, the motion of a body is complicated. This is because the forces which act on the body, and which cause it to move, are usually variable in both magnitude and direction. The motion may also have to be divided into stages, in each of which the movement satisfies some particular physical law.

In this chapter we restrict ourselves to the movement of a particle in a straight line, since this is the simplest of motions to consider.

11.1 Displacement, velocity and acceleration

Consider a particle moving along a straight line. Choose a fixed point O on the line as the origin and choose one direction of the line to be positive, the other negative. We shall always assume that the positive direction is from left to right (unless stated otherwise).

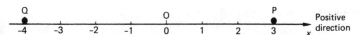

Fig. 11.1

We describe the position of the particle by giving its distance x from O (see Fig. 11.1). x is positive if the particle is to the right of O, and negative if it is to the left of O. Here we show P at $x = 3$ and Q at $x = -4$. The quantity x is called the **displacement**; it is a measure of position rather than a distance travelled.

As the particle moves, the value of x will change, so x is a function of time.

Instantaneous velocity

If the particle is moving with constant velocity, we can calculate this velocity as $\dfrac{\text{distance travelled}}{\text{time taken}}$. However, if the velocity is varying, this quotient gives the average velocity over the interval of time considered. We should like to know the instantaneous velocity of the particle, that is, its velocity at a given instant in time.

Consider a small interval of time, from t to $t + \delta t$, during which the distance from O changes from x to $x + \delta x$. The particle moves a distance δx in a time δt, and so its average velocity over this interval is $\dfrac{\delta x}{\delta t}$. As $\delta t \to 0$, $\dfrac{\delta x}{\delta t}$ gives the average velocity over a shorter and shorter time interval starting at t; we would expect the instantaneous velocity at the time t to be $\displaystyle\lim_{\delta t \to 0} \dfrac{\delta x}{\delta t}$, which is $\dfrac{dx}{dt}$, the rate of change of x (with respect to time, t).

We therefore define the velocity v of the particle to be the rate of change of distance, i.e.

$$v = \frac{dx}{dt}$$

If we know the formula giving x in terms of t, we obtain the velocity by differentiating.

If v is positive, then x is increasing, so the particle is moving in the positive direction (i.e. from left to right in Fig. 11.1). If v is negative, then x is decreasing, and the particle is moving in the negative direction (from right to left).

Acceleration

If the velocity of the particle is varying, we consider its acceleration, which is a measure of how quickly the velocity is changing. We define the average acceleration over an interval of time to be

$$\frac{\text{change in velocity}}{\text{time taken}}.$$

The acceleration a of the particle at an instant is defined to be the rate of change of velocity, i.e.

$$a = \frac{dv}{dt} = \frac{d^2x}{dt^2}$$

If we know the formula giving v in terms of t, we differentiate it to obtain a. a may be positive or negative, depending on whether v is increasing or decreasing.

Units

If the distance (x) is measured in metres, and the time (t) in seconds, then the units of velocity (v) will be metres per second $(\mathrm{m\,s^{-1}})$ and those of acceleration (a) will be metres per second per second $(\mathrm{m\,s^{-2}})$.

Example 1 A particle is moving along the x-axis. Photographs taken at times $t = 0, 1, 2, \ldots, 5$ show the positions of the particle to be as shown in Fig. 11.2.

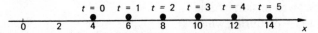

Fig. 11.2

We see that the particle is moving from left to right, and the distance travelled in each unit of time is the same. We can make a table of values for the displacement:

Time (t)	0	1	2	3	4	5
Displacement (x)	4	6	8	10	12	14

The average velocity in the period from $t = 1$ to $t = 5$ is given by

$$\frac{\text{distance travelled}}{\text{time taken}} = \frac{14-6}{5-1} = \frac{8}{4} = 2$$

and, whichever time interval we take, we shall always find that the average velocity is 2.

The particle is moving with constant velocity, $v = 2$, and, since the velocity is not changing, its acceleration is zero. If we plot a graph of x against t, we obtain a straight line with gradient 2 (Fig. 11.3). The equation of the line is

$$x = 2t + 4$$

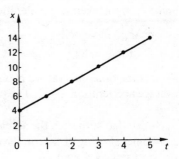

Fig. 11.3

Differentiating, we obtain

$$v = \frac{dx}{dt} = 2$$

$$a = \frac{dv}{dt} = 0$$

confirming our previous remarks. The graph of v against t is shown in Fig. 11.4.

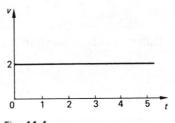

Fig. 11.4

Example 2 A particle is thrown vertically upwards, and a series of photographs taken at times $t = 0, 1, 2, \ldots, 8$ shows the positions of the particle to be as follows (see Fig. 11.5).

The particle starts at O, and reaches a maximum height $y = 16$ above O at time $t = 4$. It then falls down again, returning to O when $t = 8$.

The distance travelled in one unit of time between $t = 0$ and $t = 1$ is greater than the distance travelled in the unit of time between $t = 2$ and $t = 3$; the velocity of the particle is changing. A table of values for the height gives

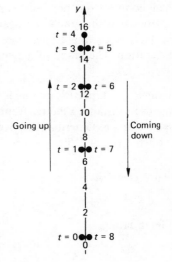

Fig. 11.5

Time (t)	0	1	2	3	4	5	6	7	8
Height (y)	0	7	12	15	16	15	12	7	0

We can estimate the velocity at the instant $t = 1$ by finding the average velocity between $t = 0$ and $t = 2$; this is $\dfrac{12-0}{2-0} = 6$.

Similarly, we estimate the velocity at the instant $t = 5$ as the average velocity between $t = 4$ and $t = 6$, which is $\dfrac{12-16}{6-4} = \dfrac{-4}{2}$

$= -2$. The negative value means that the particle is moving in the

negative direction (downwards in this case).

In this way we can build up a table of values for the velocity:

Time (t)	1	2	3	4	5	6	7
Velocity (v)	6	4	2	0	-2	-4	-6

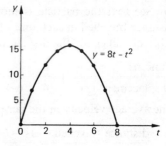

As the velocity is changing, we now consider the acceleration of the particle. For example, taking the time interval between $t = 1$ and $t = 3$, the average acceleration is

$$\frac{\text{change in velocity}}{\text{time taken}} = \frac{2-6}{3-1} = \frac{-4}{2} = -2$$

Fig. 11.6

Whichever time interval we choose, we shall always find that the average acceleration is -2. The particle has constant acceleration, $a = -2$.

Notice that, although the acceleration is constant, the particle is sometimes slowing down (on the way up), and sometimes speeding up (on the way down).

A graph of y against t looks like a quadratic curve (Fig. 11.6). In fact its equation is

$$y = 8t - t^2$$

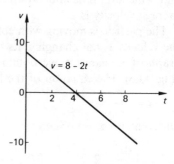

Differentiating, we obtain

$$v = \frac{dy}{dt} = 8 - 2t$$

$$a = \frac{dv}{dt} = -2$$

Fig. 11.7

which confirms our previous work.

Notice that the graph of v against t (the 'velocity–time graph': Fig. 11.7) is a straight line, whose gradient, -2, is equal to the acceleration. The graph of a against t is shown in Fig. 11.8.

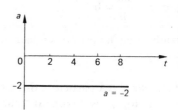

Fig. 11.8

Example 3   A particle is moving along a straight line. After t seconds its distance x metres from a point O on the line is given by $x = t^2 - 5t + 6$. Describe the position and motion of the particle (i) when $t = 2$ and (ii) when $t = 5$.

We are given that

$$x = t^2 - 5t + 6$$

Differentiating,

$$v = \frac{dx}{dt} = 2t - 5$$

and $a = \dfrac{dv}{dt} = 2$

(i)   When $t = 2$, substituting into the three formulae above, we have $x = 0$, $v = -1$, $a = 2$.

The particle is at O (since $x = 0$), and is moving with speed $1\,\text{m s}^{-1}$ from right to left (since v is negative).

Since a is positive, v is increasing. However, $v = -1$, and so its numerical value is decreasing; the particle is slowing down.

(ii)   When $t = 5$, we have $x = 6$, $v = 5$, $a = 2$.

The particle is 6 m to the right of O (since x is positive), and is moving with speed $5\,\text{m s}^{-1}$ from left to right (since v is positive).

Since a is positive, v is increasing; the particle is speeding up, with an acceleration of $2\,\text{m s}^{-2}$.

Example 4   A particle moves along a straight line, and its displacement x cm from a fixed point O after t seconds is given by $x = 30 \sin t$. Describe the motion.

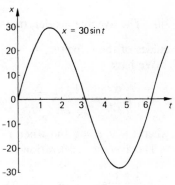

Fig. 11.9

We have

$$x = 30 \sin t$$

and thus

$$v = \frac{dx}{dt} = 30 \cos t$$

and

$$a = \frac{dv}{dt} = -30 \sin t$$

Graphs of x and v are shown in Figs 11.9 and 11.10 respectively.

The particle starts at O, moving at 30 cm s^{-1} from left to right. It comes to rest when $t = \frac{1}{2}\pi$, at a distance of 30 cm from O. It then moves from right to left until $t = \frac{3}{2}\pi$, when it is 30 cm to the left of O. It then moves from left to right. The motion of the particle is shown in Fig. 11.11.

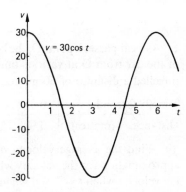

Fig. 11.10

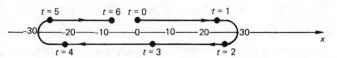

Fig. 11.11

When $t = 2\pi$ (≈ 6.28), the particle is again at O and moving at $30\,\text{cm s}^{-1}$ from left to right. The motion between $t = 2\pi$ and $t = 4\pi$ will be exactly the same as that during the first 2π seconds, and so on. We say that the motion is **periodic**, and the **period** is 2π seconds. The particle is **oscillating** between two points 30 cm on each side of O.

Notice that the acceleration, $a = -30 \sin t$, is always equal to $-x$; the displacement x satisfies the differential equation $\frac{d^2 x}{dt^2} = -x$. We shall study such oscillations later (see p. 165).

Example 5   The distance x metres travelled by a particle in t seconds is given by $x = 4\sqrt{t}$. For the interval of time from $t = 9$ to $t = 25$ find (i) the average velocity; (ii) the average acceleration.

(i)   The average velocity is $\dfrac{\text{distance travelled}}{\text{time taken}}$, so we need the positions of the particle when $t = 9$ and $t = 25$. When $t = 9$, $x = 4 \times 3 = 12$, and when $t = 25$, $x = 4 \times 5 = 20$. The average velocity is

$$\frac{20 - 12}{25 - 9} = \frac{8}{16} = \frac{1}{2} \, \mathrm{m\,s^{-1}}$$

(ii)   The average acceleration is $\dfrac{\text{change in velocity}}{\text{time taken}}$, so we need values of the velocity.
   We have

$$v = \frac{dx}{dt} = \frac{4}{2\sqrt{t}} = \frac{2}{\sqrt{t}}$$

when $t = 9$, $v = \frac{2}{3}$ and when $t = 25$, $v = \frac{2}{5}$.
   The average acceleration is

$$\frac{\frac{2}{5} - \frac{2}{3}}{25 - 9} = \frac{-\frac{4}{15}}{16} = -\frac{1}{60} \, \mathrm{m\,s^{-2}}$$

Exercise 11.1   (Motion with displacement given)

1   As a car passes a point O, the brakes are applied. The following table shows the distance (x metres) of the car from O at various times (t seconds) later. The car comes to rest after 4 seconds, having travelled a distance of 60 metres.

Time, t (seconds)	0	0.5	1	1.5	2	2.5	3	3.5	4
Distance, x (metres)	0	15	27	37	45	51	56	59	60

(i)   Find the average velocity of the car between $t = 0.5$ and $t = 1.5$. Take this value to be an approximation for the velocity when $t = 1$, and use a similar method to find approximate values for the velocity when $t = 2$ and when $t = 3$. Using these values, find the average acceleration of the car between $t = 1$ and $t = 3$.

(ii)   Draw an accurate graph of x against t. By measuring the gradient of a tangent, estimate the velocity when $t = 2$. Estimate the time taken to travel 30 metres, and hence find the average velocity while travelling the first 30 metres, and the average velocity while travelling the last 30 metres.

2   A stone is thrown vertically upwards, and after t seconds, its height, y metres, is given by $y = 50t - 5t^2$.

(i)   Make a table of values for y when $t = 0, 1, 2, \ldots$, until the stone reaches the ground again. How long is the stone in the air? Describe the motion.

(ii)   Find the average velocity of the stone while it is travelling upwards.

(iii)   By differentiation, find formulae for its velocity and acceleration in terms of t. Find the height, velocity and acceleration of the stone after 2 seconds, and after 7 seconds. In each case state whether the stone is moving up or down, and whether its speed is increasing or decreasing.

3 A particle is moving along a straight line. After t seconds, its distance x metres to the right of a fixed point O on the line is given by $x = t^3 - 8t^2 + 15t$. Describe the position and motion of the particle (i) when $t = 1$ and (ii) when $t = 4$.

In questions 4 to 7 the displacement (x or y metres) of a particle moving along a straight line is given in terms of time (t seconds). Sketch graphs of the displacement, velocity and acceleration against t, and describe the motion (for positive values of t).

4 $x = 24 - 3t$ (the positive direction is from left to right).

5 $y = 2t^2 - 8t + 6$ (the positive direction is upwards).

6 $x = 3 - 24t + 15t^2 - 2t^3$ (the positive direction is from left to right).

7 $y = t^3 - 6t^2 + 12t - 2$ (the positive direction is upwards).

8 The distance x of a particle from a fixed point O on a straight line, at time t, is given by $x = 10 \cos \pi t$.
  (i)   Find the displacement (x) and the velocity (v) when $t = 0, \frac{1}{3}, \frac{2}{3}, 1, \frac{3}{2}, 2$. Describe the motion, and sketch graphs of x and v against t.
  (ii) Express the acceleration in terms of the displacement (x).
  (iii) Find the maximum velocity and the maximum acceleration of the particle.

9 A particle is moving in a straight line so that its distance from a fixed point O on the line is $(12t - t^3)$ cm after t seconds.
  (i)   Find the distance of the particle from O when it comes (instantaneously) to rest.
  (ii) Find the velocity of the particle when it returns to O.

10 The distance x metres of a particle from a fixed point O on a straight line is given by $x = \dfrac{20t}{t+1}$, where t is the time in seconds.
  (a) Find the position and acceleration of the particle when its velocity is $5 \, \mathrm{m \, s^{-1}}$.
  (b) For the time interval from $t = 3$ to $t = 9$ find (i)   the distance travelled; (ii)   the average velocity; (iii)   the average acceleration.

11.2 Application of integration

If we know the displacement (x) of a particle in terms of time (t), we have been able to find the velocity (v) and acceleration (a) by differentiating, since $v = \dfrac{dx}{dt}$ and $a = \dfrac{dv}{dt}$. However, in practice we are more likely to know something about the acceleration, because the acceleration of a particle is proportional to the force acting upon it. If there are no forces acting on the particle, its acceleration will be zero, and it will move with constant velocity. If there is a constant force acting on the particle (a common situation), then it will move with constant acceleration.

   Suppose now that we know the acceleration a in terms of t. Since $\dfrac{dv}{dt} = a$, we can find the velocity v by integrating, $v = \int a \, dt$, but this introduces an arbitrary constant of integration. To determine the constant we need one extra piece of information, such as the velocity of the particle at the start of the motion.

   Similarly, if we know the velocity v we can find the displacement

x by integrating, since $x = \int v \, dt$, and again we need a further piece of information (such as the position of the particle at the start of the motion) in order to determine the constant of integration.

Once we have found x, v and a in terms of t, we can investigate the motion in the same way as before.

Example 1   A particle is moving along a straight line with constant acceleration. Initially the particle is at a point O and has a velocity of $8 \, \text{cm s}^{-1}$. 4 seconds later its velocity is $14 \, \text{cm s}^{-1}$. Find the time taken for the particle to travel a distance of 460 cm from O, and find its velocity after this time.

Since the acceleration is constant, we can find its value as the average acceleration over the first 4 seconds. This is

$$\frac{\text{change in velocity}}{\text{time taken}} = \frac{14 - 8}{4} = \frac{3}{2} \, \text{cm s}^{-2}$$

Thus $\dfrac{dv}{dt} = \dfrac{3}{2}$. Integrating,

$$v = \tfrac{3}{2}t + c$$

We are given that, when $t = 0$, $v = 8$, so $8 = 0 + c$, i.e. $c = 8$. Hence

$$v = \tfrac{3}{2}t + 8$$

Integrating again,

$$x = \int v \, dt = \tfrac{3}{4}t^2 + 8t + k$$

When $t = 0$, $x = 0$ (since the particle is initially at O), and so $0 = 0 + 0 + k$, i.e. $k = 0$. Thus

$$x = \tfrac{3}{4}t^2 + 8t$$

The particle is 460 cm from O when $x = 460$, i.e.

$$\tfrac{3}{4}t^2 + 8t = 460$$
i.e. $3t^2 + 32t - 1840 = 0$
i.e. $(t - 20)(3t + 92) = 0$
i.e. $t = 20$ or $t = -\tfrac{92}{3}$

The particle takes 20 seconds to travel a distance of 460 cm (the negative value of t is not relevant to the problem). To find its velocity, we substitute $t = 20$ into the formula $v = \tfrac{3}{2}t + 8$, obtaining $v = \tfrac{3}{2} \times 20 + 8 = 38$.

The velocity of the particle after 20 seconds is $38 \, \text{cm s}^{-1}$.

Example 2   A particle moving in a straight line starts at O with a velocity of $18 \, \text{m s}^{-1}$. After t seconds, its acceleration $a \, \text{m s}^{-2}$ is given by $a = 3 - 2t$. Find the position where the particle first comes to rest.

We have $a = 3 - 2t$. Integrating,

$$v = \int a \, dt = 3t - t^2 + c$$

We are given that when $t = 0$, $v = 18$, so $18 = 0 - 0 + c$, i.e. $c = 18$.
Thus
$$v = 3t - t^2 + 18$$

Integrating again,

$$x = \int v \, dt = \tfrac{3}{2}t^2 - \tfrac{1}{3}t^3 + 18t + k$$

When $t = 0$, $x = 0$ (since the particle starts at O), and so
$0 = 0 - 0 + 0 + k$, i.e. $k = 0$. Thus

$$x = \tfrac{3}{2}t^2 - \tfrac{1}{3}t^3 + 18t$$

The particle is at rest when $v = 0$, i.e.

$$3t - t^2 + 18 = 0$$
i.e. $\quad t^2 - 3t - 18 = 0$
i.e. $\quad (t - 6)(t + 3) = 0$
i.e. $\qquad\qquad t = 6 \quad (\text{or } t = -3)$

The particle first comes to rest when $t = 6$. To find the position, we
substitute into the formula for x, obtaining

$$x = \tfrac{3}{2} \times 36 - \tfrac{1}{3} \times 216 + 18 \times 6 = 90$$

The particle therefore comes momentarily to rest at a distance of
90 m from O (on the positive side).

Distance travelled

We often need to find the distance moved by a particle between two
given times, say between $t = t_1$ and $t = t_2$. Suppose that we know
the velocity v in terms of t, but we may have no information about
the actual position on the line, so we cannot find the arbitrary
constant in the integral $x = \int v \, dt$. To find the distance moved,
which is

(value of x when $t = t_2$) − (value of x when $t = t_1$)

we substitute $t = t_2$ and $t = t_1$ into the integral, and subtract the
resulting values. This is exactly what we do to evaluate the definite
integral $\displaystyle\int_{t_1}^{t_2} v \, dt$, and as we have seen (p. 72), the arbitrary constant
will cancel out. Hence, the distance moved by the particle between
$t = t_1$ and $t = t_2$ is

$$\int_{t_1}^{t_2} v \, dt$$

We can derive this important result by another method.
   Consider a small interval of time from t to $t + \delta t$. If the velocity at
time t is v, the distance travelled during this short interval is
approximately velocity × time = $v \, \delta t$. The total distance travelled

between $t = t_1$ and $t = t_2$ is therefore

$$\sum_{t=t_1}^{t=t_2} v \, \delta t$$

Taking the limit as $\delta t \to 0$, the distance travelled is

$$\lim_{\delta t \to 0} \sum_{t=t_1}^{t=t_2} v \, \delta t$$

which is $\displaystyle \int_{t_1}^{t_2} v \, dt$

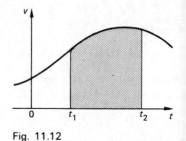

Fig. 11.12

by the definition of a definite integral as the limit of a sum.

It follows that the distance moved can be represented by the area under the velocity–time curve (Fig. 11.12).

The average velocity between $t = t_1$ and $t = t_2$ is

$$\frac{\text{distance travelled}}{\text{time taken}} = \frac{1}{t_2 - t_1} \int_{t_1}^{t_2} v \, dt$$

This is in agreement with our previous work on average values (see p. 82).

Similarly we can calculate the average acceleration between $t = t_1$ and $t = t_2$ as

$$\frac{1}{t_2 - t_1} \int_{t_1}^{t_2} a \, dt$$

and the average displacement as

$$\frac{1}{t_2 - t_1} \int_{t_1}^{t_2} x \, dt$$

Example 1   A particle moving with constant retardation slows down from $30 \, \text{m s}^{-1}$ to $20 \, \text{m s}^{-1}$ in a distance of $200 \, \text{m}$. Find the deceleration. (Note that a negative acceleration may be called a deceleration or a retardation.)

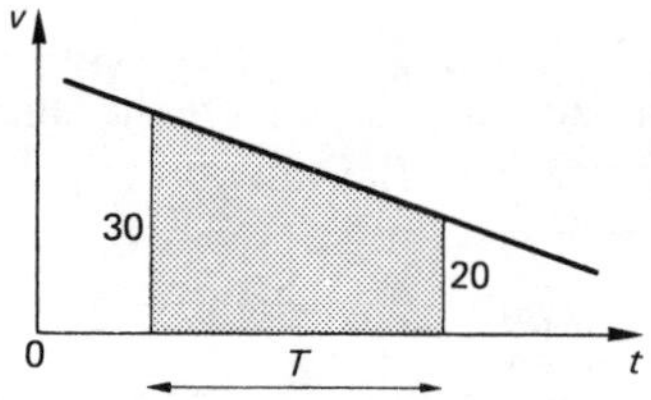

We consider the velocity–time graph (Fig. 11.13). Since the retardation is constant, this is a straight line.

Suppose that the particle takes T seconds to slow down from $30 \, \text{m s}^{-1}$ to $20 \, \text{m s}^{-1}$. We know that the area under this part of the graph is equal to 200 (the distance travelled). Thus

Fig. 11.13

$$\tfrac{1}{2}(30 + 20) \times T = 200$$
i.e. $25T = 200$
i.e. $T = 8$

The acceleration $a = \dfrac{dv}{dt}$ is the gradient of the line, thus

$$a = \frac{20 - 30}{T} = \frac{-10}{8} = -1.25$$

Hence the deceleration of the particle is $1.25 \, \text{m s}^{-2}$.

Example 2 After t seconds, the velocity $v\,\mathrm{m\,s^{-1}}$ of a particle moving in a straight line is given by $v = 10 - 3\sqrt{t}$. Find the distance moved by the particle in the period between 1 second and 9 seconds after the start, and the average velocity during this period.

Between $t = 1$ and $t = 9$, the distance moved is

$$\int_1^9 v\,dt = \int_1^9 (10 - 3\sqrt{t})\,dt = \int_1^9 (10 - 3t^{\frac{1}{2}})\,dt$$

$$= \left[10t - 2t^{\frac{3}{2}} \right]_1^9 = (90 - 54) - (10 - 2) = 28 \text{ m}$$

The average velocity during this period is

$$\frac{1}{9-1}\int_1^9 v\,dt = \frac{1}{8} \times 28 = 3.5 \text{ m s}^{-1}$$

Notice that the 'distance moved', given by $\int_{t_1}^{t_2} v\,dt$, is really the difference between the finishing and starting positions of the particle; its value is not affected by the exact way in which the particle has moved in between. It could be described more accurately as the change in position of the particle.

If v takes both positive and negative values during the motion (so that the particle moves to and fro), the integral $\int_{t_1}^{t_2} v\,dt$ will *not* give the total distance actually travelled by the particle.

Example 3 A particle starts from rest, and moves in a straight line. After t seconds, its acceleration $a\,\mathrm{m\,s^{-2}}$ is given by $a = 3t^2 - 6t + 2$. Find the total distance actually travelled by the particle during the first three seconds.

We have $\quad a = 3t^2 - 6t + 2$

Thus $\quad v = \int a\,dt = t^3 - 3t^2 + 2t + c$

When $t = 0$, $v = 0$ (since the particle starts from rest), and so $0 = 0 - 0 + 0 + c$, i.e. $c = 0$. Hence

$$v = t^3 - 3t^2 + 2t$$

The change in position during the first 3 seconds is

$$\int_0^3 v\,dt = \int_0^3 (t^3 - 3t^2 + 2t)\,dt$$

$$= \left[\frac{1}{4}t^4 - t^3 + t^2 \right]_0^3 = \left(\frac{81}{4} - 27 + 9 \right) - (0) = \frac{9}{4}$$

This means that after 3 seconds, the particle is at a distance of $\frac{9}{4}$ m from its starting point.

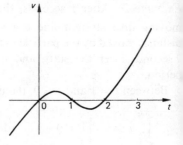

Fig. 11.14

However, the velocity v is zero when

$$t^3 - 3t^2 + 2t = 0$$

i.e. $t(t^2 - 3t + 2) = 0$

i.e. $t(t - 1)(t - 2) = 0$

i.e. $t = 0, 1$ or 2

When we draw the graph of v (Fig. 11.14) we see that v takes both positive and negative values between $t = 0$ and $t = 3$. We consider these separately.

Between $t = 0$ and $t = 1$, the distance moved is

$$\int_0^1 (t^3 - 3t^2 + 2t)\, dt = \left[\frac{1}{4}t^4 - t^3 + t^2 \right]_0^1$$

$$= (\tfrac{1}{4} - 1 + 1) - (0) = \tfrac{1}{4}$$

Between $t = 1$ and $t = 2$, the distance moved is

$$\int_1^2 (t^3 - 3t^2 + 2t)\, dt = \left[\frac{1}{4}t^4 - t^3 + t^2 \right]_1^2$$

$$= (4 - 8 + 4) - (\tfrac{1}{4} - 1 + 1) = -\tfrac{1}{4}$$

The negative value means that the particle has moved in the negative direction.

Between $t = 2$ and $t = 3$, the distance moved is

$$\int_2^3 (t^3 - 3t^2 + 2t)\, dt = \left[\frac{1}{4}t^4 - t^3 + t^2 \right]_2^3$$

$$= \left(\frac{81}{4} - 27 + 9 \right) - (4 - 8 + 4) = \frac{9}{4}$$

Assuming that the positive direction is from left to right, the particle moves a distance $\frac{1}{4}$ m to the right, then $\frac{1}{4}$ m to the left, and then $\frac{9}{4}$ m to the right. The total distance travelled is therefore $\frac{1}{4} + \frac{1}{4} + \frac{9}{4} = \frac{11}{4}$ m.

Exercise 11.2   (Motion with acceleration or velocity given)

1   A particle travels in a straight line through a point O, passing through O at zero time. Its velocity is $(6t^2 - 16t + 6)$ m s^{-1} at time t seconds. Find its acceleration and its distance from O at time t seconds. Find also the times at which the particle passes through O.   (O and C)

2   If a particle moves in a straight line with acceleration $(2t + 1)$ m s^{-2} where t is the time in seconds, find the velocity acquired 2 seconds after starting from rest, and the distance travelled in that time.   (O and C)

3   The speed v m s^{-1} of a train, t seconds after leaving a station, is given by $v = 2t - 0.03t^2$. Find how soon the train reaches its maximum speed, and how far it travels in the first minute.   (O and C)

4 Three points A, B and C in a straight line are such that $AB = 25$ cm and $BC = 135$ cm. A particle moving with constant acceleration in the line passes A with a velocity of $20 \, \text{cm} \, \text{s}^{-1}$ and B with a velocity of $30 \, \text{cm} \, \text{s}^{-1}$. Calculate its acceleration, and the number of seconds it takes to travel from B to C. (MEI)

5 A particle moves along a straight line with constant acceleration a. At time $t = 0$, it is passing through a point O with velocity u. If x is its distance from O, and v its velocity, after time t show by integration that (i) $v = u + at$ (ii) $x = ut + \frac{1}{2}at^2$. By eliminating a, deduce that (iii) $x = \frac{1}{2}(u + v)t$. By eliminating t between (i) and (iii), show that (iv) $v^2 = u^2 + 2ax$. (These formulae are used frequently in mechanics and physics, but they apply only when the acceleration is constant.)

6 A particle is moving along a straight line with a velocity $v \, \text{m} \, \text{s}^{-1}$ given by $v = 3t^2 - t^3$, where t is the time in seconds that has elapsed since the particle was at a fixed point O on the given line. At what time after leaving O does the particle first come to instantaneous rest?

If the particle is then at P, calculate:
(i)   the distance OP;
(ii) the average speed between O and P;
(iii) the maximum acceleration between O and P.

7 A particle moves in a straight line, and at time $t = 0$ it is passing through a point O with velocity $2 \, \text{m} \, \text{s}^{-1}$. It experiences a varying force, which at time t gives it an acceleration of $(2 \sin 3t) \, \text{m} \, \text{s}^{-2}$. Find an expression for the distance of the particle from O after time t. Hence find the position and velocity of the particle after 10 seconds.

8 A particle starts from rest at a point O. Its speed, $v \, \text{m} \, \text{s}^{-1}$, t seconds later is given by the formula $v = 10\sqrt{t}$. Calculate:
(i)   the distance travelled by the particle in 9 seconds;
(ii) the average velocity in the ninth second;
(iii) the acceleration after nine seconds;
(iv) the average acceleration in the ninth second.

9 A particle moves in a straight line so that at time t, measured in seconds, the acceleration, in $\text{m} \, \text{s}^{-2}$, of the particle is $(2 + 6t)$. The particle travels 11 m in the interval from $t = 1$ to $t = 2$. Calculate the speed of the particle when $t = 0$. (L)

10 The velocity $v \, \text{m} \, \text{s}^{-1}$, at time t seconds, of a particle is given by $v = \dfrac{10}{t+1}$. Make a table of values of v for $t = 0, 1, 2, \ldots, 6$, and use the trapezium rule to estimate the distance travelled by the particle in the first 6 seconds.

11.3 Motion in stages

Most journeys are complicated, and it is unlikely that a single formula could give the displacement at all times. We can consider more realistic journeys by dividing the motion into several stages, with different formulae applying to each stage.

We usually illustrate such a problem by drawing the velocity–time graph, because most aspects of the motion can easily be derived from it: the height of the curve gives the velocity (and the particle is at rest when the curve crosses the t-axis); the gradient of the curve gives the acceleration; and the area under the curve gives the distance travelled. For any stage in which the acceleration is constant, the velocity–time graph is a straight line.

Example   A car starting from rest moves in a straight line for 5 seconds with an acceleration of $(10 - 2t)\,\mathrm{m\,s^{-2}}$, where t is the time in seconds. Calculate its speed at the end of the 5 seconds and the distance it has gone in this time.

It then continues to move at this speed until such time as its average speed, from the start, has attained the value of $20\,\mathrm{m\,s^{-1}}$. The brakes are then applied and it is brought to rest with constant retardation in 10 seconds. Calculate the total distance gone. (O)

Stage 1   (for the first 5 seconds)   We are given that the acceleration is $a = 10 - 2t$. Integrating,

$$v = 10t - t^2 + c$$

When $t = 0$, $v = 0$ (since the car starts from rest), and so $c = 0$. Thus $v = 10t - t^2$.

To find the speed after 5 seconds, we substitute $t = 5$, then $v = 50 - 25 = 25\,\mathrm{m\,s^{-1}}$.

The distance travelled in the first 5 seconds is

$$\int_0^5 v\,dt = \int_0^5 (10t - t^2)\,dt$$

$$= \left[5t^2 - \frac{1}{3}t^3\right]_0^5 = \left(125 - \frac{125}{3}\right) - 0$$

$$= \frac{250}{3} = 83\tfrac{1}{3}\,\mathrm{m}$$

Stage 2   The car now moves at a constant velocity of $25\,\mathrm{m\,s^{-1}}$. This continues until the average velocity reaches $20\,\mathrm{m\,s^{-1}}$ (Fig. 11.15).

If the car moves with constant velocity for T seconds, it covers a distance of $25T$ metres; the total distance travelled from the start is $\frac{250}{3} + 25T$ metres; and the total time since the start is $5 + T$ seconds.

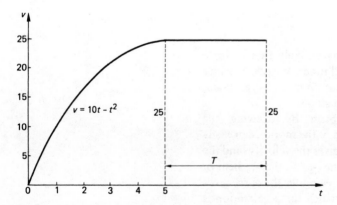

Fig. 11.15

The average velocity is $20\ \mathrm{m\,s^{-1}}$ when

$$\frac{\dfrac{250}{3} + 25T}{5 + T} = 20$$

i.e. $\qquad \dfrac{250}{3} + 25T = 100 + 20T$

i.e. $\qquad 5T = \dfrac{50}{3}$

i.e. $\qquad T = \dfrac{10}{3} = 3\tfrac{1}{3}$

The car moves with constant velocity for $3\tfrac{1}{3}$ seconds, and covers a distance of $25 \times \frac{10}{3} = 83\tfrac{1}{3}$ metres in this time.

Stage 3 The car decelerates uniformly from $25\ \mathrm{m\,s^{-1}}$ to rest in 10 seconds.

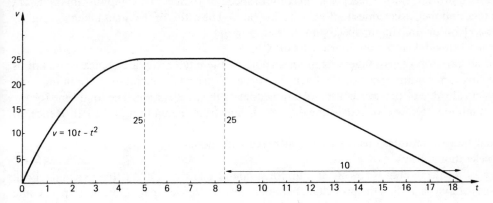

Fig. 11.16

The velocity–time graph here is a straight line (see Fig. 11.16), and the distance travelled in decelerating is the area under this section, i.e.

$\tfrac{1}{2} \times 10 \times 25 = 125\ \mathrm{m}$

The total distance travelled by the car is thus

$83\tfrac{1}{3} + 83\tfrac{1}{3} + 125 = 291\tfrac{2}{3}\ \mathrm{m}$

Do not forget that the car is moving along a straight line. A series of photographs taken at one second intervals would show the actual motion as in Fig. 11.17.

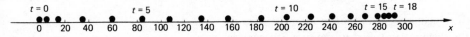

Fig. 11.17

Exercise 11.3   (Motion in stages)

1   A car sets off from rest on a straight road with a constant acceleration of $2 \, \mathrm{m \, s^{-2}}$ which continues until the car reaches a velocity of $72 \, \mathrm{km \, h^{-1}}$, which velocity it then maintains. How far does it travel before it reaches this velocity and what is its average velocity, in $\mathrm{km \, h^{-1}}$, for the first half minute of its motion? Find also its average acceleration in $\mathrm{m \, s^{-2}}$ over this period.

Some time later the driver has to make an emergency stop and brings the car to rest with uniform retardation in a distance of 40 m. Calculate the retardation, in $\mathrm{m \, s^{-2}}$, and the time taken to come to rest.   (O)

2   A saloon car travels with constant speed $15 \, \mathrm{m \, s^{-1}}$. Just as it is passing a sports car which is at rest, the sports car starts to accelerate at a uniform rate of $2 \, \mathrm{m \, s^{-2}}$ for 12 s, and then continues to travel at the constant speed it has reached after 12 s. Calculate:
  (i)   the distance travelled by the saloon car in the t seconds after it overtakes the sports car;
  (ii)   the distance travelled by the sports car during the same t seconds, (a) for $t \leqslant 12$ and
       (b) for $t > 12$;
  (iii)   the value of t when the sports car overtakes the saloon car;
  (iv)   the distance they have both travelled in that time.   (O)

3   The distance between two underground railway stations X and Y is 450 m. A train starts from X and accelerates uniformly until it reaches its maximum speed of $10 \, \mathrm{m \, s^{-1}}$. It maintains this constant speed for a time and then decelerates uniformly to stop at Y. The ratios of the times taken during the periods of acceleration, maximum speed and deceleration are 2:3:1.

Find the total time taken for the journey from X to Y.   (C)

4   A monorail car to take passengers from an airport to the city centre, a distance of 12 km, is designed to have a maximum speed of $120 \, \mathrm{km \, h^{-1}}$. Acceleration and deceleration are both uniform, and the numerical value of the deceleration is double that of the acceleration. The total time for the journey is $7\frac{1}{2}$ minutes. Draw a velocity–time diagram to illustrate the motion of the car. Hence, or otherwise, find
  (i)   for what time, in minutes, the car travels at maximum speed;
  (ii)   the acceleration, in $\mathrm{m \, s^{-2}}$.   (MEI)

5   The crew of a rowing boat is rowing at a uniform rate of one stroke every three seconds. Each stroke consists of two parts, first the propelling part and then the recovery part. During the propelling part of each stroke the velocity $v \, \mathrm{m \, s^{-1}}$ of the boat is given by the formula
$$v = 2 + 4t - 2t^2$$
where t is the time in seconds measured from the beginning of the stroke until the time that the velocity reaches a maximum. After this time the boat retards uniformly for the recovery part of the stroke until the velocity reaches $2 \, \mathrm{m \, s^{-1}}$. The stroke is then repeated.
  (i)   Calculate the times for the propelling and the recovery parts of each stroke.
  (ii)   Sketch the velocity–time graph for the boat for a period of nine seconds.
  (iii)   Calculate the distance travelled by the boat during (a) the propelling and (b) the recovery part of
       each stroke.
  (iv)   Calculate the average velocity of the boat.   (O)

6   An electric train accelerates from rest for 30 seconds at a rate $\left(1 - \dfrac{t}{30}\right) \mathrm{m s^{-2}}$, where t is the time from starting measured in seconds. It then travels at constant speed for 2 minutes and decelerates at a constant rate of $0.75 \, \mathrm{m \, s^{-2}}$ to rest at the next station. It waits 30 seconds at that station before starting again.
  (a)   Calculate the constant speed attained after 30 seconds.
  (b)   Calculate the distance travelled in the first 30 seconds.
  (c)   Calculate the time and distance involved in decelerating.
  (d)   Calculate the average speed for the whole cycle.   (O)

11.4 Oscillations

The force acting on a particle often depends on where the particle is, in which case the acceleration can be expressed in terms of the position of the particle. We shall consider a particle moving along a straight line such that its acceleration is proportional to its distance from a fixed point O on the line, and the acceleration is always directed towards O. This is called **simple harmonic motion**. It is important because it occurs frequently (for example, a mass moving on the end of a spring, the bob of a pendulum, and the vibrating ends of a tuning fork).

If x is the distance of the particle from O at time t, then the acceleration a is proportional to x; we have $a = -\omega^2 x$, where ω is a constant (the negative sign ensures that the acceleration is always directly towards O).

Thus

$$\frac{d^2 x}{dt^2} = -\omega^2 x$$

We cannot simply integrate this, because the acceleration is given in terms of x (instead of t). However, we know (see p. 140) that the general solution of this differential equation is

$$x = A \sin \omega t + B \cos \omega t$$

We have also seen (on p. 153, Example 4) that $x = 30 \sin t$ gives an oscillatory motion between two points at a distance of 30 on each side of O, and the period for one complete oscillation is 2π.

Similarly, if $x = A \sin \omega t$, and hence the velocity is $v = A\omega \cos \omega t$, the particle is oscillating between the points $x = A$ and $x = -A$ (see Fig. 11.18). The time for the first complete oscillation is given by $0 \leqslant \omega t \leqslant 2\pi$, i.e. $0 \leqslant t \leqslant \dfrac{2\pi}{\omega}$, so the **period** of oscillations is $\dfrac{2\pi}{\omega}$.

When $t = 0$, $x = 0$ and $v = A\omega$, so the particle is passing through O with its maximum velocity of $A\omega$.

O is called the **centre of oscillation**. The end-points, $x = A$ and $x = -A$, are called the **extremes** of the motion. The maximum distance, A, of the particle from the centre, is called the **amplitude** (see Fig. 11.19).

Now consider the motion given by $x = B \cos \omega t$ and thus $v = -B\omega \sin \omega t$.

The particle is oscillating with period $\dfrac{2\pi}{\omega}$. The centre of oscillation is O, and the amplitude is B (see Fig. 11.20).

When $t = 0$, $x = B$ and $v = 0$, so the particle is instantaneously at rest at an extreme of the motion (see Fig. 11.21).

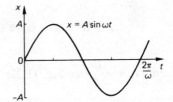

Fig. 11.18

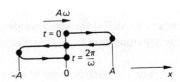

Fig. 11.19

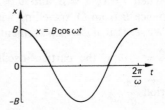

Fig. 11.20

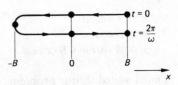

Fig. 11.21

We now consider the most general motion which satisfies

$$\frac{d^2x}{dt^2} = -\omega^2 x, \text{ i.e.}$$

$$x = A \sin \omega t + B \cos \omega t$$

and thus

$$v = A\omega \cos \omega t - B\omega \sin \omega t$$

We may write

$$x = A \sin \omega t + B \cos \omega t$$

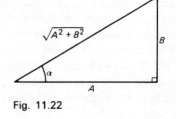

Fig. 11.22

$$= \sqrt{A^2 + B^2} \left[(\sin \omega t) \left(\frac{A}{\sqrt{A^2 + B^2}} \right) + (\cos \omega t) \left(\frac{B}{\sqrt{A^2 + B^2}} \right) \right]$$

$$= \sqrt{A^2 + B^2} (\sin \omega t \cos \alpha + \cos \omega t \sin \alpha)$$

where $\tan \alpha = \dfrac{B}{A}$ (see Fig. 11.22)

$$= \sqrt{A^2 + B^2} \sin (\omega t + \alpha)$$

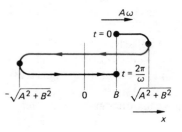

Fig. 11.23

(where α is called the **phase angle**) and so the graph of x is similar to the previous two cases (see Fig. 11.23).

Again, the particle is oscillating with period $\dfrac{2\pi}{\omega}$ about the centre O. The amplitude is now $\sqrt{A^2 + B^2}$.

When $t = 0$, $x = B$ and $v = A\omega$, so the particle starts at a distance B from O, travelling with velocity $A\omega$ (see Fig. 11.24).

We see that, if the displacement satisfies the differential equation $\dfrac{d^2x}{dt^2} = -\omega^2 x$, the particle oscillates about the centre O, with period $\dfrac{2\pi}{\omega}$.

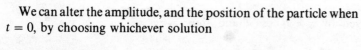

Fig. 11.24

We can alter the amplitude, and the position of the particle when $t = 0$, by choosing whichever solution

$$x = A \sin \omega t$$
$$x = B \cos \omega t$$
$$x = A \sin \omega t + B \cos \omega t$$

is most suited to our problem.

Example 1 A particle is executing simple harmonic motion, making 5 oscillations per second. The maximum speed of the particle during the motion is 400 cm s^{-1}. Find (i) the amplitude of the oscillations; (ii) the time taken for the particle to accelerate from rest at an extreme position to a speed of 200 cm s^{-1}, and the distance travelled in this time.

We are free to measure time from any point we choose. As we have to find a time from an extreme position, we choose so that the particle is at an extreme position when $t = 0$; i.e. let $x = B \cos \omega t$, then $v = -B\omega \sin \omega t$.

The time for one oscillation is $\frac{1}{5}$ seconds; thus $\frac{2\pi}{\omega} = \frac{1}{5}$, i.e. $\omega = 10\pi$. The maximum speed is $B\omega$ (when $\sin \omega t = \pm 1$), thus $B\omega = 400$; i.e. $B = \frac{400}{10\pi} = \frac{40}{\pi}$. Hence

$$x = \frac{40}{\pi} \cos(10\pi t) \quad \text{and} \quad v = -400 \sin(10\pi t)$$

(i) The amplitude is $\frac{40}{\pi} \approx 12.7 \text{ cm}$.

(ii) When the particle moves from its extreme position at $t = 0$, its velocity $v = -400 \sin(10\pi t)$ is negative. We require the first value of t for which $v = -200$,

i.e. $-400 \sin(10\pi t) = -200$

i.e. $\sin(10\pi t) = \frac{1}{2}$

i.e. $10\pi t = \frac{1}{6}\pi$

i.e. $t = \dfrac{1}{60}$

The particle takes $\dfrac{1}{60}$ s to accelerate from rest to 200 cm s^{-1}.

To find the distance travelled, we substitute into the expression $x = \frac{40}{\pi}\cos(10\pi t)$.

When $t = 0$,

$$x = \frac{40}{\pi} \approx 12.73$$

When $t = \dfrac{1}{60}$,

$$x = \frac{40}{\pi} \cos(\tfrac{1}{6}\pi) = \frac{40}{\pi} \times \frac{\sqrt{3}}{2} \approx 11.03$$

The distance travelled is $12.73 - 11.03 = 1.7 \text{ cm}$.

Example 2   The displacement x cm of a particle at time t seconds satisfies the equation $\dfrac{d^2x}{dt^2} = -4x$. When $t = \tfrac{1}{4}\pi$, the particle is 5 cm to the right of O, and is moving towards O at $3\,\mathrm{cm\,s^{-1}}$. Find x in terms of t.

Calculate (i) the period of the oscillations; (ii) the amplitude of the oscillations; (iii) the position, velocity and acceleration of the particle when $t = 0$.

The equation $\dfrac{d^2x}{dt^2} = -4x$ is of the form $\dfrac{d^2x}{dt^2} = -\omega^2x$, with $\omega = 2$. Since we are given precise initial conditions, we must use the general solution

$$x = A\sin 2t + B\cos 2t$$

and then

$$v = 2A\cos 2t - 2B\sin 2t$$

Now $x = 5$ when $t = \tfrac{1}{4}\pi$, so $5 = A + 0$, i.e. $A = 5$. Also $v = -3$ (since the particle is moving from right to left) when $t = \tfrac{1}{4}\pi$, so $-3 = 0 - 2B$, i.e. $B = \tfrac{3}{2}$. Hence

$$x = 5\sin 2t + \tfrac{3}{2}\cos 2t$$

(i)   For the period, we only need to compare the equation $\dfrac{d^2x}{dt^2} = -4x$ with $\dfrac{d^2x}{dt^2} = -\omega^2x$. The period is $\dfrac{2\pi}{\omega} = \dfrac{2\pi}{2} = \pi$ seconds.

(ii)   The amplitude is

$$\sqrt{A^2 + B^2} = \sqrt{5^2 + (\tfrac{3}{2})^2}$$

$$= \sqrt{\frac{109}{4}} \approx 5.22\ \text{cm}$$

(iii)   We have $x = 5\sin 2t + \tfrac{3}{2}\cos 2t$. When $t = 0$, $x = \tfrac{3}{2}$. Also $v = 10\cos 2t - 3\sin 2t$. When $t = 0$, $v = 10$. The acceleration is

$$a = \frac{d^2x}{dt^2} = -4x. \text{ When } t = 0,\ a = -4 \times \tfrac{3}{2} = -6.$$

The particle is $\tfrac{3}{2}$ cm to the right of O, travelling with velocity $10\ \mathrm{cm\,s^{-1}}$ (from left to right); and its acceleration is $-6\ \mathrm{cm\,s^{-2}}$ (i.e. the particle is slowing down).

Exercise 11.4   (Oscillations)

In questions 1 to 4, the displacement x of an oscillating particle is given in terms of time t. For each question:
 (i)   sketch graphs of the displacement and of the velocity against time;
 (ii) state the amplitude and the period of the oscillations;
 (iii) express the acceleration in terms of the displacement;
 (iv) state the maximum velocity of the particle;
 (v)   state the displacement and the velocity when $t = 0$.

1 $x = 4 \cos t$ **2** $x = 3 + \sin 2t$ **3** $x = 8 \sin (4\pi t + \frac{1}{3}\pi)$ **4** $x = 3 \sin 8t - 4 \cos 8t$

In questions 5 to 10 assume that the particle is executing simple harmonic motion.

5   The amplitude is 40 cm and the period is 6 seconds. Find the position, velocity and acceleration of the particle 1 second after it has passed through the centre of oscillation.

6   The amplitude is 60 mm and the particle is making 12 oscillations per second. Find the maximum velocity and the maximum acceleration.

7   The amplitude is 2 m and the period is 10 seconds. O is the centre of the oscillations. At a certain instant the particle is travelling away from O and is passing through a point Q, which is 1.6 m from O. Find the time that elapses before the particle next returns to Q (travelling back towards O).

8   The amplitude is 50 cm and the maximum speed is 20 cm s^{-1}. Find the period of oscillation. Find the time taken for the speed to decrease from 20 cm s^{-1} to 16 cm s^{-1}, and the distance travelled during this time.

9   The particle takes 2 seconds to travel from an extreme position to a point mid-way between the centre of oscillation and the other end of the motion. Find the period of oscillation.

10   The maximum velocity is 4 m s^{-1}. Find the average velocity of the particle as it moves from one extreme position to the other.

11   If the displacement x of a particle at time t is given by $x = A \sin \omega t$, and its velocity is v, show that $v^2 = \omega^2 (A^2 - x^2)$.

   A particle is executing simple harmonic motion. P is a point 6 cm from the centre of oscillation, and Q is 8 cm from the centre, on the same side as P. The particle has velocity 40 cm s^{-1} when it passes through P, and 30 cm s^{-1} when it passes through Q. Find the amplitude and period of the oscillations. Find also the time taken to travel directly from P to Q.

12   A particle moves along a straight line, and its displacement x from a fixed point O at time t satisfies the differential equation $\dfrac{d^2 x}{dt^2} = -16x$. If $x = 7$ and $\dfrac{dx}{dt} = 96$ when $t = 0$, find x in terms of t, and state the period and amplitude of the oscillations. Find the smallest (positive) value of t for which the particle is at rest.

12

Rates of change

12.1 Dependent and independent variables

We frequently meet equations which express some physical law or relationship between different quantities. For example, the area of a circle, A, is given in terms of the radius, r, by the equation $A = \pi r^2$. We calculate A by substituting a value for r; we say that A **depends** on r. A is said to be the **dependent variable** and r the **independent variable**.

When we draw a graph to illustrate such a relationship, the dependent variable is always placed on the vertical axis, and the independent variable on the horizontal axis. The independent variable will frequently be time.

Rates of change

We can differentiate the equation $A = \pi r^2$ with respect to r, to obtain $\dfrac{dA}{dr} = 2\pi r$.

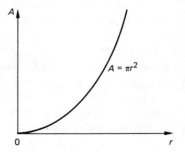

Fig. 12.1

This gives the gradient of the graph of A against r (Fig. 12.1).

Also, if a change in the radius of δr causes a change in the area of δA, then by the definition of the derivative,

$$\frac{dA}{dr} = \lim_{\delta r \to 0} \frac{\delta A}{\delta r} = \lim \left(\frac{\text{change in area}}{\text{change in radius}} \right)$$

This indicates how a change in r affects the value of A, and $\dfrac{dA}{dr}$ is called the **rate of change of A with r**.

In general, if a quantity W depends on another quantity p, $\dfrac{dW}{dp}$ is the rate of change of W with p.

If a change δp in p causes a change δW in W, then we have

$$\frac{dW}{dp} = \lim_{\delta p \to 0} \frac{\delta W}{\delta p} = \lim \left(\frac{\text{change in } W}{\text{change in } p} \right)$$

A change is positive if it is an increase, and negative if it is a decrease. If an increase in p causes a decrease in W, then δW is negative when δp is positive, and so the rate of change $\dfrac{dW}{dp}$ will be negative.

170

The units of a rate of change $\dfrac{\mathrm{d}W}{\mathrm{d}p}$ are 'units of W per unit of p'.

For example, if W is measured in kilograms and p in degrees Centigrade, then the units of $\dfrac{\mathrm{d}W}{\mathrm{d}p}$ are kilograms per degree Centigrade.

Example 1 A circular oil slick is increasing in size. Find the rate of change of its area with the radius when the radius is (i) 10 m (ii) 40 m.

If the radius is r m and the area A m², we have $A = \pi r^2$. Then $\dfrac{\mathrm{d}A}{\mathrm{d}r} = 2\pi r$.

(i)   when $r = 10$, the rate of change is $2\pi \times 10 = 20\pi$ m² per m.
(ii) when $r = 40$, the rate of change is $2\pi \times 40 = 80\pi$ m² per m.

Example 2 When a potential difference of 200 volts is applied across a resistance of R ohms, the current flowing, I amps, is given by $I = \dfrac{200}{R}$. Find the rate of change of current with resistance when the resistance is 40 ohms.

The rate of change is $\dfrac{\mathrm{d}I}{\mathrm{d}R} = -\dfrac{200}{R^2}$. When $R = 40$, $\dfrac{\mathrm{d}I}{\mathrm{d}R} = -\dfrac{200}{1600}$ $= -\frac{1}{8}$ amps per ohm. The negative value indicates that an increase in resistance will cause a decrease in the current.

Application to circumference and surface area

It may seem odd that, for the area of a circle $A = \pi r^2$, the rate of change $\dfrac{\mathrm{d}A}{\mathrm{d}r} = 2\pi r$ is equal to the circumference of the circle.

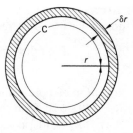

Fig. 12.2

However, consider a small increase of δr in the radius. This changes the area of the circle by adding a thin band to it (shown shaded in Fig. 12.2). This band has width δr and length approximately C, the circumference of the circle (provided that δr is sufficiently small). The increase in the area is

δA = area of band
$\approx C\,\delta r$

Thus $\dfrac{\delta A}{\delta r} \approx C$. Taking the limit as $\delta r \to 0$, the approximation becomes more and more accurate, and we obtain $\dfrac{\mathrm{d}A}{\mathrm{d}r} = C$, the circumference.

We can use a similar method to derive the formula for the surface area of a sphere. Let the radius of a sphere be r, its surface area S, and its volume V.

Consider a small increase of δr in the radius (Fig. 12.3). This changes the volume by adding to it a shell of thickness δr. The inside of the shell has area S, and the outside area is approximately the same, so its volume is approximately $S\,\delta r$.

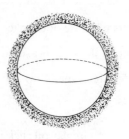

Hence the increase in volume $\delta V \approx S\delta r$, giving $\dfrac{\delta V}{\delta r} \approx S$, and

taking the limit as $\delta r \to 0$, we obtain $\dfrac{\mathrm{d}V}{\mathrm{d}r} = S$.

We have established that the volume of a sphere is $V = \frac{4}{3}\pi r^3$ (see p. 87 where we considered it as a solid of revolution). The surface

area is therefore $S = \dfrac{\mathrm{d}V}{\mathrm{d}r} = \frac{4}{3}\pi(3r^2)$, i.e. $S = 4\pi r^2$.

Fig. 12.3

Exercise 12.1   (Rates of change)

1   A circle of radius r has circumference C and area A. Find the rate of change of A with r when the radius is (i) 25 cm and (ii) 100 km.

Find also the rate of change of C with r for the same values of the radius.

Give the units in each case.

2   A sphere of radius r has surface area S and volume V. Find formulae for the rates of change of S with r, and of V with r, in terms of r. Evaluate both of these when the radius is 6 m, giving the units in each case.

3   The volume V cm^3 of a cylinder of radius r cm and height h cm is given by $V = \pi r^2 h$.

(i)   If the height is constant at 20 cm, find the rate of change of the volume with the radius when the radius is 5 cm.

(ii)   If the radius is constant at 5 cm, find the rate of change of the volume with the height when the height is 20 cm.

4   The force F newtons of attraction between two charged particles placed a distance x metres apart is given by $F = \dfrac{20}{x^2}$. Find the rate of change of the force with the distance when the particles are 2 m apart.

5   The range R m of a projectile fired with speed 80 m s^{-1} at an angle θ radians to the horizontal is given by $R = 640 \sin 2\theta$. Find the rate of change of the range with the angle of projection when this angle is (i) $\frac{1}{8}\pi$ (ii) $\frac{1}{4}\pi$ (iii) $\frac{1}{3}\pi$.

6   When an object is placed at a distance u cm from a lens of focal length 5 cm, the image is v cm from the lens, where $v = \dfrac{5u}{u-5}$. Find the rate of change of image distance with object distance when the object is 15 cm from the lens.

7   A container is in the form of a hollow inverted cone, with perpendicular height 30 cm and the radius at the top is 30 cm. If it contains water to a depth of h cm, find in terms of h (i) the radius of the surface of the water; (ii) the area S cm^2 of the water surface; (iii) the volume V cm^3 of the water.

Find the rate of change of volume with depth. What do you notice?

8   When the depth of water in a container is h, the volume of water is V and the area of the surface of the water is S. Consider a small increase of δh in the depth, and express the corresponding increase in the volume in terms of S and δh. Hence show that $S = \dfrac{\mathrm{d}V}{\mathrm{d}h}$.

For a certain container, when the depth is h cm, the volume of water is $\frac{1}{3}\pi h^2(15 - h)$ cm^3. Find the area of the surface of the water when the depth is 3 cm.

12.2   Maxima and minima

We have already used rates of change to solve problems involving maxima and minima (see p. 43). We now have a wider variety of functions at our disposal, and so we can tackle more complicated problems.

Remember that the quantity to be maximised (or minimised), say W, must first be expressed in terms of one variable only, say p. Then stationary values will occur when $\dfrac{\mathrm{d}W}{\mathrm{d}p} = 0$.

Example 1   A rectangular framework ABCD is stabilised by a strut of length 10 cm joining A to the mid-point of CD. Find the dimensions of the rectangle ABCD when (i) its area is a maximum; (ii) its perimeter is a maximum.

Let the lengths of the sides AD and AB be x cm and y cm (see Fig. 12.4). By Pythagoras' theorem,

$$x^2 + (\tfrac{1}{2}y)^2 = 10^2$$

i.e. $\qquad y = 2\sqrt{100 - x^2}$

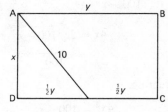

Fig. 12.4

(i)   The area of the rectangle is $S = xy$. Substituting for y, $S = 2x\sqrt{100 - x^2}$. This is now in terms of one variable only.

Rather than differentiate this expression, it is easier to consider S^2. This will have a maximum value when S does.

Let

$$W = S^2 = 4x^2(100 - x^2)$$
$$= 400x^2 - 4x^4$$

Then $\qquad \dfrac{\mathrm{d}W}{\mathrm{d}x} = 800x - 16x^3$

and $\qquad \dfrac{\mathrm{d}^2 W}{\mathrm{d}x^2} = 800 - 48x^2$

For a stationary value, $\dfrac{\mathrm{d}W}{\mathrm{d}x} = 0$, so

$$800x - 16x^3 = 0$$
i.e. $\quad 16x(50 - x^2) = 0$
i.e. $\qquad\qquad x = 0 \quad \text{or} \quad \sqrt{50}$

When $x = 0$, $\dfrac{\mathrm{d}^2 W}{\mathrm{d}x^2} = 800 > 0$, so this gives a minimum. When $x = \sqrt{50}$, $\dfrac{\mathrm{d}^2 W}{\mathrm{d}x^2} = 800 - (48 \times 50) < 0$, so this gives a maximum.

The area is a maximum when $x = \sqrt{50}$, and $y = 2\sqrt{100 - x^2} = 2\sqrt{50}$, i.e. when the dimensions are AD $\approx$ 7.07 cm and AB $\approx$ 14.14 cm.

(ii) This is a separate problem, with a different quantity (the perimeter) to be maximised.

The perimeter of the rectangle ABCD is $P = 2x + 2y$. Substituting $y = 2\sqrt{100 - x^2}$, we have

$$P = 2x + 4\sqrt{100 - x^2}$$

Thus $\dfrac{dP}{dx} = 2 + 4 \times \tfrac{1}{2}(100 - x^2)^{-\frac{1}{2}}(-2x)$

$$= 2 - \frac{4x}{\sqrt{100 - x^2}}$$

For a stationary value, $\dfrac{dP}{dx} = 0$, i.e.

$$\frac{4x}{\sqrt{100 - x^2}} = 2$$

i.e. $4x = 2\sqrt{100 - x^2}$

i.e. $2x = \sqrt{100 - x^2}$

Squaring,

$$4x^2 = 100 - x^2$$

i.e. $5x^2 = 100$

i.e. $x = \sqrt{20}$

Now $\dfrac{d^2P}{dx^2} = \dfrac{d}{dx}\left[2 - 4x(100 - x^2)^{-\frac{1}{2}} \right]$

$$= -4x(-\tfrac{1}{2})(100 - x^2)^{-\frac{3}{2}}(-2x) - 4(100 - x^2)^{-\frac{1}{2}}$$
$$= -4x^2(100 - x^2)^{-\frac{3}{2}} - 4(100 - x^2)^{-\frac{1}{2}}$$

So $\dfrac{d^2P}{dx^2} < 0$ when $x = \sqrt{20}$. Hence the perimeter is a maximum

when $x = \sqrt{20}$ and $y = 2\sqrt{100 - x^2} = 2\sqrt{80}$, i.e. when the dimensions are AD ≈ 4.47 cm and AB ≈ 17.89 cm.

Example 2   A circular cone is inscribed in a sphere of radius R. Find the maximum possible value for the curved surface area of the cone.

Suppose the radius of the base of the cone is r, and the slant height is l (see Fig. 12.5). The curved surface area is $S = \pi r l$.

It is not easy to obtain the relationship between r and l, but we can express both r and l in terms of the semi-vertical angle θ (see Figs. 12.6 and 12.7).

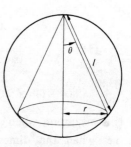

Fig. 12.5

Since the angle in a semicircle is a right angle, we have

$$l = 2R \cos \theta$$

Then $\quad r = l \sin \theta$
$$= 2R \cos \theta \sin \theta$$

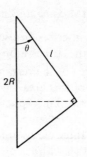

Thus $\quad S = \pi r l = \pi (2R \cos \theta \sin \theta)(2R \cos \theta)$
$$= 4\pi R^2 \sin \theta \cos^2 \theta$$

This is now expressed in terms of a single variable θ (π and R are constants). We have

Fig. 12.6

$$\frac{dS}{d\theta} = 4\pi R^2 \sin \theta (2 \cos \theta)(-\sin \theta) + 4\pi R^2 \cos \theta \cos^2 \theta$$

$$= 4\pi R^2 \cos \theta (\cos^2 \theta - 2 \sin^2 \theta)$$

For a stationary value, $\dfrac{dS}{d\theta} = 0$, i.e.

$$\cos \theta = 0 \quad \text{or} \quad 2 \sin^2 \theta = \cos^2 \theta$$
i.e. $\qquad\qquad\qquad 2 \tan^2 \theta = 1$

i.e. $\qquad\qquad\qquad \tan \theta = \dfrac{1}{\sqrt{2}}$

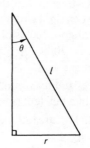

When $\cos \theta = 0$ (i.e. $\theta = \frac{1}{2}\pi$), we have $S = 0$, so this clearly does not give the maximum value.

Fig. 12.7

When $\tan \theta = \dfrac{1}{\sqrt{2}}$, we have $\sin \theta = \dfrac{1}{\sqrt{3}}$ and $\cos \theta = \dfrac{\sqrt{2}}{\sqrt{3}}$ (see Fig. 12.8) and hence

$$S = 4\pi R^2 \sin \theta \cos^2 \theta$$

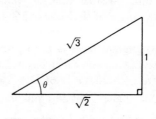

$$= 4\pi R^2 \times \frac{1}{\sqrt{3}} \times \frac{2}{3} = \frac{8\pi R^2}{3\sqrt{3}}$$

Fig. 12.8

Now $\quad \dfrac{d^2 S}{d\theta^2} = 4\pi R^2 \cos \theta [2 \cos \theta(-\sin \theta) - 4 \sin \theta \cos \theta] + 4\pi R^2 (-\sin \theta)(\cos^2 \theta - 2 \sin^2 \theta)$

$$= 4\pi R^2 (2 \sin^3 \theta - 7 \sin \theta \cos^2 \theta)$$

when $\tan \theta = \dfrac{1}{\sqrt{2}}$,

$$\frac{d^2 S}{d\theta^2} = 4\pi R^2 \left(2 \times \frac{1}{3\sqrt{3}} - 7 \times \frac{1}{\sqrt{3}} \times \frac{2}{3} \right)$$

$$= -\frac{16\pi R^2}{\sqrt{3}} < 0$$

so this gives a maximum value of S.

Hence the maximum curved surface area is $\dfrac{8\pi R^2}{3\sqrt{3}}$.

Exercise 12.2   (Maxima and minima)

1   A rectangle is inscribed in a semicircle of radius a so that one of its sides, of length $2x$, lies along the diameter of the semicircle. Show that the area of the rectangle is $2x\sqrt{a^2 - x^2}$, and find the maximum value of this area.

2   A cone is inscribed in a sphere of radius R. Find the maximum possible volume of the cone.

3   A farmer has three pieces of fencing each 10 m long and wishes to use these, together with a hedge, to make a sheep pen as shown in Fig. 12.9.

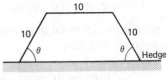

Fig. 12.9

Prove that the area of the sheep pen is $100\sin\theta\,(1 + \cos\theta)$. Find the value of θ which gives the maximum area of the sheep pen, and give the value of this area in this case.   (O)

4   The volume of a right circular cone is 18π cm^3. Find the minimum possible slant height of the cone.

5   When a certain product is offered for sale at a price of p pounds, the number which can be sold is q articles, where $q = \dfrac{3000}{4 + p^{\frac{3}{2}}}$. The income obtained from selling the product is then (qp) pounds. Find the price which gives the maximum income, and the value of this maximum income.

6   The canvas of a bell tent is in the shape of the curved surface of a cone of height h metres and base radius r metres. The area of canvas of the tent is $\pi\sqrt{48}$ m^2. Prove that
$$r^4 + h^2 r^2 - 48 = 0$$

The tent is designed in such a way that the volume contained by the canvas and the ground is the maximum possible. Find the value of r and the corresponding value of h and prove that the semi-vertical angle of the cone is $35°$ to the nearest degree.   (O)

7   A pyramid has a square base and four isosceles triangular faces. If the total area of the base and the four triangular faces is 18 m^2, find the maximum possible volume of the pyramid.

8   ABCD is a kite which has AC as its axis of symmetry. $\hat{BAD}$ is a right angle, and the lengths BC and DC are each 20 cm. If $\hat{BCD} = 2\theta$, show that the area of the kite is
$$S = 200(1 - \cos 2\theta + \sin 2\theta)\ \text{cm}^2$$

Hence find the value of θ which makes S a maximum, and find this maximum area.

9   A cylinder just fits underneath a hemispherical shell of radius 10 cm, as shown in Fig. 12.10. Find the maximum curved surface area of such a cylinder.

10   A solid cylinder is such that the sum of its base diameter and its height is 48 cm. Find the dimensions of the cylinder when (i) its volume is a maximum; (ii) its curved surface area is a maximum; (iii) its total surface area is a maximum.

12.3   Rates of change with time

The most common situation in which rates of change are used is when a quantity, say W, varies with time (t).

The rate of change of W with t, $\dfrac{dW}{dt}$, is called simply the **rate of change of W**.

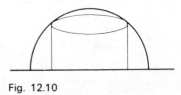

Fig. 12.10

If the value of W is increasing, $\dfrac{dW}{dt}$ is positive; and if W is decreasing, the rate of change $\dfrac{dW}{dt}$ is negative.

We have already encountered, in Chapter 11, the rate of change of displacement $\left(\text{called velocity, } v = \dfrac{dx}{dt}\right)$ and the rate of change of velocity $\left(\text{called acceleration, } a = \dfrac{dv}{dt}\right)$.

Example 1 The potential difference V volts between the terminals of an AC mains socket is given by $V = 340 \sin(100\pi t)$ where t is the time in seconds. Find the rate of change of the voltage (i) after 4 milliseconds; (ii) after 8 milliseconds.

Since $V = 340 \sin(100\pi t)$, the rate of change of V is

$$\frac{dV}{dt} = 340 \times 100\pi \cos(100\pi t)$$

$$= 34\,000\pi \cos(100\pi t)$$

(i) When $t = 0.004$ seconds,

$$\frac{dV}{dt} = 34\,000\pi \cos(0.4\pi) \approx 33\,010$$

The voltage is increasing at a rate of $33\,010$ volts per second.
(ii) When $t = 0.008$ seconds,

$$\frac{dV}{dt} = 34\,000\pi \cos(0.8\pi) \approx -86\,410$$

Since $\dfrac{dV}{dt}$ is negative, the voltage is decreasing at a rate of $86\,410$ volts per second.

Suppose now that W and p are two related quantities. If one of these varies with time, so will the other. Using the 'chain rule', we have

$$\frac{dW}{dt} = \frac{dW}{dp} \times \frac{dp}{dt}$$

This gives the relationship between the rate of change of W, $\dfrac{dW}{dt}$, and the rate of change of p, $\dfrac{dp}{dt}$. If we know one of these rates of change, we can find the other.

Example 2 The radius of a circular oil slick is increasing at a constant rate of 0.2 metres per second. Calculate the rate at which its area is increasing when the radius is (i) 10 metres and (ii) 30 metres.

If the radius is r m and the area A m^2, we have $A = \pi r^2$. Thus

$$\frac{dA}{dt} = \frac{dA}{dr} \times \frac{dr}{dt} = 2\pi r \frac{dr}{dt}$$

$\frac{dr}{dt}$ is the rate of change of the radius; we are given that $\frac{dr}{dt} = 0.2$.

(i)   When $r = 10$,

$$\frac{dA}{dt} = 2\pi r \frac{dr}{dt} = 2\pi \times 10 \times 0.2 = 4\pi$$

The area is increasing at 4π square metres per second.
(ii)   when $r = 30$,

$$\frac{dA}{dt} = 2\pi \times 30 \times 0.2 = 12\pi$$

The area is increasing at 12π square metres per second.
   Notice that, although the radius is increasing at a constant rate, the rate at which the area is increasing is not constant.

Example 3   Air is escaping from a spherical balloon. When the radius is 5 cm, air is leaving the balloon at the rate of 12π cm^3 per second. Calculate the rate at which the radius is decreasing at that instant.

If the radius is r cm and the volume V cm^3, we have $V = \frac{4}{3}\pi r^3$. Thus

$$\frac{dV}{dt} = \frac{dV}{dr} \times \frac{dr}{dt} = 4\pi r^2 \frac{dr}{dt}$$

We are given $r = 5$. Also the volume of the balloon is decreasing at 12π cm^3 per second, so $\dfrac{dV}{dt} = -12\pi$. Hence

$$-12\pi = 4\pi \times 25 \times \frac{dr}{dt}$$

i.e. $$\frac{dr}{dt} = -0.12$$

The radius is decreasing at the rate of 0.12 cm per second.

Example 4   A trough is 3 m long. Its cross-section is an isosceles triangle with base uppermost (see Fig. 12.11). The width at the top is 1.2 m, and the perpendicular height is 0.8 m. Water is poured into the trough at a constant rate of 2.4 m^3 per minute.
   Show that the depth of the water in the trough increases at a rate which is inversely proportional to the depth. Find the rate at which the water level is rising when the depth is 0.4 m.

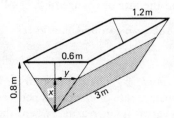

Fig. 12.11

Let V m^3 be the volume of the water in the trough when the depth is x m. We must first find the relationship between V and x.

If y m is half the width of the surface of the water, we have, by similar triangles, $\dfrac{y}{0.6} = \dfrac{x}{0.8}$, i.e. $y = \tfrac{3}{4}x$.

The cross-sectional area of the water is $yx = \tfrac{3}{4}x^2$, hence its volume is

$$V = (\tfrac{3}{4}x^2) \times 3 = \tfrac{9}{4}x^2$$

Thus $\quad \dfrac{\mathrm{d}V}{\mathrm{d}t} = \dfrac{\mathrm{d}V}{\mathrm{d}x} \times \dfrac{\mathrm{d}x}{\mathrm{d}t} = \dfrac{9x}{2}\dfrac{\mathrm{d}x}{\mathrm{d}t}$

Now the volume of water is increasing at 2.4 m^3 per minute, and so $\dfrac{\mathrm{d}V}{\mathrm{d}t} = 2.4$. Hence

$$2.4 = \dfrac{9x}{2}\dfrac{\mathrm{d}x}{\mathrm{d}t}$$

i.e. $\quad \dfrac{\mathrm{d}x}{\mathrm{d}t} = \dfrac{4.8}{9x} = \dfrac{8}{15x} = \dfrac{8}{15} \times \dfrac{1}{x}$

$\dfrac{\mathrm{d}x}{\mathrm{d}t}$ is proportional to $\dfrac{1}{x}$, i.e. inversely proportional to x.

When $x = 0.4$, $\dfrac{\mathrm{d}x}{\mathrm{d}t} = \dfrac{8}{15 \times 0.4} = \dfrac{4}{3}$, i.e. the depth is increasing at a rate of $\tfrac{4}{3}$ metres per minute (which is $\tfrac{4}{3} \times \tfrac{100}{60}$ cm per second, i.e. $2\tfrac{2}{9}$ cm per second).

Angular velocity

Suppose AB is a line which is rotating (i.e. changing its direction). If AB makes an angle θ with some fixed direction (see Fig. 12.12), then θ will vary with time.

The rate of change of θ, $\dfrac{\mathrm{d}\theta}{\mathrm{d}t}$, is called the **angular velocity** of AB.

It is measured in units such as radians per second, or revolutions per minute (one revolution $= 2\pi$ radians).

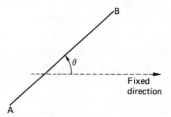

Fig. 12.12

Example 1   After t seconds, a wheel has turned through an angle θ radians, where $\theta = \tfrac{1}{2}t + \tfrac{1}{10}t^2$. Find the angular velocity after 5 seconds.

The angular velocity is $\dfrac{\mathrm{d}\theta}{\mathrm{d}t} = \tfrac{1}{2} + \tfrac{1}{5}t$

When $t = 5$, $\dfrac{\mathrm{d}\theta}{\mathrm{d}t} = \tfrac{1}{2} + 1 = \tfrac{3}{2}$. The angular velocity is $\tfrac{3}{2}$ radians per second.

Example 2   An observer O is watching a boat race from the river bank. The boat B is moving on a straight course with speed 10 m s^{-1}. A is the point on the boat's track which is nearest to O, and OA = 5 m. Find the angular velocity of OB at the instant when $\hat{AOB} = \frac{1}{3}\pi$.

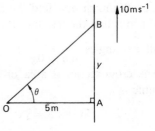

Fig. 12.13

Let $\hat{AOB} = \theta$ radians, and AB = y metres (see Fig. 12.13). The rate of change of y is the velocity of B, so we know that $\dfrac{dy}{dt} = 10$. Now $y = 5\tan\theta$, thus

$$\frac{dy}{dt} = \frac{dy}{d\theta} \times \frac{d\theta}{dt} = 5\sec^2\theta\frac{d\theta}{dt}$$

Hence $10 = 5\sec^2\theta\dfrac{d\theta}{dt}$

i.e. $\dfrac{d\theta}{dt} = \dfrac{2}{\sec^2\theta} = 2\cos^2\theta$

When $\theta = \frac{1}{3}\pi$

$$\frac{d\theta}{dt} = 2 \times (\tfrac{1}{2})^2 = \tfrac{1}{2}$$

The angular velocity of OB is $\frac{1}{2}$ radian per second.

Exercise 12.3   (Rates of change with time)

1   The depth h metres of water in a harbour is given by $h = 5 + 2\cos\left(\dfrac{\pi t}{6}\right)$ where t is the time in hours since high water. Find the rate at which the water level is rising or falling (i) 5 hours after high water; (ii) $10\frac{1}{2}$ hours after high water.

2   A disc is set spinning, and after t seconds it has turned through an angle θ radians, where $\theta = 2t - \frac{1}{10}t^2$. Find the angular velocity of the disc after 6 seconds.

3   The dimensions of a rectangle are varying. The length is initially 30 cm and decreases at a constant rate of 1 cm per second. The width is initially 12 cm and increases at a constant rate of 2 cm per second. State the length and the width of the rectangle after t seconds, and hence obtain an expression for its area. Find the rate at which the area is increasing when the figure becomes a square.

Show that there is a maximum value of the area, and find this value.

4   The radius of a circle is increasing at the rate of 3 m per second. Find the rate at which its circumference is increasing.

5   A sphere is decreasing in size. At the instant when its radius is 0.5 m, its surface area is decreasing at a rate of 0.4 m^2 per second. Calculate the rate at which the radius is decreasing at this instant, and hence find the rate at which the volume of the sphere is decreasing.

6   When sand is being poured on to a level surface the heap takes on a conical shape where the semi-vertical angle is 45°. Calculate the volume of sand when the height is h metres.

The sand is being poured at a constant rate of 1 cubic metre per minute. Calculate the rate at which the height of the heap is increasing in centimetres per second when the height is $\frac{1}{2}$ metre.   (O)

7   The cross-section of a trough, 5 metres long, is an isosceles right-angled triangle ABC, with the right-angle at A and the side BC being horizontal and uppermost. The trough is empty and water is

poured in at a rate of $2.5 \text{ m}^3 \text{s}^{-1}$. Calculate the rate, in cm s^{-1}, at which the water level is rising 10 seconds after the start. (O)

8 Under certain conditions the pressure $p \text{ N m}^{-2}$ and the volume $v \text{ m}^3$ of a gas are related by the equation $pv^{4/3} = \text{constant}$.

When $p = 10$, $v = 13$; calculate the value of the constant.

At the instant the pressure is 10 N m^{-2}, the volume is increasing at the rate $2 \text{ m}^3 \text{s}^{-1}$. Calculate the rate at which the pressure is decreasing. (O)

9 A hollow right circular cone, with base radius 12 cm and perpendicular height 18 cm, is held vertex downwards beneath a dripping tap. If the water level is x cm above the vertex, and the radius of the water surface is y cm, show that $\dfrac{y}{x} = \dfrac{2}{3}$ and that the volume of water is $\dfrac{4}{27}\pi x^3 \text{ cm}^3$.

If the tap is dripping at the rate of 2 cm^3 per second, calculate the rate at which the water level is rising when it is 6 cm above the vertex.

10 When an object is placed a distance u cm in front of a curved mirror, its image appears to be at a distance v cm behind the mirror, where $v = \dfrac{100u}{100 - u}$. If the object is 60 cm in front of the mirror and is moving towards the mirror at 5 cm s^{-1}, find the speed at which the image appears to move.

11 A ship S is anchored 1 km from the nearest point A of a straight shore. A searchlight on the ship illuminates a point P on the shore. If $A\hat{S}P = \theta$, show that $AP = \tan\theta$ km. The searchlight revolves at a rate of 2 revolutions per minute. Express this angular velocity in radians per second, and find the velocity of P along the shore when (i) $\theta = 0$ (ii) $\theta = 60°$ (iii) $\theta = 80°$.

12 A rod AB of length 10 m moves in a vertical plane with the end A in contact with a horizontal floor, and the end B in contact with a vertical wall. The floor and the wall meet at O. If $O\hat{B}A = \theta$, express OA and OB in terms of θ. At the instant when OA is 6 m, the velocity of A is 2 m s^{-1}. Find, at this instant, (i) the angular velocity of the rod; (ii) the velocity of the end B.

12.4 Differential equations

If W is a quantity which varies with time (t), a statement such as $\dfrac{dW}{dt} = 3t^2 - 2$ may be regarded as a formula from which to calculate the rate of change of W; so far in this chapter we have been calculating rates of change.

On the other hand, the statement may represent given information about the rate of change, when we are really interested in finding values of W. In this case, we regard $\dfrac{dW}{dt} = 3t^2 - 2$ as a differential equation, which we can 'solve' by integrating to obtain $W = t^3 - 2t + C$. This introduces an arbitrary constant C, and to find it we shall need an extra piece of information (such as the value of W when $t = 0$). We then have the relationship between W and t.

We have encountered similar problems in Chapter 11; given the velocity $\dfrac{dx}{dt}$, we integrated to find the displacement x, and given the acceleration $\dfrac{dv}{dt}$, we integrated to find the velocity v.

Example 1   While a tap is being turned on, the water issues from it at a rate of $(3t^2 + 6t)\,\text{cm}^3$ per second, where t is the time in seconds. This formula holds until the rate of flow reaches $45\,\text{cm}^3$ per second, when the tap is fully open. Find the total volume of water which has issued from the tap after 10 seconds.

Let V be the volume of water which has issued from the tap after t seconds. Then $\dfrac{dV}{dt}$, which is the rate at which V is increasing, is the rate of flow. We therefore have $\dfrac{dV}{dt} = 3t^2 + 6t$, until the tap becomes fully open. This occurs when $\dfrac{dV}{dt} = 45$, i.e.

$$3t^2 + 6t = 45$$

i.e. $t^2 + 2t - 15 = 0$

i.e. $(t - 3)(t + 5) = 0$

i.e. $t = 3$ or -5

Hence the tap is fully open after 3 seconds.

For $0 \leqslant t \leqslant 3$, we have $\dfrac{dV}{dt} = 3t^2 + 6t$. Integrating with respect to t,

$$V = t^3 + 3t^2 + C$$

When $t = 0$, $V = 0$, and so $C = 0$. Thus

$$V = t^3 + 3t^2$$

Putting $t = 3$, the volume of water issued during the first 3 seconds is $V = 27 + (3 \times 9) = 54\,\text{cm}^3$.

For $3 \leqslant t \leqslant 10$, the rate of flow is constant at $45\,\text{cm}^3$ per second, and so the volume of water issued during this time is $45 \times 7 = 315\,\text{cm}^3$.

Hence the total volume of water issued after 10 seconds is $54 + 315 = 369\,\text{cm}^3$.

If we draw a graph showing the rate of flow $\dfrac{dV}{dt}$ as a function of t, the volume V is represented by the area under this curve (see Fig. 12.14).

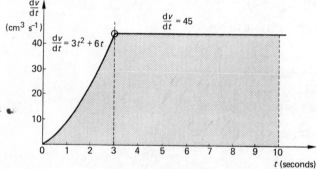

Fig. 12.14

The rate of change of a quantity W may be given in terms of W, for example

$$\frac{dW}{dt} = 2W^2$$

To solve this, we rewrite it as

$$\frac{dt}{dW} = \frac{1}{\dfrac{dW}{dt}} = \frac{1}{2W^2}$$

and then integrate with respect to W, to obtain

$$t = -\frac{1}{2W} + C$$

Again we shall need an extra piece of information in order to find C.

Example 2   A candle is burning, so that when its height is h cm, it is burning down at a rate of $3\sqrt{h}$ cm per hour. If the initial height of the candle is 36 cm, how long will it take to burn down completely?

The height h is decreasing, and so $\dfrac{dh}{dt}$ is negative; we are given that $\dfrac{dh}{dt} = -3\sqrt{h}$. Thus

$$\frac{dt}{dh} = -\frac{1}{3\sqrt{h}}$$

and, integrating with respect to h,

$$t = -\tfrac{2}{3}\sqrt{h} + C$$

When $t = 0$, $h = 36$; so $0 = -\tfrac{2}{3} \times 6 + C$, giving $C = 4$. Hence

$$t = -\tfrac{2}{3}\sqrt{h} + 4$$

The candle has burnt down completely when $h = 0$; then $t = 4$. It therefore takes 4 hours.

Exercise 12.4   (Differential equations)

1   While a river is flooding, the water level rises at the rate of $(42 - 9t)$ cm per hour, where t is the number of hours that have elapsed since the flooding started.
  (i)   By how much has the water level risen after 2 hours?
  (ii)   How long will it take for the water level to rise 90 cm?
  (iii)   What is the maximum rise in the water level, and when does this occur?

2   An international incident shakes confidence in the pound sterling. t minutes after the incident, the value of the pound is falling at a rate of $\dfrac{40}{(t+1)^2}$ cents per minute. If the value of the pound at the time

of the incident is 220 cents, find its value 9 minutes later.

3   A particle is moving along a straight line away from a fixed point O. When it is at a distance x m from O, its speed is $\dfrac{2}{x}\,\mathrm{m\,s^{-1}}$. If the particle is initially at a distance of 4 m from O, find (i) the relationship between x and t (the time in seconds); (ii) the distance travelled by the particle in the first 45 seconds.

4   An organism is growing; when its mass is m milligrams, this is increasing at the rate of $(125-m)^{\frac{1}{3}}$ milligrams per day. If its initial mass is 61 milligrams, how long will it take to grow (i) to 124 milligrams; (ii) to 125 milligrams?

5   When the volume of a balloon is $V\,\mathrm{cm^3}$, air is escaping at the rate of $(10V)^{\frac{1}{4}}\,\mathrm{cm^3}$ per second. How long will it take for the balloon to deflate completely from an initial volume of $1000\,\mathrm{cm^3}$?

12.5   Formation of differential equations

Many laws of nature refer to rates of change. For example, from mechanics, Newton's second law can be expressed as

  force = mass × acceleration

and acceleration is the rate of change of velocity.

Also, in chemistry, 'the rate at which a chemical reaction proceeds is proportional to the product of the masses of the reactants remaining'.

When these laws are expressed mathematically, we obtain differential equations. The following examples show how a differential equation can be derived from a given practical situation. This is the first step in analysing such a problem (the next step is to solve the equation).

Example 1   A population is increasing at a rate which is proportional to its size.

If N is the size of the population at time t, then $\dfrac{dN}{dt}$ is the rate of change of size, which is the rate at which the population is increasing.

We are given that $\dfrac{dN}{dt}$ is proportional to N, i.e. $\dfrac{dN}{dt}=kN$ where k is the constant of proportionality. In order to answer numerical questions about the population, we would need to know the value of k, but in this example we have not been given any information which will enable us to calculate it.

All we can say is that N satisfies the differential equation

$$\frac{dN}{dt}=kN \quad \text{(where } k \text{ is a constant)}$$

Notice that, if we attempt to solve this equation, by writing

$$\frac{dt}{dN}=\frac{1}{kN} \quad \text{and integrating, } t=\int\frac{1}{kN}dN=\frac{1}{k}\int\frac{1}{N}dN, \text{ we encoun-}$$

ter the problem of having to integrate $\dfrac{1}{N}$ with respect to N. Further discussion of this important example must wait until later (see p. 335).

Example 2 The rate at which radioactive waste decays is proportional to the mass of active material remaining. When the active mass is 20 kg, it decays at a rate of 2 kg per year.

Let m kg be the mass of active material remaining. The rate of change of m, $\dfrac{dm}{dt}$, is negative (since m is decreasing) and is proportional to m; thus

$$\frac{dm}{dt} = -km$$

In this case we can find k, because we are given that when $m = 20$, $\dfrac{dm}{dt} = -2$; so $-2 = -k \times 20$ and $k = \frac{1}{10}$.

Hence the differential equation is

$$\frac{dm}{dt} = -\tfrac{1}{10}m$$

Example 3 When an object is heating up in an oven, its temperature rises at a rate proportional to the square of the difference between its temperature and that of the oven, which remains constant at 350°C.

Suppose that its temperature at time t is θ°C. $\dfrac{d\theta}{dt}$ is the rate at which its temperature is rising. The difference between the temperatures of the object and the oven is $(350 - \theta)$°C; we are given that $\dfrac{d\theta}{dt}$ is proportional to $(350 - \theta)^2$, i.e.

$$\frac{d\theta}{dt} = k\,(350 - \theta)^2$$

Example 4 When the power to a high speed train is turned off, the effect of air resistance is to produce a deceleration which is proportional to the square of the train's velocity. When the train is travelling at 50 m s^{-1}, the deceleration is 0.2 m s^{-2}.

Let the velocity of the train after t seconds be v m s^{-1}. Then $\dfrac{dv}{dt}$ is the acceleration; $\dfrac{dv}{dt}$ is negative (since the train is decelerating) and proportional to v^2, thus

$$\frac{dv}{dt} = -kv^2$$

When $v = 50$, $\dfrac{dv}{dt} = -0.2$, so $-0.2 = -k \times 50^2$ i.e.

$k = 8 \times 10^{-5}$

Hence

$$\frac{dv}{dt} = -(8 \times 10^{-5})v^2$$

Sometimes the rate of change of one quantity (say W) is given in terms of a different, but related, quantity (say p). A differential equation involving $\dfrac{dW}{dt}$ cannot be solved if it contains any variables other than W and t; in such cases we must use the relationship between W and p to eliminate one of these variables.

Example 5   The area of a circle is being increased at a rate inversely proportional to its radius.

Suppose the circle has area A and radius r at time t. $\dfrac{dA}{dt}$ is the rate at which the area is increasing. We are given that $\dfrac{dA}{dt}$ is inversely proportional to r, which means that it is proportional to $\dfrac{1}{r}$, i.e.

$$\frac{dA}{dt} = k \times \frac{1}{r}$$

This equation cannot be solved, because it contains three variables (A, r and t).

However, we know that $A = \pi r^2$. We can use this in either of two ways:

(i)   We can eliminate r. From $A = \pi r^2$, we have $r = \sqrt{\dfrac{A}{\pi}}$, and thus

$$\frac{dA}{dt} = k\sqrt{\frac{\pi}{A}}$$

This is a differential equation involving A and t.

(ii)   We can eliminate A. Using the chain rule

$$\frac{dA}{dt} = \frac{dA}{dr} \times \frac{dr}{dt} = 2\pi r \frac{dr}{dt}$$

Substituting into $\dfrac{dA}{dt} = k \times \dfrac{1}{r}$, we obtain

$$2\pi r \frac{dr}{dt} = \frac{k}{r}$$

i.e. $$\frac{dr}{dt} = \frac{k}{2\pi r^2}$$

This is a differential equation involving r and t and it is probably the better approach for the problem. Not only is the equation 'simpler' (it does not involve square roots), but we are usually more interested in the radius of a circle than we are in its area.

Example 6   The volume of a sphere is being reduced at a rate proportional to the square of its surface area.

Suppose the volume of the sphere is V, and its surface area is S. $\dfrac{dV}{dt}$ is the rate of change of the volume; this is negative, and proportional to S^2. Thus

$$\frac{dV}{dt} = -kS^2$$

In this case, rather than finding the relationship between V and S, it is simpler to express both $\dfrac{dV}{dt}$ and S in terms of the radius r.

We have $V = \frac{4}{3}\pi r^3$, and so

$$\frac{dV}{dt} = \frac{dV}{dr} \times \frac{dr}{dt} = 4\pi r^2 \frac{dr}{dt}$$

Also $S = 4\pi r^2$

Substituting into $\dfrac{dV}{dt} = -kS^2$, we obtain

$$4\pi r^2 \frac{dr}{dt} = -k(4\pi r^2)^2$$

i.e. $$\frac{dr}{dt} = -4\pi k r^2$$

Exercise 12.5   (Forming differential equations)

For each of the following situations, choose a suitable variable, and give the differential equation which it satisfies.

Find the constant of proportionality where possible, but make no attempt to solve the equations.

1   A growing organism increases its mass at a rate which is proportional to its mass at that instant.
2   In an electrical circuit, the current is reducing at a rate proportional to the current flowing at that time. When the current is 40 milliamps, it is falling at a rate of $\frac{1}{2}$ milliamp per second.
3   A hot object cools at a rate which is proportional to the difference between its temperature and that of the surrounding air. The surroundings remain at a constant temperature of $15°C$, and when the temperature of the object is $150°C$ it is cooling at a rate of $12°C$ per minute.
4   A boy is running away from a trouble spot, and his speed is always inversely proportional to his distance from the spot. He starts 2 m from the spot, running at $6\,m\,s^{-1}$.
5   When the engine is turned off, the deceleration of a boat is proportional to the square root of its velocity.
6   The surface area of a sphere is being increased at a rate inversely proportional to its radius.

7   A tank made of porous material has a square base of side 2 metres and vertical sides. The tank contains water which seeps out through the base and the sides at a rate proportional to the total area in contact with the water. When the depth is 3 metres, it is observed to be falling at a rate of 0.2 metres per hour.

8   Money invested increases at a rate of 12% of its current value per year.

9   The rate at which a substance is deposited from a solution to form a crystal is proportional to the product of the mass already deposited and the mass remaining in the solution. The solution originally contained 20 grams of the substance. [Hint. If the amount already deposited is m grams, then the mass remaining in solution is $(20 - m)$ grams].

10   Air is escaping from a spherical balloon at a rate proportional to its surface area. When the radius is 20 cm, the air is escaping at 8 cm^3 per second.

Show that the radius is decreasing at a constant rate. Hence find the time taken for the radius to change from 20 cm to 15 cm.

12.6   Problems involving differential equations

To solve problems about the kind of situations encountered in the previous section, we first have to form a differential equation, and then we solve it.

Example 1   A hollow inverted cone, of radius 20 cm and perpendicular height 50 cm, is filled with water. The water flows out from a tap at the vertex at a rate which is proportional to the depth of the water remaining in the cone. The initial rate of flow of water is 100 cm^3 per second. Find (i) the rate at which the depth of water is decreasing initially; (ii) the time taken for the cone to empty.

Suppose that, after time t seconds, the depth of the water is x cm, and the volume V cm^3 (see Fig. 12.15). The rate of flow of water is equal to the rate of change of the volume. $\dfrac{dV}{dt}$ is negative, and proportional to x, i.e. $\dfrac{dV}{dt} = -kx$.

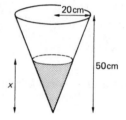

Fig. 12.15

At this stage we can find k. When $x = 50$, $\dfrac{dV}{dt} = -100$, so $-100 = -k \times 50$, i.e. $k = 2$. Thus

$$\frac{dV}{dt} = -2x$$

We must now find the relationship between V and x. The radius r of the surface of the water is given by $\dfrac{r}{20} = \dfrac{x}{50}$ (see Fig. 12.16), i.e. $r = \dfrac{2}{5}x$. The volume of water is therefore

$$V = \tfrac{1}{3}\pi r^2 x = \tfrac{1}{3}\pi(\tfrac{4}{25}x^2)x = \tfrac{4}{75}\pi x^3$$

Fig. 12.16

Thus $\dfrac{dV}{dt} = \dfrac{dV}{dx} \times \dfrac{dx}{dt} = \dfrac{4}{25}\pi x^2 \dfrac{dx}{dt}$

and, substituting into $\dfrac{dV}{dt} = -2x$, we obtain the differential equation

$$\frac{4}{25}\pi x^2 \frac{dx}{dt} = -2x$$

i.e. $\dfrac{dx}{dt} = -\dfrac{25}{2\pi x}$

(i) To find the initial rate at which the depth is decreasing, we substitute $x = 50$; then $\dfrac{dx}{dt} = -\dfrac{1}{4\pi} \approx -0.08$. Initially, the depth is decreasing at 0.08 cm per second.

(ii) To find the time taken to empty, we need to find t, and so we integrate the equation. We have

$$\frac{dt}{dx} = -\frac{2\pi x}{25}$$

Integrating with respect to x,

$$t = -\frac{\pi x^2}{25} + C$$

When $t = 0$, $x = 50$, so $0 = -\dfrac{\pi \times 50^2}{25} + C$, i.e. $C = 100\pi$. Hence

$$t = 100\pi - \frac{\pi x^2}{25}$$

The cone is empty when $x = 0$; then $t = 100\pi \approx 314$. It takes 314 seconds for the cone to empty.

Sometimes the constant of proportionality cannot be found until we have solved the differential equation.

Example 2   A tree grows to a maximum height of 25 m. While it is growing, its height increases at a rate proportional to the square root of the difference between its present height and its eventual maximum height. The tree grows from nothing to its maximum height in 120 years. Find the time taken for the tree to reach a height of 16 m, and the rate at which the height is then increasing.

Suppose the height of the tree after t years is h metres. The rate at which the height is increasing is $\dfrac{dh}{dt}$. The difference between its present height and the maximum height is $(25 - h)$. We are given that $\dfrac{dh}{dt}$ is proportional to $\sqrt{25 - h}$, i.e.

$$\frac{dh}{dt} = k\sqrt{25 - h}$$

We are not given any information about $\dfrac{dh}{dt}$, and so we cannot find k at this stage. We leave k in the equation, and solve it. We have

$$\frac{dt}{dh} = \frac{1}{k\sqrt{25-h}}$$

To find t, we integrate with respect to h:

$$t = -\frac{2}{k}\sqrt{25-h} + C$$

When $t = 0$, $h = 0$, so $0 = -\dfrac{10}{k} + C$.

Also, when $t = 120$, $h = 25$, so $120 = 0 + C$.

Thus $C = 120$ and $k = \dfrac{10}{C} = \dfrac{1}{12}$, and hence

$$t = 120 - 24\sqrt{25-h}$$

Putting $h = 16$, the time taken to reach a height of 16 m is

$$t = 120 - 24\sqrt{25-16} = 120 - 72 = 48 \text{ years}$$

The height is increasing at a rate $\dfrac{dh}{dt} = k\sqrt{25-h}$ and we now know that $k = \frac{1}{12}$. When $h = 16$, the height is increasing at a rate of

$$\frac{dh}{dt} = \frac{1}{12}\sqrt{25-h} = \frac{1}{12}\sqrt{25-16} = 0.25 \text{ metres per year}$$

Exercise 12.6   (Problems involving differential equations)

1   A girl runs to school, and her speed is proportional to the square root of the distance she has still to go. If she starts running at 6 m s^{-1} when she is 900 m from the school, how long will it take her to get to school?

2   The volume of a spherical soap bubble is being increased at a rate inversely proportional to its radius. Initially the bubble has radius 5 mm and its volume is increasing at 400 mm^3 per second. The bubble bursts when the radius is 12 mm. Find the time that elapses before it bursts.

3   An object cools at a rate proportional to the square of the difference between its temperature and that of the surrounding air, which remains constant at 20°C. Initially the object is at 220°C, and is cooling at 10°C per second. How long does it take for the object to cool to 21°C?

4   A cylindrical tank has a horizontal circular base of radius 20 cm, and vertical sides of height 64 cm. Water flows out of the tank at a rate proportional to the cube root of the depth of water remaining in the tank. When the tank is full the rate of flow is 800 cm^3 per second. Find the time taken for the tank to empty.

5   A car has a maximum speed of 20 m s^{-1} in first gear. While it is accelerating, its acceleration is proportional to the square of the difference between its present speed and its maximum speed. When starting from rest, its initial acceleration is 12 m s^{-2}. How long does it take to reach a speed of 18 m s^{-1}? Also find its speed 1 second after starting.

6   The area of a circle is being increased at a rate inversely proportional to the square of its radius. Initially the radius is 4 cm, and the *radius* is increasing at 0.1 cm per second. Find the radius of the circle after 2 minutes.

7 The thickness of ice on a pond is increasing at a rate inversely proportional to the thickness already present. If it takes 6 hours for the thickness to increase from 3 cm to 4 cm, find how long it takes to increase from 4 cm to 5 cm. At what rate is the thickness increasing when the ice is 5 cm thick?

8 A boat is travelling at 12 m s^{-1} when the engine is turned off. The deceleration of the boat is then proportional to the square of its speed. 10 seconds later its speed is 8 m s^{-1}. Find the speed and the deceleration of the boat 30 seconds after the engine was turned off.

9 Money invested in a certain enterprise increases its value at a rate proportional to the square root of the cube of its current value. An investment of £400 quadruples to £1600 in 20 months. How long will it take to quadruple from £1600 to £6400?

10 An organism takes 25 days to grow from a mass of 5 grams to its maximum mass of 30 grams. While it is growing, its mass increases at a rate proportional to the square root of the difference between its present mass and its maximum mass. Find its mass after 15 days, and the rate at which its mass is then increasing.

11 When the tide is coming in, the water level rises at a rate proportional to $\sin(\pi t/6)$, where t hours is the time elapsed since the tide turned. After 6 hours, the water has risen 4.8 m. Find the rise in the water level, and the rate at which the water is rising, 4 hours after the tide has turned.

12 A hemispherical bowl of radius 10 cm contains water to a maximum depth of x cm. You are given that the volume of water is $\pi (10x^2 - \frac{1}{3}x^3)$ cm^3, and the area of the bowl in contact with the water is $20\pi x$ cm^2. The bowl is made of porous material, and water seeps through it at a rate proportional to the area of the bowl in contact with the water. When the bowl is full, the water level is seen to be dropping at 0.4 cm per hour. How long will it take for the bowl to empty?

13 When a spherical balloon is deflating, its volume decreases at a rate proportional to the square of its surface area. If it takes 3 hours for the radius to decrease from 20 cm to 10 cm, how long will it take to decrease from 10 cm to 5 cm?

14 When a large cube of ice is suspended in the air, its volume decreases as it melts, at a rate proportional to its surface area at any instant. If x cm is the length of its side after t days, find a differential equation involving x and t. If the cube initially has a side of 40 cm, and it takes 5 days to melt completely, find the relationship between x and t.

15 A V-shaped trough (similar to that shown in Fig. 12.11, p. 178) is 100 cm long and 36 cm wide at the top. Its perpendicular height is 36 cm. Water flows out at a rate proportional to the square root of the depth of the water in the trough; when the trough is full, the rate of flow is 300 cm^3 per second. Find how long it takes for the depth to fall to 4 cm, and find the rate at which the depth is then decreasing.

12.7 Small changes

If W and p are related quantities, we have considered $\dfrac{dW}{dp}$, the rate

of change of W with p By the definition of a derivative,

$\dfrac{dW}{dp} = \lim_{\delta p \to 0} \dfrac{\delta W}{\delta p}$, where δW and δp are corresponding changes in W and p.

Provided that δp is small enough, $\dfrac{\delta W}{\delta p}$ will be close to its limiting

value, i.e. $\dfrac{\delta W}{\delta p} \approx \dfrac{dW}{dp}$, and so we have

$$\delta W \approx \frac{dW}{dp}\delta p$$

If we are given a small change in one of the variables, we can use this result to find (approximately) the corresponding change in the other one.

We can interpret this approximation on a graph of W against p (see Fig. 12.17). $\dfrac{\delta W}{\delta p}$ is the gradient of the chord AB, and $\dfrac{dW}{dp}$ is the gradient of AD, the tangent at A. $\dfrac{dW}{dp}\delta p$ is the length of CD, whereas the true value of δW is the length of CB.

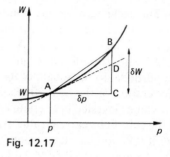

Fig. 12.17

Example 1   The radius of a circular oil slick increases from 32 m to 32.5 m. Find the approximate increase in its area.

If the radius is r and the area A, we have $A = \pi r^2$. Thus

$$\delta A \approx \frac{dA}{dr}\delta r = 2\pi r\,\delta r$$

We are given $r = 32$ and the change in r, $\delta r = 0.5$, hence

$$\delta A \approx 2\pi \times 32 \times 0.5 = 32\pi \approx 100.5$$

The approximate increase in area is $100.5\,\text{m}^2$.
[Note that this problem could also be solved directly:
when $r = 32$, $A = \pi \times 32^2 = 1024\pi$
when $r = 32.5$, $A = \pi \times 32.5^2 = 1056.25\pi$
Hence the exact increase in area is $32.25\pi \approx 101.3\,\text{m}^2$.

This numerical approach may appear to be simpler. However, the application of calculus methods to small changes is an important idea which we continue to develop.]

Example 2   The volume $V\,\text{cm}^3$ of water in a vessel is given by the formula $V = \frac{4}{75}\pi x^3$, where $x\,\text{cm}$ is the depth of the water. When the depth is 40 cm, an extra 120 cm³ of water is added to the vessel. Find the approximate increase in the depth of the water.

We have $V = \frac{4}{75}\pi x^3$, and so

$$\delta V \approx \frac{dV}{dx}\delta x = \frac{4}{25}\pi x^2\,\delta x$$

We are given $x = 40$ and $\delta V = 120$, hence

$$120 \approx \frac{4}{25}\pi \times 40^2\,\delta x$$

i.e. $\delta x \approx \dfrac{15}{32\pi} \approx 0.15$

The approximate increase in depth is 0.15 cm.

Exercise 12.7   (Small changes)

1   The radius of a sphere is increased from 10 to 10.2 cm. Find the approximate increases in (i) its volume and (ii) its surface area.

2   A spherical balloon has radius 16 cm. The volume is then increased by 100 cm³. Find the approximate increase in the radius. Hence also find the corresponding approximate increase in the surface area.

3   The time taken by a planet to complete its orbit round the sun is given by $T = kR^{\frac{3}{2}}$ where R is its mean distance from the sun and k is a constant. For the earth $R = 1.5 \times 10^8$ km and $T = 365$ days. If earth's mean distance from the sun were to be increased by 1% (i.e. by 1.5×10^6 km), find approximately how much longer the year would become. Give your answer in days correct to one decimal place.

4   When paint is kept in a hemispherical dish of radius 3 cm the relation between its volume V cm³ and its depth x cm is given by the formula $V = \frac{1}{3}\pi x^2(9 - x)$.

When the depth is 1 cm, 0.2 cm³ of paint is taken out. Find the approximate decrease in the depth.

Also find approximately what volume of paint is required to increase the depth from 2 cm to 2.1 cm.

5   A cone has fixed slant height 10 cm, but a variable semi-vertical angle θ. Show that the volume of the cone is $V = \frac{1000}{3}\pi \sin^2 \theta \cos \theta$. When $\theta = \frac{1}{6}\pi$, find approximately the change in θ which will cause a change of 5 cm³ in the volume.

6   Find approximately the radius of a sphere given that an increase of 0.1 cm in the radius causes an increase in the volume of 20 cm³.

7   The volume V m³ of water in a container is given by $V = x^3 + 4x$, where x m is the depth of the water. Show that when $x = 3$, $V = 39$. By considering a change of 1 m³ in the volume, find the approximate depth of the water when the volume is 40 m³.

8   A point source of light is fixed at a distance h above a horizontal table. A metal plate of area A is parallel to the table and is placed between the light and the table at a distance x above the table. If S is the area of the shadow of the plate on the table, find a formula for S in terms of A, h and x, and give an expression for the change δS in S when the plate is raised a small distance δx.

12.8   Proportionate and percentage changes

Suppose a quantity W changes by an amount δW. Often we are not interested in the actual amount of the change, but wish to know how the change δW compares with the original value W.

We define the **proportionate change** in W as $\dfrac{\delta W}{W}$. This has no units, and, in fact, proportionate changes are usually given as percentages.

The **percentage change** in W is $\dfrac{\delta W}{W} \times 100$.

Example 1   The radius of a sphere is increased by 2%. Find the approximate percentage increase in the volume.

Let the radius be r and the volume V. Then $V = \frac{4}{3}\pi r^3$, so

$$\delta V \approx \frac{dV}{dr}\,\delta r = 4\pi r^2\,\delta r$$

The percentage increase in V is

$$\frac{\delta V}{V} \times 100 \approx \frac{4\pi r^2\,\delta r}{V} \times 100 = \frac{4\pi r^2\,\delta r}{\frac{4}{3}\pi r^3} \times 100 = \frac{3\delta r}{r} \times 100 = 3\left(\frac{\delta r}{r} \times 100\right)$$

i.e. the percentage increase in V is 3 times the percentage increase in r. Hence the volume increases by approximately 6%.

Example 2   The force F of gravitational attraction between two bodies of masses M and m, at a distance apart of x, is given by $F = \dfrac{GMm}{x^2}$, where G is a constant. If the distance between the bodies is changed by 1%, find the approximate percentage change in the force.

Since the masses M and m remain constant, we regard $F = \dfrac{GMm}{x^2}$ as a function of x. Then

$$\frac{dF}{dx} = GMm\left(-\frac{2}{x^3}\right)$$

Thus $\delta F \approx \dfrac{dF}{dx}\delta x = -\dfrac{2GMm}{x^3}\delta x$

and the percentage change in F is

$$\frac{\delta F}{F} \times 100 = \frac{-\dfrac{2GMm}{x^3}\delta x}{\dfrac{GMm}{x^2}} \times 100 = -\frac{2GMm}{x^3}\delta x \times \frac{x^2}{GMm} \times 100$$

$$= -2\left(\frac{\delta x}{x} \times 100\right)$$

i.e. the percentage change in F is twice the percentage change in x (the minus sign means that an increase in x causes a decrease in F).

Hence, if the distance x is changed by 1%, the force changes by approximately 2%.

Example 3   The cost E pence of producing one unit of electricity is given by $E = 1.5 + 0.1\,P$, where P is the price of fuel expressed in pounds per tonne. If the price of fuel increases by $z\%$, find an expression for the percentage increase in E. Hence find the percentage rise in the cost of electricity due to a 10% increase in the cost of fuel (i) when $P = 20$; (ii) when $P = 40$.

We have

$$\delta E \approx \frac{dE}{dP}\delta P = 0.1\,\delta P$$

so the percentage increase in E is

$$\frac{\delta E}{E} \times 100 \approx \frac{0.1\,\delta P}{1.5 + 0.1P} \times 100 = \frac{0.1}{1.5 + 0.1P} \times P \times \left(\frac{\delta P}{P} \times 100\right)$$

$$= \left(\frac{P}{15 + P}\right)z$$

In this example the percentage increase in E depends on the value of P as well as the percentage increase in P.

(i)   when $z = 10$ and $P = 20$, the percentage rise in the cost of electricity is $\left(\dfrac{20}{15+20}\right) \times 10 \approx 5.7\%$.

(ii)   when $z = 10$ and $P = 40$, the percentage rise in the cost of electricity is $\left(\dfrac{40}{15+40}\right) \times 10 \approx 7.3\%$.

Application to elasticity of demand

Consider a company which manufactures articles for sale. The number of articles which can be sold (the 'demand' for the product) will depend on the price at which they are offered. A change in the price will affect the demand. Suppose, for example, that when the price is £20, the demand is 40 000 articles; when the price is increased to £21, the demand falls to 37 500 articles.

The percentage change in the price is $\dfrac{1}{20} \times 100 = 5\%$, and this causes a change in the demand of $\dfrac{2500}{40\,000} \times 100 = 6.25\%$.

The **elasticity** of the demand is defined as

$$\eta = \frac{\text{percentage change in demand}}{\text{percentage change in price}}$$

$$= \frac{6.25}{5} = 1.25$$

In this case a 5% increase in price causes a higher percentage (6.25%) reduction in demand. The price rise will in fact reduce the total sales income received by the company.

If q is the demand when the price is p, and a change δp in the price causes a change δq in the demand, the elasticity is

$$\eta = \frac{\text{percentage change in } q}{\text{percentage change in } p} = \frac{\dfrac{\delta q}{q} \times 100}{\dfrac{\delta p}{p} \times 100}$$

$$= \frac{\delta q}{\delta p} \times \frac{p}{q}$$

If we have some idea of the precise relationship between q and p, we can consider the effect of very small changes in price $\left(\text{as } \delta p \to 0, \dfrac{\delta q}{\delta p} \to \dfrac{dq}{dp}\right)$, and we calculate η as

$$\eta = \frac{dq}{dp} \times \frac{p}{q}$$

Example For a certain product, the demand q thousand, when he price is p pence, is given by $q = 720 - 6p$ (this is the 'demand schedule'). Calculate the elasticity of the demand when the price is (i) 30 pence, and (ii) 90 pence and comment on these values.

With reference to Fig. 12.18, the elasticity is

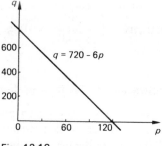

$$\eta = \frac{dq}{dp} \times \frac{p}{q} = (-6) \times \frac{p}{720 - 6p}$$

i.e. $\eta = \dfrac{p}{120 - p}$

Fig. 12.18

(We ignore the minus sign, which always occurs in these examples, since an increase in price causes a decrease in the demand.)

(i) when $p = 30$, $\eta = \dfrac{30}{120 - 30} = \dfrac{1}{3}$. For example, a price increase of 1% would reduce the demand by only $\frac{1}{3}\%$, and so the sales income would be increased.

(ii) when $p = 90$, $\eta = \dfrac{90}{120 - 90} = 3$. Now it would be advantageous to reduce the price (for example, a 1% price reduction will increase the demand by 3%).

Note that the maximum sales income is achieved when the elasticity is 1; in this example, when $\dfrac{p}{120 - p} = 1$, i.e. $p = 60$. A price of 60 pence gives the maximum income.

Exercise 12.8 (Percentage changes)

1 The radius of a circle is increased by 1%. Find the approximate percentage increase in (i) the circumference (ii) the area.

2 The surface area of a sphere is increased by 3%. Find the approximate percentage increase in the radius.

3 The side of a cube is decreased by a small amount. Show that the percentage change in the volume is approximately three times the percentage change in the length of the side.

4 The quantity T is calculated from x by means of the formula $T = k\sqrt{x}$, where k is a constant. Find the approximate percentage change in the value of T when x increases by 0.2%.

5 The power P watts generated when a potential difference of V volts is applied across a resistance of R ohms, is given by $P = \dfrac{V^2}{R}$. Find the approximate percentage change in the power when (i) the potential difference is increased by 2%; (ii) the resistance is increased by 2%.

6 When a current I amps flows through an inductance of L henrys, the energy stored in the magnetic field, E joules, is given by $E = \frac{1}{2}LI^2$. What percentage decrease in the current is required to reduce the energy by 5%?

7 The cost, Q million pounds, of building an oil installation with a target production of p thousand barrels per day is given by $Q = 2p + 0.1p^2$. If the production target is increased by $z\%$, find an expression (in terms of p and z) for the approximate percentage increase in the building cost.

Hence find the percentage increase in cost due to a 4% increase in production targets when (i) the original target was 5000 barrels per day; (ii) the original target was 20 000 barrels per day.

8   When the price of an article is p pounds, the demand is for q thousand articles. Market research shows that when $p = 3$, $q = 63$ and when $p = 9$, $q = 3$.
(i)   Assuming that the relationship between q and p is linear, find the demand schedule (i.e. express q in terms of p). Obtain an expression for the elasticity of the demand $\left(\eta = \dfrac{dq}{dp} \times \dfrac{p}{q} \right)$ and hence find the price which will produce the highest income.
(ii)   Further market research shows that the demand schedule is more accurately given by $q = p^2 - 22p + 120$ (for $p \leqslant 10$). Find the elasticity of the demand at the price found in part (i). Should the price be increased or decreased?

12.9   Application to errors

Very often one quantity is used to calculate another by means of a formula (for example, the radius r of a circle may be used to calculate its area, using the formula $A = \pi r^2$). If the first quantity contains an error then so will the value of the second quantity calculated from it. Provided the error is small, we can use the method of small changes to estimate the resulting error in the second quantity.

Example 1   The radius of a circle is 6.40 cm, but this is incorrectly measured as 6.42 cm. Find the error in the area if it is calculated from this measurement.

If r is the radius and A the area, we have $A = \pi r^2$, and so $\dfrac{dA}{dr} = 2\pi r$.

Now $r = 6.40$ and the error in r is $\delta r = 0.02$. Using $\delta A \approx \dfrac{dA}{dr}\, \delta r$,

the error in the area is

$\qquad \delta A \approx 2\pi r\, \delta r = 2\pi \times 6.40 \times 0.02$

$\qquad \qquad = 0.256\pi \approx 0.804\ \text{cm}^2$

If a quantity W contains an error δW, we define the **relative error** in W to be $\dfrac{\delta W}{W}$. This is usually given as a percentage; the **percentage error** in W is $\dfrac{\delta W}{W} \times 100$.

Example 2   A boat (B) is observed from the top of a cliff of known height h. The angle of depression (θ) of the boat is measured and is recorded as $34.2°$. The distance of the boat from the bottom of the cliff is then calculated using the formula $x = h \cot \theta$.
   It is subsequently discovered that an error of $0.8°$ has been made in the measurement of θ. What is the percentage error in the distance calculated?

With reference to Fig. 12.19, we have $x = h \cot \theta$, and so $\dfrac{dx}{d\theta} = -h \operatorname{cosec}^2\theta$. If the error in θ is $\delta\theta$ radians, the error in x is given by $\delta x \approx \dfrac{dx}{d\theta}\, \delta\theta = -h \operatorname{cosec}^2\theta\, \delta\theta$ (the negative sign indicates that an increase in θ causes a decrease in x). So the percentage error in x is

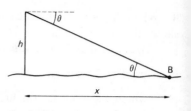

Fig. 12.19

$$\frac{\delta x}{x} \times 100 \approx \frac{-h \operatorname{cosec}^2 \theta\, \delta\theta}{h \cot \theta} \times 100 = -\frac{\delta\theta \times 100}{\sin \theta \cos \theta}$$

Now $\theta = 34.2°$, and $\delta\theta = 0.8° = 0.8 \times \dfrac{\pi}{180} = 0.014$ radians, so the percentage error in the distance is

$$(-)\frac{1.4}{\sin (34.2°) \cos (34.2°)} = 3.0\%$$

If a quantity has been obtained by measurement, or from an experiment, it is unlikely to be exactly correct, and so it contains an error. We do not know precisely what the error is (otherwise we would know the exact value of the quantity), but we can usually state the maximum possible error in the value. If we are given, say, $W = 5.83 \pm 0.12$, this means that the error δW in W, although unknown, lies between -0.12 and $+0.12$. Thus the true value of W is somewhere between 5.71 and 5.95.

We are sometimes given the maximum error as a percentage. For example, $W = 36.0 \pm 2\%$ means that the maximum error is 2% of 36.0, and so this is equivalent to $W = 36.0 \pm 0.72$.

A number given correct to a certain number of decimal places should be regarded as containing an error. For example, if $x = 4.76$ (to 2 decimal places), the true value of x might be anywhere between 4.755 and 4.765, and so the value 4.76 contains a possible error of ± 0.005. Thus $x = 4.76$ (to 2 d.p.) is equivalent to $x = 4.76 \pm 0.005$.

Example 3   The side of a cube is measured as 6 ± 0.05 cm. Calculate the volume, giving the range of possible error.

If the side of the cube is x, then its volume is $V = x^3$, and so $\dfrac{dV}{dx} = 3x^2$. If the error in x is δx, the error in V is given by

$$\delta V \approx \frac{dV}{dx} \delta x = 3x^2\, \delta x$$

Putting $x = 6$ and $\delta x = 0.05$, the maximum possible error in V is $\delta V \approx 3 \times 36 \times 0.05 = 5.4$. When $x = 6$, $V = 216$, so the volume is 216 ± 5.4 cm^3.

Example 4   The time T for one complete swing of a simple pendulum is given by $T = 2\pi \sqrt{\dfrac{l}{g}}$ where l is the length of the pendulum, and g is a known constant. If l is known to within $\pm 3\%$, find the maximum percentage error in T.

We have $T = 2\pi \sqrt{\dfrac{l}{g}} = \dfrac{2\pi}{\sqrt{g}} l^{\frac{1}{2}}$

so $\dfrac{\mathrm{d}T}{\mathrm{d}l} = \dfrac{\pi}{\sqrt{g}} l^{-\frac{1}{2}} = \dfrac{\pi}{\sqrt{gl}}$

If the error in l is δl, the error in T is given by

$$\delta T \approx \dfrac{\mathrm{d}T}{\mathrm{d}l}\delta l = \dfrac{\pi}{\sqrt{gl}}\delta l$$

and the percentage error in T is

$$\dfrac{\delta T}{T} \times 100 \approx \dfrac{\dfrac{\pi}{\sqrt{gl}}\delta l}{2\pi\sqrt{\dfrac{l}{g}}} \times 100$$

$$= \dfrac{\pi}{\sqrt{gl}}\delta l \times \dfrac{1}{2\pi} \times \sqrt{\dfrac{g}{l}} \times 100 = \dfrac{1}{2}\left(\dfrac{\delta l}{l} \times 100\right)$$

Now the maximum value of $\left(\dfrac{\delta l}{l} \times 100\right)$ is 3%; hence the maximum percentage error in T is $1\frac{1}{2}\%$.

Exercise 12.9   (Errors)

1   The side of a square 6 m long is incorrectly measured as 6.11 m. Find the resulting error in the calculation of the area of the square.

2   The area of a triangle is calculated using the formula $A = \frac{1}{2}bc\sin\theta$, where the sides b and c are known exactly, and θ is measured as $65.3°$. Subsequently it is discovered that an error of $1.0°$ has been made in the measurement of θ. Calculate the percentage error in the area.

3   The radius of a sphere is measured as 5 ± 0.04 cm. Calculate the volume and the surface area, giving the range of possible error in each case.

4   The side of a cube is given as 2.84 cm (correct to 2 d.p.). What is the maximum percentage error in the volume?

5   The height of a tower is calculated from the formula $h = d\tan\theta$, where d is a known distance, and θ is given as $21.4° \pm 2\%$. Find the maximum percentage error in the height.

6   If the radius of a sphere is subject to a small error, show that the percentage error in the volume is three times the percentage error in the radius.

7   W is calculated from x by means of the formula $W = kx^n$, where k and n are known constants. Show that the percentage error in W is n times the percentage error in x.

8   A sphere of radius a is filled with water to a depth x $(0 < x < a)$. Calculate the area of the surface of the water in terms of a and x.

If the sphere is of radius 10 cm and x is measured as 7 cm, use the method of small differences to calculate, correct to two significant figures, the maximum possible error in the calculated value of the surface area of the water if x may be in error by up to 0.5%.   (O)

Revision questions C

Revision paper C1

1 (i) Find the gradient of each of the following graphs at the point for which $x = 4$.

(a) $y = \sqrt{x}$ (b) $y = \dfrac{4}{x^2}$ (c) $y = \dfrac{x+2}{x+3}$ (d) $y = \dfrac{\sin x}{\cos x}$

(ii) Find the x coordinates of the points on the graph $y = x(x+2)^4$ for which the gradient is zero.   (O)

2 (i) Use the fact that $\sec x = \dfrac{1}{\cos x}$ to differentiate $\sec x$ with respect to x. Hence differentiate $\sec^2 x$. Hence or otherwise evaluate

$$\int_0^{\frac{1}{4}\pi} \tan x \sec^2 x \, dx$$

(ii) Evaluate (a) $\displaystyle\int_0^2 (1 + \sqrt{x})^2 \, dx$ (b) $\displaystyle\int_1^4 \left(1 + \frac{1}{x^2}\right) dx$ (O)

3 (i) Find the area between the curve $y = x^2$, the y-axis and the line $y = 1$.

(ii) Find dy/dx when $y = x \sin x$. Hence find the area bounded by the curve $y = x \cos x$, the x-axis and the ordinates $x = 0$ and $x = \frac{1}{3}\pi$. Give your answer correct to one decimal place.   (O)

4 A girl on a pony leaves a long straight road AB across a level moor at the point A and rides at right angles to the road for 3 km. She suddenly realises that she is due at B, 5 km from A, in 20 minutes. She can ride across the moor at 16 km h^{-1} and along the road at 20 km h^{-1}. She rides directly towards a point C on the road, x km from A, and then along the road from C to B. Find the value of x such that she gets to B in the least possible time. Find also how many minutes late she is on reaching B.   (O)

5 A train starts from station A to travel to station B. For the first 100 seconds it moves with velocity $v = \frac{1}{50}t^{\frac{3}{2}}$ where t is measured in seconds and v in metres per second. It then travels for 3 minutes at constant speed. The train is then 200 m from B. The brakes are immediately applied and the train is retarded uniformly to come to rest at B. Sketch the velocity–time graph for the journey. Calculate the total time of the journey and the distance from A to B.   (O)

6 While a rubber balloon is being inflated, it is in the shape of a hollow cylinder of radius x cm and length $3x$ cm which has hemispherical ends as shown in Fig. C.1.

Fig. C.1

200

Air is pumped into this balloon at a rate of $65\pi\,\text{cm}^3\,\text{s}^{-1}$. The maximum value of x before the balloon bursts is $x = 6$. How many seconds elapse before the balloon bursts? At what rate is x increasing when the balloon is about to burst? Calculate also the rate of increase of surface area at this instant. (O)

Revision paper C2

1 (i) Differentiate the following functions with respect to x:

(a) $\dfrac{x}{x+2}$ (b) $x^2(x^3+1)^2$

(ii) Let $y = \cos^2 x$. Calculate $\dfrac{d^2y}{dx^2}$ and prove that

$$\frac{d^2y}{dx^2} + 4y = 2 \tag{O}$$

2 (i) Evaluate the following integrals, giving your answers, where appropriate, correct to two significant figures:

(a) $\displaystyle\int_1^2 x^{\frac12}\,dx$ (b) $\displaystyle\int_{\frac16\pi}^{\frac13\pi} \sin 2x\,dx$ (c) $\displaystyle\int_0^{\sqrt2} x(x^2+1)^2\,dx$

(ii) The curve $x^2 + y^2 = 25$ is a circle, centre $(0,0)$ and radius 5. Use this fact to evaluate the integral

$$\int_0^5 \sqrt{(25-x^2)}\,dx \tag{O}$$

3 A glass is made in the shape formed by rotating through two right angles about the y-axis that part of the curve $y = \frac34 x^2$ which lies between $x = -4$ and $x = 4$. Given that the units on each axis are centimetres, calculate the volume of the glass in litres.

Calculate also the depth of the liquid when the glass is exactly half full. What is the radius of the surface of the liquid in this case? (O)

4 Two cyclists A and B are moving, one on each of two straight roads, with uniform speeds towards the cross-roads where the two roads intersect at right angles. When they are first observed A is 300 m and B is 400 m from the cross-roads; A has a speed of $16\,\text{m s}^{-1}$ and B has a speed of $12\,\text{m s}^{-1}$. Prove that their distance apart t seconds later is given by the expression

$$s^2 = 2.5 \times 10^5 - 1.92 \times 10^4\,t + 4 \times 10^2\,t^2$$

By considering the minimum value of s^2, or otherwise, find when the cyclists are the shortest distance apart and show that this distance is 140 m. (O)

5 The speed $v\,\text{m s}^{-1}$ of a stopping train travelling between two stations is given by the formula $v = \dfrac{t(60-t)}{60}$ where t is the time in seconds. The formula is valid for $0 \leqslant t \leqslant 60$.

(i) Find a formula for the acceleration of the train in terms of t.
(ii) Find the maximum speed of the train.
(iii) If the train is at the station when $t = 0$ and $t = 60$, find the distance between the stations.
(iv) What is the average speed of the train as it travels between the stations? (O)

6 A hemispherical bowl of radius a cm is initially full of water. The water runs out of a small hole at the bottom of the bowl at a constant rate which is such that it would empty the bowl in 24 s. Given that, when the depth of the water is x cm, the volume of water is $\frac13\pi x^2(3a-x)\,\text{cm}^3$, prove that the depth is decreasing at a rate of

$$\frac{a^3}{[36x(2a-x)]}\ \text{cm s}^{-1}$$

Find after what time the depth of water is $\frac{1}{2}a$ cm, and the rate at which the water level is then decreasing. (O and C)

Miscellaneous revision questions: paper C3

1 (i) Differentiate $(1-x^2)^5$ with respect to x. Using the product rule, or otherwise, differentiate

$$(1+x)(1-x^2)^5$$

with respect to x. For what values of x is this latter derivative zero?

(ii) Differentiate $\dfrac{x}{x^2+4}$ with respect to x. (O)

2 (i) Differentiate with respect to x

(a) $\sqrt{1+4x^2}$ (b) $\sin^2 5x$

(ii) If $y = \sqrt{\dfrac{x}{2x+1}}$

find the value of $\dfrac{dy}{dx}$ when $x = 4$. (O and C)

3 (i) Show that, when k is constant, the curve

$$y = 3x^4 - 8x^3 - 6x^2 + 24x + k$$

has a stationary point when $x = 1$ and find the values of x at the other two stationary points on the curve.

Find the values of k for which the curve touches the x-axis.

(ii) A spherical balloon is being inflated so that its volume increases at a constant rate. Show that the rate of increase of surface area is inversely proportional to the radius. (L)

4 Given that

$$y = \sqrt{3}\cos x + \sin x$$

where $0 \leqslant x \leqslant 2\pi$, find the maximum and minimum values of y.

Sketch the graph of y.

Calculate the values of x for which $\dfrac{dy}{dx} = -1$.

Show that the mean value of y^2 with respect to x in the interval $0 \leqslant x \leqslant \dfrac{\pi}{2}$ is

$$2(\pi + \sqrt{3})/\pi$$ (JMB)

5 A right circular cone has a semi-vertical angle θ and slant edge of length 10 cm. Find an expression for its volume, V, in terms of θ. Hence show that the maximum volume of the cone as θ varies occurs when $\tan\theta = \sqrt{2}$, and find the maximum volume, rounding your answer to three significant figures.

[Hint: When $\tan\theta = \sqrt{2}$, $\sin\theta = \sqrt{2}/\sqrt{3}$ and $\cos\theta = 1/\sqrt{3}$.] (O)

6 Verify that the curves $y = \cos\frac{1}{2}\pi x$ and $y = \frac{1}{2}(x^3 - 3x^2 + 2)$ intersect when $x = 0$, $x = 1$ and $x = 2$ and calculate the values of the gradients of both curves at these three points of intersection.

Calculate to one significant figure the percentage error in using

(i)   the value of $\frac{1}{2}(x^3 - 3x^2 + 2)$ when $x = \frac{1}{2}$ as the true value of $\cos \frac{1}{2}\pi x$ when $x = \frac{1}{2}$;

(ii)   the area under the graph of $\frac{1}{2}(x^3 - 3x^2 + 2)$ in place of the area under the graph of $y = \cos \frac{1}{2}\pi x$ between $x = 0$ and $x = 1$. **(O)**

7   Find the point A at which the curve $y = 1 + \sqrt{x}$ meets the y-axis. Find also the equation of the tangent to the curve at B(4, 3) and show that it passes through C(8, 4).

O is the origin. The curve formed by the line OA, arc AB and line BC is rotated through an angle 2π about the x-axis to form a vase. Find the volume of this vase, giving your answer as a multiple of π. **(O)**

8   (i)   Integrate $x(x^2 + 1)^3$ with respect to x.

(ii)   Find the volume of the solid formed when the finite area bounded by the curve $y = 5x - x^2$ and the straight line $y = x$ is rotated about the x-axis through four right angles. Give your answer as a multiple of π. **(O)**

9   (i)   A spherical balloon is being inflated, and when its radius is 10 cm the radius is increasing at 0.1 centimetres per second. Find the rate at which the volume is increasing. (Leave π in your answer.)

When the radius reaches 25 cm the surface area is increasing at 8π square centimetres per second. Find the rate at which the radius is then increasing. **(O)**

(ii)   The magnetising force H inside a long solenoid is given by the formula $H = \dfrac{knI}{l}$ where n is the number of turns of the wire, l the length of the solenoid and I the current in the wire (in suitable units) with k constant. Find the approximate percentage change in H if

(a)   I is decreased by 2%, keeping n and l constant;

(b)   l is increased by 2%, keeping I and n constant.

(iii)   The graph of $y = x^2 - 50$ cuts the x-axis near $x = 7$. Starting from the point $(7, -1)$ use the relation $\delta y \approx \dfrac{dy}{dx}\,\delta x$ to find an approximation to $\sqrt{50}$. **(O)**

10   Particle A moves along a straight line such that its distance s metres from a fixed point O on the line at time t seconds is given by $x = t^3 - 5t^2 + 3t + \frac{1}{2}$. Particle B moves along the same straight line with acceleration a in $\mathrm{m\,s^{-2}}$ at time t seconds given by $a = 2t + 1$. Initially B is 1 metre from O on the same side of O as A and moving towards O at $2\,\mathrm{m\,s^{-1}}$.

After what time is the acceleration of A greater than the acceleration of B? At what times will the velocities of the two particles be the same? On the same axes sketch the velocity–time graphs of the two particles for the period $0 \leqslant t \leqslant 6$. **(MEI)**

11   A particle moves on a straight line through a fixed point O so that at time t its displacement from O is x and its equation of motion is

$$d^2x/dt^2 = -16x$$

Given that $x = -12$ and $dx/dt = 20$ when $t = \pi/4$, find

(i)   the position and velocity of the particle when $t = \pi/2$;

(ii)   the least positive value of t for which $x = 0$, giving two decimal places in your answer. **(JMB)**

12   A car of mass one tonne was started from rest and given continuous acceleration for 35 seconds. Measurements of the resultant propulsive force were recorded and are shown in the following table correct to the nearest ten newtons. $\left[\text{The acceleration} = \dfrac{\text{force}}{1000} \text{ in this case.} \right]$

Time(s)	0	5	10	15	20	25	30	35
Force(N)	1200	1020	870	740	620	500	390	290

Use the trapezium rule to calculate an estimate of the speed of the car in metres per second at the end of the 35 second period. **(JMB)**

13

Curves

13.1 Stationary points

We recall (from Chapter 3, p. 38) that a stationary point on a curve $y = f(x)$ is a point where the gradient $\frac{dy}{dx}$ is zero; the value of y at such a point is called a stationary value. A stationary point may be a maximum, a minimum, or a point of inflexion.

If $\frac{d^2y}{dx^2}$ is negative, we have a maximum.

If $\frac{d^2y}{dx^2}$ is positive, we have a minimum.

If $\frac{d^2y}{dx^2} = 0$, it is usually best to examine the sign of $\frac{dy}{dx}$ on each side of the stationary point.

Example 1   Find and classify the stationary points on the curve $y = \frac{x}{1+x^2}$.

We have

$$\frac{dy}{dx} = \frac{(1+x^2)(1) - x(2x)}{(1+x^2)^2} = \frac{1-x^2}{(1+x^2)^2}$$

$\frac{dy}{dx} = 0$ when $1 - x^2 = 0$, i.e. when $x = \pm 1$.

When $x = -1, y = -\frac{1}{2}$ and when $x = 1, y = \frac{1}{2}$; so the stationary points are $(-1, -\frac{1}{2})$ and $(1, \frac{1}{2})$.

We have

$$\frac{d^2y}{dx^2} = \frac{d}{dx}[(1-x^2)(1+x^2)^{-2}]$$

$$= (1-x^2)(-2)(1+x^2)^{-3}(2x) + (-2x)(1+x^2)^{-2}$$

$$= \frac{-4x(1-x^2)}{(1+x^2)^3} - \frac{2x}{(1+x^2)^2}$$

When $x = -1, \frac{d^2y}{dx^2} = 0 - \frac{(-2)}{(1+1)^2} > 0$, so $(-1, -\frac{1}{2})$ is a minimum.

When $x = 1, \frac{d^2y}{dx^2} = 0 - \frac{2}{(1+1)^2} < 0$, so $(1, \frac{1}{2})$ is a maximum.

The following examples show how the methods of calculus can be used to establish certain inequalities.

Example 2   Show that $\sin x < x$ when $x > 0$ and $\sin x > x$ when $x < 0$.

Consider the curve $y = x - \sin x$. Then

$$\frac{dy}{dx} = 1 - \cos x$$

There are stationary points when $\cos x = 1$, i.e. $x = 0$, $\pm 2\pi$, $\pm 4\pi, \ldots$ At all other points, $\cos x < 1$, and so $\dfrac{dy}{dx}$ is positive. Hence all the stationary points are points of inflexion, and y is an increasing function of x.

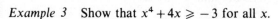

Fig. 13.1

When $x = 0$, $y = 0$. We now sketch the curve (Fig. 13.1). If $x > 0$, then $y > 0$, i.e. $x - \sin x > 0$, i.e. $\sin x < x$. If $x < 0$, then $y < 0$, i.e. $x - \sin x < 0$, i.e. $\sin x > x$.

Example 3   Show that $x^4 + 4x \geqslant -3$ for all x.

Consider the curve $y = x^4 + 4x$. Then

$$\frac{dy}{dx} = 4x^3 + 4$$

$\dfrac{dy}{dx} = 0$ when $x = -1$, and then $y = 1 - 4 = -3$. So there is only one stationary point: $(-1, -3)$.

We have

$$\frac{d^2 y}{dx^2} = 12x^2$$

which is positive when $x = -1$, so $(-1, -3)$ is a minimum. The curve is a continuous one, and so $y \geqslant -3$ for all x, i.e. $x^4 + 4x \geqslant -3$.

Exercise 13.1   (Stationary points)

Find and classify the stationary points on the following curves (questions 1 to 12).

1 $y = 4x^2 - \dfrac{1}{x}$

2 $y = \dfrac{x - 3}{x^2 + 7}$

3 $y = \dfrac{x}{(x + 1)^2}$

4 $y = \dfrac{(x + 1)(2x - 1)}{x - 1}$

5 $y = \dfrac{x^2 + 2x + 3}{x^2 - 2x - 3}$

6 $y = \dfrac{2x^2 + 4x + 3}{x^2 - 1}$

7 $y = x^2 + \dfrac{1}{x^2}$

8 $y = (x + 2)^6$

9 $y = 3x^4 - 4x^3$

10 $y = \sin x + \cos x$

11 $y = 8 \cos x + \cot x$

12 $y = \sec x + \csc x$

13 Show that the curve $y = \tan x - x$ has a point of inflexion at the origin, and deduce that

$$\tan x > x \text{ when } 0 < x < \tfrac{1}{2}\pi$$

and $\tan x < x$ when $-\tfrac{1}{2}\pi < x < 0$.

14 Show that $6x - x^6 \leqslant 5$ for all x.

15 Using the results $\sin x < x$ when $x > 0$ and $\sin x > x$ when $x < 0$ (see p. 205, Example 2), show that the curve $y = \cos x - 1 + \frac{1}{2}x^2$ has a minimum when $x = 0$. Deduce that $\cos x > 1 - \frac{1}{2}x^2$ for all non-zero values of x.

13.2 Points of inflexion

We have already seen that a stationary point such as A on Fig. 13.2 is called a point of inflexion.

The gradient $\dfrac{dy}{dx}$ is positive to the left of A, zero at A, and positive to the right of A; so $\dfrac{dy}{dx}$ has a minimum value (of zero) at A.

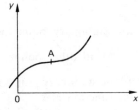

Fig. 13.2

Generally, a **point of inflexion** on a curve is any point at which $\dfrac{dy}{dx}$ has a (local) maximum or minimum value (it need not be a stationary point). At such a point the graph straightens out and starts to curve the other way. Figs. 13.3, 13.4 and 13.5 show some more examples.

To find the points of inflexion on a curve, we find the points where $\dfrac{dy}{dx}$ has a maximum or minimum, using the same methods as before. We need $\dfrac{d}{dx}\left(\dfrac{dy}{dx}\right) = 0$, i.e. $\dfrac{d^2 y}{dx^2} = 0$. We then consider $\dfrac{d^2}{dx^2}\left(\dfrac{dy}{dx}\right) = \dfrac{d^3 y}{dx^3}$. If $\dfrac{d^3 y}{dx^3} \neq 0$, then we definitely have a point of inflexion $\left(\text{because if } \dfrac{d^3 y}{dx^3} > 0, \dfrac{dy}{dx} \text{ has a minimum value, and if } \dfrac{d^3 y}{dx^3} < 0, \dfrac{dy}{dx} \text{ has a maximum value}\right)$. If however $\dfrac{d^3 y}{dx^3} = 0$, then further investigation will be necessary $\left(\dfrac{d^2 y}{dx^2} \text{ must have opposite signs before and after a point of inflexion}\right)$.

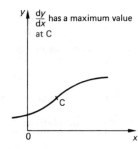

Fig. 13.3

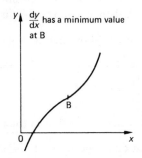

Fig. 13.4

Example 1 Find the points of inflexion on the curve $y = 16 \sin x + \tan x$.

We have $\dfrac{dy}{dx} = 16 \cos x + \sec^2 x$

and $\dfrac{d^2 y}{dx^2} = -16 \sin x + 2 \sec x (\sec x \tan x)$

$= -16 \sin x + 2 \sec^2 x \tan x$

$\dfrac{d^2 y}{dx^2} = 0$ when $16 \sin x = 2 \sec^2 x \tan x$, i.e. $16 \sin x = \dfrac{2 \sin x}{\cos^3 x}$,

so either

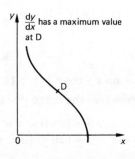

Fig. 13.5

$$\sin x = 0, \text{ i.e. } x = 0, \pi, 2\pi \text{ (for } 0 \leqslant x \leqslant 2\pi)$$

or $\cos^3 x = \dfrac{1}{8}$, i.e. $\cos x = \dfrac{1}{2}$, i.e. $x = \dfrac{\pi}{3}, \dfrac{5\pi}{3}$

But $\dfrac{d^3 y}{dx^3} = -16 \cos x + 2 \sec^2 x (\sec^2 x) + (4 \sec x \sec x \tan x)\tan x$

$$= -16 \cos x + 2 \sec^4 x + 4 \sec^2 x \tan^2 x$$

which is non-zero for all the above values of x (for example, when

$x = \dfrac{\pi}{3}, \dfrac{d^3 y}{dx^3} = -16 \times \dfrac{1}{2} + 2 \times 16 + 4 \times 4 \times 3 = 72$).

When $x = 0, \pi$ or $2\pi, y = 0$.

When $x = \dfrac{\pi}{3}, y = 16 \times \dfrac{\sqrt{3}}{2} + \sqrt{3} = 9\sqrt{3}$.

When $x = \dfrac{5\pi}{3}, y = 16 \times \left(-\dfrac{\sqrt{3}}{2}\right) - \sqrt{3} = -9\sqrt{3}$.

So the points of inflexion are $(0, 0)$, $\left(\dfrac{\pi}{3}, 9\sqrt{3}\right)$, $(\pi, 0)$,

$\left(\dfrac{5\pi}{3}, -9\sqrt{3}\right)$, $(2\pi, 0)$, and other points obtained by adding any integer multiple of 2π to these values of x.

Example 2 Find the points of inflexion on the curve $y = x^5 - 5x^4 + x$.

We have $\dfrac{dy}{dx} = 5x^4 - 20x^3 + 1$

$$\dfrac{d^2 y}{dx^2} = 20x^3 - 60x^2 = 20x^2(x - 3)$$

and so $\dfrac{d^2 y}{dx^2} = 0$ when $x = 0$ or 3

Also, $\dfrac{d^3 y}{dx^3} = 60x^2 - 120x$

When $x = 3, \dfrac{d^3 y}{dx^3} = 540 - 360 \neq 0$, so we have a point of inflexion.

When $x = 0$, $\dfrac{d^3 y}{dx^3} = 0$, and we must consider the sign of $\dfrac{d^2 y}{dx^2} = 20x^2(x - 3)$ on each side of $x = 0$. If x is close to zero, $20x^2$ is positive and $(x - 3)$ is negative, so $\dfrac{d^2 y}{dx^2}$ is negative on both sides of $x = 0$. Therefore $x = 0$ does not give a point of inflexion.

There is only one point of inflexion, when $x = 3$ and $y = 243 - 405 + 3 = -159$, i.e. at $(3, -159)$.

Exercise 13.2   (Points of inflexion)

Find the points of inflexion on the following curves.

1 $y = x^4 - 4x^3 + 7x$

2 $y = \dfrac{1}{1+x^2}$

3 $y = \dfrac{1}{1+x+x^2}$

4 $y = x^5 - 7x + 2$

5 $y = 3x + 5x^4 - 2x^6$

6 $y = 8x^2 + \dfrac{1}{x}$

7 $y = 4\sin x - \sin 2x$

8 $y = \dfrac{\sin x}{2\cos x - 1}$

13.3   Curve sketching

When sketching a curve $y = f(x)$ we usually consider the problem under the following six headings. (With experience we may not need to consider them all.)

(I)   Values of x for which y is not defined

A quotient $y = \dfrac{u}{v}$ is not defined when $v = 0$. If $v = 0$ when $x = a$, there is no value of y, and hence no point on the curve, corresponding to $x = a$. Provided $u \neq 0$, y becomes numerically large as x approaches a, and so $x = a$ is a vertical asymptote of the curve (see p. 109). [See p. 234 for a discussion of the case $u = v = 0$ when $x = a$.]

Example 1

$$y = \frac{x}{(x-1)(x+3)}$$

The denominator is zero when $x = 1$ or -3; thus $x = 1$ and $x = -3$ are vertical asymptotes of the curve.

Example 2

$$y = \frac{x}{1+x^2}$$

The denominator $(1 + x^2)$ is never zero, so y is defined for all values of x, and this curve has no vertical asymptotes.

Example 3

$$y = \sin x + \tan x$$

This may be written as $y = \sin x + \dfrac{\sin x}{\cos x}$ so y is not defined when $\cos x = 0$. There are vertical asymptotes at $x = \pm\dfrac{\pi}{2}, \ \pm\dfrac{3\pi}{2}, \ \pm\dfrac{5\pi}{2}, \ldots$

There are other reasons (apart from a zero denominator) why y may not be defined. For example, we cannot have the square root of a negative number.

Example 4

$$y = \sqrt{x^2 - 4}$$

y is not defined when $(x^2 - 4)$ is negative, i.e. when $-2 < x < 2$. The curve does not exist between $x = -2$ and $x = 2$.

Example 5

$$y = \frac{x + 2}{x\sqrt{x - 3}}$$

y is not defined when $(x - 3)$ is negative, i.e. when $x < 3$.

$x = 3$ makes the denominator zero, so $x = 3$ is a vertical asymptote of the curve. (Note that $x = 0$, although it makes the denominator zero, is not a vertical asymptote because it lies in the region where the curve does not exist.)

Example 6

$$y = \sin^{-1}(2x - 3)$$

This is meaningful only if $-1 \leqslant 2x - 3 \leqslant 1$, i.e. $2 \leqslant 2x \leqslant 4$, i.e. $1 \leqslant x \leqslant 2$. y is not defined (and the curve does not exist) when $x < 1$ or $x > 2$.

(II) Behaviour as $x \to \pm \infty$

We consider first the behaviour of x^n as $x \to \pm \infty$.

If n is positive, $x^n \to +\infty$ as $x \to +\infty$.

If n is an even integer, $x^n \to +\infty$ as $x \to -\infty$.

If n is an odd integer, $x^n \to -\infty$ as $x \to -\infty$.

If n is negative, $x^n \to 0$ as $x \to \pm \infty$.

To examine the behaviour of a function as $x \to \pm \infty$, the general method is to take out as factors the highest powers of x occurring in the numerator and the denominator.

Example 1

$$y = x^5 - 5x^4 + x$$

We have

$$y = x^5 \left(1 - \frac{5}{x} + \frac{1}{x^4} \right)$$

As $x \to \pm \infty$, $-\dfrac{5}{x}$ and $\dfrac{1}{x^4}$ tend to zero; so y behaves like x^5. Thus

$$y \to +\infty \text{ as } x \to +\infty$$
$$y \to -\infty \text{ as } x \to -\infty$$

Example 2

$$y = \frac{x+2}{2x^2 - x + 5}$$

We have

$$y = \frac{x\left(1 + \dfrac{2}{x}\right)}{2x^2\left(1 - \dfrac{1}{2x} + \dfrac{5}{2x^2}\right)} = \frac{1}{2x} \times \frac{1 + \dfrac{2}{x}}{1 - \dfrac{1}{2x} + \dfrac{5}{2x^2}}$$

As $x \to \pm\infty$, $\dfrac{2}{x}$, $-\dfrac{1}{2x}$ and $\dfrac{5}{2x^2}$ all tend to zero; so y behaves like

$\dfrac{1}{2x}$. Thus $y \to 0$ as $x \to \pm\infty$. The line $y = 0$ (the x-axis) is a horizontal asymptote to this curve.

(Alternatively, we could say that as $x \to \pm\infty$, the highest powers of x will dominate, and so $y \approx \dfrac{x}{2x^2} = \dfrac{1}{2x}$. The above working justifies this method.)

Example 3

$$y = \frac{1 + 3x^2 - x^3}{x^{\frac{2}{3}} + 1}$$

We have

$$y = \frac{-x^3\left(-\dfrac{1}{x^3} - \dfrac{3}{x} + 1\right)}{x^{\frac{2}{3}}(1 + x^{-\frac{2}{3}})} = -x^{\frac{7}{3}} \times \frac{1 - \dfrac{1}{x^3} - \dfrac{3}{x}}{1 + x^{-\frac{2}{3}}}$$

As $x \to \pm\infty$, y behaves like $-x^{\frac{7}{3}} = -(x^{\frac{1}{3}})^7$. Thus

$$y \to -\infty \text{ as } x \to +\infty$$
$$y \to +\infty \text{ as } x \to -\infty$$

Example 4

$$y = \frac{x^2}{2 - 3x^2}$$

We have

$$y = \frac{x^2}{-3x^2\left(-\dfrac{2}{3x^2} + 1\right)} = -\frac{1}{3} \times \frac{1}{1 - \dfrac{2}{3x^2}}$$

Thus $y \to -\frac{1}{3}$ as $x \to \pm\infty$. The line $y = -\frac{1}{3}$ is a horizontal asymptote.

Example 5

$$y = \frac{x}{\sqrt{x+1}}$$

y is not defined when $x < -1$, so we only need to consider $x \to +\infty$. We have

$$y = \frac{x}{\sqrt{x\left(1+\dfrac{1}{x}\right)}} = \frac{x}{\sqrt{x}\sqrt{1+\dfrac{1}{x}}} = x^{\frac{1}{2}} \times \frac{1}{\sqrt{1+\dfrac{1}{x}}}$$

Thus $y \to +\infty$ as $x \to +\infty$.

If y is expressed in terms of circular functions only $\Bigg($ for example,

$y = \dfrac{\sin x - 4 \tan 2x}{1 - 2 \cos x} \Bigg)$, there is no special behaviour as $x \to \pm \infty$, because the curve just repeats itself at fixed intervals of x (in this example, at intervals of 2π).

However, if we have a mixture of algebraic and circular functions, it is sensible to examine the behaviour as $x \to \pm \infty$.

Example 6

$$y = \frac{\sin x}{x}$$

We have

$$y = \frac{1}{x} \times \sin x$$

$\sin x$ always lies between -1 and 1, so y lies between $-\dfrac{1}{x}$ and $\dfrac{1}{x}$.

As $x \to \pm \infty$, $\dfrac{1}{x} \to 0$, and hence $y \to 0$. (We shall sketch this curve later; see p. 234.)

(III) Where *y* is zero, positive or negative

If y can be expressed as a product or quotient of simple functions, we have already seen (in Chapter 9) how to investigate whether y is zero, positive or negative.

(IV) Easy points

Note any points on the curve which can be found easily. For example, substituting $x = 0$ gives the point where the curve crosses the y-axis.

(V) Symmetry

Note any symmetry of the curve.

(a)   If the value of y is unchanged when x is replaced by $-x$, the curve is symmetrical about the y-axis. This is because whenever (x, y) is on the curve, so is $(-x, y)$.

(b)   If, when x is replaced by $-x$, y is replaced by $-y$, the curve has rotational symmetry about the point $(0, 0)$. This is because whenever (x, y) is on the curve, so is $(-x, -y)$.

(VI) Stationary points

Find the stationary points on the curve, and determine the nature of each one. This often involves tedious calculations, and it may be possible to sketch the curve without finding the precise positions of the stationary points.

Examples of curve sketching

Example 1

$$y = \frac{x}{1 + x^2}$$

(I)   The denominator $(1 + x^2)$ is never zero, and so y is defined for all values of x and the curve is continuous.

(II)   For the behaviour as $x \to \pm\infty$,

$$y = \frac{x}{x^2\left(\dfrac{1}{x^2} + 1\right)} = \frac{1}{x} \times \frac{1}{1 + \dfrac{1}{x^2}}$$

Hence $y \to 0$ as $x \to \pm\infty$.

(III) $y = 0$ when $x = 0$. Since $(1 + x^2)$ is always positive, y is negative when $x < 0$ and positive when $x > 0$.

(V)   When x is replaced by $-x$, y is replaced by

$$\frac{-x}{1 + (-x)^2} = \frac{-x}{1 + x^2} = -y$$

$\left(\text{i.e. } \dfrac{x}{1 + x^2} \text{ is an odd function}\right)$ so the curve has rotational

symmetry about the point $(0, 0)$.

(VI)   The stationary points (see p. 204, Example 1) are $(-1, -\frac{1}{2})$, which is a minimum, and $(1, \frac{1}{2})$, which is a maximum.

The curve is sketched in Fig. 13.6.

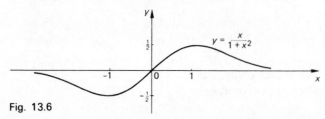

Fig. 13.6

Example 2

$$y = x + \frac{4}{x}$$

(I)   When $x = 0$, y is not defined and the curve has a vertical asymptote.

(II)   As $x \to \pm \infty$, $\frac{4}{x} \to 0$, and so y behaves like x, i.e. $y \to +\infty$ as $x \to +\infty$, and $y \to -\infty$ as $x \to -\infty$.

(III) y is negative when $x < 0$, and positive when $x > 0$.

(V)   When x is replaced by $-x$, y is replaced by $-x - \frac{4}{x} = -y$ so the curve has rotational symmetry about the origin.

(VI) $\frac{dy}{dx} = 1 - \frac{4}{x^2}$. For stationary points, $\frac{dy}{dx} = 0$, i.e. $x = -2$ or 2.

When $x = -2$, $y = -4$, and when $x = 2$, $y = 4$.

$\frac{d^2y}{dx^2} = \frac{8}{x^3}$. When $x = -2$, $\frac{d^2y}{dx^2}$ is negative, and so $(-2, -4)$ is a

maximum. When $x = 2$, $\frac{d^2y}{dx^2}$ is positive, and so $(2, 4)$ is a minimum.

The curve is sketched in Fig. 13.7.

Example 3

$$y = \frac{3x^2 - 12x + 1}{x^2 - 3x - 4}$$

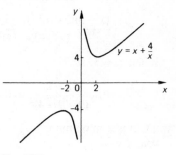

Fig. 13.7

(I)   The denominator is zero when $x^2 - 3x - 4 = 0$, i.e. $(x + 1)(x - 4) = 0$, i.e. $x = -1$ or 4. $x = -1$ and $x = 4$ are vertical asymptotes of the curve. We begin our sketch by putting in these asymptotes.

(II)   For the behaviour as $x \to \pm \infty$,

$$y = \frac{3x^2\left(1 - \dfrac{4}{x} + \dfrac{1}{3x^2}\right)}{x^2\left(1 - \dfrac{3}{x} - \dfrac{4}{x^2}\right)} = 3 \times \frac{1 - \dfrac{4}{x} + \dfrac{1}{3x^2}}{1 - \dfrac{3}{x} - \dfrac{4}{x^2}}$$

Thus $y \to 3$ as $x \to \pm \infty$, i.e. $y = 3$ is a horizontal asymptote of the curve. We put this on the sketch.

(III) $y = 0$ when $3x^2 - 12x + 1 = 0$,

i.e. $x = \dfrac{12 \pm \sqrt{144 - 12}}{6} \approx 0.1$ or 3.9. We mark these points on the sketch.

We sketch the graphs of $3x^2 - 12x + 1$ and $x^2 - 3x - 4$ (see Fig. 13.8). The sign of the quotient $y = \dfrac{3x^2 - 12x + 1}{x^2 - 3x - 4}$ is indicated on the x-axis.

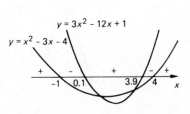

Fig. 13.8

(IV)   When $x = 0$, $y = -\frac{1}{4}$, so the curve crosses the y-axis at $(0, -\frac{1}{4})$. We mark this point on the sketch.

(VI) $\dfrac{dy}{dx} = \dfrac{(x^2 - 3x - 4)(6x - 12) - (3x^2 - 12x + 1)(2x - 3)}{(x^2 - 3x - 4)^2}$

$= \dfrac{3x^2 - 26x + 51}{(x^2 - 3x - 4)^2} = \dfrac{(x - 3)(3x - 17)}{(x^2 - 3x - 4)^2}$

For stationary points, $\dfrac{dy}{dx} = 0$, i.e. $x = 3$ or $\frac{17}{3}$. When $x = 3$, $y = 2$, and when $x = \frac{17}{3}$, $y = 2.64$. We mark these points.

We have

$\dfrac{d^2y}{dx^2} = \dfrac{d}{dx}\left[(3x^2 - 26x + 51)(x^2 - 3x - 4)^{-2}\right]$

$= (3x^2 - 26x + 51)(-2)(x^2 - 3x - 4)^{-3}(2x - 3) + (6x - 26)(x^2 - 3x - 4)^{-2}$

$= \dfrac{(x - 3)(3x - 17)(-2)(2x - 3)}{(x^2 - 3x - 4)^3} + \dfrac{6x - 26}{(x^2 - 3x - 4)^2}$

When $x = 3$,

$\dfrac{d^2y}{dx^2} = 0 + \dfrac{18 - 26}{(9 - 9 - 4)^2} < 0$

so $(3, 2)$ is a maximum.

When $x = \frac{17}{3}$,

$\dfrac{d^2y}{dx^2} = 0 + \dfrac{34 - 26}{(\frac{289}{9} - 17 - 4)^2} > 0$

so $(\frac{17}{3}, 2.64)$ is a minimum.

The curve is sketched in Fig. 13.9.

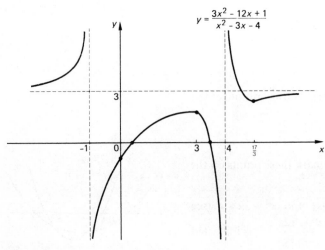

$$y = \frac{3x^2 - 12x + 1}{x^2 - 3x - 4}$$

Fig. 13.9

Example 4

$$y^2 = \frac{2x}{x-2}$$

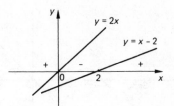

Fig. 13.10

We first sketch the curve $y = \dfrac{2x}{x-2}$.

(I) $x = 2$ is a vertical asymptote.

(II)   For the behaviour as $x \to \pm \infty$,

$$y = \frac{2x}{x\left(1 - \dfrac{2}{x}\right)} = \frac{2}{1 - \dfrac{2}{x}}$$

and so $y \to 2$ as $x \to \pm \infty$. $y = 2$ is a horizontal asymptote.

(III) $y = 0$ when $x = 0$.

We sketch the graphs of $2x$ and $(x - 2)$ (see Fig. 13.10). The sign of $\dfrac{2x}{x-2}$ is indicated on the x-axis.

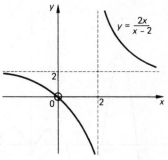

Fig. 13.11

(VI)   If $y = \dfrac{2x}{x-2}$, then

$$\frac{dy}{dx} = \frac{(x-2)(2) - (2x)(1)}{(x-2)^2} = \frac{-4}{(x-2)^2}$$

$\dfrac{dy}{dx}$ is never zero, and so there are no stationary points. In fact the gradient of the curve is always negative (see Fig. 13.11).

The required curve $y^2 = \dfrac{2x}{x-2}$ is $y = \pm\sqrt{\dfrac{2x}{x-2}}$. When $\dfrac{2x}{x-2}$ is negative (i.e. for $0 < x < 2$) there is no value of y and the curve does not exist. When $\dfrac{2x}{x-2}$ is positive, there are two values of y for each value of x. The curve is symmetrical about the x-axis (see Fig. 13.12).

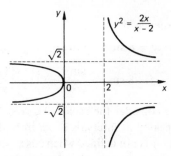

Fig. 13.12

Example 5

$$y = x \sin x$$

(III) $y = 0$ when $x = 0, \pm\pi, \pm 2\pi, \ldots$

We sketch the graphs of x and $\sin x$ (see Fig. 13.13.). The sign of their product $y = x \sin x$ has been indicated on the x-axis.

(V)   When x is replaced by $-x$, y is replaced by $(-x)\sin(-x)$ $= x \sin x = y$, so the curve is symmetrical about the y-axis.

Since $\sin x$ always lies between -1 and $+1$, the curve $y = x \sin x$ lies between the lines $y = -x$ and $y = x$. It touches these lines when $\sin x = \pm 1$, i.e. $x = \pm\frac{1}{2}\pi, \pm\frac{3}{2}\pi, \ldots$ (for example, when $x = \frac{1}{2}\pi$, $y = \frac{1}{2}\pi$; when $x = \frac{3}{2}\pi$, $y = -\frac{3}{2}\pi$). The curve is sketched in Fig. 13.14.

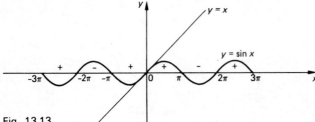

Fig. 13.13

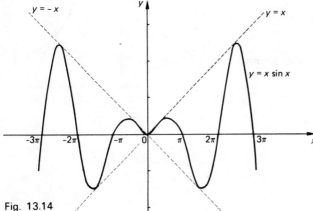

Fig. 13.14

Example 6

$$y = 8 \sin x - \tan x$$

$$= 8 \sin x - \frac{\sin x}{\cos x} = \frac{\sin x (8 \cos x - 1)}{\cos x}$$

We consider the region $0 \leqslant x \leqslant 2\pi$. This portion of the curve is repeated at intervals of 2π in both directions.

(I) y is not defined when $\cos x = 0$, i.e. when $x = \frac{1}{2}\pi$ or $\frac{3}{2}\pi$. $x = \frac{1}{2}\pi$ and $x = \frac{3}{2}\pi$ are vertical asymptotes of the curve.

(III) y is zero when $\sin x = 0$, i.e. $x = 0$, π, or 2π, and when $\cos x = \frac{1}{8}$, which gives $x \approx 1.45$ or $x \approx 4.84$.

To investigate where y is positive or negative, we sketch the graphs of $\sin x$, $8 \cos x - 1$, and $\cos x$. The sign of $y = \dfrac{\sin x (8 \cos x - 1)}{\cos x}$ is indicated on the x-axis (Fig. 13.15).

(VI) $\dfrac{dy}{dx} = 8 \cos x - \sec^2 x$

and $\dfrac{d^2y}{dx^2} = -8 \sin x - 2 \sec^2 x \tan x$

For stationary points,

$$8 \cos x = \sec^2 x$$

i.e. $\cos^3 x = \frac{1}{8}$

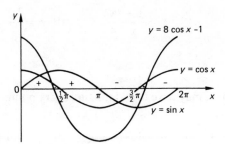

Fig. 13.15

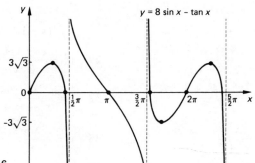

Fig. 13.16

i.e. $\cos x = \frac{1}{2}$
i.e. $x = \frac{1}{3}\pi$ or $\frac{5}{3}\pi$

When $x = \frac{1}{3}\pi$, $y = 3\sqrt{3}$ and $\dfrac{d^2y}{dx^2} = -12\sqrt{3}$, so $(\frac{1}{3}\pi, 3\sqrt{3})$ is a maximum.

When $x = \frac{5}{3}\pi$, $y = -3\sqrt{3}$ and $\dfrac{d^2y}{dx^2} = 12\sqrt{3}$, so $(\frac{5}{3}\pi, -3\sqrt{3})$ is a minimum.

We can now sketch the curve as shown in Fig. 13.16 (where we have also shown the continuation up to $x = \frac{5}{2}\pi$).

Exercise 13.3   (Sketching curves)

Write down the equations of any vertical asymptotes of the following curves (questions 1 to 10). Also state any other values of x for which y is not defined.

1   $y = 9x + \dfrac{1}{x}$ 　2 $y = \dfrac{(x-2)^2}{x+1}$ 　3 $y = \dfrac{x-3}{x^2+7}$ 　4 $y = \dfrac{2x^2 - 4x}{x^2 - 4x + 3}$

5   $y = \dfrac{\sqrt{x+1}}{x}$ 　6 $y = \dfrac{1}{x\sqrt{x-1}}$ 　7 $y = 8\cos x + \cot x$ 　8 $y = \sec x + \csc x$

9   $y = \dfrac{\sin x}{2\cos x - 1}$ 　10 $y = \cos^{-1} 5x$

Investigate the behaviour of the following curves as $x \to \pm\infty$ (questions 11 to 20).

11   $y = \dfrac{1}{1+x^2}$ 　12 $y = 4x^2 - \dfrac{1}{x}$ 　13 $y = \dfrac{x-4}{3-x}$ 　14 $y = \dfrac{2 + 3x^3}{x^{\frac{1}{3}} - 1}$

15 $y = \dfrac{(x+1)(2x-1)}{x-1}$ **16** $y = \dfrac{x}{(1+x)^2}$ **17** $y = \dfrac{1-2x^4}{x-4}$ **18** $y = \dfrac{2x^2+4x+3}{x^2-1}$

19 $y = \dfrac{\sqrt{x+1}}{x}$ **20** $y = \dfrac{\cos x}{x^2}$

Sketch the following curves (questions 21 to 40). You may have found the stationary points for some of them in Exercise 13.1.

21 $y = \dfrac{1}{1+x^2}$ **22** $y = 9x + \dfrac{1}{x}$ **23** $y = 4x^2 - \dfrac{1}{x}$

24 $y = \dfrac{x-4}{3-x}$ **25** $y = \dfrac{(x-2)^2}{x+1}$ **26** $y = \dfrac{x}{(x+1)^2}$

27 $y = x^2 + \dfrac{1}{x^2}$ **28** $y = \dfrac{(x+1)(2x-1)}{x-1}$ **29** $y^2 = \dfrac{(x+1)(2x-1)}{x-1}$

30 $y = \dfrac{x-3}{x^2+7}$ **31** $y = \dfrac{2x^2+4x+3}{x^2-1}$ **32** $y = \dfrac{2x^2-4x}{x^2-4x+3}$

33 $y = \dfrac{x^2+2x+3}{x^2-2x-3}$ **34** $y = \dfrac{x^2}{x^2+4}$ **35** $y = \dfrac{\sqrt{x+1}}{x}$

36 $y^2 = \sin x$ **37** $y = 8\cos x + \cot x$ **38** $y = \sec x + \operatorname{cosec} x$

39 $y = \dfrac{\sin x}{2\cos x - 1}$ **40** $y = x \cos x$

13.4   Implicit relations

For many curves it is difficult (or even impossible) to express y in terms of x and it is more convenient to give the equation in a form such as $x^2 - 2xy + 4y^2 = 28$. In this case we say that the curve is defined **implicitly**.

To find points on the curve, we can substitute a value for x, and solve the resulting equation for y. For example, when $x = 4$, $16 - 8y + 4y^2 = 28$, i.e. $y^2 - 2y - 3 = 0$, i.e. $y = 3$ or -1. This means that $(4, 3)$ and $(4, -1)$ are two points on the curve. If we repeat this process for several values of x, we can plot the curve (see Fig. 13.17).

This is not the graph of a function, because there are sometimes two values of y corresponding to a given value of x. It does not make sense to speak of the gradient of the curve when $x = 4$. However, we can consider the gradient at the point $(4, 3)$; when we do this, we are restricting our attention to the top part of the curve, so that y is a function of x (Fig. 13.18).

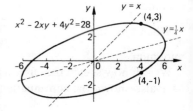

Fig. 13.17

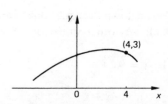

Fig. 13.18

Differentiation of implicit relations

To find the gradient, we differentiate each term in the equation with respect to x, assuming that y is a function of x.

Example   Find the gradient of the curve $x^2 - 2xy + 4y^2 = 28$
(i) at the point $(4, 3)$; (ii) at the point $(4, -1)$.

We differentiate each term with respect to x:

$$\frac{d}{dx}(x^2) = 2x$$

$2xy$ is a product, so

$$\frac{d}{dx}(2xy) = 2x\frac{d}{dx}(y) + \frac{d}{dx}(2x)y$$

$$= 2x\frac{dy}{dx} + 2y$$

$4y^2$ is a function of y, and y is a function of x, so

$$\frac{d}{dx}(4y^2) = \frac{d}{dy}(4y^2) \times \frac{dy}{dx} = 8y\frac{dy}{dx}$$

28 is a constant, so $\dfrac{d}{dx}(28) = 0$.

We therefore have

$$2x - \left(2x\frac{dy}{dx} + 2y\right) + 8y\frac{dy}{dx} = 0$$

i.e. $$2x - 2y = (2x - 8y)\frac{dy}{dx}$$

Hence $$\frac{dy}{dx} = \frac{x - y}{x - 4y}$$

This is a general expression for the gradient $\dfrac{dy}{dx}$ at any point (x, y)

on the curve.

(i)   At the point $(4, 3)$,

$$\frac{dy}{dx} = \frac{4 - 3}{4 - 12} = -\frac{1}{8}$$

(ii)   At the point $(4, -1)$,

$$\frac{dy}{dx} = \frac{4 + 1}{4 + 4} = \frac{5}{8}$$

Notes

(1)   The same formula $\dfrac{dy}{dx} = \dfrac{x - y}{x - 4y}$ can be used at both the points
$(4, 3)$ and $(4, -1)$, even though y represents different functions at
these points.

(2)   The formula $\dfrac{dy}{dx} = \dfrac{x - y}{x - 4y}$ has no meaning if (x, y) is not a
point on the curve.

(3)   When $x - y = 0$, $\dfrac{dy}{dx} = 0$, so stationary points occur where the line $y = x$ meets the curve.

(4)   When $x - 4y = 0$, $\dfrac{dy}{dx}$ is not defined. When it meets the line $y = \frac{1}{4}x$, the curve is parallel to the y-axis. (The lines $y = x$ and $y = \frac{1}{4}x$ are shown on Fig. 13.17.)

Tangents and normals

If the gradient of a curve at a point (x_1, y_1) is m_1, then the equation of the tangent at this point is

$$y - y_1 = m_1(x - x_1)$$

and the equation of the normal is

$$y - y_1 = -\frac{1}{m_1}(x - x_1)$$

Example   Find the equation of the normal to the curve $x^3 - y^3 = 3xy^2 + 1$ at the point $(2, 1)$.

Differentiating each term in the equation $x^3 - y^3 = 3xy^2 + 1$ with respect to x,

$$3x^2 - 3y^2\frac{dy}{dx} = (3x)\left(2y\frac{dy}{dx}\right) + (3)(y^2) + 0$$

At the point $(2, 1)$,

$$12 - 3\frac{dy}{dx} = 12\frac{dy}{dx} + 3$$

so $\dfrac{dy}{dx} = \dfrac{3}{5}$. The normal has gradient $-\frac{5}{3}$, and so its equation is

$$y - 1 = -\tfrac{5}{3}(x - 2)$$

i.e. $5x + 3y = 13$

Stationary points

Example   Show that $(1, 2)$ is a stationary point on the curve $x^2 - xy + y^3 = 7$, and determine its nature.

We first check that the point $(1, 2)$ does lie on the curve $x^2 - xy + y^3 = 7$.   Substituting $x = 1$ and $y = 2$, LHS $= 1 - 2 + 8 = 7 =$ RHS. Differentiating each term in the equation with respect to x,

$$2x - \left[x\frac{dy}{dx} + (1)y\right] + 3y^2\frac{dy}{dx} = 0$$

i.e. $2x - x\dfrac{dy}{dx} - y + 3y^2\dfrac{dy}{dx} = 0$

At the point (1, 2),

$$2 - \frac{dy}{dx} - 2 + 12\frac{dy}{dx} = 0$$

so $\frac{dy}{dx} = 0$. Hence (1, 2) is a stationary point on the curve.

To find $\frac{d^2y}{dx^2}$, we differentiate each term in the equation

$$2x - x\frac{dy}{dx} - y + 3y^2\frac{dy}{dx} = 0$$

For example, $3y^2\frac{dy}{dx}$ is a product, and

$$\frac{d}{dx}\left(3y^2\frac{dy}{dx}\right) = 3y^2\frac{d}{dx}\left(\frac{dy}{dx}\right) + \frac{d}{dx}(3y^2)\frac{dy}{dx}$$

$$= 3y^2\frac{d^2y}{dx^2} + \left(6y\frac{dy}{dx}\right)\frac{dy}{dx}$$

$$= 3y^2\frac{d^2y}{dx^2} + 6y\left(\frac{dy}{dx}\right)^2$$

We obtain

$$2 - \left[x\frac{d^2y}{dx^2} + (1)\frac{dy}{dx}\right] - \frac{dy}{dx} + \left[3y^2\frac{d^2y}{dx^2} + 6y\left(\frac{dy}{dx}\right)^2\right] = 0$$

i.e. $$2 - x\frac{d^2y}{dx^2} - 2\frac{dy}{dx} + 3y^2\frac{d^2y}{dx^2} + 6y\left(\frac{dy}{dx}\right)^2 = 0$$

At the point (1, 2), $x = 1$, $y = 2$ and $\frac{dy}{dx} = 0$; thus

$$2 - \frac{d^2y}{dx^2} - 0 + 12\frac{d^2y}{dx^2} + 0 = 0$$

i.e. $\frac{d^2y}{dx^2} = -\frac{2}{11}$ which is negative. Hence (1, 2) is a maximum.

Exercise 13.4   (Implicit relations)

For the following curves (questions 1 to 5), express $\frac{dy}{dx}$ in terms of x and y.

1   $x^2 + 3y^2 = 7$ 2 $4x^2 - y^3 + 2x + 3y = 0$ 3 $3x^2 + 4xy - 5y^2 = 20$
4   $xy^3 + x^3y = x - y$ 5 $\sin x + 2\cos y = 1$

6   Find the gradient of the curve $x^3 - 2y^3 = 3xy$ at the point (2, 1).
7   Find the equation of the tangent to the curve $x^2 + xy + 4y^2 = 16$ at the point (3, 1).
8   Find the equation of the normal to the curve $x^2y^2 = x^2 + 5y^2$ at the point $(3, \frac{3}{2})$.
9   Show that the tangents to the curve $y^2 - 16x - 2y = 47$, at the points $(-2, -3)$ and $(13, 17)$, are perpendicular, and find the coordinates of their point of intersection.

10   For the curve $4x^2 + y^3 = 2x + 7y$, find the values of $\dfrac{dy}{dx}$ and $\dfrac{d^2y}{dx^2}$ at the point $(-1, 2)$.

11   For the curve $x^2 + 4y^2 - 4x - 8y + 4 = 0$, express $\dfrac{dy}{dx}$ in terms of x and y. What can you say about the points $(2, 2)$ and $(0, 1)$ on the curve?

12   Show that $(-1, 3)$ and $(0, 0)$ are stationary points on the curve $3x^2 + 2xy - 5y^2 + 16y = 0$, and determine their nature.

13   Show that $(2^{\frac{1}{3}}, 2^{\frac{2}{3}})$ is a stationary point on the curve $x^3 + y^3 = 3xy$, and determine its nature.

14   Show that stationary points on the curve $x^2 + 4xy - 2y^2 + 24 = 0$ occur when $x = -2y$. Hence find and classify the stationary points.

15   If $y^2 + ay + b = x$, show that $\dfrac{d^2y}{dx^2} + 2\left(\dfrac{dy}{dx}\right)^3 = 0$.   (O and C)

13.5   The conics

Certain curves occur so frequently that they should be recognised
from their equations. We give a selection of examples.

Parabolas

(1)   Basic equation $y = kx^2$.

k positive

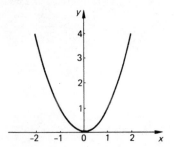

Vertex O; $k = 1$
   $y = x^2$

k negative

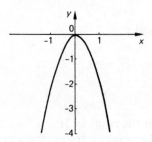

Vertex O; $k = -2$
   $y = -2x^2$

(2)   Basic equation $y^2 = lx$.

l positive

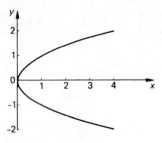

Vertex O; $l = 1$
   $y^2 = x$

l negative

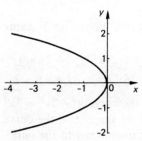

Vertex O; $l = -1$
   $y^2 = -x$

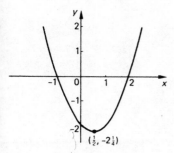

Vertex $(\frac{1}{2}, -2\frac{1}{4})$; $k = 1$
$$y + 2\frac{1}{4} = (x - \frac{1}{2})^2$$
or $y = x^2 - x - 2$
or $y = (x + 1)(x - 2)$

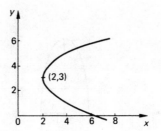

Vertex $(2, 3)$; $l = 2$
$$(y - 3)^2 = 2(x - 2)$$
or $y^2 - 6y + 9 = 2x - 4$
or $y^2 - 2x - 6y + 13 = 0$

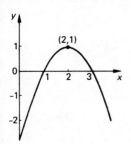

Vertex $(2, 1)$; $k = -1$
$$y - 1 = -(x - 2)^2$$
or $y = -x^2 + 4x - 3$
or $y = (1 - x)(x - 3)$

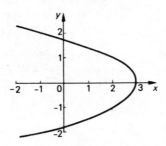

Vertex $(3, 0)$; $l = -1$
$$y^2 = -(x - 3)$$
or $y^2 = 3 - x$

Ellipses

The basic equation is $\dfrac{x^2}{a^2} + \dfrac{y^2}{b^2} = 1$

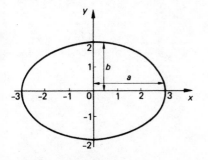

Centre O; $a = 3$, $b = 2$
$$\frac{x^2}{3^2} + \frac{y^2}{2^2} = 1$$
or $4x^2 + 9y^2 = 36$

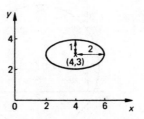

Centre $(4, 3)$; $a = 2$, $b = 1$
$$\frac{(x - 4)^2}{2^2} + \frac{(y - 3)^2}{1^2} = 1$$
or $x^2 - 8x + 16 + 4y^2 - 24y + 36 = 4$
or $x^2 + 4y^2 - 8x - 24y + 48 = 0$

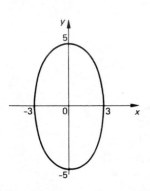

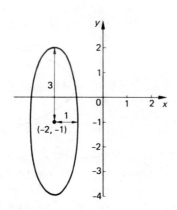

Centre O; $a = 3$, $b = 5$

$$\frac{x^2}{3^2} + \frac{y^2}{5^2} = 1$$

or $25x^2 + 9y^2 = 225$

Centre $(-2, -1)$; $a = 1$, $b = 3$

$$\frac{(x+2)^2}{1^2} + \frac{(y+1)^2}{3^2} = 1$$

or $9x^2 + 36x + 36 + y^2 + 2y + 1 = 9$

or $9x^2 + y^2 + 36x + 2y + 28 = 0$

In the special case $a = b$, the ellipse becomes a circle with radius a. The basic equation is then

$$\frac{x^2}{a^2} + \frac{y^2}{a^2} = 1$$

i.e. $x^2 + y^2 = a^2$

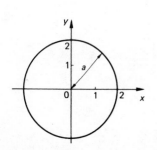

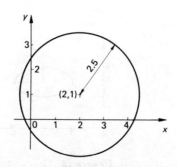

Centre O; radius 2

$$x^2 + y^2 = 4$$

Centre $(2, 1)$; radius 2.5

$$(x-2)^2 + (y-1)^2 = (2.5)^2$$

or $x^2 - 4x + 4 + y^2 - 2y + 1 = 6.25$

or $x^2 + y^2 - 4x - 2y - 1.25 = 0$

Hyperbolas

The basic equations are

$$\frac{x^2}{a^2} - \frac{y^2}{b^2} = 1 \qquad\qquad \frac{x^2}{a^2} - \frac{y^2}{b^2} = -1$$

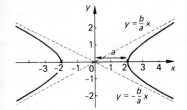

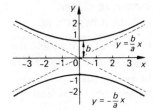

Centre O; $a = 2$, $b = 1$

$$\frac{x^2}{2^2} - \frac{y^2}{1^2} = 1$$

or $x^2 - 4y^2 = 4$
The asymptotes are $y = \pm\frac{1}{2}x$

Centre O; $a = 2$, $b = 1$

$$\frac{x^2}{2^2} - \frac{y^2}{1^2} = -1$$

or $x^2 - 4y^2 + 4 = 0$
The asymptotes are $y = \pm\frac{1}{2}x$

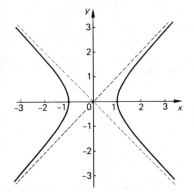

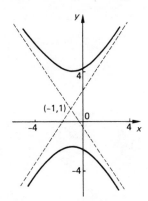

Centre O; $a = 1$, $b = 1$
$$x^2 - y^2 = 1$$
The asymptotes $y = \pm x$ are at
right angles. This is called a
rectangular hyperbola.

Centre $(-1, 1)$; $a = 2$, $b = 3$
$$\frac{(x+1)^2}{2^2} - \frac{(y-1)^2}{3^2} = -1$$

or $9x^2 + 18x + 9 - 4y^2 + 8y - 4 = -36$
or $9x^2 - 4y^2 + 18x + 8y + 41 = 0$
The asymptotes are
$$(y-1) = \pm\tfrac{3}{2}(x+1)$$
i.e. $3x - 2y + 5 = 0$ and $3x + 2y + 1 = 0$

Sketching curves of the form $Ax^2 + By^2 + Px + Qy + C = 0$

Parabolas, ellipses, circles and hyperbolas are called **conics**. All the
conics we have considered have equations of this form. Conversely,
if we are given such an equation, we can tell which type of conic it
represents by looking at the terms $Ax^2 + By^2$.

(i) If both A and B are zero, the equation, $Px + Qy + C = 0$, is that of a straight line.

(ii) If one of A and B is zero, we have a parabola:
    if $A = 0$, the equation can be rearranged as $y^2 = lx + \ldots$
    if $B = 0$, the equation can be rearranged as $y = kx^2 + \ldots$

(iii) If $A = B$, we have a circle.

(iv) If A and B have the same sign, we have an ellipse.

(v) If A and B have opposite signs, we have a hyperbola.

For further information about the curve, we can 'complete the squares' (see the following examples). This will give us the centre (of an ellipse, circle or hyperbola), or the vertex (of a parabola).

Also, we can find where the curve crosses the x-axis and the y-axis by substituting $y = 0$ and $x = 0$.

Example 1

$$2x^2 + 4x - 3y + 8 = 0$$

There is no y^2 term, and so this is a parabola. The equation is $y = \frac{2}{3}x^2 + \ldots$, hence the parabola looks like Fig. 13.19.
To find the vertex, we have

$$3y - 8 = 2x^2 + 4x$$
$$= 2(x^2 + 2x)$$

Completing the square,

Fig. 13.19

$$3y - 8 + 2 \times 1^2 = 2(x^2 + 2x + 1^2)*$$
i.e. $\qquad 3y - 6 = 2(x + 1)^2$
i.e. $\qquad y - 2 = \frac{2}{3}(x + 1)^2$

and so the vertex is at the point $(-1, 2)$.
Also, when $x = 0$, $y = \frac{8}{3}$.
The curve is sketched in Fig. 13.20.

Example 2

$$y^2 = (4 + x)(6 - x)$$

We have
$$y^2 = 24 + 2x - x^2$$
The equation is $x^2 + y^2 + \ldots = 0$, and so this is a circle.
To find the centre, we have
$$x^2 - 2x + y^2 = 24$$
Thus
$$x^2 - 2x + 1^2 + y^2 = 24 + 1^2$$
i.e. $\qquad (x - 1)^2 + y^2 = 25$

Hence the centre is $(1, 0)$, and the radius is 5 (see Fig. 13.21).

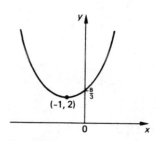

Fig. 13.20

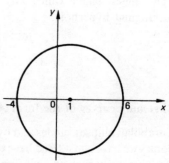

* To complete the square, starting with $x^2 + mx$, we need to add $(\frac{1}{2}m)^2$. We then have $x^2 + mx + (\frac{1}{2}m)^2$, which is $(x + \frac{1}{2}m)^2$.

Fig. 13.21

Example 3

$$6x^2 - 2y^2 + 6x + 4y - 7 = 0$$

From $6x^2 - 2y^2 + \ldots = 0$, we see that this is a hyperbola, and the asymptotes have gradients $\pm\sqrt{\frac{6}{2}}$, i.e. $\pm\sqrt{3}$.

To find the centre, we have

$$6(x^2 + x) - 2(y^2 - 2y) = 7$$

Thus

$$6[x^2 + x + (\tfrac{1}{2})^2] - 2[y^2 - 2y + 1^2] = 7 + 6 \times (\tfrac{1}{2})^2 - 2 \times 1^2$$

i.e.
$$6(x + \tfrac{1}{2})^2 - 2(y - 1)^2 = \tfrac{13}{2}$$

i.e.
$$\tfrac{12}{13}(x + \tfrac{1}{2})^2 - \tfrac{4}{13}(y - 1)^2 = 1$$

i.e.
$$\frac{(x + \tfrac{1}{2})^2}{\left(\sqrt{\dfrac{13}{12}}\right)^2} - \frac{(y - 1)^2}{\left(\sqrt{\dfrac{13}{4}}\right)^2} = 1$$

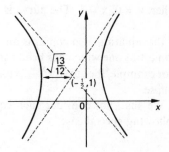

Hence the centre is $(-\tfrac{1}{2}, 1)$, and the branches of the hyperbola lie to the left and right of this (see Fig. 13.22).

Fig. 13.22

Example 4

$$4x^2 + y^2 - 6y + 8 = 0$$

From $4x^2 + y^2 + \ldots = 0$, we see that this is an ellipse, and, since the coefficient of x^2 is greater than that of y^2, it looks like Fig. 13.23.

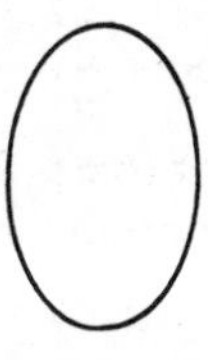

Fig. 13.23

To find the centre, we have

$$4x^2 + y^2 - 6y = -8$$

Thus

$$4x^2 + y^2 - 6y + 3^2 = -8 + 3^2$$
i.e.
$$4x^2 + (y - 3)^2 = 1$$

i.e.
$$\frac{x^2}{(\tfrac{1}{2})^2} + \frac{(y - 3)^2}{1^2} = 1$$

Hence the centre is $(0, 3)$. The curve is sketched in Fig. 13.24.

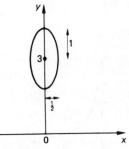

Fig. 13.24

Example 5

$$(2 + y)(1 - y) = 4x$$

We have

$$2 - y - y^2 = 4x$$

There is no x^2 term, and so this is a parabola. From $y^2 = -4x + \ldots$, the parabola looks like Fig. 13.25.

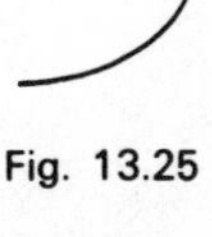

Fig. 13.25

To find the vertex,

$$y^2 + y = -4x + 2$$

Thus

$$y^2 + y + (\tfrac{1}{2})^2 = -4x + 2 + (\tfrac{1}{2})^2$$
i.e. $\qquad (y + \tfrac{1}{2})^2 = -4x + \tfrac{9}{4}$
i.e. $\qquad (y + \tfrac{1}{2})^2 = -4(x - \tfrac{9}{16})$

Hence the vertex is at $(\tfrac{9}{16}, -\tfrac{1}{2})$. Also, when $x = 0$, $y = -2$ or 1; and when $y = 0$, $x = \tfrac{1}{2}$. The curve is shown in Fig. 13.26.

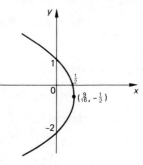

Fig. 13.26

If the equation also contains an xy term, then it still represents a conic, but one which has been rotated from the standard position. For example, the curve $x^2 - 2xy + 4y^2 = 28$, drawn on p. 218, is an ellipse.

We shall not consider such equations in detail, except for the following special case.

Rectangular hyperbolas

A rectangular hyperbola $\dfrac{x^2}{a^2} - \dfrac{y^2}{a^2} = 1$ has asymptotes $y = \pm x$ which are perpendicular (see p. 225). If the curve is rotated through an angle $\tfrac{1}{4}\pi$ (about the origin), then it will have the x-axis and the y-axis as asymptotes. The equation then takes a different form.

The basic equation is

$$xy = k$$

k positive

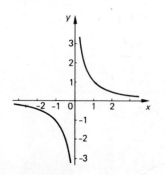

k negative

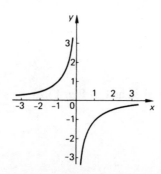

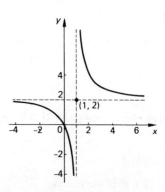

Centre O; $k = 1$
$$xy = 1$$
or $\quad y = \dfrac{1}{x}$

Centre O; $k = -1$
$$xy = -1$$
or $\quad y = -\dfrac{1}{x}$

Centre $(1, 2)$; $k = 2$
$$(x - 1)(y - 2) = 2$$
or $\quad xy - 2x - y = 0$

The equation $Hxy + Px + Qy + C = 0$ represents a rectangular hyperbola.

Example   Sketch the curve

$$xy + x - 3y - 1 = 0$$

We have

$$xy + x - 3y = 1$$

Thus

$$x(y+1) - 3y - 3 = 1 - 3$$
i.e. $x(y+1) - 3(y+1) = -2$
i.e. $(x-3)(y+1) = -2$

This is a rectangular hyperbola, with centre $(3, -1)$, and asymptotes $x = 3$ and $y = -1$.
   Also, when $x = 0$, $y = -\frac{1}{3}$; and when $y = 0$, $x = 1$. The curve is sketched in Fig. 13.27.

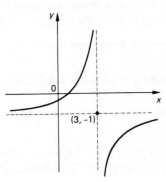

Fig. 13.27

Exercise 13.5   (Conics)

Sketch the following curves (questions 1 to 18).

1   $xy - 2 = 0$

2   $y^2 = 9 - x^2$

3   $16x^2 - 9y^2 = 144$

4   $x^2 + 4y^2 = 4$

5   $y = 4 - x^2$

6   $49x^2 + 16y^2 = 784$

7   $y^2 = 9 - x$

8   $y = x(x-4)$

9   $y^2 = 36 + x^2$

10   $(y-1)^2 = x$

11   $y(x-2) = 16$

12   $y^2 = (8-x)(x+2)$

13   $y = x^2 + 2x - 4$

14   $x^2 + y^2 - 2x + 6y - 6 = 0$

15   $y^2 - 6x + 4y - 12 = 0$

16   $x^2 - 4y^2 - 2x - 8y = 0$

17   $4x^2 + y^2 - 8x - 4y + 4 = 0$

18   $xy - 4x + 3y = 9$

19   Find the equation of the tangent at the point $(2, -1)$ to the ellipse

$$3x^2 + 5y^2 - 2x + 8y - 5 = 0$$

20   The normal to the parabola $y^2 = 2(x-1)$ at the point $P(3, 2)$ meets the parabola again at Q. Find the coordinates of Q.

21   Find the volumes of the solids generated when the area enclosed by the ellipse $x^2 + 4y^2 = 36$ is rotated (i) about the x-axis and (ii) about the y-axis.

22   Sketch the curve $4y^2 = (x-1)(5-x)$. The area enclosed by the curve is rotated about the x-axis. Find the volume of the solid so formed.   (O and C)

23   The area lying between the curves $x^2 + y^2 = 1$ and $x^2 + 4y^2 = 4$ for which x is positive is rotated about the x-axis. Find the volume of the solid thus formed.   (O and C)

24   Find the area enclosed between the curve $y^2 = 2x - 6$ and the line $x = 5$.

25   Find the area enclosed between the curves $y^2 = 8x$ and $x^2 = 8y$.

13.6 Limits by calculator

We have already encountered the idea of a limit, in differentiation and integration. Suppose now that we are interested in the behaviour of a function $y = f(x)$ in the neighbourhood of a particular value of x, say $x = a$. If $f(x)$ approaches a limiting value as x is taken closer and closer to a, we write this limiting value as $\lim\limits_{x \to a} f(x)$, said 'the limit, as x tends to a, of $f(x)$'.

Investigation using a calculator

Example 1 Investigate $\lim\limits_{x \to 2} (5x - x^2)$.

Let $y = 5x - x^2$. We consider values of x close to 2 (both less than and greater than 2), and calculate the corresponding values of y.

x	1.9	1.99	1.999	2.1	2.01	2.001
$y = 5x - x^2$	5.89	5.9899	5.998 999	6.09	6.0099	6.000 999

It appears that, as x becomes closer and closer to 2, y approaches a limiting value of 6. We write $y \to 6$ as $x \to 2$, or

$$\lim_{x \to 2} (5x - x^2) = 6$$

In this case we can obtain the limiting value by simply substituting $x = 2$ into $5x - x^2$, which gives $y = 5 \times 2 - 2^2 = 6$.

Example 2 Investigate $\lim\limits_{x \to 9} \left(\dfrac{\sqrt{x} - 3}{x - 9} \right)$.

We cannot just substitute $x = 9$ into this, because we would then have $\dfrac{3 - 3}{9 - 9} = \dfrac{0}{0}$, which is meaningless.

When x is close to 9, we obtain the following values

x	8.9	8.99	8.999	9.1	9.01	9.001
$\dfrac{\sqrt{x} - 3}{x - 9}$	0.167 132	0.166 713	0.166 671	0.166 206	0.166 620	0.166 662

It appears that $\lim\limits_{x \to 9} \left(\dfrac{\sqrt{x} - 3}{x - 9} \right) = 0.1667$ (to 4 decimal places).

Example 3 Investigate $\lim\limits_{x \to \frac{1}{2}\pi} \left(\dfrac{\cos x}{1 - \sin x} \right)$.

When x is close to $\frac{1}{2}\pi$ ($= 1.570796$), we obtain the following values (remember that we must work in radians).

x	1.5	1.57	1.5707	1.6	1.58	1.571
$\dfrac{\cos x}{1 - \sin x}$	28.24	2511.6	20804.9	-68.48	-217.3	-9825.0

There is no limiting value, and so $\lim\limits_{x \to \frac{1}{2}\pi} \left(\dfrac{\cos x}{1 - \sin x} \right)$ does not exist.

Consider the curve $y = \dfrac{\cos x}{1 - \sin x}$ in the neighbourhood of $x = \frac{1}{2}\pi$.

If $x < \frac{1}{2}\pi$, y becomes large and positive as x is taken close to $\frac{1}{2}\pi$; we write $y \to +\infty$ as $x \to \frac{1}{2}\pi$ from below.

If $x > \frac{1}{2}\pi$, y becomes large and negative as x is taken close to $\frac{1}{2}\pi$; we write $y \to -\infty$ as $x \to \frac{1}{2}\pi$ from above.

The curve has a vertical asymptote at $x = \frac{1}{2}\pi$ (see Fig. 13.28).

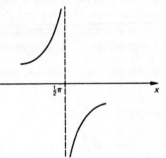

Fig. 13.28

Example 4 Investigate $\lim\limits_{\substack{x \to -1 \\ (x > -1)}} \left(\dfrac{x^2 - 1}{\sqrt{x + 1}} \right)$.

The restriction $x > -1$ means that we are only to consider values of x greater than -1 (this avoids having the square root of a negative number).

We obtain the following values.

x	-0.9	-0.99	-0.999
$\dfrac{x^2 - 1}{\sqrt{x + 1}}$	-0.6008	-0.1990	-0.0632

It appears that $\lim\limits_{\substack{x \to -1 \\ (x > -1)}} \left(\dfrac{x^2 - 1}{\sqrt{x + 1}} \right) = 0$

We may also write $\dfrac{x^2 - 1}{\sqrt{x + 1}} \to 0$ as $x \to -1$ from above.

Exercise 13.6 (Limits by calculator)

Use your calculator to investigate the following limits.

1 $\lim\limits_{x \to 3} \dfrac{x^2 - 9}{x - 3}$

2 $\lim\limits_{\substack{x \to 1 \\ (x > 1)}} \dfrac{\sqrt{x} - 1}{\sqrt{x - 1}}$

3 $\lim\limits_{x \to 0} \dfrac{\tan x}{x}$

4 $\lim\limits_{x \to 0} \dfrac{\sin x}{x^2}$

5 $\lim\limits_{x \to \frac{1}{2}\pi} (\sec x - \tan x)$

6 $\lim\limits_{x \to 0} \dfrac{1 - \cos x}{x}$

7 $\lim\limits_{x \to 0} \dfrac{\sin x - x \cos x}{x^3}$

8 $\lim\limits_{x \to \pi} \dfrac{\tan x}{\sin x}$

9 $\lim\limits_{x \to 0} \dfrac{4^x - 1}{x}$

10 $\lim\limits_{\substack{x \to 0 \\ (x > 0)}} x^x$

13.7 Evaluating limits

The use of a calculator can give us a good idea of what the value of a particular limit might be, but this does not constitute a proper proof.

To establish the value of a limit, $\lim_{x \to a} f(x)$, we must rearrange the formula for $f(x)$ in such a way that its behaviour near $x = a$ can easily be seen. If $f(x)$ is an algebraic function, we usually achieve this by dividing the numerator and the denominator by a suitable power of $(x - a)$. If $f(x)$ contains circular functions, we can use trigonometrical identities; remember also that $\dfrac{\sin x}{x} \to 1$ as $x \to 0$.

Example 1

$$\lim_{x \to 2} \frac{x^3 - 8}{x - 2}$$

Let

$$y = \frac{x^3 - 8}{x - 2} = \frac{(x - 2)(x^2 + 2x + 4)}{x - 2}$$

$$= x^2 + 2x + 4$$

As $x \to 2$, we can now obtain the limit by simply substituting $x = 2$; thus $y \to 2^2 + 2 \times 2 + 4 = 12$. Hence

$$\lim_{x \to 2} \frac{x^3 - 8}{x - 2} = 12$$

Example 2

$$\lim_{x \to 0} \frac{x^3 + x}{x^{\frac{4}{3}}}$$

Let

$$y = \frac{x^3 + x}{x^{\frac{4}{3}}} = \frac{x(x^2 + 1)}{x^{\frac{4}{3}}}$$

$$= \frac{x^2 + 1}{x^{\frac{1}{3}}}$$

As $x \to 0$, $x^2 + 1 \to 1$, and so y behaves like $\dfrac{1}{x^{\frac{1}{3}}}$. Thus

$$y \to -\infty \text{ as } x \to 0 \text{ from below}$$
$$y \to +\infty \text{ as } x \to 0 \text{ from above}$$

and

$$\lim_{x \to 0} \frac{x^3 + x}{x^{\frac{4}{3}}} \text{ does not exist}$$

Example 3

$$\lim_{x \to 9} \left(\frac{\sqrt{x} - 3}{x - 9} \right)$$

Let

$$y = \frac{\sqrt{x} - 3}{x - 9} = \frac{\sqrt{x} - 3}{x - 9} \times \frac{\sqrt{x} + 3}{\sqrt{x} + 3} = \frac{x - 9}{(x - 9)(\sqrt{x} + 3)} = \frac{1}{\sqrt{x} + 3}$$

As $x \to 9$, $y \to \dfrac{1}{3 + 3} = \dfrac{1}{6}$. Hence

$$\lim_{x \to 9} \left(\frac{\sqrt{x} - 3}{x - 9} \right) = \frac{1}{6}$$

Example 4

$$\lim_{\substack{x \to -1 \\ (x > -1)}} \left(\frac{x^2 - 1}{\sqrt{x + 1}} \right)$$

Let

$$y = \frac{x^2 - 1}{\sqrt{x + 1}} = \frac{(x + 1)(x - 1)}{\sqrt{x + 1}}$$

$$= (x + 1)^{\frac{1}{2}}(x - 1)$$

As $x \to -1$, $y \to 0 \times (-2) = 0$. Hence

$$\lim_{\substack{x \to -1 \\ (x > -1)}} \left(\frac{x^2 - 1}{\sqrt{x + 1}} \right) = 0$$

Example 5

$$\lim_{x \to \pi} \left(\frac{1 + \cos x}{\sin^2 x} \right)$$

Let

$$y = \frac{1 + \cos x}{\sin^2 x} = \frac{1 + \cos x}{1 - \cos^2 x} = \frac{1 + \cos x}{(1 + \cos x)(1 - \cos x)} = \frac{1}{1 - \cos x}$$

As $x \to \pi$, $y \to \dfrac{1}{1 - (-1)} = \dfrac{1}{2}$. Hence

$$\lim_{x \to \pi} \left(\frac{1 + \cos x}{\sin^2 x} \right) = \frac{1}{2}$$

Example 6

$$\lim_{x \to 0} \frac{\sin 2x \tan x}{x^2}$$

Let

$$y = \frac{\sin 2x \tan x}{x^2} = \frac{\sin 2x \sin x}{x^2 \cos x}$$

$$= \frac{\sin 2x}{2x} \times \frac{\sin x}{x} \times \frac{2}{\cos x}$$

As $x \to 0$, $y \to 1 \times 1 \times \frac{2}{1} = 2$. Hence

$$\lim_{x \to 0} \frac{\sin 2x \tan x}{x^2} = 2$$

Example 7

$$\lim_{\substack{x \to 0 \\ (x > 0)}} \frac{\sqrt{x}}{\sin x}$$

Let

$$y = \frac{\sqrt{x}}{\sin x} = \frac{x}{\sin x} \times \frac{1}{\sqrt{x}}$$

As $x \to 0$, $\dfrac{x}{\sin x} \to 1$, and so y behaves like $\dfrac{1}{\sqrt{x}}$. Thus $y \to +\infty$

as $x \to 0$ from above, and

$$\lim_{\substack{x \to 0 \\ (x > 0)}} \frac{\sqrt{x}}{\sin x} \text{ does not exist.}$$

Sketching curves

Example   Sketch the curve $y = \dfrac{\sin x}{x}$.

(I) y is not defined when $x = 0$. As the numerator and denominator are both zero, we need to consider the limit.

As $x \to 0$, $y = \dfrac{\sin x}{x} \to 1$, and so the curve approaches the point $(0, 1)$.

(III) $y = 0$ when $\sin x = 0$ and $x \neq 0$, i.e. when $x = \pm\pi$, $\pm 2\pi$, ... We find the sign of $y = \dfrac{\sin x}{x}$ (indicated on Fig. 13.29 on the x-axis) by sketching the graphs of $\sin x$ and x.

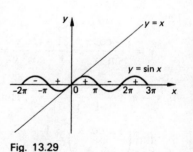

Fig. 13.29

(V) When x is replaced by $-x$, y is replaced by $\dfrac{\sin(-x)}{-x}$

$= \dfrac{-\sin x}{-x} = \dfrac{\sin x}{x} = y$, so the curve is symmetrical about the

y-axis.

Since $\sin x$ always lies between -1 and $+1$, the curve $y = \dfrac{\sin x}{x}$

is between the curves $y = -\dfrac{1}{x}$ and $y = \dfrac{1}{x}$. It touches these curves

when $\sin x = \pm 1$, i.e. $x = \pm\frac{1}{2}\pi,\ \pm\frac{3}{2}\pi,\ \ldots$ (see Fig. 13.30).

Strictly, the curve $y = \dfrac{\sin x}{x}$ has a gap in it when $x = 0$, since y is

then not defined. It is reasonable to fill this gap by defining $y = 1$
when $x = 0$.

The curve given by

$$y = \begin{cases} \dfrac{\sin x}{x} & \text{if } x \neq 0 \\[2mm] 1 & \text{if } x = 0 \end{cases}$$

is then continuous.

Fig. 13.30

Exercise 13.7   (Limits)

Evaluate the following limits (questions 1 to 12) where possible; if the limit does not exist, explain why.

1 $\displaystyle\lim_{x \to 1} \frac{x^2 - 1}{x - 1}$

2 $\displaystyle\lim_{x \to 3} \frac{x^3 - 8x - 3}{x - 3}$

3 $\displaystyle\lim_{\substack{x \to -2 \\ (x > -2)}} \frac{x^2 + x - 2}{\sqrt{x + 2}}$

4 $\displaystyle\lim_{\substack{x \to 3 \\ (x > 3)}} \frac{\sqrt{x^2 - 9}}{x - 3}$

5 $\displaystyle\lim_{x \to 0} \frac{\tan x}{x}$

6 $\displaystyle\lim_{x \to \frac{1}{2}\pi} \frac{\sec x}{\tan x}$

7 $\displaystyle\lim_{x \to 0} \frac{\sin x}{x^2}$

8 $\displaystyle\lim_{x \to 0} \frac{\tan 3x \sin 2x}{x^2}$

9 $\displaystyle\lim_{x \to 0} \frac{1 - \cos x}{x^2}$ $\left(\text{multiply by } \dfrac{1 + \cos x}{1 + \cos x}\right)$

10 $\displaystyle\lim_{x \to \pi} \frac{\sin 4x + \sin 2x}{3 \sin 3x}$

11 $\displaystyle\lim_{x \to 0} \frac{\tan x - \sin x}{x^3}$

12 $\displaystyle\lim_{x \to \frac{1}{2}\pi} \frac{1 - \sin x}{\cos x}$

Sketch the following curves.

13 $y = \dfrac{\tan x}{x}$ (for $-\frac{1}{2}\pi < x < \frac{1}{2}\pi$)

14 $y = \dfrac{\sin x}{x^2}$ (for $-4\pi \leqslant x \leqslant 4\pi$)

15 $y = \dfrac{1 - \sin x}{\cos x}$ (for $0 \leqslant x \leqslant 2\pi$)

13.8 Continuity

We have often said that a curve is continuous. Intuitively, this means that the curve has no breaks in it. We make the idea of continuity precise by using limits. We consider the continuity of a curve at a particular value of x.

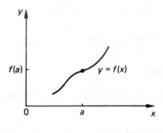

A curve $y = f(x)$ is said to be **continuous at** $x = a$ if $\lim\limits_{x \to a} f(x)$ exists, and is equal to $f(a)$. Then $y \to f(a)$ as $x \to a$, and the curve approaches the point $(a, f(a))$ from both sides (see Fig. 13.31).

Fig. 13.31

The following examples illustrate the types of discontinuity which we have already met.

Example 1

$$y = \frac{1}{x-2}$$

y is not defined, and the curve has a vertical asymptote, when $x = 2$. Also,

$y \to -\infty$ as $x \to 2$ from below
$y \to +\infty$ as $x \to 2$ from above

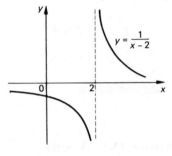

and so $\lim\limits_{x \to 2} \left(\dfrac{1}{x-2} \right)$ does not exist. This curve is not continuous at $x = 2$.

However, the curve is continuous at all other values of x (see Fig. 13.32).

Fig. 13.32

Example 2

$$y = f(x) = \begin{cases} 1 & \text{if } x \leqslant 2 \\ 3 & \text{if } x > 2 \end{cases}$$

We have

$y \to 1$ as $x \to 2$ from below
$y \to 3$ as $x \to 2$ from above

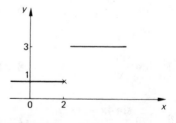

and so $\lim\limits_{x \to 2} f(x)$ does not exist. The curve is not continuous at $x = 2$. This is called a 'jump discontinuity'.

Again, the curve is continuous at all other values of x (see Fig. 13.33).

Fig. 13.33

It can be shown that, if u and v are continuous functions (i.e. their graphs are continuous curves) then the functions $u + v$, $u - v$, and uv are also continuous. The quotient $\dfrac{u}{v}$ is continuous except at values of x which make $v = 0$. If y is a continuous function of u, and u is a continuous function of x, then the composite function is also continuous.

The basic functions x, x^2, x^3, . . . , $\sin x$ and $\cos x$, are continuous at all values of x. Using the results stated above, we can investigate the continuity of any function made up from these basic ones.

Example 3

$$y = \frac{x^2 \sin x - 7x^3 + 12}{x(x-1) \cos (x^2)}$$

$x^2 \sin x$ is continuous
(since it is the product of two continuous functions).

$x^2 \sin x - 7x^3 + 12$ is continuous
(since it is a sum of continuous functions).

$$\cos (x^2) = \cos u \text{ where } u = x^2$$

This is the composite of two continuous functions, and is therefore continuous.

$x(x-1) \cos (x^2)$ is continuous
(since it is a product of continuous functions).

$$y = \frac{x^2 \sin x - 7x^3 + 12}{x(x-1) \cos (x^2)}$$

is the quotient of two functions, each of which is continuous at all values of x.

We can therefore assert that this quotient is continuous at all values of x except those which make the denominator zero.

Hence the given curve has discontinuities at $x = 0$, 1, $\pm \sqrt{\tfrac{1}{2}\pi}$, $\pm \sqrt{\tfrac{3}{2}\pi}$, . . . , but it is continuous at all other values of x.

Example 4 Find constants a and b so that the function defined by

$$f(x) = \begin{cases} 1 - x^2 & \text{if } x < 2 \\ 9 + ax + bx^2 & \text{if } x \geqslant 2 \end{cases}$$

shall be continuous, and shall have a continuous derivative, for all values of x.   (W)

Since $1 - x^2$ and $9 + ax + bx^2$ are continuous functions, the only possible point of discontinuity of $f(x)$ is at $x = 2$ (where the definition changes).

As $x \to 2$ from below,

$$f(x) = 1 - x^2 \to 1 - 4 = -3$$

As $x \to 2$ from above,

$$f(x) = 9 + ax + bx^2 \to 9 + 2a + 4b$$

If $f(x)$ is continuous, these two limits must be the same, i.e.

$$9 + 2a + 4b = -3$$

i.e. $2a + 4b = -12$

The derivative is

$$f'(x) = \begin{cases} -2x & \text{if } x < 2 \\ a + 2bx & \text{if } x > 2 \end{cases}$$

As $x \to 2$ from below,

$$f'(x) = -2x \to -4$$

As $x \to 2$ from above,

$$f'(x) = a + 2bx \to a + 4b$$

If $f'(x)$ is continuous, we must have

$$a + 4b = -4$$

Solving the simultaneous equations

$$2a + 4b = -12$$
$$a + 4b = -4$$

we obtain $a = -8$ and $b = 1$.

Exercise 13.8   (Continuity)

State any values of x at which the following curves are discontinuous (questions 1 to 5).

1 $y = \dfrac{x-2}{x(x+3)}$

2 $y = \cos x - \sec x$

3 $y = \dfrac{4x - 3\cos 2x}{(x^2 - 4)\sin x}$

4 $y = \begin{cases} 2 & \text{if } x \leqslant 3 \\ 5 & \text{if } x > 3 \end{cases}$

5 $y = \begin{cases} x^2 & \text{if } x \leqslant 1 \\ 2x - 1 & \text{if } 1 < x \leqslant 4 \\ 9 - x & \text{if } x > 4 \end{cases}$

6 If $f(x) = x^{1/3}$, investigate the continuity of the functions $f(x)$ and $f'(x)$.

7 The function $f(x)$ is defined by

$$f(x) = \begin{cases} x+4 & \text{if } x \leqslant 2 \\ 5x - x^2 & \text{if } 2 < x \leqslant 3 \\ 6 & \text{if } x > 3 \end{cases}$$

Investigate the continuity of $f(x)$ and of its derivative.

8 The function $f(x)$ is defined by

$$f(x) = \begin{cases} x^2 + 6x + 5 & \text{if } x \leqslant -1 \\ ax^2 + bx + c & \text{if } -1 < x < 0 \\ x^2 + 2x + d & \text{if } x \geqslant 0 \end{cases}$$

Determine the constants a, b, c, d so that $f(x)$ and $f'(x)$ are continuous for all x.   (W)

14

Parametric equations

14.1 Sketching curves from parametric equations

In Chapter 11 we described the motion of a particle along a straight line by giving its x-coordinate (the displacement) in terms of time. We now consider a particle moving in two dimensions, in the plane containing the x- and y-axes. In order to describe the motion, we need to know both its x-coordinate and its y-coordinate in terms of time.

Example 1 A particle moves so that, at time t, it is at the point (x, y) where

$$x = t^2$$
and $y = 2t^3 + 3$

Here it is convenient to allow t to take both positive and negative values; the time $t = 0$ is now just some reference point during the motion, and it no longer represents the start of the motion.

We first make a table of values.

t	-2	-1.5	-1	-0.5	0	0.5	1	1.5	2
$x = t^2$	4	2.25	1	0.25	0	0.25	1	2.25	4
$y = 2t^3 + 3$	-13	-3.75	1	2.75	3	3.25	5	9.75	19

At time $t = -2$, the particle is at the point $(4, -13)$; at time $t = -1.5$, it is at $(2.25, -3.75)$; and so on. If we plot these points, and join them up as in Fig. 14.1, we see that the particle is moving along a curve.

This curve consists of all points (x, y) for which $x = t^2$ and $y = 2t^3 + 3$, as t takes all possible values.

If we are just interested in the curve, there is no need to think of t as representing time. x and y are given in terms of a third variable t, called a **parameter** (of course, any other letter could be used instead of t). We say that the curve is defined **parametrically**.

When $t = 2$ we have $x = 4$ and $y = 19$. The point $(4, 19)$ on the curve may be called 'the point $t = 2$'.

Similarly, the general point $(t^2, 2t^3 + 3)$ on the curve is called 'the point t'.

We can obtain the ordinary (Cartesian) equation of the curve by eliminating t, in which case

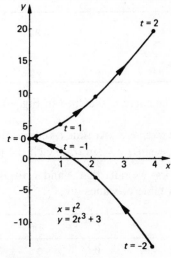

Fig. 14.1

239

$$(y-3)^2 = (2t^3)^2 = 4t^6 = 4x^3$$

The equation of the curve is

$$(y-3)^2 = 4x^3$$

but this is not particularly useful. In this chapter we shall show how the properties of the curve (such as its gradient) may be found directly from the parametric equations.

When sketching a curve which is given parametrically we usually need to make a table of values. We consider carefully any values of t for which x or y is not defined (this enables us to find any asymptotes of the curve). We also investigate the behaviour of x and y as $t \to \pm \infty$.

Example 2 Sketch the curve defined by $x = \dfrac{1}{t-1}$, $y = \dfrac{t}{t+1}$.

x is not defined when $t = 1$. As $t \to 1$, x becomes arbitrarily large, and $y \to \dfrac{1}{1+1} = \frac{1}{2}$. Thus $y = \frac{1}{2}$ is an asymptote.

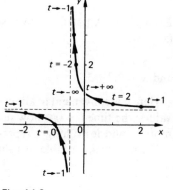

Fig. 14.2

y is not defined when $t = -1$. As $t \to -1$, $x \to \dfrac{1}{-1-1} = -\frac{1}{2}$, and y becomes arbitrarily large. Thus $x = -\frac{1}{2}$ is an asymptote.

As $t \to \pm\infty$, we have $x = \dfrac{1}{t-1} = \dfrac{1/t}{1-1/t} \to 0$ and $y = \dfrac{t}{t+1}$

$= \dfrac{1}{1+1/t} \to 1$. So the curve approaches the point $(0, 1)$. Strictly the point $(0, 1)$ itself does not lie on the curve, since it does not correspond to any real value of t.

A table of values gives

t	-2	-1.5	-0.5	0	0.5	1.5	2
$x = \dfrac{1}{t-1}$	-0.33	-0.4	-0.67	-1	-2	2	1
$y = \dfrac{t}{t+1}$	2	3	-1	0	0.33	0.6	0.67

The curve is sketched in Fig. 14.2.

Example 3 At time t, the position (x, y) of a particle is given by $x = \sin t$, $y = \sin 2t$. Sketch the path of the particle.

First we note that x and y always lie between -1 and $+1$.
A table of values gives

t	0	$\frac{1}{6}\pi$	$\frac{1}{4}\pi$	$\frac{1}{3}\pi$	$\frac{1}{2}\pi$	π	$\frac{3}{2}\pi$	2π
$x = \sin t$	0	0.5	0.71	0.87	1	0	-1	0
$y = \sin 2t$	0	0.87	1	0.87	0	0	0	0

Replacing t by $\pi - t$, (x, y) is replaced by $(\sin(\pi - t), \sin(2\pi - 2t))$, which is $(\sin t, -\sin 2t)$, or $(x, -y)$; so the curve is symmetrical about the x-axis. Also, the section for $\frac{1}{2}\pi \leqslant t \leqslant \pi$ is the reflection in the x-axis of the section for $0 \leqslant t \leqslant \frac{1}{2}\pi$.

Replacing t by $\pi + t$, the point (x, y) is replaced by $(\sin(\pi + t), \sin(2\pi + 2t))$, which is $(-\sin t, \sin 2t)$, or $(-x, y)$; so the curve is symmetrical about the y-axis. The section of curve for $\pi \leqslant t \leqslant 2\pi$ is the reflection in the y-axis of the section for $0 \leqslant t \leqslant \pi$.

Replacing t by $2\pi + t$, (x, y) is unchanged, so the particle retraces the path at intervals of 2π (see Fig. 14.3).

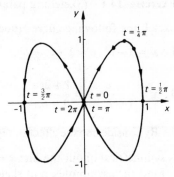

Fig. 14.3

Parametric equations for conics

The parametric equations most commonly used for the standard conics are given below.

The ellipse $\left(\dfrac{x^2}{a^2} + \dfrac{y^2}{b^2} = 1\right)$

$$x = a\cos t$$
$$y = b\sin t$$

The parabola $(y^2 = 4ax)$

$$x = at^2$$
$$y = 2at$$

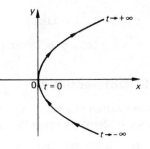

The circle $(x^2 + y^2 = a^2)$

$$x = a\cos t$$
$$y = a\sin t$$

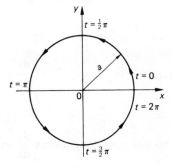

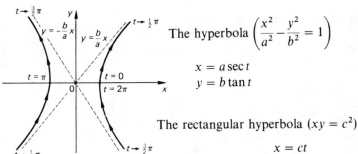

The hyperbola $\left(\dfrac{x^2}{a^2} - \dfrac{y^2}{b^2} = 1\right)$

$$x = a\sec t$$
$$y = b\tan t$$

The rectangular hyperbola $(xy = c^2)$

$$x = ct$$
$$y = \frac{c}{t}$$

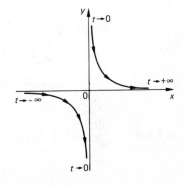

Exercise 14.1   (Sketching parametric curves)

Sketch the following curves (questions 1 to 6).

1 $x = 1 - t^3, y = 1 + t^2$ **2** $x = \sin 2t, y = \cos t$ **3** $x = 2t, y = 4 - \dfrac{3}{t}$

4 $x = \dfrac{1 + t}{1 - 2t}, y = \dfrac{1 + 2t}{1 - t}$ **5** $x = \dfrac{4t}{1 + t}, y = \dfrac{4t^2}{1 - t}$ **6** $x = \sin t, y = \tan t$

7   By considering the effect of replacing t by $-t$, show that the curve defined by $x = \dfrac{t}{2 + t}, y = \dfrac{t}{2 - t}$, is symmetrical about the line $y = -x$.
   Find the asymptotes, and sketch the curve.

Sketch the following conics (questions 8 to 15).

8 $x = 3t^2, y = 6t$ **9** $x = 2t, y = \dfrac{2}{t}$ **10** $x = 4\cos t, y = 2\sin t$

11 $x = 3\sec t, y = \tan t$ **12** $x = \cos t, y = 3\sin t$ **13** $x = -2t^2, y = 4t$

14 $x = 2 + \cos t, y = 3 + \sin t$ **15** $x = 2t + 3, y = \dfrac{2}{t} - 1$

14.2   Differentiation

If the equation of a curve is given parametrically in terms of t, we find its gradient by using the chain rule. We have

$$\frac{dy}{dx} = \frac{dy}{dt} \times \frac{dt}{dx} = \frac{dy}{dx} \times \frac{1}{\dfrac{dx}{dt}}$$

Hence

$$\boxed{\frac{dy}{dx} = \frac{\dfrac{dy}{dt}}{\dfrac{dx}{dt}}}$$

This will give the gradient $\dfrac{dy}{dx}$ in terms of the parameter t. Also,

$$\frac{d^2y}{dx^2} = \frac{d}{dx}\left(\frac{dy}{dx}\right) = \frac{d}{dt}\left(\frac{dy}{dx}\right) \times \frac{dt}{dx}$$

Thus

$$\boxed{\frac{d^2y}{dx^2} = \frac{\dfrac{d}{dt}\left(\dfrac{dy}{dx}\right)}{\dfrac{dx}{dt}}}$$

and **not** $\dfrac{\dfrac{d^2y}{dt^2}}{\dfrac{d^2x}{dt^2}}$

Example For the curve $x = t^2$, $y = 2t^3 + 3$, find $\dfrac{dy}{dx}$ and $\dfrac{d^2y}{dx^2}$ in terms of t. Evaluate both at the point $(4, -13)$.

We have

$$\frac{dy}{dx} = \frac{\dfrac{dy}{dt}}{\dfrac{dx}{dt}} = \frac{6t^2}{2t} = 3t$$

$$\frac{d^2y}{dx^2} = \frac{\dfrac{d}{dt}(3t)}{\dfrac{dx}{dt}} = \frac{3}{2t}$$

The point $(4, -13)$ corresponds to $t = -2$, so then $\dfrac{dy}{dx} = -6$ and $\dfrac{d^2y}{dx^2} = -\dfrac{3}{4}$.

(Note that, when $t = 0$, the gradient $\dfrac{dy}{dx}$ is zero. This confirms the shape of the curve as shown in the sketch on p. 239.)

Stationary points

Example Find and classify the stationary points on the curve $x = \sin t$, $y = \sin 2t$.

We have

$$\frac{dy}{dx} = \frac{\dfrac{dy}{dt}}{\dfrac{dx}{dt}} = \frac{2\cos 2t}{\cos t}$$

(Notice how this confirms the shape of the curve, drawn on p. 241. When $t = \frac{1}{2}\pi$ and $t = \frac{3}{2}\pi$, $\dfrac{dy}{dx}$ is not defined and the curve is parallel to the y-axis. At the origin, when $t = 0$ the gradient is 2 and when $t = \pi$ the gradient is -2.)

For stationary points, $\dfrac{dy}{dx} = 0$, i.e. $\cos 2t = 0$, i.e. $2t = \frac{1}{2}\pi,\ \frac{3}{2}\pi,\ \frac{5}{2}\pi, \ldots$

Thus $t =$	$\frac{1}{4}\pi,$	$\frac{3}{4}\pi,$	$\frac{5}{4}\pi,$	$\frac{7}{4}\pi,$	$\frac{9}{4}\pi, \ldots$
Then $x = \sin t =$	$\dfrac{1}{\sqrt{2}}$	$\dfrac{1}{\sqrt{2}}$	$-\dfrac{1}{\sqrt{2}}$	$-\dfrac{1}{\sqrt{2}}$	$\dfrac{1}{\sqrt{2}}$
and $y = \sin 2t =$	1	-1	1	-1	1

The stationary points are $\left(\dfrac{1}{\sqrt{2}}, 1\right), \left(\dfrac{1}{\sqrt{2}}, -1\right), \left(-\dfrac{1}{\sqrt{2}}, 1\right)$ and $\left(-\dfrac{1}{\sqrt{2}}, -1\right)$.

Now

$$\frac{d^2y}{dx^2} = \frac{\dfrac{d}{dt}\left(\dfrac{2\cos 2t}{\cos t}\right)}{\dfrac{dx}{dt}}$$

$$= \frac{\dfrac{(\cos t)(-4\sin 2t) + (2\cos 2t)(\sin t)}{\cos^2 t}}{\cos t}$$

$$= \frac{-4\sin 2t}{\cos^2 t} + \frac{2\cos 2t \sin t}{\cos^3 t}$$

$$= \frac{-4\sin 2t}{\cos^2 t} \quad \text{at stationary points (since } \cos 2t = 0)$$

When $t = \frac{1}{4}\pi$,

$$\frac{d^2y}{dx^2} = \frac{-4(1)}{\frac{1}{2}}$$

which is negative, so $\left(\dfrac{1}{\sqrt{2}}, 1\right)$ is a maximum.

When $t = \frac{3}{4}\pi$,

$$\frac{d^2y}{dx^2} = \frac{-4(-1)}{\frac{1}{2}}$$

which is positive, so $\left(\dfrac{1}{\sqrt{2}}, -1\right)$ is a minimum.

When $t = \frac{5}{4}\pi$,

$$\frac{d^2y}{dx^2} = \frac{-4(1)}{\frac{1}{2}}$$

which is negative, so $\left(-\dfrac{1}{\sqrt{2}}, 1\right)$ is a maximum.

When $t = \frac{7}{4}\pi$,

$$\frac{d^2y}{dx^2} = \frac{-4(-1)}{\frac{1}{2}}$$

which is positive, so $\left(-\dfrac{1}{\sqrt{2}}, -1\right)$ is a minimum.

Exercise 14.2   (Differentiation)

For the following curves (questions 1 to 4) find $\dfrac{dy}{dx}$ and $\dfrac{d^2y}{dx^2}$ in terms of t.

1 $x = 3t^2$, $y = 6t$ **2** $x = 2t$, $y = \dfrac{2}{t}$ **3** $x = 4\cos t$, $y = 3\sin t$ **4** $x = 3\sec t$, $y = \tan t$

5   Find the values of $\dfrac{dy}{dx}$ and $\dfrac{d^2y}{dx^2}$ at the point $(2, 2)$ on the curve $x = 1 - t^3$, $y = 1 + t^2$.

6   Find the gradient of the curve $x = \dfrac{1+t}{1-2t}$, $y = \dfrac{1+2t}{1-t}$ at the point $(1, 1)$.

7   Find the points on the curve $x = t(t - 4)$, $y = t^2(t - 4)$ at which the gradient is $\frac{3}{2}$. Find also the point at which the curve is parallel to the y-axis.

8   Find the points on the curve $x = 3\cos t$, $y = 8\sin t$ at which the gradient is -2.

9   Find the points on the curve $x = 3t^2 + 5$, $y = 2t^3 + 6t$ at which $\dfrac{d^2y}{dx^2}$ is zero.

10   If x and y are given in terms of a parameter t, explain how to find $\dfrac{dx}{dy}$ and $\dfrac{d^2x}{dy^2}$. For the curve $x = 2t$, $y = 4 - \dfrac{3}{t}$, express $\dfrac{dy}{dx}, \dfrac{d^2y}{dx^2}, \dfrac{dx}{dy}$ and $\dfrac{d^2x}{dy^2}$ in terms of t.

11   Find and classify the stationary points on the curve $x = \dfrac{t}{1+t}$, $y = \dfrac{t^2}{1+t}$.

12   If $x = a\cos^3 t$, $y = a\sin^3 t$, find $\dfrac{dy}{dx}$ in terms of t, and sketch the curve.

14.3   Tangents and normals

Once we know the gradient of a curve, we can apply the usual methods to find the equations of tangents and normals to the curve.

Example 1   Find the equation of the tangent to the curve $x = 2t + t^2$, $y = 2t^2 + t^3$ at the point $(3, -9)$.

We have

$$\frac{dy}{dx} = \frac{\dfrac{dy}{dt}}{\dfrac{dx}{dt}} = \frac{4t + 3t^2}{2 + 2t}$$

The point $(3, -9)$ corresponds to $t = -3$; then

$$\frac{dy}{dx} = \frac{-12 + 27}{2 - 6} = -\frac{15}{4}$$

The equation of the tangent is

$$y + 9 = -\frac{15}{4}(x - 3)$$

i.e. $15x + 4y = 9$

Example 2 Find the equation of the normal to the curve $x = t + \dfrac{1}{t}, y = t - \dfrac{1}{t}$ at the point $t = 2$, and find the point where this normal meets the curve again.

We have

$$\frac{dy}{dx} = \frac{\dfrac{dy}{dt}}{\dfrac{dx}{dt}} = \frac{1 + \dfrac{1}{t^2}}{1 - \dfrac{1}{t^2}} = \frac{t^2 + 1}{t^2 - 1}$$

When $t = 2$, $x = \frac{5}{2}$, $y = \frac{3}{2}$ and $\dfrac{dy}{dx} = \frac{5}{3}$. The gradient of the normal is $-\frac{3}{5}$, and so its equation is

$$y - \frac{3}{2} = -\frac{3}{5}\left(x - \frac{5}{2}\right)$$

i.e. $3x + 5y = 15$

To find where this line meets the curve, we substitute $x = t + \dfrac{1}{t}$ and $y = t - \dfrac{1}{t}$. Thus

$$3\left(t + \frac{1}{t}\right) + 5\left(t - \frac{1}{t}\right) = 15$$

i.e. $3t^2 + 3 + 5t^2 - 5 = 15t$

i.e. $8t^2 - 15t - 2 = 0$

i.e. $(t - 2)(8t + 1) = 0$

The line meets the curve at the point $t = 2$ (where it is the normal), and at the point $t = -\frac{1}{8}$. When $t = -\frac{1}{8}$, $x = -\frac{1}{8} - 8 = -8\frac{1}{8}$ and $y = -\frac{1}{8} + 8 = 7\frac{7}{8}$, so the normal meets the curve again at the point $(-8\frac{1}{8}, 7\frac{7}{8})$.

Example 3 Find the equation of the tangent to the curve $x = 3t^2$, $y = 4t + 1$ at the point t. Hence find the equations of the tangents which pass through the point $(6, -5)$, and state their points of contact with the curve.

We have

$$\frac{dy}{dx} = \frac{\dfrac{dy}{dt}}{\dfrac{dx}{dt}} = \frac{4}{6t} = \frac{2}{3t}$$

The point t is $(3t^2, 4t + 1)$, and the gradient at this point is $\dfrac{2}{3t}$. The equation of the tangent is therefore

$$y - (4t + 1) = \frac{2}{3t}(x - 3t^2)$$

$$3ty - 12t^2 - 3t = 2x - 6t^2$$

$$2x - 3ty + 6t^2 + 3t = 0$$

This tangent passes through the point $(6, -5)$ if $x = 6$ and $y = -5$ satisfies its equation, i.e.

$$12 + 15t + 6t^2 + 3t = 0$$
$$t^2 + 3t + 2 = 0$$
$$(t + 1)(t + 2) = 0$$
$$t = -1 \quad \text{or} \quad -2$$

When $t = -1$, the equation of the tangent is $2x + 3y + 6 - 3 = 0$, or $2x + 3y + 3 = 0$, and its point of contact is the point $t = -1$, i.e. $(3, -3)$. When $t = -2$, the tangent is $2x + 6y + 24 - 6 = 0$, or $x + 3y + 9 = 0$ and its point of contact is the point $t = -2$, i.e. $(12, -7)$.

Exercise 14.3 (Tangents and normals)

1 Find the equation of the tangent to the curve $x = 1 - t^3$, $y = 1 + t^2$ at the point $(0, 2)$.

2 Find the equation of the normal to the curve $x = \dfrac{t}{1 - t}$, $y = \dfrac{t^2}{1 - t}$ at the point $t = \frac{1}{2}$.

3 Find the equation of the tangent to the curve $x = \sec t$, $y = \tan t$ at the point $t = \frac{1}{4}\pi$.

4 Find the equation of the tangent to the curve $x = 2t - 1$, $y = t^3$ at the point t.

5 Find the equation of the normal to the curve $x = 2\cos t$, $y = \sin t$ at the point t.

6 Find the equations of the tangent and the normal to the parabola $x = at^2$, $y = 2at$ at the point t.

7 Find the equation of the normal to the curve $x = a\cos^3 t$, $y = a\sin^3 t$ at the point t.

8 Find the equations of the tangents to the curve $x = \dfrac{2}{t}$, $y = 4 - 3t$ which have gradient 6, and give their points of contact with the curve.

9 Find the equation of the tangent to the curve $x = 2t^2$, $y = 2t^3$ at the point $t = 2$, and find the point where this tangent meets the curve again.

10 Find the equation of the tangent to the curve $x = 3t$, $y = \dfrac{3}{t}$ at the point t. Hence find the equations of the tangents which pass through the point $(7, -1)$, and give their points of contact with the curve.

11 Find the equation of the normal to the curve $x = 2t^2$, $y = 4t$ at the point t. Hence show that there are three normals which pass through the point $(12, 0)$, and give their equations.

12 The line $4x + 3y = 27$ meets the curve $x = t^2 - 2t$, $y = 1 + 4t$ at P and Q. Show that the tangents to the curve at P and Q are perpendicular, and find their point of intersection.

14.4 Areas

We now look at the problem of finding the area between a curve and the x-axis, when the curve is given in terms of a parameter t.

Consider a section of the curve corresponding to values of t between t_1 and t_2 (Fig. 14.4). If two points on the curve correspond to t and $(t + \delta t)$, the change δt in t will cause a change δx in x, and the area of the shaded rectangle is $y\,\delta x$. Provided that δt is small,

$$\frac{\delta x}{\delta t} \approx \frac{dx}{dt} \quad \left(\text{since } \frac{dx}{dt} = \lim_{\delta t \to 0} \frac{\delta x}{\delta t} \right)$$

and so

$$\delta x \approx \frac{dx}{dt} \times \delta t$$

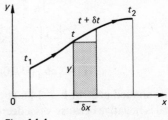

Fig. 14.4

The total area under this section of the curve is approximately

$$\sum_{t=t_1}^{t=t_2} y \, \delta x \approx \sum_{t=t_1}^{t=t_2} \left(y \frac{dx}{dt} \right) \delta t$$

As $\delta t \to 0$, we obtain the area between the curve and the x-axis as

$$\int_{t_1}^{t_2} \left(y \frac{dx}{dt} \right) dt$$

Before integrating, we must express y and $\dfrac{dx}{dt}$ in terms of t. The definite integral will give the numerical value of the area, provided that the curve does not cross the x-axis, nor turn back on itself, between $t = t_1$ and $t = t_2$.

Example 1   Find the area between the x-axis and the curve $x = t^2 - 4t + 8$, $y = t^3$ from $t = 2$ to $t = 3$.

We have

$$\int_2^3 \left(y \frac{dx}{dt} \right) dt = \int_2^3 t^3 (2t - 4) \, dt$$

$$= \int_2^3 (2t^4 - 4t^3) \, dt$$

$$= \left[\frac{2}{5} t^5 - t^4 \right]_2^3$$

$$= \left(\frac{486}{5} - 81 \right) - \left(\frac{64}{5} - 16 \right) = \frac{97}{5}$$

The required area is therefore $\frac{97}{5}$ (see Fig. 14.5).

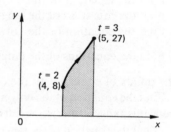

Fig. 14.5

Example 2   Find the area enclosed by the ellipse $\dfrac{x^2}{a^2} + \dfrac{y^2}{b^2} = 1$.

The ellipse is given parametrically by $x = a \cos t$, $y = b \sin t$. The top half of the ellipse corresponds to values of t between 0 and π (see Fig. 14.6). The area between this section of the curve and the x-axis is given by the numerical value of

$$\int_0^\pi y \frac{dx}{dt} \, dt = \int_0^\pi b \sin t \, (-a \sin t) \, dt = -ab \int_0^\pi \sin^2 t \, dt$$

$$= -ab \int_0^\pi \frac{1}{2}(1 - \cos 2t) \, dt$$

$$= -\frac{1}{2} ab \left[t - \frac{1}{2} \sin 2t \right]_0^\pi$$

$$= -\tfrac{1}{2} ab \{ (\pi - 0) - (0 - 0) \} = -\tfrac{1}{2} \pi ab$$

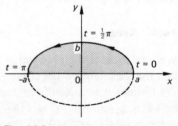

Fig. 14.6

The shaded area in Fig. 14.6 is therefore $\frac{1}{2}\pi ab$ (the definite integral is negative because x is a decreasing function of t between $t = 0$ and $t = \pi$). The area enclosed by the ellipse is double this, i.e. πab.

Exercise 14.4 (Areas)

Find the area between the x-axis and the following curves (questions 1 to 5), between the given values of t.

1   $x = 3t^2$, $y = 6t$, from $t = 1$ to $t = 2$.   2 $x = 1 - t^3$, $y = 1 + t^2$, from $t = 0$ to $t = 2$.

3   $x = t(2 - t)$, $y = \dfrac{1 + t}{1 - t}$, from $t = -1$ to $t = 0$.

4   $x = 3 \cos t$, $y = 2 \sin t$, from $t = 0$ to $t = \frac{1}{3}\pi$.
5   $x = a(t - \sin t)$, $y = a(1 - \cos t)$, from $t = 0$ to $t = 2\pi$.

6   Write down the values of t where the curve $x = 2t^3$, $y = t(t + 2)$ crosses the x-axis, and find the area enclosed between the curve and the x-axis.
7   Find the total area enclosed by the curve $x = \sin t$, $y = \sin 2t$. [For a sketch, see p. 241. Use $2 \sin A \cos B = \sin (A + B) + \sin (A - B)$.]
8   Show that the area between the y-axis and a curve given parametrically, from $t = t_1$ to $t = t_2$, is given by

$$\int_{t_1}^{t_2} x \frac{dy}{dt}\, dt$$

Sketch the curve $x = t(1 - t^2)$, $y = 2 - t^2$, and find the area of its loop.

14.5   Length of a curve

The speed of a particle

We now return to the idea of a particle moving in two dimensions, with its x- and y-coordinates given in terms of time, t. Consider the velocity of the particle. $\dfrac{dx}{dt}$ is the velocity in the x-direction, and $\dfrac{dy}{dt}$ is the velocity in the y-direction (see Fig. 14.7). The actual speed of the particle, say v, is found by taking the resultant of these two components. The speed is

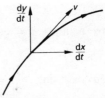

Fig. 14.7

$$v = \sqrt{\left(\frac{dx}{dt}\right)^2 + \left(\frac{dy}{dt}\right)^2}$$

and the direction of motion is along the tangent to the curve.

Alternatively, consider a small interval of time δt, and suppose x and y change by δx and δy (Fig. 14.8). The distance travelled is approximately $\sqrt{(\delta x)^2 + (\delta y)^2}$, hence the speed is approximately

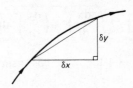

Fig. 14.8

$$\frac{\sqrt{(\delta x)^2 + (\delta y)^2}}{\delta t} = \sqrt{\left(\frac{\delta x}{\delta t}\right)^2 + \left(\frac{\delta y}{\delta t}\right)^2}$$

As $\delta t \to 0$, we obtain the speed,

$$v = \sqrt{\left(\frac{dx}{dt}\right)^2 + \left(\frac{dy}{dt}\right)^2}$$

as before.

The distance travelled

If the speed of the particle is v, the distance travelled in a short time δt is approximately $v\delta t$, and so the total distance travelled between times $t = t_1$ and $t = t_2$ is approximately $\sum\limits_{t=t_1}^{t=t_2} v\delta t$. Taking the limit as $\delta t \to 0$, the distance travelled is

$$\int_{t_1}^{t_2} v\,dt = \int_{t_1}^{t_2} \sqrt{\left(\frac{dx}{dt}\right)^2 + \left(\frac{dy}{dt}\right)^2}\,dt$$

Length of arc

If we now consider the curve along which the particle is moving, the distance travelled is equal to the length of the curve. It follows that, for a curve given parametrically in terms of t, the length of the curve between the points $t = t_1$ and $t = t_2$ is

$$\int_{t_1}^{t_2} \sqrt{\left(\frac{dx}{dt}\right)^2 + \left(\frac{dy}{dt}\right)^2}\,dt$$

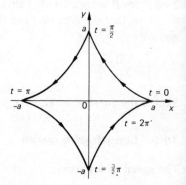

Fig. 14.9

Example 1   Find the total length of the astroid $x = a\cos^3 t$, $y = a\sin^3 t$.

Since the curve is symmetrical, we shall find the total length as four times the length of the curve between $t = 0$ and $t = \frac{1}{2}\pi$ (Fig. 14.9). Now

$$x = a\cos^3 t \quad \text{and} \quad y = a\sin^3 t$$

so

$$\frac{dx}{dt} = -3a\cos^2 t\,\sin t \qquad \frac{dy}{dt} = 3a\sin^2 t\,\cos t$$

The total length is

$$4\int_0^{\frac{1}{2}\pi} \sqrt{\left(\frac{dx}{dt}\right)^2 + \left(\frac{dy}{dt}\right)^2}\,dt$$

$$= 4\int_0^{\frac{1}{2}\pi} \sqrt{9a^2\cos^4 t\,\sin^2 t + 9a^2\sin^4 t\,\cos^2 t}\,dt$$

$$= 4\int_0^{\frac{1}{2}\pi} \sqrt{9a^2\cos^2 t\,\sin^2 t\,(\cos^2 t + \sin^2 t)}\,dt$$

$$= 4\int_0^{\frac{1}{2}\pi} 3a\cos t\,\sin t\,dt = \int_0^{\frac{1}{2}\pi} 6a\sin 2t\,dt$$

$$= \left[-3a\cos 2t\right]_0^{\frac{1}{2}\pi} = (3a) - (-3a)$$

$$= 6a$$

Example 2 A circle of radius a is rolling along the x-axis, and P is a point marked on its circumference. The path of P is called a **cycloid**. If P is initially at the origin, show that, when the circle has rotated through t radians, the coordinates of P are

$$x = a(t - \sin t)$$
$$y = a(1 - \cos t)$$

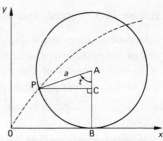

Find the length of one arch of the cycloid.

The arc PB has rolled along OB (Fig. 14.10), and so OB $=$ arc PB $= at$. The x-coordinate of P is

Fig. 14.10

$$x = OB - PC = at - a \sin t$$
$$= a(t - \sin t)$$

The y-coordinate of P is

$$y = BC = AB - AC = a - a \cos t$$
$$= a(1 - \cos t)$$

The cycloid is shown in Fig. 14.11.

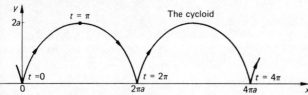

Fig. 14.11

Now $\quad \dfrac{dx}{dt} = a(1 - \cos t)$

and $\quad \dfrac{dy}{dt} = a \sin t$

The length of one arch is

$$\int_0^{2\pi} \sqrt{\left(\frac{dx}{dt}\right)^2 + \left(\frac{dy}{dt}\right)^2}\ dt = \int_0^{2\pi} \sqrt{a^2(1 - \cos t)^2 + a^2 \sin^2 t}\ dt$$

$$= \int_0^{2\pi} a\sqrt{1 - 2\cos t + \cos^2 t + \sin^2 t}\ dt$$

$$= \int_0^{2\pi} a\sqrt{2 - 2\cos t}\ dt$$

$$= \int_0^{2\pi} a\sqrt{4 \sin^2 \tfrac{1}{2}t}\ dt \quad (\text{since } 1 - \cos t = 2 \sin^2 \tfrac{1}{2}t)$$

$$= \int_0^{2\pi} 2a \sin \tfrac{1}{2}t\ dt$$

$$= \left[-4a \cos \frac{1}{2}t \right]_0^{2\pi} = (4a) - (-4a)$$

$$= 8a$$

Exercise 14.5   (Speed, and length of curves)

1   At time t, the position (x, y) of a moving particle is given by $x = t(2+t)$, $y = t(3-t)$. Find the speed of the particle (i) when $t = 0$ and (ii) when $t = 2$. Find the position of the particle when its speed is a minimum, and find this minimum speed.

2   A particle moves so that its position (x, y) at time t is given by $x = 4\cos t$, $y = 3\sin t$. Show that its speed is $\sqrt{9+7\sin^2 t}$. Hence find the maximum and minimum speeds, and the corresponding positions of the particle.

3   At time t, a particle is at the point $(a\cos t, a\sin t)$. Find the distance moved by the particle between $t = 0$ and $t = 2\pi$, and comment on your answer.

4   By using the parametric form $x = t$, $y = 2t^{3/2}$, find the length of the curve $y = 2x^{3/2}$ between $x = 0$ and $x = 11$.

5   Find the length of the curve $x = t^2$, $y = t - \frac{1}{3}t^3$ between the points $t = -3$ and $t = 3$.

6   Find the length of that part of the curve $x = 4t + 2t^{3/2}$, $y = 4t - 2t^{3/2}$, which lies above the x-axis (give your answer to 3 decimal places).

7   Sketch the curve $x = \cos 2t$, $y = 4(\sin t + \cos t)$. Show that

$$\left(\frac{dx}{dt}\right)^2 + \left(\frac{dy}{dt}\right)^2 = 4(2 - \sin 2t)^2$$

and hence find the total length of the curve.

15

Integration II

Introduction

To integrate a function $f(x)$, we have to find a function which, when differentiated, yields $f(x)$. In this chapter we study some methods for doing this.

Everything we know about differentiation tells us something about integration. For example, the fact that $\dfrac{d}{dx}(x^3) = 3x^2$ tells us that $\int 3x^2\, dx = x^3 + C$; the fact that $\dfrac{d}{dx}(x\sin x) = x\cos x + \sin x$ tells us that $\int (x\cos x + \sin x)\, dx = x\sin x + C$. Each 'rule' for differentiation will give rise to some 'rule' for integration. However, this does not mean that we can guarantee to integrate any given function. Nor is it always obvious which method will provide the easiest solution, or indeed a solution at all.

If we are asked for an indefinite integral, we give the most general answer by adding an arbitrary constant $+C$. If we omit this, then the answer we give is only one of the infinite set of possible answers, and this could lead to misunderstandings. Sometimes two different methods for finding the same integral will give apparently different 'answers'. In fact the two answers will differ by a constant.

An integral can always be checked by differentiating it, whereupon the original function should be obtained. This check must always be made if there is any doubt.

15.1 Integration by substitution

Suppose y is a function of x which we wish to integrate. Let $Q = \int y\, dx$ be the required integral; then $\dfrac{dQ}{dx} = y$. We try to simplify the integral by introducing a new variable, say t, which is related to x.

Using the chain rule,

$$\frac{dQ}{dt} = \frac{dQ}{dx} \times \frac{dx}{dt} = y\,\frac{dx}{dt}$$

and so

$$Q = \int y\,\frac{dx}{dt}\,dt$$

Thus $\int y\,dx = \int y\dfrac{dx}{dt}\,dt$

This is **integration by substitution** (or integration using a change of variable). It follows from the chain rule of differentiation.

We first express $y\dfrac{dx}{dt}$ in terms of t. Integrating this (with respect to t), we obtain a function of t. Finally, we use the relationship between x and t to express the integral in terms of x.

Example 1   Integrate $x(2x-3)^4$.

This will be simplified by writing $2x-3=t$. Differentiating this equation with respect to t, $2\dfrac{dx}{dt}=1$, so $\dfrac{dx}{dt}=\frac{1}{2}$. We have $x=\frac{1}{2}(t+3)$, and thus

$$\int x(2x-3)^4\,dx = \int x\,(2x-3)^4\dfrac{dx}{dt}\,dt = \int \tfrac{1}{2}(t+3)t^4(\tfrac{1}{2})dt$$

$$= \int(\tfrac{1}{4}t^5+\tfrac{3}{4}t^4)\,dt = \tfrac{1}{24}\,t^6+\tfrac{3}{20}\,t^5+C$$

$$= \tfrac{1}{24}(2x-3)^6+\tfrac{3}{20}(2x-3)^5+C$$

We can check this by differentiating it. We obtain

$$\tfrac{1}{24}\times 6(2x-3)^5\times 2+\tfrac{3}{20}\times 5(2x-3)^4\times 2$$
$$= \tfrac{1}{2}(2x-3)^5+\tfrac{3}{2}(2x-3)^4$$
$$= \tfrac{1}{2}(2x-3)^4[(2x-3)+3] = x(2x-3)^4$$

Example 2   Integrate $\dfrac{x}{\sqrt{4+x^2}}$.

We put $4+x^2=t^2$

Then $2x\dfrac{dx}{dt}=2t$

i.e. $x\dfrac{dx}{dt}=t$

Thus $\displaystyle\int\frac{x}{\sqrt{4+x^2}}\,dx = \int\frac{x}{\sqrt{4+x^2}}\dfrac{dx}{dt}\,dt$

$$= \int\frac{t}{t}\,dt = \int 1\,dt = t+C$$

$$= \sqrt{4+x^2}+C$$

We check that

$$\frac{d}{dx}(\sqrt{4+x^2}+C) = \frac{d}{dx}[(4+x^2)^{\frac{1}{2}}+C]$$

$$= \tfrac{1}{2}(4+x^2)^{-\frac{1}{2}} \times (2x)$$

$$= \frac{x}{\sqrt{4+x^2}}$$

In practice, the setting out of these examples can be shortened as follows.

(1) $\displaystyle\int x(2x-3)^4\,dx$

Put $2x - 3 = t$; then $2\dfrac{dx}{dt} = 1$, so $dx = \tfrac{1}{2}dt$.

Then

$$\int x(2x-3)^4\,dx = \int \tfrac{1}{2}(t+3)t^4(\tfrac{1}{2}dt)$$

$$= \int (\tfrac{1}{4}t^5 + \tfrac{3}{4}t^4)dt = \tfrac{1}{24}t^6 + \tfrac{3}{20}t^5 + C$$

$$= \tfrac{1}{24}(2x-3)^6 + \tfrac{3}{20}(2x-3)^5 + C$$

(2) $\displaystyle\int \frac{x}{\sqrt{4+x^2}}\,dx$

Put $4 + x^2 = t^2$; then $2x\,dx = 2t\,dt$. (Here we have differentiated the LHS with respect to x, and multiplied the result by dx. We have differentiated the RHS with respect to t, and multiplied the result by dt.) Then

$$\int \frac{x}{\sqrt{4+x^2}}\,dx = \int \frac{1}{t}t\,dt \qquad \text{(replacing `$x\,dx$' by `$t\,dt$')}$$

$$= \int 1\,dt = t + C$$

$$= \sqrt{4+x^2} + C$$

Of course this is just a shorter way of doing exactly the same thing as before. The statement $dx = \tfrac{1}{2}dt$ simply means that in the first integral, the 'dx' may be replaced by '$\tfrac{1}{2}dt$'. It does not imply that we can attach any meaning to dx or dt as separate quantities.

This method will only lead to a successful integration if the substitution produces a function of t which we can integrate. Finding a substitution which will work for a given integral is an art which requires much practice.

Of course any letter (instead of t) can be used to represent the new variable. Other letters commonly used are u, y, and θ.

Example 3

$$\int (1 - 3x)^5 \, dx$$

This will be simplified if we replace $1 - 3x$ by a new variable. We put

$$1 - 3x = t$$

then

$$-3 \, dx = dt$$

$$\int (1 - 3x)^5 \, dx = \int t^5 (-\tfrac{1}{3} dt) = -\tfrac{1}{18} t^6 + C$$

$$= -\tfrac{1}{18} (1 - 3x)^6 + C$$

Example 4

$$\int \frac{x}{\sqrt{1 + 4x}} \, dx$$

This is made difficult by the presence of $\sqrt{1 + 4x}$. We replace this by a new variable. Put

$$1 + 4x = t^2$$

then

$$4 \, dx = 2t \, dt$$

$$\int \frac{x}{\sqrt{1 + 4x}} \, dx = \int \frac{\tfrac{1}{4}(t^2 - 1)}{t} (\tfrac{1}{2} t \, dt)$$

$$= \int (\tfrac{1}{8} t^2 - \tfrac{1}{8}) dt = \tfrac{1}{24} t^3 - \tfrac{1}{8} t + C$$

$$= \tfrac{1}{24} \sqrt{(1 + 4x)^3} - \tfrac{1}{8} \sqrt{1 + 4x} + C$$

Example 5

$$\int \frac{\sin x}{(1 + \cos x)^2} \, dx$$

Here we notice that $\sin x$ is (apart from sign) the derivative of $(1 + \cos x)$. We put

$$1 + \cos x = u$$

then

$$-\sin x \, dx = du$$

$$\int \frac{\sin x}{(1 + \cos x)^2} \, dx = \int \frac{1}{u^2} (-du)$$

$$= \int (-u^{-2}) du = u^{-1} + C$$

$$= \frac{1}{1 + \cos x} + C$$

Example 6

$$\int x^3 \sqrt{1-x^2}\, dx$$

We can write x^3 as $x^2 \times x$; x is a multiple of the derivative of $1-x^2$.
We put

$$1-x^2 = t^2$$

then

$$-2x\, dx = 2t\, dt$$

$$\int x^3 \sqrt{1-x^2}\, dx = \int x^2 \sqrt{1-x^2}\,(x\, dx) = \int (1-t^2)t(-t\, dt)$$

$$= \int (-t^2 + t^4)dt = -\tfrac{1}{3}t^3 + \tfrac{1}{5}t^5 + C$$

$$= -\tfrac{1}{3}\sqrt{(1-x^2)^3} + \tfrac{1}{5}\sqrt{(1-x^2)^5} + C$$

Example 7

$$\int \sin^2 x \cos^3 x\, dx$$

We have $\quad \sin^2 x \cos^3 x = \sin^2 x \cos^2 x \cos x$

$$= \sin^2 x(1 - \sin^2 x)\cos x$$

and $\cos x$ is the derivative of $\sin x$. We put

$$\sin x = t$$

then

$$\cos x\, dx = dt$$

$$\int \sin^2 x \cos^3 x\, dx = \int \sin^2 x(1 - \sin^2 x)\cos x\, dx$$

$$= \int t^2(1-t^2)\, dt = \int (t^2 - t^4)\, dt$$

$$= \tfrac{1}{3}t^3 - \tfrac{1}{5}t^5 + C = \tfrac{1}{3}\sin^3 x - \tfrac{1}{5}\sin^5 x + C$$

Example 8

$$\int \frac{1}{4+x^2}\, dx$$

The identity $1 + \tan^2 \theta = \sec^2 \theta$ suggests that we should try the substitution

$$x = 2\tan\theta$$

then

$$dx = 2\sec^2 \theta\, d\theta$$

$$\int \frac{1}{4+x^2}\, dx = \int \frac{1}{4 + 4\tan^2\theta}\,(2\sec^2\theta\, d\theta) = \int \frac{2\sec^2\theta}{4\sec^2\theta}\, d\theta$$

$$= \int \tfrac{1}{2}\, d\theta = \tfrac{1}{2}\theta + C = \tfrac{1}{2}\tan^{-1}(\tfrac{1}{2}x) + C$$

Example 9

$$\int \frac{1}{\sqrt{5-4x-x^2}}\,dx$$

We first complete the square:

$$5 - 4x - x^2 = 9 - (4 + 4x + x^2) = 9 - (x+2)^2$$

Since $1 - \sin^2\theta = \cos^2\theta$, we put

$$x + 2 = 3\sin\theta$$

then

$$dx = 3\cos\theta\,d\theta$$

$$\int \frac{1}{\sqrt{5-4x-x^2}}\,dx = \int \frac{1}{\sqrt{9-(x+2)^2}}\,dx = \int \frac{1}{\sqrt{9-9\sin^2\theta}}\,(3\cos\theta\,d\theta)$$

$$= \int \frac{3\cos\theta}{\sqrt{9\cos^2\theta}}\,d\theta = \int \frac{3\cos\theta}{3\cos\theta}\,d\theta$$

$$= \int 1\,d\theta = \theta + C = \sin^{-1}\left(\frac{x+2}{3}\right) + C$$

Exercise 15.1 (Integration by substitution)

In questions 1 to 23, use the given substitution to integrate the function.

1 $(2x-7)^5$ $[2x-7=t]$

2 $(3-4x)^9$ $[3-4x=t]$

3 $\dfrac{1}{\sqrt{1-5x}}$ $[1-5x=t^2]$

4 $x(x+2)^3$ $[x+2=t]$

5 $x(2x+3)^{\frac{3}{2}}$ $[2x+3=t^2]$

6 $\dfrac{x}{(2x-1)^3}$ $[2x-1=t]$

7 $x(2+x^2)^3$ $[2+x^2=t]$

8 $\dfrac{x^2}{(1+2x^3)^2}$ $[1+2x^3=t]$

9 $\dfrac{x}{\sqrt{x^2-1}}$ $[x^2-1=t^2]$

10 $\sin^2 x \cos x$ $[\sin x = t]$

11 $\tan^2 x \sec^2 x$ $[\tan x = t]$

12 $\dfrac{\cos x}{\sqrt{1+\sin x}}$ $[1+\sin x = t^2]$

13 $\dfrac{1}{\sqrt{1-16x^2}}$ $[4x = \sin\theta]$

14 $\dfrac{1}{9+25x^2}$ $[5x = 3\tan\theta]$

15 $\dfrac{1}{\sqrt{3+2x-x^2}}$ $[x-1=2\sin\theta]$

16 $x\sqrt{x+1}$ $[x+1=u^2]$

17 $\cos^3 x$ $[\sin x = u]$

18 $\sin^3 x \cos^2 x$ $[\cos x = t]$

19 $x^3 \cos(x^4)$ $[x^4 = u]$

20 $\sec^4 x$ $[\tan x = t]$

21 $\sec^3 x \tan^3 x$ $[\sec x = u]$

22 $\dfrac{x^2}{1+4x^6}$ $[2x^3 = \tan\theta]$

23 $\dfrac{\cos x}{\sqrt{4-\sin^2 x}}$ $[\sin x = 2\sin\theta]$

24 Use the substitution $x = \sec\theta$ to find

$$\int \frac{1}{\sqrt{x^2(x^2-1)}}\, dx$$

25 Find $\int \sin x \cos x\, dx$ (i) by using the substitution $\sin x = u$; (ii) by using the substitution $\cos x = t$; (iii) by first writing $\sin x \cos x = \frac{1}{2}\sin 2x$. Are your answers the same?

15.2 Definite integrals by substitution

To evaluate a definite integral by substitution, we could carry out the process of the previous section, and put in the limits of x at the end. However, it is usually quicker to transform the required integral into a definite integral involving the new variable, which is then evaluated with its own limits, without returning to the original variable x.

Suppose we wish to evaluate $\int_a^b y\, dx$. If we introduce a new variable t, we have $\int y\, dx = \int y \dfrac{dx}{dt}\, dt$, and this gives us a function of t. Instead of expressing this in terms of x, and then substituting $x = b$ and $x = a$, we can simply substitute $t = \beta$ and $t = \alpha$, where β and α are the values of t which correspond to the values b and a of x. Thus

$$\int_a^b y\, dx = \int_\alpha^\beta y \frac{dx}{dt}\, dt$$

where $t = \alpha$ when $x = a$ and $t = \beta$ when $x = b$.

If there is more than one value of t corresponding to a given value of x, we remove any ambiguity by placing suitable restrictions on t at the outset. For example, when we use a substitution like $1 + 4x = t^2$, we assume that $t \geqslant 0$. When we put $x = \sin\theta$, we assume that $-\frac{1}{2}\pi \leqslant \theta \leqslant \frac{1}{2}\pi$. When we put $x = \cos\theta$, we assume that $0 \leqslant \theta \leqslant \pi$. When we put $x = \tan\theta$, we assume that $-\frac{1}{2}\pi < \theta < \frac{1}{2}\pi$.

We also check that any substitution we make in an integral is valid for all values of x between the limits of integration.

Example 1

$$\int_0^3 \frac{1}{\sqrt{x+1}}\, dx$$

Put $x + 1 = t^2$ (where $t \geqslant 0$)

then $dx = 2t\, dt$

When $x = 0$, $t = 1$; and when $x = 3$, $t = 2$.

Then $\displaystyle\int_0^3 \frac{1}{\sqrt{x+1}}\, dx = \int_1^2 \frac{1}{t}(2t\, dt) = \int_1^2 2\, dt$

$$= \Big[\, 2t\, \Big]_1^2 = 4 - 2 = 2$$

The substitution gives

$$\int_0^3 \frac{1}{\sqrt{x+1}}\,dx = \int_1^2 2\,dt$$

We shall now investigate this numerically.

$\int_0^3 \frac{1}{\sqrt{x+1}}\,dx$ is the area under the curve $y = \frac{1}{\sqrt{x+1}}$ from $x = 0$ to $x = 3$. We divide this area into six strips of width 0.5. The area of each strip is found approximately as the area of a rectangle with height equal to that of the curve at the mid-point of the strip (see Fig. 15.1).

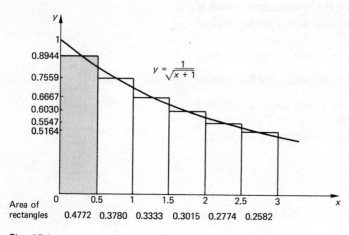

Area of rectangles   0.4772 0.3780 0.3333 0.3015 0.2774 0.2582

Fig. 15.1

Using the substitution $x + 1 = t^2$, the division points are transformed as follows:

x	0	0.5	1	1.5	2	2.5	3
t	1	1.225	1.414	1.581	1.732	1.871	2

We wish to find a curve such that the area between this new curve and the t-axis is equal to the area under the corresponding part of the original curve. Accordingly, on the t-axis, we construct rectangles having the same area as the corresponding rectangles on the x-axis (see Fig. 15.2).

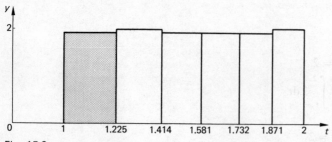

Fig. 15.2

Rectangle between	Width	Area	Height
$t = 1$ and $t = 1.225$	0.225	0.4472	$\dfrac{0.4472}{0.225} = 1.988$
$t = 1.225$ and $t = 1.414$	0.189	0.3780	$\dfrac{0.3780}{0.189} = 2.000$
$t = 1.414$ and $t = 1.581$	0.167	0.3333	$\dfrac{0.3333}{0.167} = 1.996$
$t = 1.581$ and $t = 1.732$	0.151	0.3015	$\dfrac{0.3015}{0.151} = 1.997$
$t = 1.732$ and $t = 1.871$	0.139	0.2774	$\dfrac{0.2774}{0.139} = 1.996$
$t = 1.871$ and $t = 2$	0.129	0.2582	$\dfrac{0.2582}{0.129} = 2.002$

It appears that the required curve is the line $y = 2$; the area under the curve $y = \dfrac{1}{\sqrt{x+1}}$ (between $x = 0$ and $x = 3$) is equal to the area under the line $y = 2$ between the corresponding values of t ($t = 1$ and $t = 2$). Thus

$$\int_0^3 \frac{1}{\sqrt{x+1}} \, dx = \int_1^2 2 \, dt$$

Example 2

$$\int_0^4 x\sqrt{2x+1} \, dx$$

Put $\qquad\qquad 2x + 1 = t^2 \quad \text{(where } t \geqslant 0)$

Then $\qquad\qquad 2 \, dx = 2t \, dt$

When $x = 0$, $t = 1$; and when $x = 4$, $t = 3$. Then

$$\int_0^4 x\sqrt{2x+1} \, dx = \int_1^3 \tfrac{1}{2}(t^2 - 1)t(t \, dt)$$

$$= \int_1^3 (\tfrac{1}{2}t^4 - \tfrac{1}{2}t^2) \, dt$$

$$= \left[\frac{1}{10}t^5 - \frac{1}{6}t^3 \right]_1^3$$

$$= \left(\frac{243}{10} - \frac{27}{6} \right) - \left(\frac{1}{10} - \frac{1}{6} \right) = \frac{298}{15}$$

Example 3

$$\int_0^{\frac{\sqrt{3}}{2}} \frac{1}{9+4x^2}dx$$

We have

$$9 + 4x^2 = 3^2 + (2x)^2$$

So we put

$$2x = 3\tan\theta \quad (\text{where } -\tfrac{1}{2}\pi < \theta < \tfrac{1}{2}\pi)$$

Then $2\,dx = 3\sec^2\theta\,d\theta$

When $x = 0$, $\theta = 0$; and when $x = \dfrac{\sqrt{3}}{2}$, $\tan\theta = \dfrac{\sqrt{3}}{3} = \dfrac{1}{\sqrt{3}}$,

so $\theta = \tfrac{1}{6}\pi$. Hence

$$\begin{aligned}
\int_0^{\frac{\sqrt{3}}{2}} \frac{1}{9+4x^2}dx &= \int_0^{\frac{1}{6}\pi} \frac{1}{9+9\tan^2\theta}\,(\tfrac{3}{2}\sec^2\theta\,d\theta) \\
&= \int_0^{\frac{1}{6}\pi} \frac{3}{2} \times \frac{\sec^2\theta}{9\sec^2\theta}\,d\theta \\
&= \int_0^{\frac{1}{6}\pi} \frac{1}{6}d\theta = \left[\frac{1}{6}\theta\right]_0^{\frac{1}{6}\pi} \\
&= \frac{1}{36}\pi
\end{aligned}$$

Example 4

$$\int_0^{5\pi} \cos^4 x \sin x \, dx$$

Put $\qquad \cos x = u$

Then $\qquad -\sin x\,dx = du$

When $x = 0$, $u = 1$; and when $x = 5\pi$, $u = -1$. Then

$$\int_0^{5\pi} \cos^4 x \sin x \, dx = \int_1^{-1} u^4(-du)$$

$$= \left[-\frac{1}{5}u^5\right]_1^{-1} = (\tfrac{1}{5}) - (-\tfrac{1}{5})$$

$$= \tfrac{2}{5}$$

Note that, in $\int_1^{-1}(-u^4)\,du$, the upper limit of integration is less than the lower one. This does not affect the method of evaluation: after integrating, we substitute first $u = -1$, and then $u = 1$.

Exercise 15.2   (Definite integrals by substitution)

In questions 1 to 13, use the given substitutions to evaluate the definite integrals

1   $\displaystyle\int_0^2 x(3-x)^4\,dx$ $[3-x=t]$

2   $\displaystyle\int_0^4 \frac{x}{\sqrt{2x+1}}\,dx$ $[2x+1=t^2]$

3   $\displaystyle\int_3^{11} x^2\sqrt{x-2}\,dx$ $[x-2=t^2]$

4   $\displaystyle\int_0^1 x^3\sqrt{1+x^4}\,dx$ $[1+x^4=t^2]$

5   $\displaystyle\int_{-\frac{1}{2}\pi}^{\frac{1}{2}\pi} \sin^4 x\cos x\,dx$ $[\sin x=t]$

6   $\displaystyle\int_0^{9\pi} \cos^6 x\sin x\,dx$ $[\cos x=t]$

7   $\displaystyle\int_0^{\frac{1}{4}\pi} \tan^3 x\sec^2 x\,dx$ $[\tan x=t]$

8   $\displaystyle\int_0^1 \frac{1}{\sqrt{4-x^2}}\,dx$ $[x=2\sin\theta]$

9   $\displaystyle\int_0^{\frac{2}{3}} \frac{1}{4+9x^2}\,dx$ $[3x=2\tan\theta]$

10   $\displaystyle\int_0^{\frac{1}{2}} \frac{x^2}{\sqrt{1-x^2}}\,dx$ $[x=\sin\theta]$

11   $\displaystyle\int_{-1}^1 \sqrt{1-x^2}\,dx$ $[x=\sin\theta]$

12   $\displaystyle\int_0^{\pi} \sin^3 x\cos^4 x\,dx$ $[\cos x=u]$

13   $\displaystyle\int_0^{\frac{1}{3}\pi} \frac{\sin x}{\sqrt[3]{\cos x}}\,dx$ $[\cos x=u^3]$

14   Show that, if $x>1$, the substitution $x=\sec\theta$ (where $0<\theta<\frac{1}{2}\pi$) reduces the integral

$\displaystyle\int \frac{1}{x\sqrt{x^2-1}}\,dx$ to $\int 1\,d\theta$. Hence evaluate $\displaystyle\int_{2/\sqrt{3}}^2 \frac{1}{x\sqrt{x^2-1}}\,dx$.

If $x<-1$, show that the substitution $x=\sec\theta$ (where $\frac{1}{2}\pi<\theta<\pi$) reduces the integral

$\displaystyle\int \frac{1}{x\sqrt{x^2-1}}\,dx$ to $\int(-1)\,d\theta$. Hence evaluate $\displaystyle\int_{-2}^{-\sqrt{2}} \frac{1}{x\sqrt{x^2-1}}\,dx$.

15   Explain the fallacy in the following evaluation of $I=\int_{-1}^1 x^2\,dx$.
'Put $x^2=u$

Then $2x\,dx=du$

When $x=-1$, $u=1$; and when $x=1$, $u=1$. Thus

$$I=\int_{-1}^1 (x)(x\,dx)$$

$$=\int_1^1 \sqrt{u}\,(\tfrac{1}{2}\,du)=\left[\frac{1}{3}u^{\frac{3}{2}}\right]_1^1$$

$$=\tfrac{1}{3}-\tfrac{1}{3}=0'$$

[Hint: is the replacement of x by $\sqrt{u}$ valid throughout the range of integration? Remember that $\sqrt{u}$ is the positive square root of u.]

15.3   Integration by inspection

Integration by substitution can be a lengthy process, but after a little practice the general form of the integral can often be seen without actually carrying out the substitution. This is particularly true when we can recognise a function occurring with its derivative. In such cases it is quicker to determine the integral by a trial differentiation (which could be done mentally).

Example 1   Integrate $x\sqrt{1-x^2}$.

x is a constant multiple of the derivative of $(1-x^2)$, so the integral is a multiple of $(1-x^2)^{\frac{3}{2}}$. Now

$$\frac{d}{dx}\left[(1-x^2)^{\frac{3}{2}}\right] = \tfrac{3}{2}(1-x^2)^{\frac{1}{2}}(-2x) = -3x\sqrt{1-x^2}$$

Hence $\displaystyle\int x\sqrt{1-x^2}\,dx = -\tfrac{1}{3}(1-x^2)^{\frac{3}{2}}+C = -\tfrac{1}{3}\sqrt{(1-x^2)^3}+C$

Example 2   Integrate $\sin^3 x \cos x$.

$\cos x$ is the derivative of $\sin x$, so the integral is a multiple of $\sin^4 x$. Now

$$\frac{d}{dx}(\sin^4 x) = 4\sin^3 x \cos x$$

Hence $\int \sin^3 x \cos x\,dx = \tfrac{1}{4}\sin^4 x + C$

Example 3   Integrate $\dfrac{(\tan^{-1} x)^2}{1+x^2}$.

$\dfrac{1}{1+x^2}$ is the derivative of $\tan^{-1} x$, so the integral is a multiple of $(\tan^{-1} x)^3$. Now

$$\frac{d}{dx}\left[(\tan^{-1} x)^3\right] = 3(\tan^{-1} x)^2\left(\frac{1}{1+x^2}\right)$$

Hence $\displaystyle\int\frac{(\tan^{-1} x)^2}{1+x^2}\,dx = \tfrac{1}{3}(\tan^{-1} x)^3 + C$

As a matter of notation, integrals such as

$$\int\frac{1}{1+x^2}\,dx \quad\text{and}\quad \int\frac{x}{\sqrt{1-x^2}}\,dx$$

are often written

$$\int\frac{dx}{1+x^2} \quad\text{and}\quad \int\frac{x\,dx}{\sqrt{1-x^2}}$$

Exercise 15.3   (Integration by inspection)

In questions 1 to 20, integrate the given functions.

1 $x(x^2-1)^4$ **2** $x(1-x^2)^{\frac{3}{2}}$ **3** $x^2\sqrt{1-2x^3}$

4 $\dfrac{x}{(1+x^2)^2}$ **5** $\dfrac{x^2}{\sqrt{x^3-2}}$ **6** $\dfrac{x^5}{(x^6-4)^3}$

7 $\dfrac{x^3}{\sqrt{1-x^4}}$ **8** $\dfrac{x-1}{\sqrt{x^2-2x+4}}$ **9** $\dfrac{(1+2\sqrt{x})^{\frac{2}{3}}}{\sqrt{x}}$

10 $\sin^4 x\cos x$ **11** $\cos^5 x\sin x$ **12** $\tan x\sec^2 x$

13 $\sin x\sqrt{1-\cos x}$ **14** $\sec^2 x\sqrt{\tan x}$ **15** $\dfrac{\cos x-\sin x}{\sqrt{\sin x+\cos x}}$

16 $x\cos(x^2)$ **17** $x^2\sin(x^3+1)$ **18** $\dfrac{1}{\sqrt{x}}\sin\sqrt{x}$

19 $\dfrac{\sin^{-1}x}{\sqrt{1-x^2}}$ **20** $\dfrac{(\tan^{-1}x)^{\frac{3}{2}}}{1+x^2}$

In questions 21 to 25, evaluate the definite integrals.

21 $\displaystyle\int_0^1 x(x^2+1)^3\,dx$ **22** $\displaystyle\int_{-1}^0 \dfrac{x^2\,dx}{(1-x^3)^2}$ **23** $\displaystyle\int_0^\pi \cos^2 x\sin x\,dx$

24 $\displaystyle\int_0^{\frac{1}{2}\pi} \dfrac{\cos x\,dx}{\sqrt{1+\sin x}}$ **25** $\displaystyle\int_0^{\frac{1}{2}} \dfrac{(\sin^{-1}x)^2}{\sqrt{1-x^2}}\,dx$

15.4   Integration by parts

We now consider the integration of a product of two functions, such as $x\cos x$. This occurs as one of the terms in the derivative of $x\sin x$; in fact

$$\frac{d}{dx}(x\sin x) = x\cos x + \sin x$$

Thus

$$\int(x\cos x+\sin x)dx = x\sin x+C$$

or $\displaystyle\int x\cos x\,dx+\int \sin x\,dx = x\sin x+C$

and so

$$\int x\cos x\,dx = x\sin x-\int \sin x\,dx+C$$
$$= x\sin x+\cos x+C$$

We check that

$$\frac{d}{dx}(x\sin x+\cos x+C) = x\cos x+\sin x-\sin x$$

$$= x\cos x$$

The following method generalises this process.

If u and v are functions of x, then the product rule of differentiation gives

$$\frac{d}{dx}(uv) = u\frac{dv}{dx} + \frac{du}{dx}v$$

Integrating both sides of this equation,

$$uv = \int u\frac{dv}{dx}dx + \int \frac{du}{dx}v\,dx$$

or
$$\boxed{\int u\frac{dv}{dx}dx = uv - \int \frac{du}{dx}v\,dx}^{*}$$

This is called **integration by parts**. It follows from the product rule of differentiation, and it is valid whenever u, v, $\dfrac{du}{dx}$ and $\dfrac{dv}{dx}$ are continuous functions.

Example 1

$\int x \cos x \, dx$

Put $u = x$, $\dfrac{dv}{dx} = \cos x$

Then $\dfrac{du}{dx} = 1$, $v = \sin x$

Then $\int x \cos x \, dx = x \sin x - \int 1 \times \sin x \, dx$

$$= x \sin x + \cos x + C$$

Example 2

$\displaystyle\int_{0}^{\frac{1}{2}\pi} x \sin 2x \, dx$

Put $u = x$, $\dfrac{dv}{dx} = \sin 2x$

Then $\dfrac{du}{dx} = 1$, $v = -\tfrac{1}{2}\cos 2x$

---

* We may learn this by saying $\displaystyle\int 1\text{st} \times 2\text{nd} = 1\text{st} \times \int 2\text{nd} - \int \frac{d}{dx}(1\text{st}) \times \int 2\text{nd}$

or 1st function times the integral of the 2nd minus the integral of the differential of the 1st times the integral of the 2nd.

Then $\displaystyle\int_0^{\frac{1}{2}\pi} x \sin 2x\, dx = \left[x(-\tfrac{1}{2}\cos 2x) \right]_0^{\frac{1}{2}\pi} - \int_0^{\frac{1}{2}\pi} (1)(-\tfrac{1}{2}\cos 2x)\, dx$

$$= \left[-\tfrac{1}{2}x \cos 2x + \tfrac{1}{4}\sin 2x \right]_0^{\frac{1}{2}\pi}$$

$$= \{ -\tfrac{1}{2}(\tfrac{1}{2}\pi)(-1) + 0 \} - \{0\}$$

$$= \tfrac{1}{4}\pi$$

The function to be integrated is expressed as the product of two 'parts', one of which is differentiated and the other integrated. If we can integrate the product of these two new functions, then the method is successful.

The function v may be any integral of $\dfrac{dv}{dx}$. There is no need to include an arbitrary constant at this stage, and putting it in would just complicate the process unnecessarily. For example, when finding $\int x \cos x\, dx$, suppose we put

$$u = x, \qquad \frac{dv}{dx} = \cos x$$

$$\frac{du}{dx} = 1, \qquad v = \sin x + k$$

then $\int x \cos x\, dx = x(\sin x + k) - \int (1)(\sin x + k)\, dx$
$$= x \sin x + kx + \cos x - kx + C$$
$$= x \sin x + \cos x + C \quad \text{as before}$$

Some harder examples are now given.

Example 3

$$\int x^2 \cos x\, dx$$

Differentiating x^2 and integrating $\cos x$ will lead to the integral $\int 2x \sin x\, dx$. We then repeat the process; differentiating $2x$ and integrating $\sin x$ will lead to $\int 2(-\cos x)\, dx$, which we can integrate. First we put

$$u = x^2, \qquad \frac{dv}{dx} = \cos x$$

$$\frac{du}{dx} = 2x, \qquad v = \sin x$$

$$\int x^2 \cos x\, dx = x^2 \sin x - \int 2x \sin x\, dx$$

Now put $\quad u = 2x, \quad \dfrac{dv}{dx} = \sin x$

$$\dfrac{du}{dx} = 2, \quad v = -\cos x$$

Then $\int x^2 \cos x \, dx = x^2 \sin x - [2x(-\cos x) - \int 2(-\cos x)dx]$

$$= x^2 \sin x + 2x \cos x - \int 2 \cos x \, dx$$

$$= x^2 \sin x + 2x \cos x - 2 \sin x + C$$

Example 4

$\int \sin^{-1} x \, dx$

$\sin^{-1} x$ does not appear to be a product, but we may regard it as $(\sin^{-1} x)\,(1)$. Differentiating $\sin^{-1} x$ and integrating 1 leads to

$\displaystyle\int \dfrac{x}{\sqrt{1-x^2}} dx$, which we can integrate by inspection. So we put

$$u = \sin^{-1} x, \quad \dfrac{dv}{dx} = 1$$

$$\dfrac{du}{dx} = \dfrac{1}{\sqrt{1-x^2}}, \quad v = x$$

$$\int \sin^{-1} x \, dx = (\sin^{-1} x)\,(x) - \int \dfrac{x}{\sqrt{1-x^2}} dx$$

$$= x \sin^{-1} x + \sqrt{1-x^2} + C$$

Exercise 15.4   (Integration by parts)

In questions 1 to 10, integrate the given functions with respect to x.

1   $x \sin x$ 2 $x \sin 3x$ 3 $x \cos 4x$ 4 $x \sin \frac{1}{3}x$ 5 $(3 - 2x) \cos 3x$

6   $x^2 \cos x$ 7 $(x^2 - 2x) \sin 3x$ 8 $\cos^{-1} x$ 9 $\sin^{-1} 3x$

10   $x \sin^2 x$ [First write $\sin^2 x = \frac{1}{2}(1 - \cos 2x)$.]

In questions 11 to 15, evaluate the definite integrals.

11   $\displaystyle\int_0^{\pi} x \sin x \, dx$ 12 $\displaystyle\int_0^{2\pi} x \cos \frac{1}{2}x \, dx$ 13 $\displaystyle\int_0^{\frac{1}{2}\pi} x^2 \sin 2x \, dx$

14   $\displaystyle\int_0^{\frac{1}{4}} \sin^{-1}(2x) \, dx$ 15 $\displaystyle\int_0^{\pi} x^3 \cos x \, dx$

16   Show that replacing v by $v + k$ in the general formula for integration by parts does not affect the result.

15.5 Methods of integration

We have encountered the following methods which can be used in integration.

1) Recognition of a standard integral.
2) Integration by inspection.
3) Integration by the use of a substitution.
4) Integration involving the use of an identity (usually trigonometric).
5) Integration of a product, by parts.

Standard integrals

Before making a list of 'standard integrals' we integrate $\dfrac{1}{\sqrt{a^2 - x^2}}$

and $\dfrac{1}{a^2 + x^2}$ (where a is a positive constant).

For $\displaystyle\int \frac{1}{\sqrt{a^2 - x^2}}\, dx$, we put

$$x = a \sin \theta$$

Then
$$dx = a \cos \theta\, d\theta$$

and
$$\int \frac{1}{\sqrt{a^2 - x^2}}\, dx = \int \frac{1}{\sqrt{a^2 - a^2 \sin^2 \theta}}\, a \cos \theta\, d\theta$$

$$= \int \frac{a \cos \theta}{a \cos \theta}\, d\theta$$

$$= \int 1\, d\theta = \theta + C$$

$$= \sin^{-1} \frac{x}{a} + C$$

For $\displaystyle\int \frac{1}{a^2 + x^2}\, dx$, we put

$$x = a \tan \theta$$

Then
$$dx = a \sec^2 \theta\, d\theta$$

and
$$\int \frac{1}{a^2 + x^2}\, dx = \int \frac{1}{a^2 + a^2 \tan^2 \theta}\, a \sec^2 \theta\, d\theta$$

$$= \int \frac{a \sec^2 \theta}{a^2 \sec^2 \theta}\, d\theta$$

$$= \int \frac{1}{a}\, d\theta$$

$$= \frac{1}{a}\theta + C = \frac{1}{a}\tan^{-1} \frac{x}{a} + C$$

We now have

$$\int x^n \, dx = \frac{x^{n+1}}{n+1} + C \quad \text{(provided } n \neq -1\text{)}$$

$$\int \sin x \, dx = -\cos x + C$$

$$\int \cos x \, dx = \sin x + C$$

$$\int \frac{1}{\sqrt{a^2 - x^2}} \, dx = \sin^{-1} \frac{x}{a} + C$$

$$\int \frac{1}{a^2 + x^2} \, dx = \frac{1}{a} \tan^{-1} \frac{x}{a} + C$$

Example 1

$$\int (3x + 7)^4 \, dx$$

This is of the form $\int x^n \, dx$, with $n = 4$ and x replaced by the linear function $(3x + 7)$. We divide by the coefficient of x (which is 3). Then

$$\int (3x + 7)^4 \, dx = \tfrac{1}{3} \times \tfrac{1}{5}(3x + 7)^5 + C = \tfrac{1}{15}(3x + 7)^5 + C$$

Example 2

$$\int \frac{1}{\sqrt{25 - 4x^2}} \, dx = \int \frac{1}{\sqrt{5^2 - (2x)^2}} \, dx$$

This is of the form $\int \frac{1}{\sqrt{a^2 - x^2}} \, dx$, with $a = 5$ and x replaced by $2x$.

Then

$$\int \frac{1}{\sqrt{25 - 4x^2}} \, dx = \frac{1}{2} \sin^{-1} \left(\frac{2x}{5} \right) + C$$

Example 3

$$\int \frac{1}{4 + 9x^2} \, dx = \int \frac{1}{2^2 + (3x)^2} \, dx$$

This is of the form $\int \frac{1}{a^2 + x^2} \, dx$, with $a = 2$ and x replaced by $3x$.

Then $\int \frac{1}{4 + 9x^2} \, dx = \tfrac{1}{3} \times \tfrac{1}{2} \tan^{-1} \left(\frac{3x}{2} \right) + C = \tfrac{1}{6} \tan^{-1} \left(\frac{3x}{2} \right) + C$

Integration by inspection

We may be able to guess the form of the integral (especially when we notice a function and its derivative). We check our guess by differentiating it, and adjust the numerical factor accordingly. Integrals of this type may alternatively be solved by substitution.

Example

$$\int \sec^2 x \tan^3 x \, dx$$

We notice that $\sec^2 x$ is the derivative of $\tan x$, and we guess that the integral is a multiple of $\tan^4 x$. Now

$$\frac{d}{dx}(\tan^4 x) = 4 \tan^3 x \sec^2 x$$

Hence $\quad \int \sec^2 x \tan^3 x \, dx = \frac{1}{4}\tan^4 x + C$

Integration by the use of a substitution

(i) If we notice a function $f(x)$ and its derivative $f'(x)$, then we can substitute $f(x) = t$ (or $f(x) = t^2$).

(ii) If the integral contains $\sqrt{1-x^2}$, put $x = \sin\theta$, since $1 - \sin^2\theta = \cos^2\theta$. Similarly, if it contains, for example, $\sqrt{4-9x^2}$, substitute $3x = 2\sin\theta$, and so on.

(iii) If the integral contains $\sqrt{1+x^2}$, put $x = \tan\theta$, since $1 + \tan^2\theta = \sec^2\theta$.

(iv) If it contains $\sqrt{x}$ and $\sqrt{1-x}$, put $x = \sin^2\theta$.

(v) If it contains $\sqrt{1+x}$ and $\sqrt{1-x}$, put $x = \cos 2\theta$, since $1 + \cos 2\theta = 2\cos^2\theta$ and $1 - \cos 2\theta = 2\sin^2\theta$.

(vi) For integrals of the type $\int \sin^m x \cos^n x \, dx$, we put $\cos x = u$ if m is odd, or $\sin x = u$ if n is odd. (If m and n are both odd, either substitution will do. If m and n are both even, we need to use identities such as $\sin^2 x = \frac{1}{2}(1 - \cos 2x)$; see the next section.)

Example 1

$$\int \frac{x^3}{\sqrt{4-x^4}} \, dx$$

Since x^3 is a multiple of the derivative of $4 - x^4$, we put

$$4 - x^4 = t^2$$

Then $\quad -4x^3 \, dx = 2t \, dt$

and $\quad \displaystyle\int \frac{x^3}{\sqrt{4-x^4}} \, dx = \int \frac{1}{t}(-\tfrac{1}{2}t \, dt)$

$$= \int(-\tfrac{1}{2}) \, dt = -\tfrac{1}{2}t + C$$

$$= -\tfrac{1}{2}\sqrt{4-x^4} + C$$

Example 2

$$\int \frac{x}{\sqrt{4-x^4}}\,dx = \int \frac{x}{\sqrt{4-(x^2)^2}}\,dx$$

and x is a multiple of the derivative of x^2, so we put

$$x^2 = t$$
Then $\qquad 2x\,dx = dt$

and $\displaystyle\int \frac{x}{\sqrt{4-x^4}}\,dx = \int \frac{1}{\sqrt{4-t^2}}\,(\tfrac{1}{2}\,dt) = \tfrac{1}{2}\int \frac{1}{\sqrt{2^2-t^2}}\,dt$

$$= \tfrac{1}{2}\sin^{-1}\frac{t}{2}+C \quad \text{(using the standard integral for } \int\frac{1}{\sqrt{a^2-t^2}}\,dt\text{, with } a = 2)$$

$$= \tfrac{1}{2}\sin^{-1}\left(\frac{x^2}{2}\right)+C$$

Example 3

$$\int \sqrt{\frac{1-x}{1+x}}\,dx$$

Put $\qquad\qquad x = \cos 2\theta$

Then $\qquad\qquad dx = -2\sin 2\theta\,d\theta$

and $\displaystyle\int \sqrt{\frac{1-x}{1+x}}\,dx = \int \sqrt{\frac{2\sin^2\theta}{2\cos^2\theta}}\,(-2\sin 2\theta\,d\theta)$

$$= -\int \frac{\sin\theta}{\cos\theta}\,(4\sin\theta\cos\theta)\,d\theta$$

$$= -\int 4\sin^2\theta\,d\theta = -\int 2\,(1-\cos 2\theta)\,d\theta$$

$$= -2\theta+\sin 2\theta+C = -\cos^{-1}x+\sqrt{1-x^2}+C$$

Alternatively, multiplying by $\dfrac{\sqrt{1-x}}{\sqrt{1-x}}$,

$$\int \sqrt{\frac{1-x}{1+x}}\,dx = \int \frac{1-x}{\sqrt{1-x^2}}\,dx$$

$$= \int \frac{1}{\sqrt{1-x^2}}\,dx - \int \frac{x}{\sqrt{1-x^2}}\,dx$$

$$= \sin^{-1}x+\sqrt{1-x^2}+C \quad \text{(by inspection)}$$

This solution looks slightly different from the previous one, but it should be remembered that $\sin^{-1}x = \tfrac{1}{2}\pi-\cos^{-1}x$.

Example 4

$\int \sin^3 x \cos^4 x \, dx$

Put $\qquad\qquad \cos x = u$

Then $\qquad -\sin x \, dx = du$

and $\quad \int \sin^3 x \cos^4 x \, dx = \int \sin^2 x \cos^4 x \sin x \, dx$

$\qquad\qquad\qquad = \int (1 - u^2) u^4 (-du) = \int (u^6 - u^4) du$

$\qquad\qquad\qquad = \frac{1}{7} u^7 - \frac{1}{5} u^5 + C$

$\qquad\qquad\qquad = \frac{1}{7} \cos^7 x - \frac{1}{5} \cos^5 x + C$

Example 5

$\int \cos^5 x \, dx$

Put $\qquad\qquad \sin x = u$

Then $\qquad \cos x \, dx = du$

and $\quad \int \cos^5 x \, dx = \int \cos^4 x \cos x \, dx$

$\qquad\qquad\qquad = \int (1 - u^2)^2 du = \int (1 - 2u^2 + u^4) du$

$\qquad\qquad\qquad = u - \frac{2}{3} u^3 + \frac{1}{5} u^5 + C$

$\qquad\qquad\qquad = \sin x - \frac{2}{3} \sin^3 x + \frac{1}{5} \sin^5 x + C$

Integrals which involve the use of an identity

Identities such as

$\qquad \sin^2 x = \frac{1}{2}(1 - \cos 2x)$

$\qquad \cos^2 x = \frac{1}{2}(1 + \cos 2x)$

$\quad \sin x \cos x = \frac{1}{2} \sin 2x$

$\qquad \tan^2 x = \sec^2 x - 1$

$\sin A \cos B = \frac{1}{2}[\sin(A + B) + \sin(A - B)]$

may be used to rewrite an expression in terms of functions which
we can integrate.

Example 1

$\int \sin^2 x \, dx = \int \frac{1}{2}(1 - \cos 2x) \, dx$

$\qquad\qquad = \frac{1}{2} x - \frac{1}{4} \sin 2x + C$

Example 2

$\int \sin^2 3x \, dx = \int \frac{1}{2}(1 - \cos 6x) \, dx$

$\qquad\qquad = \frac{1}{2} x - \frac{1}{12} \sin 6x + C$

Example 3

$$\int \tan^2 2x \, dx = \int (\sec^2 2x - 1) \, dx$$
$$= \tfrac{1}{2} \tan 2x - x + C$$

Example 4

$$\int \cos 5x \cos 2x \, dx = \int \tfrac{1}{2} (\cos 7x + \cos 3x) \, dx$$
$$(\text{using } \cos A \cos B = \tfrac{1}{2} [\cos (A+B) + \cos (A-B)])$$
$$= \tfrac{1}{14} \sin 7x + \tfrac{1}{6} \sin 3x + C$$

Example 5

$$\int \cos^4 x \, dx = \int (\cos^2 x)^2 \, dx = \int [\tfrac{1}{2}(1 + \cos 2x)]^2 \, dx$$
$$= \int (\tfrac{1}{4} + \tfrac{1}{2}\cos 2x + \tfrac{1}{4}\cos^2 2x) \, dx$$
$$= \tfrac{1}{4}x + \tfrac{1}{4} \sin 2x + \int \tfrac{1}{8} (1 + \cos 4x) \, dx$$
$$= \tfrac{1}{4}x + \tfrac{1}{4} \sin 2x + \tfrac{1}{8}x + \tfrac{1}{32} \sin 4x + C$$
$$= \tfrac{3}{8}x + \tfrac{1}{4} \sin 2x + \tfrac{1}{32} \sin 4x + C$$

Example 6

$$\int \sin^2 x \cos^2 x \, dx = \tfrac{1}{4}\int (2 \sin x \cos x)^2 \, dx = \tfrac{1}{4}\int \sin^2 2x \, dx$$
$$= \tfrac{1}{4}\int \tfrac{1}{2}(1 - \cos 4x) \, dx$$
$$= \tfrac{1}{8}x - \tfrac{1}{32} \sin 4x + C$$

Integration of a product by parts

Example Find the average value of $x \sin x$ over the interval $0 \leqslant x \leqslant \pi$.

The average value (see p. 82) is

$$\frac{1}{\pi - 0} \int_0^\pi x \sin x \, dx$$

and we shall integrate by parts.

Put $u = x,$ $\dfrac{dv}{dx} = \sin x$

$\dfrac{du}{dx} = 1,$ $v = -\cos x$

Then $\dfrac{1}{\pi} \int_0^\pi x \sin x \, dx = \dfrac{1}{\pi} \left[-x \cos x \right]_0^\pi - \dfrac{1}{\pi} \int_0^\pi (-\cos x) \, dx$

$$= \frac{1}{\pi} \left[-x \cos x + \sin x \right]_0^\pi$$

$$= \frac{1}{\pi} \left\{ (\pi + 0) - (0 + 0) \right\} = 1$$

The average value is 1.

Exercise 15.5   (Miscellaneous integration)

Integrate the following functions with respect to x (questions 1 to 30).

1 $\dfrac{3x+1}{\sqrt{x}}$

2 $\sin x \cos^4 x$

3 $\dfrac{x}{\sqrt{x^2-1}}$

4 $\dfrac{1}{(5-3x)^4}$

5 $\dfrac{x}{(5-2x)^3}$

6 $x^2 \sin(x^3)$

7 $\dfrac{1}{\sqrt{25-4x^2}}$

8 $x \sin 3x$

9 $\sec^2 x \tan^4 x$

10 $\dfrac{\cos x}{(3-2\sin x)^2}$

11 $\cos^2 5x$

12 $\sin^3 x \cos^6 x$

13 $\tan^2 3x$

14 $\sin 5x \cos 2x$

15 $\sec^2(\frac{1}{6}\pi - 2x)$

16 $x^2\sqrt{2+x^3}$

17 $\dfrac{x^3}{\sqrt{4-x^2}}$

18 $\sin^4 x$

19 $\dfrac{x^3}{\sqrt{1+x^4}}$

20 $\dfrac{x}{1+x^4}$

21 $\dfrac{1}{x^2}\sin\left(\dfrac{1}{x}\right)$

22 $\dfrac{x^2}{\sqrt{x+2}}$

23 $\dfrac{\tan^{-1}x}{1+x^2}$

24 $\dfrac{1}{49+16x^2}$

25 $\dfrac{1}{\sqrt{x(1-x)}}$ [put $x = \sin^2 \theta$]

26 $x^2 \cos 2x$

27 $x^2 \sin^2 x$ [use question 26]

28 $x \sin x \cos x$

29 $x \sin^{-1}x$ [first put $x = \sin \theta$]

30 $\dfrac{1}{\sqrt{x}(1+x)}$ [put $x = \tan^2 \theta$]

Evaluate the following definite integrals (questions 31 to 38).

31 $\displaystyle\int_0^\pi x \cos x \, dx$

32 $\displaystyle\int_0^4 3x\sqrt{x^2+9}\,dx$

33 $\displaystyle\int_{\sqrt2}^{\sqrt5} \dfrac{x^3}{\sqrt{x^2-1}}\,dx$

34 $\displaystyle\int_0^3 \dfrac{1}{9+x^2}\,dx$

35 $\displaystyle\int_0^2 \dfrac{x^2}{\sqrt{x^3+1}}\,dx$

36 $\displaystyle\int_{\frac12}^1 \sqrt{\dfrac{1-x}{x}}\,dx$

37 $\displaystyle\int_0^{\frac12\pi} \sin^3 2x \, dx$

38 $\displaystyle\int_0^1 \dfrac{1}{(1+x^2)^{\frac32}}\,dx$ [put $x = \tan \theta$]

39 Sketch the curve $y = \cos^2 x$ in the interval $-\pi \leqslant x \leqslant \pi$. Find the area of the region R bounded by the x-axis, that part of the curve for which $0 \leqslant x \leqslant \pi$ and the lines $x = 0, x = \pi$. Find also the volume of the solid body of revolution formed by rotating the region R about the x-axis.   (L)

40 The area between the curve $y = \sin x - \dfrac{2x}{\pi}$ and the x-axis, from $x = 0$ to $x = \frac12\pi$, is rotated about the x-axis. Find the volume of the solid of revolution.

41 Find the average value of $\sin x$ over the interval $0 \leqslant x \leqslant \pi$. The 'root mean square' value of a function y over an interval is the square root of the average value of y^2. Find the root mean square value of $\sin x$ over the interval $0 \leqslant x \leqslant \pi$.

42 Find the root mean square value of $x(10-x)$ over the interval $0 \leqslant x \leqslant 10$.

Solve the following differential equations.

43 $\dfrac{dy}{dx} = \dfrac{2x+3}{\sqrt{x^2+3x+6}}$ given that $y = 3$ when $x = 2$.

44 $(1+x^2)\dfrac{dy}{dx} = 3$ given that $y = 0$ when $x = 1$.

45 $\dfrac{dy}{dx} = \sqrt{9-y^2}$ given that $y = \frac{3}{2}$ when $x = 0$.

16

Numerical calculus

Introduction

In this chapter we introduce some numerical methods which are
widely used.

1) A method for solving equations which is suitable for computer
   calculation.
2) A more accurate method than the trapezium rule for evaluating definite integrals, which is again used for computer
   calculation.
3) A method for finding the numerical value of the first and
   second derivatives of a function at a specified value for x. In
   itself this method is not very useful since it is so inaccurate.
   However the formulae used give a basis for the numerical
   solution of differential equations.
4) A method for interpolating from a table of values.

16.1 The numerical solution of equations, using Newton's method

There are many equations which cannot be solved exactly (for
example, $x^5 - 3x^2 + 1 = 0$ and $\cos x = 2x$). We now describe
Newton's method, which can be applied to find an approximate
solution for any equation.

We first arrange the equation in the form $f(x) = 0$ (for example,
the equation $\cos x = 2x$ would be written as $\cos x - 2x = 0$, so that
$f(x) = \cos x - 2x$). Then a root (or solution) of the equation is a
value of x where the curve $y = f(x)$ crosses the x-axis (see
Fig. 16.1).

Suppose that $x = x_1$ is an approximate root of the equation. On
Fig. 16.1, PT is the tangent to the curve $y = f(x)$ at the point
$P(x_1, f(x_1))$. If this tangent crosses the x-axis at $x = x_2$, we can see
that x_2 is a better approximation to the root than x_1

The gradient of PT is

$$\frac{f(x_1)}{x_1 - x_2}$$

and, since it is the tangent at P, the gradient of PT is $f'(x_1)$. Thus

$$\frac{f(x_1)}{x_1 - x_2} = f'(x_1)$$

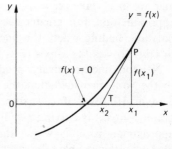

Fig. 16.1

277

giving

$$x_1 - x_2 = \frac{f(x_1)}{f'(x_1)}$$

and hence

$$x_2 = x_1 - \frac{f(x_1)}{f'(x_1)}$$

It is now possible to repeat the process, starting from the new approximation x_2. We obtain

$$x_3 = x_2 - \frac{f(x_2)}{f'(x_2)}$$

and this is a better approximation than x_2 (see Fig. 16.2).

Continuing in this way, we obtain a sequence of values x_1, x_2, x_3, ... which become closer and closer to the root. The formula for calculating the next value in the sequence is

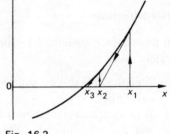

Fig. 16.2

$$\boxed{x_{n+1} = x_n - \frac{f(x_n)}{f'(x_n)}}$$

This is the **Newton–Raphson formula**.

A process such as this, where each value in the sequence is used to calculate the next one, is called an **iterative process**; each step is an **iteration**.

Before we can use the Newton–Raphson formula, we need to find a first approximation to the root. We can usually do this by sketching graphs. Alternatively, we may substitute values for x until we find two values, say $x = a$ and $x = b$, for which $f(a) < 0$ and $f(b) > 0$. We then know that there is a root between a and b, and we may take $x_1 = a$, $x_1 = b$, or $x_1 = \frac{1}{2}(a + b)$ as a first approximation. For greater accuracy, we could use linear interpolation, finding where the line joining the points $(a, f(a))$ and $(b, f(b))$ crosses the x-axis (Fig. 16.3).

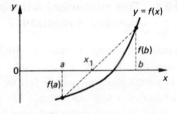

Fig. 16.3

Newton's method always produces a sequence which converges to the root, provided that the first approximation x_1 is good enough. However, if x_1 is some distance from the root and the graph $y = f(x)$ is curving rapidly, it can happen that the second approximation x_2 is worse than the first (Fig. 16.4). We can overcome this problem by starting closer to the root.

If the equation $f(x) = 0$ has several roots, we can find them all, by starting close to each root in turn.

Although Newton's method is called an 'approximate' method, any desired degree of accuracy can be obtained by using sufficient iterations.

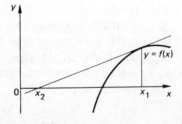

Fig. 16.4

There are methods by which cubic and quartic equations may be solved exactly, but these methods are tedious. In practice it is much quicker to use Newton's method.

Example 1 Using Newton's method, find, correct to two decimal places, the roots of the equation $12x^3 + 4x^2 - 15x - 4 = 0$.

Let $f(x) = 12x^3 + 4x^2 - 15x - 4$

Then $f(-2) = -54$; $f(-1) = 3$; $f(0) = -4$; $f(1) = -3$; $f(2) = 78$.

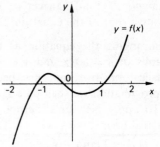

Fig. 16.5

So there is a root between -2 and -1, another between -1 and 0, and the third is between 1 and 2. We can sketch the curve as shown in Fig. 16.5.

The Newton–Raphson formula is

$$x_{n+1} = x_n - \frac{f(x_n)}{f'(x_n)}$$

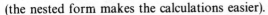

$$= x_n - \frac{12x_n^3 + 4x_n^2 - 15x_n - 4}{36x_n^2 + 8x_n - 15}$$

$$= \frac{24x_n^3 + 4x_n^2 + 4}{36x_n^2 + 8x_n - 15} = \frac{(24x_n + 4)x_n^2 + 4}{(36x_n + 8)x_n - 15}$$

(the nested form makes the calculations easier).

For the root between -2 and -1, start with $x_1 = -1$. Then

$$x_2 = \frac{-16}{13} = -1.231$$

$$x_3 = \frac{-34.71}{29.70} = -1.169$$

$$x_4 = \frac{-28.87}{24.84} = -1.162$$

$$x_5 = \frac{-28.25}{24.31} = -1.162$$

For the root between -1 and 0, start with $x_1 = 0$. Then

$$x_2 = \frac{4}{-15} = -0.267$$

$$x_3 = \frac{3.828}{-14.57} = -0.263$$

$$x_4 = \frac{3.840}{-14.61} = -0.263$$

For the root between 1 and 2, start with $x_1 = 1$. Then

$$x_2 = \frac{32}{29} = 1.103$$

$$x_3 = \frac{41.07}{37.62} = 1.092$$

$$x_4 = \frac{40.02}{36.66} = 1.092$$

Hence the roots are -1.16, -0.26, 1.09 (correct to 2 decimal places).

Example 2   Find (correct to 3 significant figures) the smallest positive root of the equation $2 \tan x + x = 0$.

Rearranging the equation as $\tan x = -\frac{1}{2}x$, and sketching the graphs of $y = \tan x$ and $y = -\frac{1}{2}x$ (Fig. 16.6), we see that the smallest positive root lies between $\frac{1}{2}\pi$ and π. So as a first approximation we shall take $x_1 = 2.4$ ($\approx \frac{3}{4}\pi$).

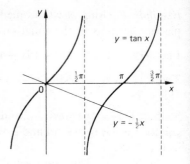

Fig. 16.6

To apply Newton's method, we require the equation in its original form $2 \tan x + x = 0$. Let

$$f(x) = 2 \tan x + x$$

Then the Newton–Raphson formula is

$$x_{n+1} = x_n - \frac{f(x_n)}{f'(x_n)}$$

$$= x_n - \frac{2 \tan x_n + x_n}{2 \sec^2 x_n + 1} = \frac{2x_n \sec^2 x_n - 2 \tan x_n}{2 \sec^2 x_n + 1}$$

$$= \frac{2x_n - 2 \sin x_n \cos x_n}{2 + \cos^2 x_n} = \frac{2x_n - \sin (2x_n)}{2 + \cos^2 x_n}$$

Taking $x_1 = 2.4$ and, remembering to work in radians, we have

$$x_2 = \frac{4.8 - \sin (4.8)}{2 + \cos^2 (2.4)} = 2.279$$

$$x_3 = \frac{4.558 - \sin (4.558)}{2 + \cos^2 (2.279)} = 2.289$$

$$x_4 = \frac{4.578 - \sin (4.578)}{2 + \cos^2 (2.289)} = 2.289$$

Hence the required root is $x = 2.29$ (correct to 3 s.f.).

Exercise 16.1   (Newton's method)

1   Show that the equation $x^3 - x - 2 = 0$ has a root between 1 and 2, and use Newton's method to find this root correct to 2 decimal places.

2   Find (correct to 2 decimal places) the three roots of the equation $x^3 - 3x^2 + 1 = 0$.

3   Find (correct to 3 decimal places), the root of the equation $x^5 = 5x - 3$ which lies between 0 and 1.

4   Find the stationary points on the curve $y = x^3 - 3x^2 - 9x - 7$, and deduce that the equation $x^3 - 3x^2 - 9x - 7 = 0$ has only one real root. Use Newton's method to find this root correct to 2 decimal places.

5   Find (correct to 2 decimal places) the smallest positive root of the equation $2 \sin x - x = 0$.

6   Find (correct to 2 decimal places) the smallest positive root of the equation $\tan x = x - 1$.

7   Show that the equation $x = 2\pi \cos x$ has a root $x = 2\pi$ and two other positive roots. Use Newton's method to find these other two positive roots, correct to 2 decimal places.

8   Show by a graphical method, or otherwise, that the equation $\sin x - \dfrac{(x+1)}{(x-1)} = 0$ has no positive real roots.

How many real roots are there between -2π and 0?

Using the Newton–Raphson iterative method find, correct to five decimal places, the real root nearest to zero.

(AEB)

16.2 Numerical integration: Simpson's rule

We may need to find an approximate value for a definite integral $\int_a^b f(x)\,dx$ if either (i) we cannot integrate $f(x)$ or (ii) we are just given a table of values for $f(x)$. We have already met the trapezium rule (see p. 96) for which the curve $y = f(x)$ was replaced by a series of straight lines.

We now replace the curve by a series of parabolic segments; this should give a better approximation.

First consider a curve $y = f(x)$, between $x = -h$ and $x = h$, passing through the points $L(-h, y_1)$, $M(0, y_2)$ and $N(h, y_3)$. Let $y = px^2 + qx + r$ be the equation of a parabola passing through L, M and N (see Fig. 16.7). The coordinates of these points must satisfy the equation, so we have

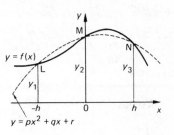

Fig. 16.7

$$y_1 = ph^2 - qh + r \qquad (1)$$
$$y_2 = r \qquad (2)$$
$$y_3 = ph^2 + qh + r \qquad (3)$$

Thus $r = y_2$, and, adding (1) and (3), $y_1 + y_3 = 2ph^2 + 2r$, giving

$$2ph^2 = y_1 + y_3 - 2r = y_1 + y_3 - 2y_2$$

Replacing the curve $y = f(x)$ by the parabola $y = px^2 + qx + r$, we obtain

$$\int_{-h}^{h} f(x)\,dx \approx \int_{-h}^{h} (px^2 + qx + r)\,dx$$

$$= \left[\tfrac{1}{3}px^3 + \tfrac{1}{2}qx^2 + rx \right]_{-h}^{h}$$

$$= (\tfrac{1}{3}ph^3 + \tfrac{1}{2}qh^2 + rh) - (-\tfrac{1}{3}ph^3 + \tfrac{1}{2}qh^2 - rh)$$

$$= \tfrac{2}{3}ph^3 + 2rh$$

$$= \tfrac{1}{3}h(2ph^2 + 6r)$$

$$= \tfrac{1}{3}h(y_1 + y_3 - 2y_2 + 6y_2)$$

$$= \tfrac{1}{3}h(y_1 + 4y_2 + y_3)$$

This gives an approximate value for the definite integral in terms of the y-coordinates at three points. The position of the origin does not matter. If x_1, x_2 and x_3 are three values of x (where $x_2 = x_1 + h$ and $x_3 = x_2 + h$), and the corresponding values of $f(x)$ are y_1, y_2 and y_3 (see Fig. 16.8), then

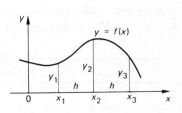

Fig. 16.8

$$\int_{x_1}^{x_3} f(x)\,dx \approx \tfrac{1}{3}h(y_1 + 4y_2 + y_3)$$

To find an approximate value for $\int_a^b f(x)\,dx$, we divide the area under the curve $y = f(x)$ into an even number of strips, each of width h. Let the dividing values of x be $x_1, x_2, x_3, \ldots, x_n$ (where $x_1 = a$, $x_n = b$), and the corresponding values of $f(x)$ be y_1, $y_2, \ldots, y_n$. Note that n must be an odd number (Fig. 16.9 overleaf).

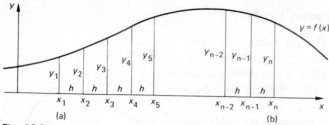

Fig. 16.9

We have

$$\int_a^b f(x)\,dx = \int_{x_1}^{x_3} f(x)\,dx + \int_{x_3}^{x_5} f(x)\,dx + \ldots + \int_{x_{n-2}}^{x_n} f(x)\,dx$$

$$\approx \tfrac{1}{3}h(y_1 + 4y_2 + y_3) + \tfrac{1}{3}h(y_3 + 4y_4 + y_5) + \ldots + \tfrac{1}{3}h(y_{n-2} + 4y_{n-1} + y_n)$$

i.e.

$$\boxed{\int_a^b f(x)\,dx \approx \tfrac{1}{3}h(y_1 + 4y_2 + 2y_3 + 4y_4 + 2y_5 + 4y_6 + \ldots + 2y_{n-2} + 4y_{n-1} + y_n)}$$

This is **Simpson's rule**. Over each pair of strips (for example, from $x = x_1$ to $x = x_3$), the curve $y = f(x)$ has been replaced by a segment of a parabola. Remember that we must use an even number of strips. The number of ordinates (i.e. values of $f(x)$, namely $y_1, y_2, \ldots, y_n$) must therefore be odd.

Example 1   Use Simpson's rule with 7 ordinates to find an approximate value for

$$\int_0^{1.5} \sqrt{1 + x^3}\,dx$$

If there are 7 ordinates, there are 6 strips; so each strip has width $h = \frac{1.5}{6} = 0.25$. The dividing values of x are $0, 0.25, 0.5, 0.75, 1, 1.25, 1.5$.

x	$y = \sqrt{1 + x^3}$	Multiplier m	my
0	1	1	1
0.25	1.0078	4	4.0311
0.5	1.0607	2	2.1213
0.75	1.1924	4	4.7697
1	1.4142	2	2.8284
1.25	1.7185	4	6.8739
1.5	2.0917	1	2.0917
			23.7161

$$\int_0^{1.5} \sqrt{1 + x^3}\,dx \approx \tfrac{1}{3}h \times 23.7161 = \tfrac{1}{3} \times 0.25 \times 23.7161 \approx 1.976$$

(After each value of y is calculated, it is multiplied by the appropriate multiplier, and the result, my, is added to the memory. Finally, the total is recalled from the memory, and multiplied by $\frac{1}{3}h$.)

Example 2   Find an approximate value for $\int_3^5 f(x)\,dx$, where $f(x)$ is the function tabulated below.

x	3	3.5	4	4.5	5
$f(x)$	5.4	4.6	4.2	4.1	4.7

We have 4 strips of width $h = 0.5$.

x	y	Multiplier m	my
3	5.4	1	5.4
3.5	4.6	4	18.4
4	4.2	2	8.4
4.5	4.1	4	16.4
5	4.7	1	4.7
			53.3

Thus $\displaystyle\int_3^5 f(x)\,dx \approx \frac{1}{3} \times 0.5 \times 53.3 \approx 8.883$

Accuracy of Simpson's rule

Simpson's rule usually gives a much better approximation than the trapezium rule using the same number of strips.

When we know the formula for $f(x)$, we can use Simpson's rule to find the value of a definite integral to any required degree of accuracy, by using enough strips. To check that our value is correct to, say, 3 decimal places, we should repeat the calculation, using more strips. If the two values agree (to 3 decimal places), we have probably achieved the required accuracy.

If $f(x)$ is a quadratic function (i.e. $f(x) = px^2 + qx + r$), the parabolic segments will coincide with the curve $y = f(x)$, and Simpson's rule will give the exact value of $\int_a^b f(x)\,dx$. Surprisingly, it also gives the exact value when $f(x)$ is a cubic function.

Exercise 16.2   (Simpson's rule)

In questions 1 to 7 use Simpson's rule to find approximate values for the definite integrals.

1 $\displaystyle\int_2^3 \frac{1}{x}\,dx$ [use 5 ordinates]

2 $\displaystyle\int_1^4 \frac{1}{\sqrt{3+x^2}}\,dx$ [use 7 ordinates]

3 $\displaystyle\int_3^7 \frac{1}{x^2-4}\,dx$ [use 5 ordinates]

4 $\displaystyle\int_0^\pi \sqrt{\sin x}\,dx$ [use 5 ordinates]

5 $\displaystyle\int_0^{\frac{1}{3}\pi} \sec x\,dx$ [use 7 ordinates]

6 $\displaystyle\int_0^3 f(x)\,dx$, where $f(x)$ is the function tabulated below:

x	0	0.5	1	1.5	2	2.5	3
$f(x)$	1.105	1.822	3.004	4.953	8.166	13.464	22.198

7 $\displaystyle\int_{-1}^1 f(x)\,dx$, where $f(x)$ is the function tabulated below:

x	-1	-0.8	-0.6	-0.4	-0.2	0	0.2	0.4	0.6	0.8	1
$f(x)$	1.45	0.94	0.60	0.58		0.92	1.60	2.09	2.37	2.23	2.02	1.52

8 Apply Simpson's rule (using 3 ordinates) to $\int_2^4 x^3\,dx$, and show that it gives the exact value of this integral.

Show generally that Simpson's rule with 3 ordinates gives the exact value of the integral $\int_a^b x^3\,dx$.
9 Apply Simpson's rule (using 3 ordinates) to $\int_{-h}^h x^4\,dx$, and show that the error in this value is $\frac{4}{15}h^5$.
10 Apply Simpson's rule to $\int_0^\pi (1+\cos x)\,dx$ (i) using 5 ordinates; (ii) using 7 ordinates. Find also the exact value of the integral.

16.3 Numerical differentiation, and its applications

In Chapter 8, we saw how the gradient of a curve at a point A may be found numerically. This is useful when the curve is defined by a table of values, or when we do not know how to differentiate the equation of the curve.

We take two points Q and R on the curve, one on each side of A, and find the gradient of QR.

In general, suppose A is a point on a curve $y = f(x)$, and take an x-step of h on each side of A (see Fig. 16.10). The two points on the curve are then $Q(x-h,\ f(x-h))$ and $R(x+h,\ f(x+h))$, and the gradient of QR is

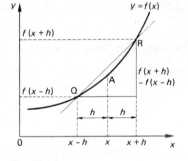

Fig. 16.10

$$\frac{f(x+h)-f(x-h)}{2h}$$

Thus

$$f'(x) \approx \frac{f(x+h)-f(x-h)}{2h}$$

Numerical differentiation involves subtraction of numbers which are close together. Consequently, any inaccuracies in tabulated values (which will certainly be present if the values have been obtained from an experiment) will have a considerable effect on the calculated values of $f'(x)$.

Similarly, if we know the formula for $f(x)$ and we try to use

$$f'(x) \approx \frac{f(x+h)-f(x-h)}{2h}$$

with extremely small values of h, our results will be affected by errors due to the calculator.

Application to differential equations

The formula we have found for $f'(x)$ can be used in the tabulation of a function, as shown in the following example.

Example Suppose the function f has values $f(0.8) = 0.5556$, $f(1.0) = 0.5000$, and satisfies the differential equation

$$f'(x) = f(x) - \frac{f(x) + 1}{x + 1}$$

$\Bigg($ Thus $y = f(x)$ is a solution of the differential equation

$\dfrac{dy}{dx} = y - \dfrac{y + 1}{x + 1}.\Bigg)$

Use a step-by-step method with step-length 0.2 to calculate $f(1.2)$, $f(1.4)$ and $f(1.6)$.

We have
$$f'(x) \approx \frac{f(x + h) - f(x - h)}{2h}$$

and
$$f'(x) = f(x) - \frac{f(x) + 1}{x + 1}$$

Thus
$$\frac{f(x + h) - f(x - h)}{2h} \approx f(x) - \frac{f(x) + 1}{x + 1}$$

We use a step length $h = 0.2$.
Putting $x = 1.0$,

$$\frac{f(1.2) - f(0.8)}{0.4} \approx f(1.0) - \frac{f(1.0) + 1}{1.0 + 1}$$

$$\frac{f(1.2) - 0.5556}{0.4} \approx 0.5000 - \frac{0.5000 + 1}{2}$$

$$= -0.2500$$

$$f(1.2) \approx (-0.2500 \times 0.4) + 0.5556$$

$$= 0.4556$$

Putting $x = 1.2$,

$$\frac{f(1.4) - f(1.0)}{0.4} \approx f(1.2) - \frac{f(1.2) + 1}{1.2 + 1}$$

$$\frac{f(1.4) - 0.5000}{0.4} \approx 0.4556 - \frac{0.4556 + 1}{2.2}$$

$$f(1.4) \approx 0.4176$$

Putting $x = 1.4$,

$$\frac{f(1.6) - f(1.2)}{0.4} \approx f(1.4) - \frac{f(1.4) + 1}{1.4 + 1}$$

$$\frac{f(1.6) - 0.4556}{0.4} \approx 0.4176 - \frac{0.4176 + 1}{2.4}$$

$$f(1.6) \approx 0.3864$$

The required values are approximately

x	1.2	1.4	1.6
f(x)	0.4556	0.4176	0.3864

Note that it can be shown that the exact solution of the differential equation is $f(x) = \dfrac{1}{x+1}$. The true values are therefore

x	1.2	1.4	1.6
f(x)	0.4545	0.4167	0.3846

An approximation for the second derivative

Suppose, for example, that we are given the following three points on a curve.

x	2	3	4
f(x)	10	15	17

We wish to find the value of the second derivative when $x = 3$, i.e. $f''(3)$.

In Fig. 16.11, the gradient of QA is $\dfrac{15-10}{1}$. This is an approximation to the gradient of the curve at $x = 2.5$, i.e.

$$f'(2.5) \approx \frac{15-10}{1}$$

Similarly, the gradient of the curve at $x = 3.5$ is approximately the gradient of AR, i.e.

$$f'(3.5) \approx \frac{17-15}{1}.$$

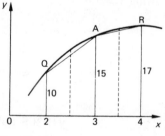

Fig. 16.11

The second derivative $f''(x)$ is the rate of change of the gradient $f'(x)$. Hence

$$f''(3) \approx \frac{f'(3.5) - f'(2.5)}{3.5 - 2.5}$$

$$= \frac{\left(\dfrac{17-15}{1}\right) - \left(\dfrac{15-10}{1}\right)}{1}$$

$$= 17 - 2 \times 15 + 10$$
$$= -3$$

We have no idea of the accuracy of this method, but it is the best we can do from the given information.

In general, if A is a point on a curve $y = f(x)$ (see Fig. 16.12), and we take an x-step of h on each side of A, we have

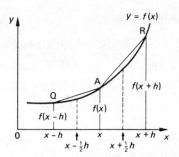

$$f'(x - \tfrac{1}{2}h) \approx \frac{f(x) - f(x - h)}{h} \quad \text{(the gradient of QA)}$$

$$f'(x + \tfrac{1}{2}h) \approx \frac{f(x + h) - f(x)}{h} \quad \text{(the gradient of AR)}$$

and

$$f''(x) \approx \frac{f'(x + \tfrac{1}{2}h) - f'(x - \tfrac{1}{2}h)}{(x + \tfrac{1}{2}h) - (x - \tfrac{1}{2}h)}$$

$$\approx \frac{\dfrac{f(x + h) - f(x)}{h} - \dfrac{f(x) - f(x - h)}{h}}{h}$$

Fig. 16.12

Hence

$$\boxed{f''(x) \approx \frac{f(x + h) - 2f(x) + f(x - h)}{h^2}}$$

Again, this formula can be used to build a table of values for a solution to a differential equation, as shown in the following example.

Example   Suppose a function satisfies the differential equation $f''(x) = 4f(x)$, and has the following values:

$f(0.20) = 1.08107$
$f(0.25) = 1.12763$

Use a step-by-step process to obtain values of $f(0.15)$, $f(0.30)$ and $f(0.35)$.

We have

$$f''(x) \approx \frac{f(x + h) - 2f(x) + f(x - h)}{h^2}$$

and

$$f''(x) = 4f(x)$$

Thus

$$\frac{f(x + h) - 2f(x) + f(x - h)}{h^2} \approx 4f(x)$$

We shall use a step length $h = 0.05$.
Putting $x = 0.20$,

$$\frac{f(0.25) - 2f(0.20) + f(0.15)}{(0.05)^2} \approx 4f(0.20)$$

$$\frac{1.12763 - 2 \times 1.081\,07 + f(0.15)}{0.0025} \approx 4 \times 1.081\,07$$

$$f(0.15) \approx 4 \times 1.081\,07 \times 0.0025 - 1.127\,63 + 2 \times 1.081\,07 = 1.045\,32$$

Putting $x = 0.25$,

$$\frac{f(0.30) - 2f(0.25) + f(0.20)}{(0.05)^2} \approx 4f(0.25)$$

$$\frac{f(0.30) - 2 \times 1.12763 + 1.08107}{0.0025} \approx 4 \times 1.12763$$

$$f(0.30) \approx 1.18547$$

Putting $x = 0.30$,

$$\frac{f(0.35) - 2f(0.30) + f(0.25)}{(0.05)^2} \approx 4f(0.30).$$

$$\frac{f(0.35) - 2 \times 1.18547 + 1.12763}{0.0025} \approx 4 \times 1.18547$$

$$f(0.35) \approx 1.25516$$

We have obtained the following values:

x	0.15	0.20	0.25	0.30	0.35
$f(x)$	1.045 32	1.081 07	1.127 63	1.185 47	1.255 16

Note that the exact solution is $f(x) = \cosh 2x$ (see Chapter 23), and the true values are

x	0.15	0.20	0.25	0.30	0.35
$f(x)$	1.045 34	1.081 07	1.127 63	1.185 47	1.255 17

Exercise 16.3   (Numerical differentiation, and differential equations)

1   The following table gives values of a function $f(x)$.

x	3.0	3.1	3.2	3.3	3.4	3.5
$f(x)$	0.720 58	0.916 91	1.116 55	1.311 54	1.494 11	1.656 99

Tabulate approximate values of $f'(x)$ and $f''(x)$ for $x = 3.1, 3.2, 3.3, 3.4$.

2   If $f(x) = \dfrac{\sin x}{x}$, use the formulae

$$f'(x) \approx \frac{f(x+h) - f(x-h)}{2h}$$

and $$f''(x) \approx \frac{f(x+h) - 2f(x) + f(x-h)}{h^2}$$

with $h = 0.01$, to find approximate values for $f'(1)$ and $f''(1)$. Check your answers by differentiation.

3   The function $f(x)$ satisfies the differential equation

$$f'(x) = x^2 - \frac{f(x)}{x}$$

and has values $f(1.0) = 0.25$ and $f(1.1) = 0.3328$. Use a step-by-step method to obtain values of $f(1.2)$, $f(1.3)$ and $f(1.4)$.

4   The function $f(x)$ satisfies the differential equation

$$f''(x) = \sin \pi x - f(x)$$

and has values $f(-0.1) = -0.26466$ and $f(0) = 0$. Evaluate the function step by step for $x = 0.1, 0.2, 0.3$, giving your answers to 5 decimal places.   (MEI)

5   The function $y = f(x)$ satisfies the differential equation

$$\frac{d^2y}{dx^2} = \frac{1}{100}[1 + (y-2)^2]$$

and it is given that $y = 3$ when $x = 0$ and $y = 3$ when $x = 1$. Use a step-by-step method, with step length 1, to obtain values of y when $x = 2, 3, 4$.

6   A function $f(x)$ satisfies the differential equation

$$f''(x) = f(x)$$

and it is given that $f(1) = 4.7008$ and $f'(1) = 6.1723$. Writing $f(0.9) = p$, $f(1.1) = q$, and using the approximate formulae for $f'(x)$ and $f''(x)$ with $x = 1$ and $h = 0.1$, obtain two simultaneous equations for p and q. Hence evaluate approximately $f(0.9)$ and $f(1.1)$. Using the formula for $f''(x)$ with $x = 1.1$ and $x = 1.2$, obtain values for $f(1.2)$ and $f(1.3)$.

7   A function $f(x)$ satisfies the differential equation

$$f''(x) = 2f(x) - 4x$$

subject to the conditions $f(0) = 0$ and $f(1) = 1$. Writing $f(0.25) = y_1$, $f(0.5) = y_2$, $f(0.75) = y_3$, and applying the formula

$$f''(x) \approx \frac{f(x+h) - 2f(x) + f(x-h)}{h^2}$$

with $h = 0.25$ and $x = 0.25, 0.5, 0.75$, obtain three simultaneous equations for y_1, y_2 and y_3. Hence calculate approximate values for $f(0.25)$, $f(0.5)$ and $f(0.75)$, and for $f'(0.25)$, $f'(0.5)$ and $f'(0.75)$.

16.4   The Lagrange interpolating formula

Here we explain one method for finding a formula to represent a tabulated function. The formula can then be used to evaluate the function at other values of x (this is called **interpolation**).

Example 1   Find the quadratic function which has the following table of values:

x	1	2	3
y	0.50	0.33	0.25

We first find a quadratic function q_1 which takes the value 0.50 when $x = 1$, and is zero when $x = 2$ and when $x = 3$. We have

$$q_1 = k_1(x-2)(x-3)$$

and, since $q_1 = 0.50$ when $x = 1$,

$$0.50 = k_1(1-2)(1-3)$$

so $k_1 = \dfrac{0.50}{(1-2)(1-3)}$ and $q_1 = 0.50\dfrac{(x-2)(x-3)}{(1-2)(1-3)}$

Similarly,

$$q_2 = 0.33\frac{(x-1)(x-3)}{(2-1)(2-3)}$$

is a quadratic function which takes the value 0.33 when $x = 2$, and is zero when $x = 1$ and $x = 3$.

Also

$$q_3 = 0.25\frac{(x-1)\,(x-2)}{(3-1)\,(3-2)}$$

takes the value 0.25 when $x = 3$, and is zero when $x = 1$ and when $x = 2$.

Now consider $q_1 + q_2 + q_3$. This is a quadratic function:

when $x = 1$, its value is $0.50 + 0 + 0 = 0.50$;
when $x = 2$, its value is $0 + 0.33 + 0 = 0.33$;
when $x = 3$, its value is $0 + 0 + 0.25 = 0.25$.

Hence this is the required function, and

$$y = q_1 + q_2 + q_3$$

$$= 0.50\frac{(x-2)(x-3)}{(1-2)(1-3)} + 0.33\frac{(x-1)(x-3)}{(2-1)(2-3)} + 0.25\frac{(x-1)(x-2)}{(3-1)(3-2)}$$

$$= 0.25(x-2)(x-3) - 0.33(x-1)(x-3) + 0.125(x-1)(x-2)$$
$$= 0.045x^2 - 0.305x + 0.76$$

In general, the quadratic function having values

x	x_1	x_2	x_3
y	y_1	y_2	y_3

is $\quad y = y_1\dfrac{(x-x_2)(x-x_3)}{(x_1-x_2)(x_1-x_3)} + y_2\dfrac{(x-x_1)(x-x_3)}{(x_2-x_1)(x_2-x_3)} + y_3\dfrac{(x-x_1)(x-x_2)}{(x_3-x_1)(x_3-x_2)}$

This is the **Lagrange interpolating formula**. We can use it to find the value of y at any given value of x (there is no need to multiply out the quadratic). Note that the values of x need not be equally spaced.

Example 2   A function $f(x)$ takes the following values

x	1	2	3
$f(x)$	0.50	0.33	0.25

Find an approximate value for $f(2.2)$.

When $x = 2.2$,

$$\frac{(x - x_2)(x - x_3)}{(x_1 - x_2)(x_1 - x_3)} = \frac{(2.2 - 2)(2.2 - 3)}{(1 - 2)(1 - 3)} = -0.08$$

$$\frac{(x - x_1)(x - x_3)}{(x_2 - x_1)(x_2 - x_3)} = \frac{(2.2 - 1)(2.2 - 3)}{(2 - 1)(2 - 3)} = 0.96$$

and $\quad \dfrac{(x - x_1)(x - x_2)}{(x_3 - x_1)(x_3 - x_2)} = \dfrac{(2.2 - 1)(2.2 - 2)}{(3 - 1)(3 - 2)} = 0.12$

Thus $\quad f(2.2) \approx (0.50)(-0.08) + (0.33)(0.96) + (0.25)(0.12) = 0.3068$

(We have effectively replaced the function $f(x)$ by a quadratic function passing through the three given points. The value we obtain for $f(2.2)$ is therefore only an approximate one.)

We can extend the method to find a cubic passing through 4 given points, or a quartic passing through 5 given points, and so on.

Example 3 A cubic function $f(x)$ is tabulated below. Evaluate $f(0.25)$.

x	0.1	0.2	0.3	0.4
$f(x)$	0.02	0.07	0.24	0.59

We proceed in a similar way to that shown in Example 2. When $x = 0.25$.

$$\frac{(x - x_2)(x - x_3)(x - x_4)}{(x_1 - x_2)(x_1 - x_3)(x_1 - x_4)} = \frac{(0.05)(-0.05)(-0.15)}{(-0.1)(-0.2)(-0.3)} = -0.0625$$

$$\frac{(x - x_1)(x - x_3)(x - x_4)}{(x_2 - x_1)(x_2 - x_3)(x_2 - x_4)} = \frac{(0.15)(-0.05)(-0.15)}{(0.1)(-0.1)(-0.2)} = 0.5625$$

$$\frac{(x - x_1)(x - x_2)(x - x_4)}{(x_3 - x_1)(x_3 - x_2)(x_3 - x_4)} = \frac{(0.15)(0.05)(-0.15)}{(0.2)(0.1)(-0.1)} = 0.5625$$

$$\frac{(x - x_1)(x - x_2)(x - x_3)}{(x_4 - x_1)(x_4 - x_2)(x_4 - x_3)} = \frac{(0.15)(0.05)(-0.05)}{(0.3)(0.2)(0.1)} = -0.0625$$

Thus $\quad f(0.25) = (0.02)(-0.0625) + (0.07)(0.5625) + (0.24)(0.5625) + (0.59)(-0.0625)$

$$= -0.001\,25 + 0.039\,375 + 0.135 - 0.036\,875$$

$$= 0.136\,25$$

Exercise 16.4   (Lagrange interpolating formula)

1   Find the quadratic function which has the following table of values, and use it to find $f(0.58)$.

x	0.50	0.60	0.70
$f(x)$	1.649	1.822	2.014

2   The following table gives values of a cubic function $f(x)$. Find an expression for $f(x)$.

x	0	1	2	3
$f(x)$	1	0	−1	7

(AEB)

3   The following table gives values of $\sqrt{x}$ tabulated against values of x. Use it to obtain a cubic polynomial approximation to $\sqrt{x}$ for $0 \leqslant x \leqslant 1.5$.

x	0	0.5	1.0	1.5
$\sqrt{x}$	0.0000	0.7071	1.0000	1.2247

(MEI)

Revision questions D

Revision paper D1

1 (i) If $x^2 = \tan y$, find $\dfrac{dy}{dx}$ in terms of x.   (L)

   (ii) Show that $\displaystyle\int_1^{\sqrt{3}} \dfrac{x+3}{\sqrt{4-x^2}}\,dx = \dfrac{\pi}{2} + \sqrt{3} - 1$.   (L)

2 (i) Find the range of values of x if $x^2 - 5x + 4 < 0$.

   (ii) Sketch the graph of $y = x^2 - 5x + 4$.

  (iii) If $x = 2$, what is the value of $|x^2 - 5x + 4|$?

  (iv) Sketch the graph of $y = |x^2 - 5x + 4|$.

   (v) If $y = mx$ is a tangent to the curve $y = |x^2 - 5x + 4|$ between $x = 1$ and $x = 4$, find the value of m, indicating clearly the reason for discarding any other values.   (S)

3 Find the coordinates of the turning point of the curve $y = \dfrac{x^2 - 4}{(x+1)^2}$ and ascertain the nature of this turning point.   (AEB)

4 (a) Prove that the tangent at the point with parameter t on the curve

$$x = \frac{\cos t}{1 + \sin t}, \quad y = \frac{\sin t}{1 + \cos t},$$

has equation

$$x(1 + \sin t) + y(1 + \cos t) = \cos t + \sin t$$

   (b) Given that $x^3 - 3xy + y^3 = k$, where k is a constant, find $\dfrac{dy}{dx}$ in terms of x and y.

Show that the curve $x^3 - 3xy + y^3 = k$ cannot possess a tangent parallel to either the x-axis or the y-axis if $k < -1$.   (MEI)

5 (i) The function f is defined by

$$f(x) = \sin x \text{ for } x \leqslant 0$$
$$f(x) = x \quad\ \text{ for } x > 0$$

Sketch the graphs of $f(x)$ and its derivative $f'(x)$ for $-\pi/2 < x < \pi/2$ and decide whether the functions f and f' are continuous at $x = 0$ or not.

   (ii) Use the substitution $x = 2\cos^2 \theta + 5\sin^2 \theta$ to evaluate

$$\int_2^5 \sqrt{(x-2)(5-x)}\,dx \tag*{(L)}$$

6   Tabulate, to three places of decimals, the values of $\sqrt{1+x^2}$ for values of x from 0 to 0.4 at intervals of 0.1. Estimate, to two places of decimals, the value of

$$\int_0^{0.4} \sqrt{1+x^2}\,dx$$

by Simpson's method using five ordinates.

By expanding $\sqrt{1+x^2}$ in powers of x as far as the term in x^4 and integrating term by term, obtain, to two places of decimals, a second estimate for the value of this integral.   (L)

Revision paper D2

1   Draw the graphs of $\sin^{-1} x$ and $\cos^{-1} x$ ($-1 \leqslant x \leqslant 1$), stating the range of values of each function.

Prove that $\dfrac{d}{dx}(\sin^{-1} x) = \dfrac{1}{\sqrt{1-x^2}}$ and find $\dfrac{d}{dx}(\cos^{-1} x)$.

Evaluate $\displaystyle\int_0^1 \sin^{-1} x\,dx$.   (O)

2   Sketch the curve $k^2 y^2 = x^2(k^2 - x^2)$, where k is a positive constant, showing its shape at the origin and at the points $(\pm k, 0)$.
(a)   Find the area enclosed by that portion of the curve lying in the first and fourth quadrants.
(b)   This portion of the curve is rotated through two right angles about the x-axis. Find the volume of the solid of revolution generated.
(c)   If the value of k is increased by 0.1%, find the approximate percentage increases in the corresponding area and volume defined above.   (O and C)

3   Find $f'(x)$ where $f(x) = \dfrac{2x}{1-x^2}$ and hence sketch the graph of $f(x)$. Show that $f(x)$ takes all real values. Prove also that it takes each of these values, with one exception, for two distinct values of x.

Show also that $g(x) = \dfrac{2x}{1+x^2}$ takes all values in the interval between -1 and $+1$ inclusive and no others.

By comparing the graphs, or otherwise, find the values of x for which $f(x) \geqslant g(x)$.   (MEI)

4   (a)   Prove that if $x = \cos\theta$, $y = \sin^5\theta$ then $\dfrac{d^2 y}{dx^2} = 5\sin 3\theta$.

(b)   Prove that if $t = \tan\frac{1}{2}\theta$ then $\dfrac{d\theta}{dt} = \dfrac{2}{1+t^2}$ and hence, or otherwise, evaluate $\displaystyle\int_0^\pi \dfrac{1}{3+2\cos\theta}\,d\theta$.

(c)   Evaluate $\displaystyle\int_0^1 \dfrac{2x+1}{\sqrt{4-x^2}}\,dx$ giving your answer correct to two decimal places.   (W)

5   Show that the equation $x^3 - 3x^2 - 2 = 0$ has a root between $x = 3$ and $x = 4$ and that this is the largest positive root of the equation.

Starting with $x = 3$, use two iterations of Newton's method to find an approximate value for this root. Justify the accuracy to which your answer is given.   (MEI)

6 Use Simpson's formula

$$\int_a^b f(x)\,dx \approx \tfrac{1}{6}(b-a)(y_0+4y_1+y_2)$$

to obtain the approximation

$$\int_0^{10} f(x)\,dx \approx \tfrac{1}{3}\left[(y_0+y_{10})+4(y_1+y_3+y_5+y_7+y_9)+2(y_2+y_4+y_6+y_8)\right]$$

where y_i is the value of $f(x)$ when $x = i$.

By considering the case when $f(x)$ is x^n, prove that, within the range of validity of this rule,

$$4(0.1^n+0.3^n+0.5^n+0.7^n+0.9^n)+2(0.2^n+0.4^n+0.6^n+0.8^n) \approx \frac{29-n}{1+n}$$

Prove by direct computation of each side that, when $n = 4$, the magnitude of the error is 0.008%.

(MEI)

Miscellaneous revision questions: paper D3

1 Differentiate with respect to x (a) $\dfrac{x-1}{x+1}$ (b) $\cos^2(3x+4)$ (c) $x(\sin x)^{\frac{1}{2}}$ (L)

2 Evaluate each of the following integrals: (i) $\displaystyle\int_1^4 \frac{2x+1}{\sqrt{x}}\,dx$ (ii) $\displaystyle\int_1^4 \frac{x}{\sqrt{2x+1}}\,dx$

(O and C)

3 An inkwell is in the form of an inverted cone of depth 3 cm and radius at the top 2 cm. Find the volume of ink when its depth is x cm.

When the depth of the ink is 1 cm, it is decided to fill the inkwell. Ink is poured in at the rate of 2π cm^3 s^{-1}.

(i) How long does it take to fill the inkwell?

(ii) What is the rate at which the depth is increasing when the depth of the ink is 2 cm?   (O)

4 A particle moves on a straight line, and its velocity v at time t is given by $v = 12\cos 3t$. If, when $t = 0$, the displacement x is 5 units, find an expression for x at time t. Show that x satisfies

$$\frac{d^2x}{dt^2} = 45-9x$$

Show that x always lies between 1 and 9 (inclusive). Find also the greatest acceleration of the particle.   (O)

5 A barrel stands 4 feet high and has a diameter of 3 feet at its centre and 2 feet diameter at its ends. It is modelled by a curve $x = a - by^2$ between suitable values of y which is then rotated through 4 right angles about the y-axis so that the centre of the barrel is at the origin.

Prove that the value of a is $1\frac{1}{2}$ and find the value of b. Calculate the volume of the barrel.   (O)

6 Find the stationary values of the function $\dfrac{2x^2+9x+6}{x^2+4x+1}$ and the values of x for which the denominator vanishes.

Make an accurate sketch of the graph of the function.   (MEI)

7 (i) Prove that if $x = \dfrac{1-t^2}{1+t^2}$ and $y = \dfrac{4t}{1+t^2}$ then $\dfrac{dy}{dx} = t-\dfrac{1}{t} = -\dfrac{4x}{y}$.

Find the stationary points on the curve and where the curve has infinite gradient.   (O and C)

(ii)   By means of the substitution $x = \sin^2\theta$, or otherwise, evaluate

$$\int_0^{\frac{1}{2}} \left(\frac{x}{1-x}\right)^{\frac{1}{2}} dx$$

giving your answer in terms of π.   (JMB)

8   Two tangents are drawn from a point at a fixed distance a from the centre of a circle of variable radius. The angle subtended at the centre of this circle by the minor arc joining the points of contact of the tangents is 2θ. Prove that the area of the region enclosed by the tangents and the minor arc is

$$a^2(\tfrac{1}{2}\sin 2\theta - \theta \cos^2\theta)$$

Show that this area is a maximum when the length of each tangent is equal to the length of the minor arc.   (MEI)

9   Obtain $\int x^2 \cos 2x\,dx$ and hence prove that

$$\int_0^{\pi} x^2 \cos 2x\,dx = \tfrac{1}{2}\pi$$

Find the area of the region enclosed by the curve $y = x \sin x$, for $0 \leqslant x \leqslant \pi$, and the x-axis.

The region is rotated through 2π radians about the x-axis. Find the volume of the solid of revolution thus generated.

From your results, or otherwise, find the volume of the solid of revolution generated by rotation of the same region through 2π radians about the line $y = -\tfrac{1}{8}$. [Leave your answers in terms of π.]
                                                                (O and C)

10   (a)   State the principal values of the functions arc sin x, arc cos x and arc tan x, and explain why these conventions are necessary.

Prove without the use of tables that

$$\text{arc cos } \sqrt{\tfrac{1}{3}} - \text{arc sin } \sqrt{\tfrac{1}{3}} = \frac{\pi}{2} - \text{arc tan } (2\sqrt{2})$$

where principal values are taken.

(b)   Evaluate $\displaystyle\int_0^{\frac{\pi}{2}} \cos x\,dx$ and $\displaystyle\int_0^1 \text{arc cos } x\,dx$.

Draw sketches to illustrate and explain the relation between the results you obtain.   (MEI)

11   Show that the equation $x^4 = 10x - 5$ has a root between 1 and 2. Use Newton's method to find this root correct to two decimal places. By writing the equation in the form $x = (10x - 5)^{\frac{1}{4}}$ and using another iterative method, find this same root to the same degree of accuracy.   (S)

12   Show that the formula

$$\int_0^{2h} f(x)\,dx \approx \frac{h}{3}[f(0) + 4f(h) + f(2h)]$$

is exact when $f(x)$ is a cubic polynomial in x.

Use the formula given above to obtain an approximate value of I where $I = \displaystyle\int_0^{\frac{1}{2}} \frac{1}{\sqrt{1-x^2}}\,dx$. Give your result to three decimal places.

Show, by direct integration, that $I = \dfrac{\pi}{6}$.   (L)

17

Infinite series and applications

Introduction

You will probably have seen the geometric series

$$1 + \tfrac{1}{2} + \tfrac{1}{4} + \tfrac{1}{8} + \ldots$$

which has a 'sum to infinity' of 2. (This means that the sum can be made as close to 2 as we please, by taking a sufficient number of terms of the series.) More generally the series

$$1 + x + x^2 + x^3 + \ldots$$

has a sum to infinity of $\dfrac{1}{1-x}$ (provided that $-1 < x < 1$), and we write

$$\frac{1}{1-x} = 1 + x + x^2 + x^3 + \ldots$$

Thus the function $\dfrac{1}{1-x}$ can be expressed as an infinite series involving powers of x. We shall see that most functions (including circular functions, and other types which we shall meet later) can be expressed in a similar way, using multiples of powers of x; for example, we shall show that

$$\sin x = x - \frac{1}{3!}x^3 + \frac{1}{5!}x^5 - \frac{1}{7!}x^7 + \ldots$$

This is an example of a **power series**, which is given in general as

$$a_0 + a_1 x + a_2 x^2 + \ldots + a_n x^n + \ldots$$

where $a_0, a_1, a_2, \ldots$ are constants. A power series 'terminates' if these constants are all zero from some point onwards; otherwise we have an infinite number of terms.

Power series are very useful, because they provide methods which can be applied to any type of function. Also they can be used to evaluate a given function to any desired degree of accuracy. When, for example, we press the 'sin' key on a calculator, an appropriate series is automatically summed within the calculator before the resulting value is displayed.

17.1 Polynomials

A polynomial is a terminating power series,

$$p(x) = a_0 + a_1 x + a_2 x^2 + \ldots + a_{n-1} x^{n-1} + a_n x^n$$

where n is a positive integer, and $a_0, a_1, \ldots, a_n$ are real numbers (called the coefficients of the polynomial). The polynomial is said to have degree n (provided $a_n \neq 0$).

A polynomial of degree 1 is a linear function $a_0 + a_1 x$ (its graph is a straight line); a polynomial of degree 2 is a quadratic function

$$a_0 + a_1 x + a_2 x^2$$

a polynomial of degree 3 is a cubic function

$$a_0 + a_1 x + a_2 x^2 + a_3 x^3$$

and so on.

Finding the coefficients of a polynomial

A polynomial of degree 3 has four coefficients, and so we shall need four items of information to find them. We can find the coefficients if, for example, we are given the values of the polynomial and of its first three derivatives when $x = 0$.

Example Find the polynomial $p(x)$ of degree 3 which is such that $p(0) = 2$, $p'(0) = -3$, $p''(0) = 1$ and $p'''(0) = 5$.

Let $p(x) = a_0 + a_1 x + a_2 x^2 + a_3 x^3$

Putting $x = 0$, we have $p(0) = a_0$, and so $a_0 = 2$. Also,

$$p'(x) = a_1 + 2a_2 x + 3a_3 x^2$$

Putting $x = 0$, $p'(0) = a_1$, so $a_1 = -3$.

$$p''(x) = 2a_2 + 6a_3 x$$

Putting $x = 0$, $p''(0) = 2a_2$, so $a_2 = \frac{1}{2}$.

$$p'''(x) = 6a_3$$

Putting $x = 0$, $p'''(0) = 6a_3$, so $a_3 = \frac{5}{6}$.

Hence the polynomial is

$$p(x) = 2 - 3x + \tfrac{1}{2}x^2 + \tfrac{5}{6}x^3$$

In general, we can find the coefficients of a polynomial $p(x)$ of degree n if we are given the values of $p(0)$, $p'(0)$, $p''(0)$, $\ldots$, $p^{(n)}(0)$, where $p^{(n)}(x)$ means the nth derivative of $p(x)$.

Let $p(x) = a_0 + a_1 x + a_2 x^2 + \ldots + a_n x^n$

Putting $x = 0$, $p(0) = a_0$.

$$p'(x) = a_1 + 2a_2 x + 3a_3 x^2 + \ldots + na_n x^{n-1}$$

Putting $x = 0$, $p'(0) = a_1$.

$$p''(x) = 2a_2 + 3 \times 2a_3 x + 4 \times 3a_4 x^2 + \ldots + n(n-1)a_n x^{n-2}$$

Putting $x = 0$, $p''(0) = 2a_2$, so $a_2 = \frac{1}{2}p''(0)$.

$$p'''(x) = 3 \times 2a_3 + 4 \times 3 \times 2a_4 x + \ldots + n(n-1)(n-2)a_n x^{n-3}$$

Putting $x = 0$, $p'''(0) = 3!\, a_3$, so $a_3 = \dfrac{1}{3!}p'''(0)$.

$$\vdots$$

$$p^{(r)}(x) = r!a_r + \ldots + n(n-1)\ldots(n-r+1)a_n x^{n-r}$$

Putting $x = 0$, $p^{(r)}(0) = r!a_r$, so $a_r = \dfrac{1}{r!}\,p^{(r)}(0)$.

Hence, $p(x) = p(0) + xp'(0) + \dfrac{x^2}{2!}p''(0) + \ldots + \dfrac{x^r}{r!}p^{(r)}(0) + \ldots + \dfrac{x^n}{n!}p^{(n)}(0)$

Approximating a function by a polynomial

Polynomials are simple to evaluate, to differentiate, and to integrate, and we therefore use them to approximate to other functions (such as $\sqrt{1+x}$ and $\cos x$). One method for finding a polynomial $p(x)$ which approximates to a function $f(x)$ is to choose certain values of x, and arrange that $p(x)$ and $f(x)$ are equal for these values (for example, for a cubic approximation, we could choose any four values of x). We would expect this to give a reasonable approximation between the chosen values of x, and we used this method when interpolating from a table of values (see p. 289.

We shall now consider a different method, where we concentrate on a single value of x, which, at first, will be $x = 0$. We find an approximating polynomial $p(x)$ by arranging that the values of $p(x)$ and of as many of its derivatives as possible, are equal to those of $f(x)$, when $x = 0$. We would expect this to give a reasonable approximation when x is close to zero.

Example 1   Consider the function $f(x) = \sqrt{1+x}$.

We shall find the 'best fitting' polynomial approximations to the curve $y = \sqrt{1+x}$ in the neighbourhood of the point $(0, 1)$, by considering the derivatives of $f(x)$ when $x = 0$ (Fig. 17.1). We have

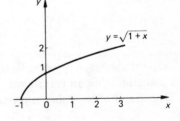

Fig. 17.1

$$f(x) = (1+x)^{\frac{1}{2}} \qquad\qquad \text{so} \qquad f(0) = 1$$

$$f'(x) = \tfrac{1}{2}(1+x)^{-\frac{1}{2}} \qquad\qquad \text{so} \qquad f'(0) = \tfrac{1}{2}$$

$$f''(x) = (\tfrac{1}{2})(-\tfrac{1}{2})(1+x)^{-\frac{3}{2}} \qquad \text{so} \qquad f''(0) = -\tfrac{1}{4}$$

$$f'''(x) = (\tfrac{1}{2})(-\tfrac{1}{2})(-\tfrac{3}{2})(1+x)^{-\frac{5}{2}} \qquad \text{so} \qquad f'''(0) = \tfrac{3}{8}$$

$$\vdots$$

$$f^{(n)}(x) = (\tfrac{1}{2})(-\tfrac{1}{2})(-\tfrac{3}{2})\ldots\left(-\frac{2n-3}{2}\right)(1+x)^{-(2n-1)/2}$$

$$\text{so} \quad f^{(n)}(0) = (-1)^{n+1}\frac{1 \times 3 \times 5 \times \ldots \times (2n-3)}{2^n}$$

The best fitting straight line passes through $(0, 1)$ and has gradient $\frac{1}{2}$, and so it is

$$y = 1 + \tfrac{1}{2}x$$

(this is the tangent at the point $(0, 1)$).

The best fitting quadratic

$$q(x) = a_0 + a_1 x + a_2 x^2$$

will have $q(0) = 1, \quad q'(0) = \tfrac{1}{2}, \quad q''(0) = -\tfrac{1}{4}$

and so it is

$$q(x) = 1 + x(\tfrac{1}{2}) + \frac{x^2}{2!}(-\tfrac{1}{4})$$

i.e. $y = 1 + \tfrac{1}{2}x - \tfrac{1}{8}x^2$

The best fitting cubic

$$c(x) = a_0 + a_1 x + a_2 x^2 + a_3 x^3$$

will have $c(0) = 1, \quad c'(0) = \tfrac{1}{2}, \quad c''(0) = -\tfrac{1}{4}, \quad c'''(0) = \tfrac{3}{8}$

and so it is

$$c(x) = 1 + x(\tfrac{1}{2}) + \frac{x^2}{2!}(-\tfrac{1}{4}) + \frac{x^3}{3!}(\tfrac{3}{8})$$

i.e. $y = 1 + \tfrac{1}{2}x - \tfrac{1}{8}x^2 + \tfrac{1}{16}x^3$

For the best fitting polynomial of degree n,

$$p(x) = a_0 + a_1 x + a_2 x^2 + \ldots + a_n x^n$$

we need

$$p(0) = 1, \quad p'(0) = \tfrac{1}{2}, \quad p''(0) = -\tfrac{1}{4}, \ldots \ p^n(0) = (-1)^{n+1}\frac{1 \times 3 \times 5 \times \ldots \times (2n-3)}{2^n}$$

and so it is

$$y = 1 + \tfrac{1}{2}x - \tfrac{1}{8}x^2 + \tfrac{1}{16}x^3 + \ldots + (-1)^{n+1}\frac{1 \times 3 \times 5 \times \ldots \times (2n-3)x^n}{2^n n!}$$

Remarkably, each polynomial is obtained from the previous one by just adding on an extra term.

The graph of the cubic approximation

$$y = 1 + \tfrac{1}{2}x - \tfrac{1}{8}x^2 + \tfrac{1}{16}x^3$$

is shown in Fig. 17.2. The fit is very good when x is close to zero (for example, when $x = 0.4$, $y = 1 + \tfrac{1}{2}x - \tfrac{1}{8}x^2 + \tfrac{1}{16}x^3$ gives $y = 1.184$ whereas $y = \sqrt{1+x}$ gives $y = 1.183$), but it deteriorates as x becomes larger. Also the cubic curve continues to give values when $x < -1$, where the original function $f(x) = \sqrt{1+x}$ is not defined.

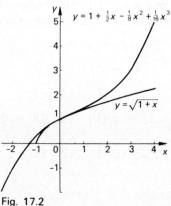

Fig. 17.2

In general, if $f(x)$ is a function, the best fitting polynomial of degree n, in the neighbourhood of $x = 0$, is

$$p(x) = f(0) + xf'(0) + \frac{x^2}{2!}f''(0) + \ldots + \frac{x^n}{n!}f^{(n)}(0)$$

Example 2 If $y = \cos x$, find the best fitting polynomial of degree 4, in the neighbourhood of the point $(0, 1)$.

Let

$$f(x) = \cos x \qquad \text{so} \qquad f(0) = 1$$

Then

$$
\begin{aligned}
f'(x) &= -\sin x & f'(0) &= 0 \\
f''(x) &= -\cos x & f''(0) &= -1 \\
f'''(x) &= \sin x & f'''(0) &= 0 \\
f^{iv}(x) &= \cos x & f^{iv}(0) &= 1
\end{aligned}
$$

The best fitting quartic is therefore

$$y = 1 + x(0) + \frac{x^2}{2!}(-1) + \frac{x^3}{3!}(0) + \frac{x^4}{4!}(1)$$

i.e. $y = 1 - \frac{1}{2}x^2 + \frac{1}{24}x^4$

Exercise 17.1 (Approximating functions by polynomials)

1 Find the polynomial $p(x)$ of degree 2 which is such that $p(0) = 3$, $p'(0) = -1$ and $p''(0) = 4$.

2 Find the polynomial $p(x)$ of degree 3 which is such that $p(0) = 2$, $p'(0) = 5$, $p''(0) = -3$ and $p'''(0) = 6$.

3 Find the polynomial $p(x)$ of degree 4 which is such that $p(0) = -3$, $p'(0) = 2$, $p''(0) = 4$, $p'''(0) = 0$ and $p^{iv}(0) = -8$.

4 If $f(x) = \dfrac{1}{(1-x)^2}$, find the values of $f(0), f'(0), f''(0)$ and $f'''(0)$. Hence find the best fitting straight line, quadratic and cubic, all in the neighbourhood of the point $(0, 1)$. Sketch the graphs of $f(x)$ and of its cubic approximation on the same diagram.

5 If $y = \sin x$, find the best fitting polynomial of degree 3 in the neighbourhood of the point $(0, 0)$. Sketch the graphs of $y = \sin x$ and of its cubic approximation on the same diagram.

6 Explain why it is not possible to find polynomial approximations to the function $f(x) = x^{\frac{1}{3}}$ in the neighbourhood of $x = 0$.

17.2 The best fitting polynomial at a general point

So far we have always considered the values of a function and of its derivatives at the point where $x = 0$, as this seemed the simplest point to consider. However we could have chosen any other point, say where $x = a$.

Suppose we wish to find the best fitting polynomial of degree n to a function $f(x)$ in the neighbourhood of $x = a$.

As we are interested in values of x close to a, so that $(x - a)$ is small, we write a general polynomial of degree n in the form

$$p(x) = b_0 + b_1(x - a) + b_2(x - a)^2 + \ldots + b_n(x - a)^n$$

Putting $x = a$,

$$p(a) = b_0$$

Then $p'(x) = b_1 + 2b_2(x - a) + 3b_3(x - a)^2 + \ldots + nb_n(x - a)^{n-1}$

Putting $x = a$,

$$p'(a) = b_1$$

$$p''(x) = 2b_2 + 3 \times 2b_3(x - a) + 4 \times 3b_4(x - a)^2 + \ldots + n(n - 1)b_n(x - a)^{n-2}$$

Putting $x = a$,

$$p''(a) = 2b_2$$

so $\quad b_2 = \frac{1}{2}p''(a)$

$$p'''(x) = 3 \times 2b_3 + 4 \times 3 \times 2b_4(x - a) + \ldots + n(n - 1)(n - 2)b_n(x - a)^{n-3}$$

Putting $x = a$,

$$p'''(a) = 3!b_3$$

so $\quad b_3 = \frac{1}{3!}p'''(a)$

$$\vdots$$

$$p^{(n)}(x) = n(n - 1)(n - 2) \ldots 2b_n$$

Putting $x = a$,

$$p^{(n)}(a) = n!b_n$$

so $\quad b_n = \frac{1}{n!}p^{(n)}(a)$

Thus $\quad p(x) = p(a) + (x - a)p'(a) + \dfrac{(x - a)^2}{2!}p''(a) + \ldots + \dfrac{(x - a)^n}{n!}p^{(n)}(a)$

We define the best fitting polynomial of degree n to the function $f(x)$, in the neighbourhood of $x = a$, to be the polynomial $p(x)$ which satisfies

$$p(a) = f(a), \quad p'(a) = f'(a)$$
$$p''(a) = f''(a), \ldots$$
$$p^{(n)}(a) = f^{(n)}(a)$$

Hence it is

$$p(x) = f(a) + (x - a)f'(a) + \frac{(x - a)^2}{2!}f''(a) + \ldots + \frac{(x - a)^n}{n!}f^{(n)}(a)$$

Example Find the quadratic polynomial which best fits the curve $y = \sqrt{1+x}$ in the neighbourhood of the point $(3, 2)$.

Let $f(x) = \sqrt{1+x}$ so $f(3) = 2$

$\quad f'(x) = \frac{1}{2}(1+x)^{-\frac{1}{2}}$ so $f'(3) = \frac{1}{4}$

$\quad f''(x) = (\frac{1}{2})(-\frac{1}{2})(1+x)^{-\frac{3}{2}}$ so $f''(3) = -\dfrac{1}{32}$

The best fitting quadratic $q(x)$ is therefore

$$q(x) = 2 + (x-3)\,(\tfrac{1}{4}) + \frac{(x-3)^2}{2!}\left(-\frac{1}{32}\right)$$

$$= 2 + \frac{1}{4}x - \frac{3}{4} - \frac{1}{64}(x^2 - 6x + 9) \;=\; \frac{71}{64} + \frac{11}{32}x - \frac{1}{64}x^2$$

Exercise 17.2 (Best fitting polynomials, at any given point)

1 Find the cubic polynomial $p(x)$ given that $p(1) = 5$, $p'(1) = -3$, $p''(1) = 4$, $p'''(1) = 6$.
2 If $f(x) = 1 + x - 3x^2 + x^3$, find the values of $f(2)$, $f'(2)$ and $f''(2)$. Hence find the best fitting quadratic in the neighbourhood of $x = 2$.
3 If $y = \dfrac{1}{x}$, find the best fitting polynomial of degree 3 in the neighbourhood of the point $(1, 1)$.
4 If $y = x^{\frac{3}{2}}$, find the best fitting quadratic in the neighbourhood of the point $(4, 8)$.
5 If $y = \tan x$, find the best fitting quadratic in the neighbourhood of the point $(\frac{1}{4}\pi, 1)$.

17.3 Maclaurin's series

We now show how successive polynomial approximations lead to the expression of a function as an infinite power series.

Example 1 Use polynomial approximations to evaluate $f(x) = \sqrt{1+x}$ when $x = 0.4$.

We have already found (see p. 299, Example 1) the polynomial approximations to $\sqrt{1+x}$ in the neighbourhood of $x = 0$. These are $1 + \frac{1}{2}x$, $1 + \frac{1}{2}x - \frac{1}{8}x^2$, $1 + \frac{1}{2}x - \frac{1}{8}x^2 + \frac{1}{16}x^3$; the quartic approximation is

$$1 + \tfrac{1}{2}x - \tfrac{1}{8}x^2 + \tfrac{1}{16}x^3 - \frac{1 \times 3 \times 5}{2^4 \times 4!}x^4 = 1 + \tfrac{1}{2}x - \tfrac{1}{8}x^2 + \tfrac{1}{16}x^3 - \frac{5}{128}x^4$$

and so on. When $x = 0.4$, the linear approximation

$$1 + \tfrac{1}{2}x$$

gives $f(0.4) \approx 1 + \frac{1}{2} \times 0.4 = 1.2$

the quadratic $1 + \frac{1}{2}x - \frac{1}{8}x^2$

gives $f(0.4) \approx 1 + \frac{1}{2} \times 0.4 - \frac{1}{8} \times 0.4^2$

$$= 1.2 - 0.02 = 1.18$$

the cubic $1 + \frac{1}{2}x - \frac{1}{8}x^2 + \frac{1}{16}x^3$

gives $f(0.4) \approx 1 + \frac{1}{2} \times 0.4 - \frac{1}{8} \times 0.4^2 + \frac{1}{16} \times 0.4^3$

$$= 1.18 + 0.004 = 1.184$$

the quartic $1 + \frac{1}{2}x - \frac{1}{8}x^2 + \frac{1}{16}x^3 - \frac{5}{128}x^4$

gives $f(0.4) \approx 1 + \frac{1}{2} \times 0.4 - \frac{1}{8} \times 0.4^2 + \frac{1}{16} \times 0.4^3 - \frac{5}{128} \times 0.4^4$

$$= 1.184 - 0.001 = 1.183$$

and so on. It appears that successive polynomial approximations give values closer and closer to the true value $f(0.4)$ (which is $\sqrt{1.4} = 1.1832\ldots$). We may achieve any desired degree of accuracy by taking sufficient terms of the series

$$1 + \frac{1}{2} \times 0.4 - \frac{1}{8} \times 0.4^2 + \frac{1}{16} \times 0.4^3 - \ldots$$

and the exact value is the sum of the **infinite** series

$$f(0.4) = 1 + \frac{1}{2} \times 0.4 - \frac{1}{8} \times 0.4^2 + \frac{1}{16} \times 0.4^3 - \ldots$$

We may obtain other values of $f(x)$ in a similar way; thus

$$\sqrt{1 + x} = 1 + \frac{1}{2}x - \frac{1}{8}x^2 + \frac{1}{16}x^3 - \ldots$$

provided that x is not too far from zero. However, if we try to put, for example, $x = 4$, the series is

$$1 + \frac{1}{2} \times 4 - \frac{1}{8} \times 4^2 + \frac{1}{16} \times 4^3 - \frac{5}{128} \times 4^4 + \ldots$$

$$= 1 + 2 - 2 + 4 - 10 + \ldots$$

and the terms become larger and larger; this series does not have a sum to infinity.

In fact, for this example, the method will work provided that $-1 < x \leqslant 1$.

In general, if $f(x)$ is a function, we have seen (on p. 301) that the best fitting polynomial of degree n, in the neighbourhood of $x = 0$, is

$$f(0) + xf'(0) + \frac{x^2}{2!}f''(0) + \ldots + \frac{x^n}{n!}f^{(n)}(0)$$

Provided x is small enough, we shall have

$$f(x) \approx f(0) + xf'(0) + \frac{x^2}{2!}f''(0) + \ldots + \frac{x^n}{n!}f^{(n)}(0)$$

and this approximation can be made as good as we please by taking n sufficiently large. $f(x)$ is given by the sum of the infinite series

$$\boxed{f(x) = f(0) + xf'(0) + \frac{x^2}{2!}f''(0) + \ldots + \frac{x^n}{n!}f^{(n)}(0) + \ldots}$$

This is the **Maclaurin series** (or expansion) for $f(x)$.

It is not easy to say how small x needs to be for this to be true. Even if the series converges (i.e. has a sum to infinity), it does not necessarily follow that its sum is equal to $f(x)$. We shall state the validity of our series without proof.

For example, the Maclaurin series for $\sqrt{1+x}$ is

$$\sqrt{1+x} = 1 + \tfrac{1}{2}x - \tfrac{1}{8}x^2 + \tfrac{1}{16}x^3 - \tfrac{5}{128}x^4 + \ldots$$

and this is true when $-1 < x \leqslant 1$.

Example 2   Find the Maclaurin series for $\sin x$, and hence evaluate $\sin(0.2)$, correct to 5 decimal places.

Let $f(x) = \sin x$,   then $f(0) = 0$

$$
\begin{array}{llll}
f'(x) = \cos x, & \text{so} & f'(0) = 1 \\
f''(x) = -\sin x, & \text{so} & f''(0) = 0 \\
f'''(x) = -\cos x, & \text{so} & f'''(0) = -1 \\
f^{iv}(x) = \sin x, & \text{so} & f^{iv}(0) = 0 \\
f^{v}(x) = \cos x, & \text{so} & f^{v}(0) = 1
\end{array}
$$

and so on. Hence the Maclaurin series is

$$\sin x = 0 + x(1) + \frac{x^2}{2!}(0) + \frac{x^3}{3!}(-1) + \frac{x^4}{4!}(0) + \frac{x^5}{5!}(1) + \ldots$$

$$= x - \frac{x^3}{3!} + \frac{x^5}{5!} - \frac{x^7}{7!} + \ldots$$

To evaluate $\sin(0.2)$ we substitute $x = 0.2$ into the series

$$\sin(0.2) = 0.2 - \frac{(0.2)^3}{6} + \frac{(0.2)^5}{120} - \frac{(0.2)^7}{5040} + \ldots$$

$$= 0.2 - 0.001\,333\,33 + 0.000\,002\,67 - \ldots$$

$$= 0.198\,669\,34 - \ldots$$

$$= 0.198\,67 \quad \text{(correct to 5 decimal places)}$$

(Enter 0.2 on the calculator, and place it in the memory.

Multiply by $\dfrac{-(0.2)^2}{2 \times 3}$, to give $\dfrac{-(0.2)^3}{3!}$, and add this to the memory.

Multiply by $\dfrac{-(0.2)^2}{4 \times 5}$, to give $\dfrac{(0.2)^5}{5!}$, and add this to the memory.

Multiply by $\dfrac{-(0.2)^2}{6 \times 7}$, and so on, until the term displayed on the calculator is sufficiently small.

Finally, recall the sum from the memory.)

If the series is to be evaluated using a computer, we would define the terms by $u_1 = x$ and the recurrence relation

$$u_n = \frac{-x^2}{(2n-2)(2n-1)} \times u_{n-1}$$

The series for $\sin x$ is true for all values of x, although when x is large, many terms of the series will be required.

For example, we can illustrate the periodicity of $\sin x$ by evaluating $\sin(2\pi + 0.2)$ and $\sin(4\pi + 0.2)$.

When $x = 2\pi + 0.2 = 6.4832$,

$$\sin(2\pi + 0.2) = 6.4832 - \frac{(6.4832)^3}{3!} + \ldots$$

$$\approx 0.1987 \quad \text{using 13 terms}$$

When $x = 4\pi + 0.2 = 12.7664$,

$$\sin(4\pi + 0.2) = 12.7664 - \frac{(12.7664)^3}{3!} + \ldots$$

$$\approx 0.1986 \quad \text{using 20 terms}$$

This demonstrates that $\sin(2\pi + 0.2)$ and $\sin(4\pi + 0.2)$ are equal to $\sin(0.2)$.

Similarly, the Maclaurin series for $\cos x$ is

$$\cos x = 1 - \frac{x^2}{2!} + \frac{x^4}{4!} - \frac{x^6}{6!} + \ldots$$

and this is also true for all values of x.

Example 3   Find the Maclaurin series for $\sec x$, up to the term in x^4, and hence find an approximate value for $\sec(0.05)$.

Let $f(x) = \sec x$ so $f(0) = 1$

$\qquad f'(x) = \sec x \tan x$ so $f'(0) = 0$

$\qquad f''(x) = \sec x \tan^2 x + \sec^3 x$ so $f''(0) = 1$

$\qquad f'''(x) = \sec x \tan^3 x + 5 \sec^3 x \tan x$ so $f'''(0) = 0$

$\qquad f^{iv}(x) = \sec x \tan^4 x + 18 \sec^3 x \tan^2 x + 5 \sec^5 x$ so $f^{iv}(0) = 5$

Hence the Maclaurin series is

$$\sec x = 1 + \frac{x^2}{2!}(1) + \frac{x^4}{4!}(5) + \ldots$$

$$= 1 + \tfrac{1}{2}x^2 + \frac{5}{24}x^4 + \ldots$$

Substituting $x = 0.05$,

$$\sec(0.05) = 1 + \tfrac{1}{2} \times (0.05)^2 + \frac{5}{24} \times (0.05)^4 + \ldots$$

$$= 1 + 0.001\,25 + 0.000\,001 + \ldots$$

$$\approx 1.001\,251$$

Example 4   Find the Maclaurin series for $(1 + x)^n$, where n need not be a positive integer.

Let $\quad f(x) = (1 + x)^n$ $\qquad\qquad$ so $\quad f(0) = 1$

$\qquad f'(x) = n(1 + x)^{n-1}$ $\qquad\quad$ so $\quad f'(0) = n$

$\qquad f''(x) = n(n - 1)(1 + x)^{n-2}$ $\qquad$ so $\quad f''(0) = n(n - 1)$

$\qquad f'''(x) = n(n - 1)(n - 2)(1 + x)^{n-3}$ so $\quad f'''(0) = n(n - 1)(n - 2)$

Hence the Maclaurin series is

$$(1 + x)^n = 1 + nx + \frac{n(n - 1)}{2!}x^2 + \frac{n(n - 1)(n - 2)}{3!}x^3 + \dots$$

This is called the **binomial series**, and it is true when $-1 < x < 1$.

For example,

$$(1 + x)^{\frac{1}{4}} = 1 + \tfrac{1}{4}x + \frac{(\frac{1}{4})(-\frac{3}{4})}{2!}x^2 + \frac{(\frac{1}{4})(-\frac{3}{4})(-\frac{7}{4})}{3!}x^3 + \dots$$

$$= 1 + \tfrac{1}{4}x - \frac{3}{32}x^2 + \frac{7}{125}x^3 + \dots$$

and $\quad (1 + x)^{-2} = 1 + (-2)x + \frac{(-2)(-3)}{2!}x^2 + \frac{(-2)(-3)(-4)}{3!}x^3 + \dots$

$$= 1 - 2x + 3x^2 - 4x^3 + \dots$$

Exercise 17.3   (Maclaurin series)

Find the Maclaurin series for the given functions, as far as the term indicated.

1   $\cos x$ (up to x^6). Hence evaluate $\cos(0.3)$.

2   $\tan x$ (up to x^3). Hence evaluate $\tan(0.02)$.

3   $(1 + x)^{\frac{3}{2}}$ (up to x^4).

4   $(1 + x)^{-\frac{2}{3}}$ (up to x^3).

5   $\sqrt{4 + x}$ (up to x^3). Hence evaluate $\sqrt{4.1}$.

6   $(3 - 2x)^{-2}$ (up to x^3).

7   $x \sin x$ (up to x^4). Verify that this series may also be obtained by multiplying the Maclaurin series for $\sin x$ by x.

8   $\sec^2 x$ (up to x^2). Verify that this series may also be obtained by differentiating the series for $\tan x$ found in question 2.

9   $\sin^2 x$ (up to x^4). Hence evaluate $\sin^2(0.25)$. **10** $\sin 2x$ (up to x^5).

17.4   Taylor's series

The Maclaurin series for a function $f(x)$ may only be valid when x is small. Even when the series *is* valid for large values of x it is not usually practical to use it to evaluate $f(x)$, as the convergence may be very slow.

So we shall consider the function $f(x)$ in the neighbourhood of a general point, say where $x = a$. (In practice we shall choose a value of a for which $f(a)$ is known.) We have seen that the best fitting polynomial of degree n to the function $f(x)$, in the neighbourhood of $x = a$, is

$$f(a) + (x - a)f'(a) + \frac{(x - a)^2}{2!}f''(a) + \dots + \frac{(x - a)^n}{n!}f^{(n)}(a)$$

Provided that x is close enough to a, we have

$$f(x) \approx f(a) + (x-a)f'(a) + \frac{(x-a)^2}{2!}f''(a) + \ldots + \frac{(x-a)^n}{n!}f^{(n)}(a)$$

and $f(x)$ is given by the sum of the infinite series

$$f(x) = f(a) + (x-a)f'(a) + \frac{(x-a)^2}{2!}f''(a) + \ldots$$

We write $x = a + h$, so that h is small (when x is close to a); then we have

$$\boxed{f(a+h) = f(a) + h\,f'(a) + \frac{h^2}{2!}f''(a) + \ldots}$$

This is the **Taylor series** (or expansion) for $f(a+h)$. If the value of $f(a)$ is known, Taylor's series can be used to evaluate $f(x)$ when x is close to a.

Example 1 Find the Taylor series for $\sin(\frac{1}{6}\pi + h)$ as far as the term in h^3 and hence evaluate $\sin 29°$, correct to 4 decimal places.

Let $\quad f(x) = \sin x \qquad$ so $\quad f\left(\frac{1}{6}\pi\right) = \frac{1}{2}$

$\qquad f'(x) = \cos x \qquad$ so $\quad f'\left(\frac{1}{6}\pi\right) = \frac{\sqrt{3}}{2}$

$\qquad f''(x) = -\sin x \quad$ so $\quad f''\left(\frac{1}{6}\pi\right) = -\frac{1}{2}$

$\qquad f'''(x) = -\cos x \quad$ so $\quad f'''\left(\frac{1}{6}\pi\right) = -\frac{\sqrt{3}}{2}$

Hence the Taylor series is

$$\sin\left(\frac{1}{6}\pi + h\right) = \frac{1}{2} + h\left(\frac{\sqrt{3}}{2}\right) + \frac{h^2}{2!}\left(-\frac{1}{2}\right) + \frac{h^3}{3!}\left(-\frac{\sqrt{3}}{2}\right) + \ldots$$

$$= \frac{1}{2} + \frac{\sqrt{3}}{2}h - \frac{1}{4}h^2 - \frac{\sqrt{3}}{12}h^3 + \ldots$$

Now $\quad 29° = 30° - 1° = \dfrac{\pi}{6} - \dfrac{\pi}{180}$ radians

so substituting $h = -\dfrac{\pi}{180}$,

$$\sin 29° = \frac{1}{2} + \frac{\sqrt{3}}{2}\times\left(-\frac{\pi}{180}\right) - \frac{1}{4}\times\left(-\frac{\pi}{180}\right)^2 - \frac{\sqrt{3}}{12}\times\left(-\frac{\pi}{180}\right)^3 + \ldots$$

$$= 0.5 - 0.015\,115 - 0.000\,076 + 0.000\,001 + \ldots$$

$$= 0.484\,810$$

$$= 0.4848 \quad \text{(to 4 decimal places)}$$

Example 2   Find the first four terms in the Taylor expansion of $\tan^{-1}(1+h)$, and hence find an approximation for $\tan^{-1}(1.1)$.

Let $f(x) = \tan^{-1}x$ so $f(1) = \frac{1}{4}\pi$

$$f'(x) = \frac{1}{1+x^2}$$ so $f'(1) = \frac{1}{2}$

$$f''(x) = \frac{-2x}{(1+x^2)^2}$$ so $f''(1) = -\frac{1}{2}$

$$f'''(x) = \frac{(1+x^2)^2(-2) + (2x)(2)(1+x^2)(2x)}{(1+x^2)^4}$$ so $f'''(1) = \frac{1}{2}$

Hence the Taylor series is

$$\tan^{-1}(1+h) = \frac{1}{4}\pi + h\left(\frac{1}{2}\right) + \frac{h^2}{2!}\left(-\frac{1}{2}\right) + \frac{h^3}{3!}\left(\frac{1}{2}\right) + \ldots$$

$$= \frac{1}{4}\pi + \frac{1}{2}h - \frac{1}{4}h^2 + \frac{1}{12}h^3 + \ldots$$

Putting $h = 0.1$,

$$\tan^{-1}(1.1) = \frac{1}{4}\pi + \frac{1}{2} \times 0.1 - \frac{1}{4} \times (0.1)^2 + \frac{1}{12} \times (0.1)^3 + \ldots$$

$$= 0.7854 + 0.05 - 0.0025 + 0.0001 + \ldots \approx 0.8330$$

The Maclaurin series is a special case of Taylor's series (with $a = 0$), and the term 'Taylor series' is often used to include both Maclaurin and Taylor series.

Exercise 17.4   (Taylor series)

Find the Taylor series for the following:

1   $\sin\left(\frac{1}{3}\pi + h\right)$ (as far as h^3). Hence evaluate $\sin 62°$.
2   $\cos\left(\frac{1}{4}\pi + h\right)$ (as far as h^3). Hence evaluate $\cos 44°$.
3   $\tan\left(\frac{1}{4}\pi + h\right)$ (as far as h^3).
4   $\sin\left(\frac{1}{2}\pi + h\right)$ (as far as h^6). Comment on the series you have obtained.
5   $\sec\left(\frac{1}{3}\pi + h\right)$ (as far as h^2).
6   $\sin^{-1}\left(\frac{1}{2} + h\right)$ (as far as h^2). Hence evaluate $\sin^{-1}(0.47)$.

17.5   Other methods for finding series

To obtain the Maclaurin series for $f(x)$ directly, we have to differentiate many times, and this is sometimes very laborious. Often we can use 'standard' series, such as

$$\sin x = x - \frac{x^3}{3!} + \frac{x^5}{5!} - \ldots$$

$$\cos x = 1 - \frac{x^2}{2!} + \frac{x^4}{4!} - \ldots$$

$$(1+x)^n = 1 + nx + \frac{n(n-1)}{2!}x^2 + \ldots$$

to derive power series expansions for more complicated functions. We assume that the methods we use are valid.

Example 1   Expand sin (x^3) as a power series.

We have $\sin x = x - \dfrac{x^3}{3!} + \dfrac{x^5}{5!} - \ldots$

and replacing x by x^3 we obtain

$$\sin(x^3) = x^3 - \frac{(x^3)^3}{3!} + \frac{(x^3)^5}{5!} - \ldots$$

$$= x^3 - \frac{x^9}{3!} + \frac{x^{15}}{5!} - \ldots$$

Example 2   Expand $\cos^2 x$ as a power series.

We have $\cos^2 x = \frac{1}{2}(1 + \cos 2x)$

and $\cos 2x = 1 - \dfrac{(2x)^2}{2!} + \dfrac{(2x)^4}{4!} - \ldots$

so $\cos^2 x = \frac{1}{2}\left(1 + 1 - \dfrac{4x^2}{2!} + \dfrac{16x^4}{4!} - \ldots\right)$

$$= 1 - \frac{2x^2}{2!} + \frac{8x^4}{4!} - \frac{32x^6}{6!} + \ldots$$

Example 3   Expand $\sqrt{1 - x^2}$ as a power series.

We have
$$(1 + x)^{\frac{1}{2}} = 1 + \tfrac{1}{2}x + \frac{(\frac{1}{2})(-\frac{1}{2})}{2!}x^2 + \ldots$$
and replacing x by $(-x^2)$ we obtain

$$(1 - x^2)^{\frac{1}{2}} = 1 + \tfrac{1}{2}(-x^2) + \frac{(\frac{1}{2})(-\frac{1}{2})}{2!}(-x^2)^2 + \frac{(\frac{1}{2})(-\frac{1}{2})(-\frac{3}{2})}{3!}(-x^2)^3 + \ldots$$

$$= 1 - \tfrac{1}{2}x^2 - \tfrac{1}{8}x^4 - \tfrac{1}{16}x^6 - \ldots$$

Example 4   Find the expansion of $\dfrac{\sin x}{(1 + x)^{\frac{5}{2}}}$ (up to the term in x^3).

We write this as

$$\sin x(1 + x)^{-\frac{5}{2}}$$

Now $\sin x = x - \dfrac{x^3}{3!} + \dfrac{x^5}{5!} - \ldots$

$$= x - \tfrac{1}{6}x^3 + \frac{1}{120}x^5 - \ldots$$

and $(1 + x)^{-\frac{5}{2}} = 1 + (-\tfrac{5}{2})x + \dfrac{(-\frac{5}{2})(-\frac{7}{2})}{2!}x^2 + \dfrac{(-\frac{5}{2})(-\frac{7}{2})(-\frac{9}{2})}{3!}x^3 + \ldots$

$$= 1 - \frac{5}{2}x + \frac{35}{8}x^2 - \frac{105}{16}x^3 + \ldots$$

Multiplying these two series together, we have

$$\sin x (1+x)^{-\frac{5}{2}} = \left(x - \frac{1}{6}x^3 + \frac{1}{120}x^5 - \cdots \right)\left(1 - \frac{5}{2}x + \frac{35}{8}x^2 - \frac{105}{16}x^3 + \cdots \right)$$

$$= x - \frac{5}{2}x^2 + \frac{35}{8}x^3 - \frac{1}{6}x^3 + \cdots$$

$$= x - \frac{5}{2}x^2 + \frac{101}{24}x^3 + \cdots$$

Example 5 Expand $\tan^{-1} x$ as a power series.

Since $\int \frac{1}{1+x^2} dx = \tan^{-1} x + C$

we shall expand $\frac{1}{1+x^2}$ using the binomial series, and then

integrate.

$$\frac{1}{1+x^2} = (1+x^2)^{-1} = 1 + (-1)(x^2) + \frac{(-1)(-2)}{2!}(x^2)^2 + \cdots$$

$$= 1 - x^2 + x^4 - x^6 + \cdots$$

Integrating,

$$\tan^{-1} x = C + x - \tfrac{1}{3}x^3 + \tfrac{1}{5}x^5 - \tfrac{1}{7}x^7 + \cdots$$

where C is the arbitrary constant of integration. Putting $x = 0$,

$$\tan^{-1} 0 = 0$$

and so $C = 0$

Hence $\tan^{-1} x = x - \tfrac{1}{3}x^3 + \tfrac{1}{5}x^5 - \tfrac{1}{7}x^7 + \cdots$

Standard series

The following are regarded as 'standard' series.

$$(1+x)^n = 1 + nx + \frac{n(n-1)}{2!}x^2 + \frac{n(n-1)(n-2)}{3!}x^3 + \cdots$$

$$\text{valid when } -1 < x < 1$$

$$\sin x = x - \frac{x^3}{3!} + \frac{x^5}{5!} - \frac{x^7}{7!} + \cdots \quad \text{valid for all } x$$

$$\cos x = 1 - \frac{x^2}{2!} + \frac{x^4}{4!} - \frac{x^6}{6!} + \cdots \quad \text{valid for all } x$$

$$\tan^{-1} x = x - \frac{x^3}{3} + \frac{x^5}{5} - \frac{x^7}{7} + \cdots \quad \text{valid when } -1 \leqslant x \leqslant 1$$

Evaluation of π

The series for $\tan^{-1}x$ can be used to evaluate π. Putting $x = 1$, we have $\tan^{-1}1 = \frac{1}{4}\pi$, and thus

$$\frac{1}{4}\pi = 1 - \frac{1}{3} + \frac{1}{5} - \frac{1}{7} + \ldots$$

However this series converges very slowly, and a large number of terms would be required to obtain a reasonable value. The result

$$\frac{1}{4}\pi = \tan^{-1}\frac{1}{2} + \tan^{-1}\frac{1}{3}$$

(see p. 144) gives a more practical method, since the series

$$\tan^{-1}\frac{1}{2} = \frac{1}{2} - \frac{1}{3}(\frac{1}{2})^3 + \frac{1}{5}(\frac{1}{2})^5 - \ldots$$

and $\quad \tan^{-1}\frac{1}{3} = \frac{1}{3} - \frac{1}{3}(\frac{1}{3})^3 + \frac{1}{5}(\frac{1}{3})^5 - \ldots$

converge fairly rapidly.

Other results which may be used are

$$\frac{1}{4}\pi = 2\tan^{-1}\frac{1}{3} + \tan^{-1}\frac{1}{7}$$

and $\quad \frac{1}{4}\pi = 4\tan^{-1}\frac{1}{5} - \tan^{-1}\frac{1}{239}$

Using differential equations

Suppose we have to find the Maclaurin series for $y = f(x)$. Rather than just differentiating y several times, it is sometimes easier to find a differential equation satisfied by y. We then differentiate this equation.

Example   If $y = (\sin^{-1}x)^2$, show that

$$(1 - x^2)\frac{d^2y}{dx^2} - x\frac{dy}{dx} = 2$$

Hence find the Maclaurin series for y, up to the term in x^4.

Let
$$y = f(x) = (\sin^{-1}x)^2 \qquad \text{When} \quad x = 0,\ y = 0$$
$$\text{i.e.} \qquad f(0) = 0$$

We have
$$\frac{dy}{dx} = \frac{2\sin^{-1}x}{\sqrt{1 - x^2}} \qquad \text{When} \quad x = 0,\ \frac{dy}{dx} = 0$$
$$\text{i.e.} \qquad f'(0) = 0$$

Thus
$$\sqrt{1 - x^2}\,\frac{dy}{dx} = 2\sin^{-1}x$$

and, differentiating,

$$\sqrt{1 - x^2}\,\frac{d^2y}{dx^2} - \frac{x}{\sqrt{1 - x^2}}\frac{dy}{dx} = \frac{2}{\sqrt{1 - x^2}}$$

Hence
$$(1-x^2)\frac{d^2y}{dx^2} - x\frac{dy}{dx} = 2$$

When $x = 0$, $\dfrac{d^2y}{dx^2} - 0 = 2$

i.e. $\dfrac{d^2y}{dx^2} = 2$

Differentiating,
$$(1-x^2)\frac{d^3y}{dx^3} - 2x\frac{d^2y}{dx^2} - x\frac{d^2y}{dx^2} - \frac{dy}{dx} = 0$$

i.e.
$$(1-x^2)\frac{d^3y}{dx^3} - 3x\frac{d^2y}{dx^2} - \frac{dy}{dx} = 0$$

When $x = 0$, $\dfrac{d^3y}{dx^3} - 0 - 0 = 0$

i.e. $\dfrac{d^3y}{dx^3} = 0$

Differentiating again,
$$(1-x^2)\frac{d^4y}{dx^4} - 2x\frac{d^3y}{dx^3} - 3x\frac{d^3y}{dx^3} - 3\frac{d^2y}{dx^2} - \frac{d^2y}{dx^2} = 0$$

i.e.
$$(1-x^2)\frac{d^4y}{dx^4} - 5x\frac{d^3y}{dx^3} - 4\frac{d^2y}{dx^2} = 0$$

When $x = 0$, $\dfrac{d^4y}{dx^4} - 0 - 4 \times 2 = 0$

i.e. $\dfrac{d^4y}{dx^4} = 8$

The Maclaurin series is therefore

$$y = 0 + x(0) + \frac{x^2}{2!}(2) + \frac{x^3}{3!}(0) + \frac{x^4}{4!}(8) + \cdots$$
$$= x^2 + \tfrac{1}{3}x^4 + \cdots$$

Exercise 17.5   (Power series)

In questions 1 to 11, find power series expansions for the given functions. Give the first three non-zero terms.

1   $\sin 2x$ 2 $\cos 3x$ 3 $\cos(x^2)$ 4 $\sin^2 x$
5   $\sin 3x \cos 4x$ [Use $2\cos A \sin B = \sin(A+B) - \sin(A-B)$]
6   $\cos 2x \cos 3x$ 7 $(1-x)^{-3}$ 8 $(1+x^2)^{-\frac{1}{2}}$
9   $(8-3x^2)^{2/3}$ [Write as $[8(1-\tfrac{3}{8}x^2)]^{2/3}$]   10 $\dfrac{\sin x}{\sqrt{1+x}}$ 11 $\dfrac{\cos}{(1-x^2)^{\frac{3}{2}}}$

12 Verify that the series for $\cos x$ may be obtained by differentiating the series

$$\sin x = x - \frac{x^3}{3!} + \frac{x^5}{5!} - \ldots$$

13 Use the binomial series to expand $\dfrac{1}{\sqrt{1-x^2}}$ as a power series, up to the term in x^6. Hence expand $\sin^{-1} x$ as a power series (up to the term in x^7). By putting $x = \frac{1}{2}$, obtain an approximate value for π.

14 Use the series for $\tan^{-1} x$ to evaluate $\tan^{-1} \frac{1}{2}$ and $\tan^{-1} \frac{1}{3}$, each correct to 5 decimal places. Using

$$\tfrac{1}{4}\pi = \tan^{-1} \tfrac{1}{2} + \tan^{-1} \tfrac{1}{3}$$

obtain an approximation for π.

15 If $y = \sqrt{\sec x + \tan x}$, show that $\cos x \dfrac{dy}{dx} = \frac{1}{2} y$. Hence find the Maclaurin series for y, up to the term in x^4, and use it to evaluate y when $x = 0.2$.

17.6 Power series for integrals

There are many functions which cannot be integrated exactly, for example $\sqrt{1+x^3}$ and $\dfrac{\sin x}{x}$. However we can express these as power series, and then integrate the series.

Example 1 Obtain a power series for the integral $\int \sqrt{1+x^3}\, dx$ and hence evaluate $\int_0^{0.5} \sqrt{1+x^3}\, dx$ correct to 4 decimal places.

Using the binomial series,

$$(1+x^3)^{1/2} = 1 + \tfrac{1}{2}x^3 + \frac{(\frac{1}{2})(-\frac{1}{2})}{2!}(x^3)^2 + \frac{(\frac{1}{2})(-\frac{1}{2})(-\frac{3}{2})}{3!}(x^3)^3 + \ldots$$

$$= 1 + \tfrac{1}{2}x^3 - \tfrac{1}{8}x^6 + \tfrac{1}{16}x^9 + \ldots$$

Thus $\displaystyle\int \sqrt{1+x^3}\, dx = C + x + \frac{1}{8}x^4 - \frac{1}{56}x^7 + \frac{1}{160}x^{10} + \ldots$

where C is the arbitrary constant of integration.

Thus $\displaystyle\int_0^{0.5} \sqrt{1+x^3}\, dx = \left[x + \frac{1}{8}x^4 - \frac{1}{56}x^7 + \frac{1}{160}x^{10} + \ldots \right]_0^{0.5}$

$$= 0.5 + \frac{1}{8} \times (0.5)^4 - \frac{1}{56} \times (0.5)^7 + \frac{1}{160} \times (0.5)^{10} + \ldots$$

$$= 0.5 + 0.007\,813 - 0.000\,140 + 0.000\,006 + \ldots$$
$$= 0.507\,679$$
$$= 0.5077 \quad \text{(to 4 decimal places)}$$

Example 2 Obtain a power series for the integral $\int \dfrac{\sin x}{x}\,dx$. (This

integral occurs in telecommunication theory.)

We have

$$\sin x = x - \frac{x^3}{3!} + \frac{x^5}{5!} - \cdots$$

so

$$\frac{\sin x}{x} = 1 - \frac{x^2}{3!} + \frac{x^4}{5!} - \cdots$$

Thus

$$\int \frac{\sin x}{x}\,dx = C + x - \frac{x^3}{3 \times 3!} + \frac{x^5}{5 \times 5!} - \cdots$$

Exercise 17.6   (Power series for integrals)

Obtain power series for the following integrals.

1 $\int \dfrac{1}{1+x}\,dx$ (up to the term in x^4). Hence evaluate $\int_0^{0.1} \dfrac{1}{1+x}\,dx$.

2 $\int \dfrac{1}{1-x^4}\,dx$ (up to x^9).   **3** $\int \sqrt[3]{1+x^2}\,dx$ (up to x^5).

4 $\int \sin(x^2)\,dx$ (up to x^{11}). Hence evaluate $\int_0^{0.5} \sin(x^2)\,dx$.

5 $\int \sec x\,dx$ (up to x^5). Hence evaluate $\int_0^{0.2} \sec x\,dx$.

17.7 Solution of differential equations

Suppose a function $y = f(x)$ satisfies a differential equation $\dfrac{dy}{dx} = g(x, y)$, and we are given the value of y when $x = a$, i.e. $f(a)$.

We can then use the differential equation to find the value of $\dfrac{dy}{dx}$

when $x = a$, i.e. $f'(a)$. By differentiating the equation, we can find the value of $\dfrac{d^2 y}{dx^2}$ when $x = a$, i.e. $f''(a)$. By differentiating again, we can obtain $f'''(a)$, and so on.

If we assume that the solution $y = f(a + h)$ can be expanded as an infinite Taylor series, we then have

$$y = f(a + h) = f(a) + hf'(a) + \frac{h^2}{2!} f''(a) + \cdots$$

and this can be used to obtain values of the solution.

Example 1   The function y satisfies the differential equation $\frac{dy}{dx} = 9.8 - y$, and $y = 0$ when $x = 0$. (This equation might occur in the study of resisted motion.) Find the Maclaurin expansion of y, and find the value of y when $x = 0.4$.

When $x = 0$ we are given that $y = 0$.

Also $\frac{dy}{dx} = 9.8 - y$

$$\text{When} \quad x = 0, \frac{dy}{dx} = 9.8$$

(since $y = 0$)

Differentiating,

$$\frac{d^2y}{dx^2} = -\frac{dy}{dx}$$

$$\text{When} \quad x = 0, \frac{d^2y}{dx^2} = -9.8$$

$$\left(\text{since} \ \frac{dy}{dx} = 9.8 \right)$$

$$\frac{d^3y}{dx^3} = -\frac{d^2y}{dx^2}$$

$$\text{When} \quad x = 0, \frac{d^3y}{dx^3} = 9.8$$

$$\frac{d^4y}{dx^4} = -\frac{d^3y}{dx^3}$$

$$\text{When} \quad x = 0, \frac{d^4y}{dx^4} = -9.8$$

and so on.

Hence the Maclaurin series for y is

$$y = 0 + x(9.8) + \frac{x^2}{2!}(-9.8) + \frac{x^3}{3!}(9.8) + \frac{x^4}{4!}(-9.8) + \dots$$

$$= 9.8x - 9.8\frac{x^2}{2!} + 9.8\frac{x^3}{3!} - 9.8\frac{x^4}{4!} + \dots$$

When $x = 0.4$,

$$y = 9.8 \times 0.4 - 9.8 \times \frac{(0.4)^2}{2} + 9.8 \times \frac{(0.4)^3}{6} - 9.8 \times \frac{(0.4)^4}{24} + \dots$$

$$= 3.92 - 0.784 + 0.10453 - 0.01045 + \dots$$
$$\approx 3.23$$

Example 2 y satisfies the differential equation $\dfrac{dy}{dx} = 2xy$, and
$y = 1$ when $x = 1$. Find a Taylor series for y, and use it to evaluate
y when $x = 1.1$.

Let $y = f(x)$.

$$\text{When} \quad x = 1 \text{ we are given that } y = 1.$$

Also

$$\frac{dy}{dx} = 2xy$$

$$\text{When} \quad x = 1, \quad \frac{dy}{dx} = 2 \times 1 \times 1 = 2$$
$$(\text{since } y = 1)$$

Differentiating,

$$\frac{d^2y}{dx^2} = 2y + 2x\frac{dy}{dx}$$

$$\text{When} \quad x = 1, \quad \frac{d^2y}{dx^2} = 2 \times 1 + 2 \times 1 \times 2 = 6$$
$$\left(\text{since } y = 1, \frac{dy}{dx} = 2\right)$$

$$\frac{d^3y}{dx^3} = 4\frac{dy}{dx} + 2x\frac{d^2y}{dx^2}$$

$$\text{When} \quad x = 1, \quad \frac{d^3y}{dx^3} = 4 \times 2 + 2 \times 1 \times 6 = 20$$

$$\frac{d^4y}{dx^4} = 6\frac{d^2y}{dx^2} + 2x\frac{d^3y}{dx^3}$$

$$\text{When} \quad x = 1, \quad \frac{d^4y}{dx^4} = 6 \times 6 + 2 \times 1 \times 20 = 76$$

So the Taylor series for $y = f(1 + h)$ is

$$y = f(1 + h) = 1 + h(2) + \frac{h^2}{2!}(6) + \frac{h^3}{3!}(20) + \frac{h^4}{4!}(76) + \dots$$

$$= 1 + 2h + 3h^2 + \frac{10}{3}h^3 + \frac{19}{6}h^4 + \dots$$

Putting $h = 0.1$, the value of y when $x = 1.1$ is

$$y = 1 + 2 \times 0.1 + 3 \times (0.1)^2 + \frac{10}{3} \times (0.1)^3 + \frac{19}{6} \times (0.1)^4 + \dots$$
$$= 1 + 0.2 + 0.03 + 0.0033 + 0.0003 + \dots$$
$$\approx 1.234$$

Exercise 17.7 (Differential equations, using Taylor's series)

1 The function y satisfies the differential equation $\dfrac{dy}{dx} = 2y$, and $y = 1$ when $x = 0$. Find the Maclaurin expansion of y, and find the value of y when $x = 0.1$.

2 The function y satisfies the differential equation $\dfrac{dy}{dx} = 3 - \frac{1}{2}y$, and $y = 0$ when $x = 0$. Find the Maclaurin expansion of y, and find the value of y when $x = -0.2$.

3 The function y satisfies the differential equation $\dfrac{dy}{dx} = x^2 + y$, and $y = 2$ when $x = 1$. Find the Taylor expansion of y, and use it to evaluate y when $x = 1.25$.

4 The function y satisfies the differential equation $\dfrac{dy}{dx} = x^2 y$, and $y = 1$ when $x = 1$. Find the Taylor expansion of y, and evaluate y when $x = 1.1$.

5 The function y satisfies the differential equation $\dfrac{dy}{dx} = x - y^2$ and $y = 1$ when $x = 0$. Find the Maclaurin expansion of y, and evaluate y when $x = 0.05$.

17.8 L'Hôpital's rule for limits

Suppose we need to find the limiting value of a quotient, say $\lim\limits_{x \to a} \dfrac{f(x)}{g(x)}$, when $f(a) = g(a) = 0$. The Taylor series for $f(x)$ and $g(x)$, in the neighbourhood of $x = a$, are

$$f(a+h) = hf'(a) + \tfrac{1}{2}h^2 f''(a) + \ldots \quad (\text{since } f(a) = 0)$$

and $\quad g(a+h) = hg'(a) + \tfrac{1}{2}h^2 g''(a) + \ldots \quad (\text{since } g(a) = 0)$

Then $\quad \lim\limits_{x \to a} \dfrac{f(x)}{g(x)} = \lim\limits_{h \to 0} \dfrac{f(a+h)}{g(a+h)}$

$$= \lim\limits_{h \to 0} \frac{hf'(a) + \tfrac{1}{2}h^2 f''(a) + \ldots}{hg'(a) + \tfrac{1}{2}h^2 g''(a) + \ldots}$$

$$= \lim\limits_{h \to 0} \frac{f'(a) + \tfrac{1}{2}hf''(a) + \ldots}{g'(a) + \tfrac{1}{2}hg''(a) + \ldots}$$

$$= \frac{f'(a)}{g'(a)} \quad \text{provided that } g'(a) \neq 0$$

This is **L'Hôpital's rule**.

If $f'(a) = g'(a) = 0$, then

$$\lim\limits_{x \to a} \frac{f(x)}{g(x)} = \lim\limits_{h \to 0} \frac{\tfrac{1}{2}h^2 f''(a) + \tfrac{1}{6}h^3 f'''(a) + \ldots}{\tfrac{1}{2}h^2 g''(a) + \tfrac{1}{6}h^3 g'''(a) + \ldots}$$

$$= \lim\limits_{h \to 0} \frac{f''(a) + 2hf'''(a) + \ldots}{g''(a) + 2hg'''(a) + \ldots}$$

$$= \frac{f''(a)}{g''(a)}$$

and so on.

$\left[\text{If } f'(a) = 0 \text{ and } g'(a) \neq 0, \text{ then } \lim_{x \to a} \dfrac{f(x)}{g(x)} = 0. \right.$

If $f'(a) \neq 0$ and $g'(a) = 0$, then as $x \to a$, $\dfrac{f(x)}{g(x)}$ behaves like

$\dfrac{f'(a)}{\frac{1}{2}(x-a)g''(a)}$, which becomes arbitrarily large.$]$

Example 1 Evaluate $\lim\limits_{x \to 2} \dfrac{x^5 - 32}{x^2 - 4}$.

Let $f(x) = x^5 - 32$, so $f(2) = 0$ and $g(x) = x^2 - 4$, so $g(2) = 0$
  $f'(x) = 5x^4$,   so $f'(2) = 80$ and $g'(x) = 2x$,   so $g'(2) = 4$

Hence $\lim\limits_{x \to 2} \dfrac{x^5 - 32}{x^2 - 4} = \dfrac{f'(2)}{g'(2)} = \dfrac{80}{4} = 20$

Example 2 Evaluate $\lim\limits_{x \to 0} \dfrac{x^2}{1 - \cos 3x}$.

Let $f(x) = x^2$, so $f(0) = 0$ and $g(x) = 1 - \cos 3x$, so $g(0) = 0$
  $f'(x) = 2x$, so $f'(0) = 0$ and $g'(x) = 3 \sin 3x$, so $g'(0) = 0$
  $f''(x) = 2$, so $f''(0) = 2$ and $g''(x) = 9 \cos 3x$, so $g''(0) = 9$

Hence $\lim\limits_{x \to 0} \dfrac{x^2}{1 - \cos 3x} = \dfrac{f''(0)}{g''(0)} = \dfrac{2}{9}$

Exercise 17.8 (L'Hôpital's rule)

Use L'Hôpital's rule to evaluate the following limits.

1 $\lim\limits_{x \to 3} \dfrac{x^4 - 81}{x^2 - 9}$ 2 $\lim\limits_{x \to -1} \dfrac{2x^2 - x - 3}{x^5 + 1}$ 3 $\lim\limits_{x \to 4} \dfrac{\sqrt{x} - 2}{x - 4}$ 4 $\lim\limits_{x \to 1} \dfrac{x + 1 - 2\sqrt{x}}{(x^4 - 1)^2}$

5 $\lim\limits_{x \to 0} \dfrac{\sin 2x}{x}$ 6 $\lim\limits_{x \to 0} \dfrac{x}{\sin 2x + \tan 3x}$ 7 $\lim\limits_{x \to 0} \dfrac{1 - \cos x}{x^2}$ 8 $\lim\limits_{x \to \pi} \dfrac{\sin x}{x^2 - \pi^2}$

9 $\lim\limits_{x \to 0} \dfrac{\tan x - x}{x - \sin x}$ 10 $\lim\limits_{x \to \pi} \dfrac{x + \sin x - \pi}{(x - \pi)^3}$

17.9 The accuracy of numerical methods

Most numerical methods involve some form of polynomial approximation. For example, in Simpson's rule for the evaluation of a definite integral $\int_a^b f(x)\,dx$, sections of the curve $y = f(x)$ are replaced by quadratic curves. In Newton's method, the next approximation to a root of the equation $f(x) = 0$ is found by replacing the curve $y = f(x)$ by its tangent (which is the linear approximation). Taylor series are useful in the theoretical study of numerical methods, and in the investigation of the errors involved.

In the following, we consider a function $f(x)$ in the neighbourhood of $x = a$. Suppose that the Taylor series is

$$f(a + h) = b_0 + b_1 h + b_2 h^2 + b_3 h^3 + \ldots$$

so that $b_0 = f(a), b_1 = f'(a), b_2 = \frac{1}{2} f''(a), b_3 = \dfrac{1}{3!} f'''(a)$, and so on.

Numerical differentiation

Taking the first 3 terms of the Taylor series, we have

$$f(a + h) \approx b_0 + b_1 h + b_2 h^2$$

Replacing h by $-h$,

$$f(a - h) \approx b_0 - b_1 h + b_2 h^2$$

Subtracting,

$$f(a + h) - f(a - h) \approx 2b_1 h$$

thus $$f'(a) = b_1 \approx \frac{f(a + h) - f(a - h)}{2h}$$

Adding the two equations,

$$f(a + h) + f(a - h) \approx 2b_0 + 2b_2 h^2$$

thus $$f''(a) = 2b_2$$

$$\approx \frac{f(a + h) - 2b_0 + f(a - h)}{h^2}$$

$$= \frac{f(a + h) - 2f(a) + f(a - h)}{h^2}$$

These are the formulae which we have already used for numerical differentiation (see p. 284). We now investigate the errors involved in these formulae.

First consider

$$f'(a) \approx \frac{f(a + h) - f(a - h)}{2h}$$

We have

$$\frac{f(a + h) - f(a - h)}{2h} = \frac{1}{2h} [(b_0 + b_1 h + b_2 h^2 + b_3 h^3 + \ldots) - (b_0 - b_1 h + b_2 h^2 - b_3 h^3 + \ldots)]$$

$$= b_1 + b_3 h^2 + b_5 h^4 + \ldots$$

whereas the true value is

$$f'(a) = b_1$$

Hence the error is

$$b_3 h^2 + b_5 h^4 + \ldots$$

Provided $b_3 \neq 0$, and h is sufficiently small, the term $b_3 h^2$ is larger than the others, and so the error is approximately

$$b_3 h^2 = \frac{h^2}{6} f'''(a)$$

The error is proportional to h^2; if the value of h is halved, the error is divided by 4, and so on.

This is the error due to the use of the formula, and we have assumed that the values of $f(a+h)$ and $f(a-h)$ are known exactly. In practice this is unlikely; if the values are found using tables or a calculator, they will contain rounding errors, and if they are obtained from an experiment, they will also contain errors.

If $f(a+h)$ and $f(a-h)$ are each subject to an error $\pm \varepsilon$, then the maximum error in the calculated value

$$f'(a) \approx \frac{f(a+h)-f(a-h)}{2h}$$

resulting from this error is

$$\frac{\varepsilon+\varepsilon}{2h} = \frac{\varepsilon}{h}$$

As h decreases, this error increases, and this is the main weakness of numerical differentiation.

Now consider

$$f''(a) \approx \frac{f(a+h)-2f(a)+f(a-h)}{h^2}$$

We have

$$\frac{f(a+h)-2f(a)+f(a-h)}{h^2} = \frac{1}{h^2}[(b_0+b_1 h+b_2 h^2+ \ldots)-2b_0+(b_0-b_1 h+b_2 h^2- \ldots)]$$
$$= 2b_2+2b_4 h^2+2b_6 h^4+ \ldots$$

whereas the true value is

$$f''(a) = 2b_2$$

Hence the error is

$$2b_4 h^2 + 2b_6 h^4 + \ldots$$

which (provided $b_4 \neq 0$) is approximately

$$2b_4 h^2 = \frac{h^2}{12} f^{iv}(a)$$

Again the error due to the use of the formula is proportional to h^2. However, if the values $f(a+h), f(a)$ and $f(a-h)$ are each subject to an error $\pm \varepsilon$, then the maximum resulting error in $f''(a)$ is

$$\frac{\varepsilon+2\varepsilon+\varepsilon}{h^2} = \frac{4\varepsilon}{h^2}$$

(which is even 'worse' than that for $f'(a)$).

Trapezium rule

To evaluate a definite integral $\int_p^q f(x)\,dx$, the area under the curve is divided into n strips of width h $\left(\text{so that } h = \dfrac{q-p}{n}\right)$. In Fig. 17.3, a typical strip, say between $x = a$ and $x = a+h$, is replaced by a trapezium, and its area is calculated as

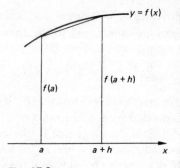

Fig. 17.3

$$\tfrac{1}{2}h[f(a)+f(a+h)] = \tfrac{1}{2}h[b_0 + b_0 + b_1 h + b_2 h^2 + \ldots]$$
$$= b_0 h + \tfrac{1}{2}b_1 h^2 + \tfrac{1}{2}b_2 h^3 + \ldots$$

whereas its true area is

$$\int_a^{a+h} f(x)\,dx = \int_a^{a+h} [b_0 + b_1(x-a) + b_2(x-a)^2 + \ldots]\,dx$$

$$= \left[b_0 x + \tfrac{1}{2}b_1(x-a)^2 + \tfrac{1}{3}b_2(x-a)^3 + \ldots \right]_a^{a+h}$$

$$= b_0 h + \tfrac{1}{2}b_1 h^2 + \tfrac{1}{3}b_2 h^3 + \tfrac{1}{4}b_3 h^4 + \ldots$$

Hence the error is

$$\tfrac{1}{6}b_2 h^3 + \tfrac{1}{4}b_3 h^4 + \ldots \approx \tfrac{1}{6}b_2 h^3$$
$$= \frac{h^3}{12} f''(a)$$

Since there are n strips, the total error is less than

$$\frac{nh^3}{12}M$$

where M is the maximum (numerical) value of $f''(x)$ over the interval $p \leqslant x \leqslant q$; and

$$\frac{nh^3}{12}M = \frac{(q-p)}{h}\frac{h^3}{12}M$$
$$= \frac{h^2(q-p)M}{12}$$

The error due to the use of the trapezium rule

$$\int_p^q f(x)\,dx \approx h(\tfrac{1}{2}y_0 + y_1 + y_2 + \ldots + y_{n-1} + \tfrac{1}{2}y_n)$$

is proportional to h^2.

If the values $y_0, y_1, \ldots y_n$ are each subject to an error $\pm\varepsilon$, there will be an additional error, which is less than

$$h(\tfrac{1}{2}\varepsilon + \varepsilon + \varepsilon + \ldots + \varepsilon + \tfrac{1}{2}\varepsilon) = hn\varepsilon$$
$$= \varepsilon(q-p)$$

Notice that this is not affected by the value of h, and so (unlike numerical differentiation) h may be taken as small as is necessary, without introducing large errors from this source.

Simpson's rule

To evaluate $\int_p^q f(x)\,dx$, the area is now divided into n strips (where n is even) of width $h = \dfrac{q-p}{n}$. The area of a typical pair of strips, say between $x = a - h$ and $x = a + h$ (see Fig. 17.4) is calculated as

$$\tfrac{1}{3}h[f(a-h) + 4f(a) + f(a+h)]$$
$$= \tfrac{1}{3}h[(b_0 - b_1 h + b_2 h^2 + \ldots) + 4b_0 + (b_0 + b_1 h + b_2 h^2 + \ldots)]$$
$$= 2b_0 h + \tfrac{2}{3}b_2 h^3 + \tfrac{2}{3}b_4 h^5 + \ldots$$

whereas the true value is

$$\int_{a-h}^{a+h} f(x)\,dx = \int_{a-h}^{a+h} [b_0 + b_1(x-a) + b_2(x-a)^2 + \ldots]\,dx$$
$$= \left[b_0 x + \tfrac{1}{2}b_1(x-a)^2 + \tfrac{1}{3}b_2(x-a)^3 + \ldots \right]_{a-h}^{a+h}$$
$$= 2b_0 h + \tfrac{2}{3}b_2 h^3 + \tfrac{2}{5}b_4 h^5 + \ldots$$

Hence the error is

$$\frac{4}{15}b_4 h^5 + \frac{8}{21}b_6 h^7 + \ldots \approx \frac{4}{15}b_4 h^5 = \frac{h^5}{90}f^{iv}(a)$$

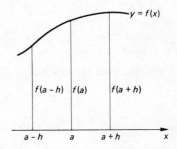

Fig. 17.4

There are $\tfrac{1}{2}n$ pairs of strips, and so the total error is less than

$$(\tfrac{1}{2}n)\frac{h^5}{90}M$$

where M is the maximum (numerical) value of $f^{iv}(x)$ over the interval $p \leqslant x \leqslant q$; and

$$(\tfrac{1}{2}n)\frac{h^5}{90}M = \tfrac{1}{2}\frac{(q-p)}{h}\frac{h^5}{90}M = \frac{h^4(q-p)M}{180}$$

This error is (approximately) proportional to h^4; if the value of h is halved, the error is divided by 16, and so on.

Again, if the ordinates are each subject to an error $\pm\varepsilon$, there is an additional error in

$$\int_p^q f(x)\,dx \approx \tfrac{1}{3}h(y_0 + 4y_1 + 2y_2 + \ldots + 4y_{n-1} + y_n)$$

which is less than

$$\tfrac{1}{3}h(\varepsilon + 4\varepsilon + 2\varepsilon + 4\varepsilon + \ldots + 4\varepsilon + \varepsilon) = \tfrac{1}{3}h(3n\varepsilon) = \varepsilon(q-p)$$

Example Values of a function $f(x)$ are tabulated, correct to 4 decimal places, for $x = 1.00, 1.01, 1.02, \ldots, 4.00$, and it is known that $|f^{iv}(x)| \leqslant 5$ whenever $1 \leqslant x \leqslant 4$. Simpson's rule is to be used to find a value for $I = \displaystyle\int_1^4 f(x)\,dx$. What is the maximum error in I due to rounding errors in the values of $f(x)$?

Suggest a strip width which could be used, if the error due to the use of Simpson's rule (the 'truncation error') is to be less than the maximum rounding error.

Each value of $f(x)$ is correct to 4 decimal places, and is therefore subject to a rounding error of up to ± 0.00005. The maximum rounding error in I, given by the formula $\varepsilon(q-p)$, is $0.00005 \times (4-1) = 0.00015$ (whatever strip width is used). If we use Simpson's rule with strip width h, the maximum truncation error is

$$\frac{h^4(q-p)\max|f^{iv}(x)|}{180} = \frac{h^4 \times 3 \times 5}{180}$$

$$= \frac{1}{12}h^4$$

This is less than the maximum rounding error when

$$\frac{1}{12}h^4 < 0.00015$$

$$h^4 < 0.0018$$
$$h < 0.206$$

The strip width must be a multiple of 0.01, and it must give an even number of strips, so we may take $h = 0.15$ (giving 20 strips between $x = 1$ and $x = 4$).

Convergence of iterative processes; Newton's method

An iterative process generates a sequence $x_1, x_2, x_3, \ldots$ where x_{n+1} is calculated as a function of x_n, say $x_{n+1} = G(x_n)$. If the sequence converges to a limit a, then we have $G(a) = a$ (since

$G(a) = \lim_{n \to \infty} G(x_n) = \lim_{n \to \infty} x_{n+1} = a$), and the Taylor series for $G(a+h)$ is

$$G(a+h) = a + b_1 h + b_2 h^2 + \ldots$$

where $b_1 = G'(a)$, $b_2 = \frac{1}{2}G''(a), \ldots$

Suppose the difference between x_n and the limit a is e_n (we call this the error in x_n). Then

$$x_n = a + e_n$$

$$x_{n+1} = a + e_{n+1}$$

and so on.
Since
$$x_{n+1} = G(x_n)$$

we have

$$a + e_{n+1} = G(a + e_n)$$
$$= a + b_1 e_n + b_2 e_n^2 + \ldots$$
$$e_{n+1} = b_1 e_n + b_2 e_n^2 + b_3 e_n^3 + \ldots$$

If $b_1 \neq 0$, we have (provided that e_n is sufficiently small)

$$e_{n+1} \approx b_1 e_n$$

At each stage the error is multiplied by a constant. This is **first order convergence**, and for the sequence to converge, we need $-1 < b_1 < 1$, i.e. $-1 < G'(a) < 1$.

If $b_1 = 0$ (i.e. $G'(a) = 0$), we have

$$e_{n+1} \approx b_2 e_n^2$$

This is **second order convergence**; the error is proportional to the square of the error at the previous step, and this gives more rapid convergence than first order convergence. The sequence always converges (provided x_1 is close enough to a). Here we have assumed that $b_2 \neq 0$. If $b_1 = b_2 = 0$ and $b_3 \neq 0$, we have third order convergence, and so on.

Newton's method for solving the equation $f(x) = 0$ uses the formula

$$x_{n+1} = x_n - \frac{f(x_n)}{f'(x_n)}$$

Thus

$$G(x) = x - \frac{f(x)}{f'(x)}$$

The sequence $x_1, x_2, x_3, \ldots$ converges to a root a, where $f(a) = 0$.

$$G'(x) = 1 - \frac{f'(x)f'(x) - f(x)f''(x)}{[f'(x)]^2}$$

$$= \frac{f(x)f''(x)}{[f'(x)]^2}$$

so $\quad G'(a) = \frac{f(a)f''(a)}{[f'(a)]^2} = 0 \quad$ since $f(a) = 0$

$$\text{(provided that } f'(a) \neq 0)$$

and $G''(x) = \dfrac{d}{dx}\left\{f(x)f''(x)[f'(x)]^{-2}\right\}$

$$= f'(x)f''(x)[f'(x)]^{-2} + f(x)\frac{d}{dx}\left\{f''(x)[f'(x)]^{-2}\right\}$$

so $\quad G''(a) = \dfrac{f''(a)}{f'(a)}$

Hence Newton's method gives second order convergence (provided that $f'(a) \neq 0$), and

$$e_{n+1} \approx \frac{f''(a)}{2f'(a)} e_n^2$$

18

The exponential function

18.1 The exponential function

'Exponent' is another word for a power; for example, in x^3, the exponent is 3. An exponential function of x is one in which x appears as a power; examples are 2^x, 3^x, and so on.

Consider the function $y = 2^x$. We first construct a table of values $\left(\text{remember that, for example, } 2^{-3} = \dfrac{1}{2^3} = \dfrac{1}{8}\right)$.

x	-3	-2	-1	0	1	2	3
$y = 2^x$	0.125	0.25	0.5	1	2	4	8

Using this table, we can draw the curve $y = 2^x$ (Fig. 18.1).

Other values of 2^x may be obtained using the 'y^x' key on a calculator. Use your calculator to find $2^{1.4}$ (the answer should be 2.639).

We can find the gradient of the curve $y = 2^x$ numerically, by taking an x-step of 0.01 to each side of the required point. For example, when $x = 1$, the points on each side are $(0.99, 2^{0.99})$ and $(1.01, 2^{1.01})$, and the gradient is approximately

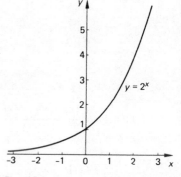

Fig. 18.1

$$\frac{2^{1.01} - 2^{0.99}}{1.01 - 0.99} = \frac{2.0139 - 1.9862}{0.02} = 1.386$$

Repeating this process for other values of x, we obtain

x	-3	-2	-1	0	1	2	3
$\dfrac{dy}{dx}$ (approx.)	0.087	0.173	0.347	0.693	1.386	2.773	5.545

We use these values to draw a graph of the gradient function (Fig. 18.2).

This is similar in shape to the original curve $y = 2^x$. A close inspection of the tables shows that, for each value of x,

$$\frac{dy}{dx} \approx 0.693 \times 2^x$$

(for example, when $x = 2$, $2.773 \approx 0.693 \times 4$).

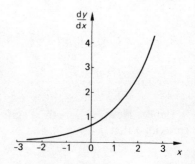

Fig. 18.2

Differentiation from first principles

We now consider the differentiation of $y = 2^x$ from first principles.
If δx and δy are corresponding changes in x and y, then

$$y + \delta y = 2^{x + \delta x}$$

Subtracting,

$$\delta y = 2^{x + \delta x} - 2^x$$
$$= 2^x \times 2^{\delta x} - 2^x$$
$$= 2^x(2^{\delta x} - 1)$$

Dividing by δx,

$$\frac{\delta y}{\delta x} = 2^x \left(\frac{2^{\delta x} - 1}{\delta x} \right)$$

As $\delta x \to 0$,

$$\frac{dy}{dx} = 2^x \lim_{\delta x \to 0} \left(\frac{2^{\delta x} - 1}{\delta x} \right)$$

$$= k \times 2^x$$

where

$$k = \lim_{\delta x \to 0} \left(\frac{2^{\delta x} - 1}{\delta x} \right) \qquad \text{This value does not depend on } x.$$

$\left(\text{Note that } k \text{ is the value of } \dfrac{dy}{dx} \text{ when } x = 0. \right)$ We have confirmed

that if $y = 2^x$, then $\dfrac{dy}{dx}$ is a multiple of 2^x.

We can use a calculator to find the value of $k = \lim\limits_{\delta x \to 0} \left(\dfrac{2^{\delta x} - 1}{\delta x} \right)$.

We take values of δx close to zero; for example, when $\delta x = 0.1$,

$$\frac{2^{\delta x} - 1}{\delta x} = \frac{2^{0.1} - 1}{0.1} = \frac{1.0718 - 1}{0.1} = 0.718$$

We obtain the following values.

δx	0.1	0.01	0.001	0.0001	-0.01	-0.001	-0.0001
$\dfrac{2^{\delta x} - 1}{\delta x}$	0.718	0.696	0.693	0.693	0.691	0.693	0.693

Thus $k = 0.693$ (correct to 3 decimal places), and if $y = 2^x$, then
$\dfrac{dy}{dx} = 0.693 \times 2^x$.

Similarly, if $y = 3^x$ then $\dfrac{dy}{dx} = k \times 3^x$, where

$$k = \lim_{\delta x \to 0} \left(\frac{3^{\delta x} - 1}{\delta x} \right)$$

In this case we find that $k = 1.099$ (correct to 3 decimal places).

The number e

More generally, suppose that a is a positive constant.

If $y = a^x$ then $\dfrac{dy}{dx} = k \times a^x$, where $k = \lim\limits_{\delta x \to 0}\left(\dfrac{a^{\delta x} - 1}{\delta x}\right)$.

When $a = 2$, we have $k = 0.693$.
When $a = 3$, we have $k = 1.099$.

It is plausible that for some number between 2 and 3, we would have $k = 1$. This number is always written e. Thus e is defined to be the number for which

$$\lim_{\delta x \to 0}\left(\frac{e^{\delta x} - 1}{\delta x}\right) = 1$$

If $y = e^x$ then $\dfrac{dy}{dx} = 1 \times e^x = e^x$.

Thus the derived function is the same as the original function.

The exponential function e^x may also be written $\exp x$.

Values of e^x can be found directly on a calculator, using the e^x key. (On some calculators we use 'inverse $\ln x$'. We shall see the reason for this in Chapter 19.) Use your calculator to find $e^{1.8}$ and $e^{-0.76}$ (the answers should be 6.0496 and 0.4677).

Exercise 18.1 (Exponential functions: numerical investigation)

1 Tabulate values of 3^x for $x = -2, -1.5, -1, \ldots, +2$, and draw a graph of the curve $y = 3^x$.

Find approximate values of $\dfrac{dy}{dx}$ numerically (by taking an x-step of 0.01 to each side) when $x = -2, -1, 0, 1, 2$. Show that $\dfrac{dy}{dx} \approx k \times 3^x$, and estimate k from your tables of values.

2 Working from first principles, show that if $y = 3^x$ then $\dfrac{dy}{dx} = k \times 3^x$, and give an expression for k (involving a limit). Use a calculator to evaluate k correct to 3 decimal places.

3 Evaluate

$$\lim_{\delta x \to 0}\left(\frac{a^{\delta x} - 1}{\delta x}\right)$$

(i) when $a = 2.7$; (ii) when $a = 2.8$. Deduce that the decimal expansion of e begins 2.7 . . . Use a similar method to find the next figure.

18.2   Sketching exponential functions

The curve $y = e^x$ (Fig. 18.3) is similar to $y = 2^x$ and $y = 3^x$. Notice that

i) e^x is always positive;
ii) e^x is an increasing function;
iii) $e^x \to 0$ as $x \to -\infty$ and $e^x \to +\infty$ as $x \to +\infty$;
iv) $e^x < 1$ when $x < 0$, $e^0 = 1$, and $e^x > 1$ when $x > 0$.

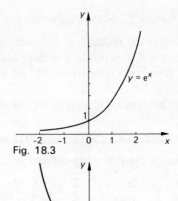

Fig. 18.3

Example 1   Sketch the curve $y = e^{-x}$.

We have

$$y = e^{-x} = \frac{1}{e^x}$$

Alternatively, this is the reflection in the y-axis of the curve $y = e^x$ (since x has been replaced by $-x$).
The curve is shown in Fig. 18.4.

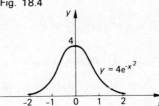

Fig. 18.4

Example 2   Sketch the curve $y = 4e^{-x^2}$.

When $x = 0$, $y = 4e^0 = 4$. Otherwise, $-x^2$ is always negative, so $e^{-x^2} < 1$, and $y < 4$.
As $x \to \pm\infty$, $-x^2 \to -\infty$, and so $y \to 0$.
The curve is shown in Fig. 18.5.

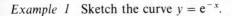

Fig. 18.5

Example 3   Sketch the curve $y = 4 - 3e^{-2x}$.

When $x = 0$, $y = 4 - 3 = 1$.
As $x \to +\infty$, $e^{-2x} \to 0$, so $y \to 4$. Since e^{-2x} is always positive, y is always less than 4. As $x \to -\infty$, $e^{-2x} \to +\infty$, so $y \to -\infty$.
The curve is shown in Fig. 18.6.

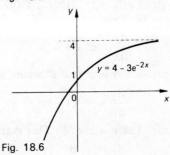

Fig. 18.6

Example 4   Sketch the curve $y = e^{\sin x}$

When $x = 0, \pi, 2\pi, \ldots$, we have $\sin x = 0$ and so $y = 1$.
When $0 < x < \pi$, $\sin x > 0$, and so $y > 1$.
When $\pi < x < 2\pi$, $\sin x < 0$, and so $0 < y < 1$.
Since $\sin x$ (Fig. 18.7) is between -1 and $+1$, y is always between e^{-1} and e.
This part of the curve (for $0 \leqslant x \leqslant 2\pi$) is repeated at intervals of 2π.
The curve is shown in Fig. 18.8.

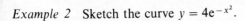

Fig. 18.7

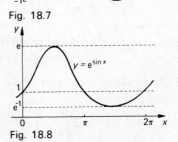

Fig. 18.8

Example 5   Sketch the curve $y = e^{-x} \sin x$.

This is the product of e^{-x} and $\sin x$ (see Fig. 18.9). $y = 0$ when $\sin x = 0$, y is positive or negative in the same regions as $\sin x$.

Since $\sin x$ is between -1 and $+1$, y always lies between $-e^{-x}$ and e^{-x}.

The curve is shown in Fig. 18.10.

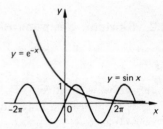

Fig. 18.9

Exercise 18.2   (Sketching exponential functions)

Sketch the following curves.

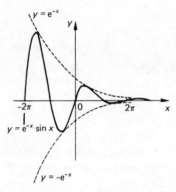

1 $y = 2e^x$	**2** $y = e^{-3x}$	**3** $y = e^{x-2}$
4 $y = 3 + e^{-x}$	**5** $y = 5 - 3e^{2x}$	**6** $y = e^{x^2}$
7 $y = e^{x^2 - 2x}$	**8** $y = e^{-\cos x}$	**9** $y = e^{-\tan^2 x}$
10 $y = e^x \cos x$	**11** $y = \sin(\pi e^{-x})$, for $x \geqslant 0$	**12** $y^2 = e^x$

18.3   Differentiation of exponential functions

If $y = e^x$ then $\dfrac{dy}{dx} = e^x$.

Example 1   Differentiate

(i) e^{3x} (ii) $e^{\cos x}$ (iii) $x^2 e^{-5x}$ (iv) $\dfrac{e^x}{e^x + 1}$

Fig. 18.10

(i)   Let $y = e^{3x}$. This is a function of a function: $y = e^u$ where $u = 3x$. Then

$$\frac{dy}{dx} = e^u \times 3 = 3e^{3x}$$

(ii)   Let $y = e^{\cos x} = e^u$ where $u = \cos x$. Then

$$\frac{dy}{dx} = e^u \times (-\sin x) = -\sin x \, e^{\cos x}$$

(iii)   Let $y = x^2 e^{-5x}$. This is the product of x^2 and e^{-5x}. Then

$$\frac{dy}{dx} = x^2 \frac{d}{dx}(e^{-5x}) + \frac{d}{dx}(x^2)e^{-5x}$$
$$= x^2(-5e^{-5x}) + 2xe^{-5x} = x(2 - 5x)e^{-5x}$$

(iv)   Let $y = \dfrac{e^x}{e^x + 1}$. Using the quotient rule,

$$\frac{dy}{dx} = \frac{(e^x + 1)e^x - e^x(e^x)}{(e^x + 1)^2} = \frac{e^x}{(e^x + 1)^2}$$

Example 2   Find the equation of the tangent at the origin to the curve $e^{-x} + e^{2y} = x + 3y + 2$

Differentiating each term of the equation with respect to x, we obtain

$$-e^{-x} + e^{2y} \times 2\frac{dy}{dx} = 1 + 3\frac{dy}{dx}$$

At the origin $(0, 0)$,

$$-1 + 2\frac{dy}{dx} = 1 + 3\frac{dy}{dx} \qquad \text{Thus} \qquad \frac{dy}{dx} = -2$$

The equation of the tangent is therefore $y = -2x$.

Exercise 18.3 (Differentiation of exponential functions)

Differentiate the following functions with respect to x (questions 1 to 20).

1 e^{-x}	**2** $3e^{2x} + 4e^{-5x}$	**3** e^{x^2}	**4** $e^{\sqrt{x}}$	**5** xe^x
6 $\cos(e^x)$	**7** $e^{\sin x}$	**8** $\dfrac{e^x}{x^2}$	**9** $e^x \sin x$	**10** $e^{\tan 2x}$
11 $\tan^{-1}(e^x)$	**12** $x^3 e^{-2x}$	**13** $\dfrac{e^x}{2x-3}$	**14** $e^{-3x} \cos 2x$	**15** e^{e^x}
16 $\sin^{-1}(e^{-\frac{1}{2}x})$	**17** $\dfrac{1}{(1+e^{2x})^3}$	**18** $e^{1/x}$	**19** $x^2 e^{-x} \cos x$	**20** $\dfrac{e^x}{\sqrt{1-x^2}}$

Find and classify the stationary points on the following curves (questions 21 to 24).

21 $y = xe^{-x}$ **22** $y = x^2 e^{3x}$ **23** $y = \dfrac{e^x}{1+x^2}$ **24** $y = e^{-x} \sin x$ (between $x = 0$ and $x = 2\pi$).

Find the points of inflexion on the following curves (questions 25 to 27).

2 $y = (1+x)e^{2x}$ **26** $y = e^{-\frac{1}{2}x^2}$ **27** $y = e^{-x} \cos x$ (between $x = 0$ and $x = 2\pi$).

28 Find the equation of the tangent to the curve $y = e^{-2x}$ at the point $(1, e^{-2})$. Find the point where this tangent crosses the x-axis.

29 For the curve $e^x + e^{2y} = 2$, find $\dfrac{dy}{dx}$ in terms of x and y. Find the equation of the normal to this curve at the origin.

30 For the curve

$$\begin{cases} x = 1 + e^t \\ y = 1 - e^{-t} \end{cases}$$

find $\dfrac{dy}{dx}$ and $\dfrac{d^2 y}{dx^2}$ in terms of t.

31 For the curve

$$\begin{cases} x = e^{-t} \\ y = te^{-t} \end{cases}$$

find the equation of the tangent at the point t. Hence show that no tangent to this curve passes through the origin.

32 By sketching the curves $y = e^x$ and $y = 2x + 3$, show that the equation $e^x = 2x + 3$ has two real roots. Use Newton's method to find these roots, each correct to 2 decimal places.

18.4 Integration of exponential functions

Since

$$\frac{d}{dx}(e^x) = e^x$$

we have

$$\int e^x \, dx = e^x + C$$

Example 1   Integrate e^{-4x}.

This is a function of a function: $e^{-4x} = e^u$ where $u = -4x$ is a linear function, so we have

$$\int e^{-4x}\,dx = -\tfrac{1}{4}e^{-4x} + C$$

Example 2   Integrate xe^{-x^2}.

x is a multiple of the derivative of $-x^2$, so we put

$$-x^2 = u$$

Then $-2x\,dx = du$

and $\int xe^{-x^2}\,dx = \int e^u(-\tfrac{1}{2}du)$
$$= -\tfrac{1}{2}e^u + C = -\tfrac{1}{2}e^{-x^2} + C$$

Example 3   Integrate $\dfrac{e^{\sqrt{x}}}{\sqrt{x}}$.

$\dfrac{1}{\sqrt{x}}$ is a multiple of the derivative of $\sqrt{x}$, so we put

$$\sqrt{x} = u$$

Then $\dfrac{1}{2\sqrt{x}}dx = du$

and $\displaystyle\int \frac{e^{\sqrt{x}}}{\sqrt{x}}dx = \int e^u(2du)$

$$= 2e^u + C = 2e^{\sqrt{x}} + C$$

Example 4

$$\int_0^1 e^x \sin(e^x)\,dx$$

Put $e^x = u$; then $e^x\,dx = du$. When $x = 0$, $u = 1$ and when $x = 1$, $u = e$. So

$$\int_0^1 e^x \sin(e^x)\,dx = \int_1^e \sin u\,du$$

$$= \Big[-\cos u\Big]_1^e$$

$$= (-\cos e) - (-\cos 1)$$

$$= \cos 1 - \cos e \approx 1.452$$

The integrals in Examples 1 to 4 could also be done 'by inspection'.

Example 5   Integrate xe^{-x}.

We use integration by parts. Put

$$u = x \qquad \frac{dv}{dx} = e^{-x}$$

$$\frac{du}{dx} = 1 \qquad v = -e^{-x}$$

Then

$$\int xe^{-x}\,dx = x(-e^{-x}) - \int (1)(-e^{-x})\,dx$$

$$= -xe^{-x} + \int e^{-x}\,dx = -xe^{-x} - e^{-x} + C$$

Example 6   Evaluate $\displaystyle\int_0^2 x^2 e^{2x}\,dx$.

This needs two applications of integration by parts. Put

$$u = x^2 \qquad \frac{dv}{dx} = e^{2x}$$

$$\frac{du}{dx} = 2x \qquad v = \tfrac{1}{2}e^{2x}$$

So

$$\int_0^2 x^2 e^{2x}\,dx = \left[x^2(\tfrac{1}{2}e^{2x}) \right]_0^2 - \int_0^2 (2x)(\tfrac{1}{2}e^{2x})\,dx$$

$$= 2e^4 - \int_0^2 xe^{2x}\,dx$$

Now put

$$u = x \qquad \frac{dv}{dx} = e^{2x}$$

$$\frac{du}{dx} = 1 \qquad v = \tfrac{1}{2}e^{2x}$$

So

$$\int_0^2 x^2 e^{2x}\,dx = 2e^4 - \left\{ \left[x(\tfrac{1}{2}e^{2x}) \right]_0^2 - \int_0^2 \tfrac{1}{2}e^{2x}\,dx \right\}$$

$$= 2e^4 - e^4 + \int_0^2 \tfrac{1}{2}e^{2x}\,dx$$

$$= e^4 + \left[\tfrac{1}{4}e^{2x} \right]_0^2 = e^4 + (\tfrac{1}{4}e^4) - (\tfrac{1}{4})$$

$$= \tfrac{5}{4}e^4 - \tfrac{1}{4} \approx 68.00$$

Example 7 Integrate $e^{-2x} \sin 3x$.

We apply integration by parts twice. It does not matter which part is chosen for differentiation and which for integration. Let

$$I = \int e^{-2x} \sin 3x \, dx$$

Put

$$u = e^{-2x} \qquad \frac{dv}{dx} = \sin 3x$$

$$\frac{du}{dx} = -2e^{-2x} \qquad v = -\tfrac{1}{3} \cos 3x$$

So $I = -\tfrac{1}{3} e^{-2x} \cos 3x - \tfrac{2}{3} \int e^{-2x} \cos 3x \, dx$

Now put

$$u = e^{-2x} \qquad \frac{dv}{dx} = \cos 3x$$

$$\frac{du}{dx} = -2e^{-2x} \qquad v = \tfrac{1}{3} \sin 3x$$

So $I = -\tfrac{1}{3} e^{-2x} \cos 3x - \tfrac{2}{3} (\tfrac{1}{3} e^{-2x} \sin 3x + \tfrac{2}{3} \int e^{-2x} \sin 3x \, dx)$

$ = -\tfrac{1}{3} e^{-2x} \cos 3x - \tfrac{2}{9} e^{-2x} \sin 3x - \tfrac{4}{9} I$

Thus

$$\tfrac{13}{9} I = -\tfrac{1}{9} e^{-2x} (3 \cos 3x + 2 \sin 3x)$$

and hence

$$\int e^{-2x} \sin 3x \, dx = I = -\tfrac{1}{13} e^{-2x} (3 \cos 3x + 2 \sin 3x) + C$$

Exercise 18.4 (Integration of exponential functions)

Integrate the following functions with respect to x (questions 1 to 20).

1 e^{5x} 2 e^{-2x} 3 $3e^{\frac{1}{2}x}$ 4 e^{7-3x} 5 $(e^x - e^{2x})^2$ 6 xe^{x^2}

7 $2x^3 e^{-x^4}$ 8 $\cos x \, e^{\sin x}$ 9 $\sin 2x \, e^{\cos 2x}$ 10 $\dfrac{e^{-2\sqrt{x}}}{\sqrt{x}}$ 11 $\sec^2 x \, e^{3 \tan x}$ 12 $\dfrac{e^x}{(4+e^x)^3}$

13 $\dfrac{e^x}{\sqrt{1-e^x}}$ 14 $\dfrac{e^x}{\sqrt{1-e^{2x}}}$ [Put $e^x = u$.]

15 $\dfrac{1}{e^x + e^{-x}}$ [Multiply numerator and denominator by e^x, then put $e^x = u$.]

16 xe^x 17 $(2x - 3)e^{-4x}$ 18 $x^2 e^{3x}$ 19 $e^{-x} \cos x$ 20 $e^{3x} \sin 4x$

Evaluate the following definite integrals (questions 21 to 25).

21 $\displaystyle\int_0^2 e^{-3x} \, dx$ 22 $\displaystyle\int_0^1 xe^{-x^2} \, dx$ 23 $\displaystyle\int_0^2 xe^{2x} \, dx$ 24 $\displaystyle\int_{-1}^1 x^2 e^{-x^3} \, dx$ 25 $\displaystyle\int_0^{\frac{1}{2}\pi} e^{-x} \sin 2x \, dx$

26 Solve the differential equation

$$e^x \frac{dy}{dx} = 1 + 2x$$

given that $y = 3$ when $x = 0$.

27 Find the average value of e^{2x} over the interval $1 \leqslant x \leqslant 3$.

28 Find the area under the curve $y = 2e^{-x}$, between $x = 0$ and $x = 2$. Also find the volume of the solid formed when this area is rotated about the x-axis.

29 Find the area enclosed by the curve $y = (1 - x)e^x$, the x-axis and the y-axis.

30 Use Simpson's rule with 7 ordinates to find an approximate value for $\int_0^3 e^{-x^2} dx$. Verify that this value is approximately $\frac{1}{2}\sqrt{\pi}$.

18.5 Differential equations of type $\dfrac{dy}{dx} = ky$

If $\quad y = e^x$

then $\quad \dfrac{dy}{dx} = e^x = y$

Also, if

$$y = Ae^x$$

where A is any constant, then

$$\frac{dy}{dx} = Ae^x = y$$

so $y = Ae^x$ is a solution of the differential equation $\dfrac{dy}{dx} = y$.

More generally, if

$$y = Ae^{kx}$$

then $\quad \dfrac{dy}{dx} = Ake^{kx} = ky$

$y = Ae^{kx}$ is the general solution of the differential equation

$$\frac{dy}{dx} = ky$$

Example 1 Find the general solution of $\dfrac{dy}{dx} = -3y$.

This is of the form

$$\frac{dy}{dx} = ky \quad \text{with } k = -3$$

So the general solution is

$$y = Ae^{-3x}$$

Example 2 Solve the differential equation $\dfrac{dy}{dx} = 2y$, given that $y = 3$ when $x = 1$.

The general solution is

$$y = Ae^{2x}$$

Substituting $x = 1$ and $y = 3, 3 = Ae^2$, so $A = 3e^{-2}$. Hence the solution is

$$y = 3e^{-2}e^{2x} = 3e^{2x-2}$$

i.e. $y = 3e^{2(x-1)}$

Differential equations of the type

$$\frac{dy}{dx} = ky$$

will arise when the rate of change of a quantity y is proportional to the value of y itself. This is applicable to many practical situations involving growth or decay (see Chapter 21).

Exercise 18.5 $\left(\text{Differential equations: } \dfrac{dy}{dx} = ky\right)$

Write down the general solution of the following differential equations (questions 1 to 4).

1 $\dfrac{dy}{dx} = 2y$ **2** $\dfrac{dy}{dx} = -5y$ **3** $3\dfrac{dy}{dx} = y$ **4** $4\dfrac{dy}{dx} + 7y = 0$

Solve the following differential equations (questions 5 to 8).

5 $\dfrac{dy}{dx} = -y$, given that $y = 3$ when $x = 0$. **6** $\dfrac{dy}{dx} = \dfrac{1}{2}y$, given that $y = -1$ when $x = 0$.

7 $\dfrac{dy}{dx} = 3y$, given that $y = 5$ when $x = 2$. **8** $5\dfrac{dy}{dx} + 2y = 0$, given that $y = -4$ when $x = 5$.

9 Verify that $y = Ae^{2x} + Be^{3x}$ is a solution of $\dfrac{d^2y}{dx^2} - 5\dfrac{dy}{dx} + 6y = 0$.

10 Verify that $y = e^x \cos 3x$ is a solution of $\dfrac{d^2y}{dx^2} - 2\dfrac{dy}{dx} + 10y = 0$.

18.6 Maclaurin series for e^x

We can expand e^x as a Maclaurin series.

Let $\quad f(x) = e^x, \qquad$ then $f(0) = 1$
$\qquad f'(x) = e^x, \qquad$ so $f'(0) = 1$
$\qquad f''(x) = e^x, \qquad$ so $f''(0) = 1$
and so on. Hence

$$e^x = 1 + x + \frac{x^2}{2!} + \frac{x^3}{3!} + \dots$$

This series is valid for all values of x, and it provides a practical method for calculating values of e^x.

Example 1 Use the series expansion to evaluate $e^{0.4}$, correct to 6 decimal places.

We have

$$e^{0.4} = 1 + 0.4 + \frac{(0.4)^2}{2!} + \frac{(0.4)^3}{3!} + \ldots$$

We enter 1 onto the calculator, and add this to the memory; multiply by 0.4, and add to the memory; multiply by $\dfrac{0.4}{2}$ $\left(\text{the display is now } \dfrac{(0.4)^2}{2!} = 0.08\right)$, and add to the memory; multiply by $\dfrac{0.4}{3}$ $\left(\text{the display is now } \dfrac{(0.4)^3}{3!}\right)$, and so on.

We obtain $e^{0.4} = 1.491\,825$ (correct to 6 decimal places).

When we program a computer to evaluate the series

$$e^x = 1 + x + \frac{x^2}{2!} + \ldots$$

we define the terms by $u_0 = 1$ and the recurrence relation

$$u_n = \frac{x}{n} \times u_{n-1}$$

In particular, taking $x = 1$, we obtain

$$e = 1 + 1 + \frac{1}{2!} + \frac{1}{3!} + \ldots$$

The first 12 terms of this series gives $e = 2.718\,281 \ldots$

The series for e^x can be used to obtain power series for related functions.

Example 2 Find power series expansions (up to the term in x^4) for (i) e^{-x} (ii) e^{6x} (iii) $e^x \sin x$.

(i) Replacing x by $-x$ in the series for e^x, we have

$$e^{-x} = 1 + (-x) + \frac{(-x)^2}{2!} + \frac{(-x)^3}{3!} + \ldots$$

$$= 1 - x + \tfrac{1}{2}x^2 - \tfrac{1}{6}x^3 + \tfrac{1}{24}x^4 - \ldots$$

(ii) $e^{6x} = 1 + (6x) + \dfrac{(6x)^2}{2!} + \dfrac{(6x)^3}{3!} + \ldots$

$$= 1 + 6x + 18x^2 + 36x^3 + 54x^4 + \ldots$$

(iii)   Multiplying the series for e^x and $\sin x$, we have

$$e^x \sin x = (1 + x + \tfrac{1}{2}x^2 + \tfrac{1}{6}x^3 + \tfrac{1}{24}x^4 + \ldots)(x - \tfrac{1}{6}x^3 + \ldots)$$
$$= x - \tfrac{1}{6}x^3 + x^2 - \tfrac{1}{6}x^4 + \tfrac{1}{2}x^3 - \tfrac{1}{12}x^5 + \tfrac{1}{6}x^4 - \tfrac{1}{36}x^6 + \ldots$$
$$= x + x^2 + \tfrac{1}{3}x^3 + \ldots$$

(the coefficient of x^4 is zero).

Example 3   Expand $e^{\sin x}$ as a power series (up to the term in x^4).

In this case, we shall find the Maclaurin series directly. Let

$$y = e^{\sin x}$$

When $x = 0$,
$y = 1$

$$\frac{dy}{dx} = e^{\sin x} \cos x = y \cos x$$

When $x = 0$,

$$\frac{dy}{dx} = 1 \times 1 = 1$$

$$\frac{d^2y}{dx^2} = -y \sin x + \frac{dy}{dx} \cos x$$

When $x = 0$,

$$\frac{d^2y}{dx^2} = (-1 \times 0) + (1 \times 1) = 1$$

$$\frac{d^3y}{dx^3} = -y \cos x - 2\frac{dy}{dx} \sin x + \frac{d^2y}{dx^2} \cos x$$

When $x = 0$,

$$\frac{d^3y}{dx^3} = (-1 \times 1) - (2 \times 1 \times 0) + (1 \times 1) = 0$$

$$\frac{d^4y}{dx^4} = y \sin x - \frac{dy}{dx} \cos x - 2\frac{dy}{dx} \cos x - 2\frac{d^2y}{dx^2} \sin x - \frac{d^2y}{dx^2} \sin x + \frac{d^3y}{dx^3} \cos x$$

When $x = 0$,

$$\frac{d^4y}{dx^4} = 0 - 1 - 2 - 0 + 0 = -3$$

The Maclaurin series is

$$e^{\sin x} = 1 + x(1) + \frac{x^2}{2!}(1) + \frac{x^3}{3!}(0) + \frac{x^4}{4!}(-3) + \ldots$$

$$= 1 + x + \tfrac{1}{2}x^2 - \tfrac{1}{8}x^4 + \ldots$$

Exercise 18.6 (Power series for e^x)

1 Use the Maclaurin series for e^x to evaluate (i) $e^{1.2}$ (ii) $e^{-0.2}$, both correct to 6 decimal places. Obtain power series for the following functions, as far as the term in the power indicated (questions 2 to 6).

2 e^{-2x} (up to x^4).

3 x^2e^{-x} (up to x^5).

4 $e^x \cos x$ (up to x^4).

5 $e^{-x}(1 + \sin x - \cos x)$ (up to x^4).

6 $e^{-2x} \sin^2 x$ (up to x^5; first write $\sin^2 x = \frac{1}{2}(1 - \cos 2x)$).

7 Let $y = \dfrac{e^x}{1+x}$. By differentiating the equation $(1+x)y = e^x$, find the values of $y, \dfrac{dy}{dx}, \dfrac{d^2y}{dx^2}, \dots, \dfrac{d^5y}{dx^5}$ when $x = 0$. Hence obtain the Maclaurin series for y (up to x^5).

8 If $y = e^{(\cos x - 1)}$, show that

$$\frac{dy}{dx} = -y \sin x$$

and obtain the Maclaurin series for y up to the term in x^4. Write down the series for $e^{\cos x}$ (up to x^4).

9 Find the Maclaurin series for $e^{\tan x}$ (up to x^3).

10 If $y = e^{2x} \sin x$, find the values of y and $\dfrac{dy}{dx}$ when $x = 0$. Also show that

$$\frac{d^2y}{dx^2} = 4\frac{dy}{dx} - 5y$$

and, by differentiating this equation, find the Maclaurin series for y (up to x^6).

11 If $y = \tan^{-1}(e^x)$, show that

$$(1 + e^{2x})\frac{dy}{dx} = e^x$$

and obtain the Maclaurin series for y (up to x^4). Use your series to evaluate $\tan^{-1}(e^{0.1})$ approximately.

12 Find the Maclaurin series for $y = \sqrt{\frac{1}{2}(1 + e^{-x})}$, up to the term in x^3. [You may find it easier to differentiate the equation in the form $y^2 = \frac{1}{2}(1 + e^{-x})$.]

18.7 Limits involving exponential functions

Behaviour as $x \to +\infty$

We know that

$$e^x \to +\infty \quad \text{as} \quad x \to +\infty$$
$$\text{and} \quad e^{-x} \to 0 \quad \text{as} \quad x \to +\infty$$

We now consider the behaviour as $x \to +\infty$ of $\dfrac{e^x}{x^n}$, where n is a positive integer. Both numerator and denominator tend to $+\infty$, and so the result is not immediately obvious. However,

$$\frac{e^x}{x^n} = \frac{1}{x^n}\left(1 + x + \frac{x^2}{2!} + \dots + \frac{x^n}{n!} + \frac{x^{n+1}}{(n+1)!} + \dots\right)$$

$$= \frac{1}{x^n} + \frac{1}{x^{n-1}} + \frac{1}{2!\,x^{n-2}} + \dots + \frac{1}{n!} + \frac{x}{(n+1)!} + \frac{x^2}{(n+2)!} + \dots$$

As $x \to +\infty$, $\dfrac{1}{x^n} \to 0$, $\dfrac{1}{x^{n-1}} \to 0, \ldots$, but $\dfrac{x}{(n+1)!} \to +\infty$,

$\dfrac{x^2}{(n+2)!} \to +\infty, \ldots$ and hence $\dfrac{e^x}{x^n} \to +\infty$.

Thus e^x, tending to $+\infty$ in the numerator, dominates x^n tending to $+\infty$ in the denominator. We may say that, as $x \to +\infty$, $e^x \to +\infty$ 'faster' than any power of x.

The argument above can be modified as follows when n is any positive number (i.e. not necessarily an integer). Let N be the greatest integer which is less than or equal to n. Then

$$\frac{e^x}{x^n} = \frac{1}{x^n}\left(1 + x + \ldots + \frac{x^N}{N!} + \frac{x^{N+1}}{(N+1)!} + \ldots \right)$$

$$= \frac{1}{x^n} + \frac{1}{x^{n-1}} + \ldots + \frac{1}{N!\,x^{n-N}} + \frac{x^{N+1-n}}{(N+1)!} + \ldots$$

Hence $\dfrac{e^x}{x^n} \to +\infty$ as $x \to +\infty$.

If n is negative, say $n = -m$, then

$$\frac{e^x}{x^n} = x^m e^x$$

As $x \to +\infty$, $x^m \to +\infty$ and $e^x \to +\infty$, hence $\dfrac{e^x}{x^n} \to +\infty$.

We have shown that, for any number n,

$$\boxed{\frac{e^x}{x^n} \to +\infty \quad \text{as} \quad x \to +\infty}$$

Taking the reciprocal, $\dfrac{x^n}{e^x} \to 0$ as $x \to +\infty$, i.e.

$$\boxed{x^n e^{-x} \to 0 \quad \text{as} \quad x \to +\infty}$$

Sketching curves

Example 1　Sketch the curve $y = xe^x$.

We have $y = 0$ when $x = 0$; $y > 0$ when $x > 0$; $y < 0$ when $x < 0$.
　As $x \to +\infty$, $y \to +\infty$.
　As $x \to -\infty$, put $x = -z$ (so that $z \to +\infty$); then $y = -ze^{-z} \to 0$.
　To find any stationary points,

$$\frac{dy}{dx} = xe^x + e^x = (x+1)e^x$$

$$\frac{d^2y}{dx^2} = xe^x + e^x + e^x = (x+2)e^x$$

Thus $\dfrac{dy}{dx} = 0$ when $x = -1$, and then $\dfrac{d^2y}{dx^2}$ is positive; hence there is a minimum at $(-1, -e^{-1})$.

The curve is sketched in Fig. 18.11.

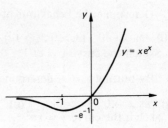

$y = xe^x$

Fig. 18.11

Example 2 Sketch the curve $y = x^4 e^{-3x}$.

We have $y = 0$ when $x = 0$. Otherwise, y is always positive.

As $x \to +\infty$, $y = x^4 e^{-3x} \to 0$.

As $x \to -\infty$, put $x = -z$; then $y = (-z)^4 e^{3z} \to +\infty$.

To find the stationary points,

$$\frac{dy}{dx} = x^4(-3e^{-3x}) + 4x^3 e^{-3x}$$

$$= (-3x + 4)x^3 e^{-3x}$$

which is zero when $x = 0$ or $\frac{4}{3}$. $x = 0$ gives a minimum, and $x = \frac{4}{3}$ gives a maximum.

The curve is sketched in Fig. 18.12.

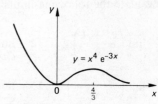

$y = x^4 e^{-3x}$

Fig. 18.12

Other limits

Example Evaluate $\lim\limits_{x \to 0} \left(\dfrac{e^{2x} - 1}{x} \right)$.

Using the series

$$e^{2x} = 1 + (2x) + \frac{(2x)^2}{2!} + \cdots$$

we obtain

$$\frac{e^{2x} - 1}{x} = \frac{(1 + 2x + 2x^2 + \cdots) - 1}{x}$$

$$= \frac{2x + 2x^2 + \cdots}{x} = 2 + 2x + \cdots$$

As $x \to 0$, $\dfrac{e^{2x} - 1}{x} \to 2$, i.e.

$$\lim_{x \to 0} \left(\frac{e^{2x} - 1}{x} \right) = 2$$

Exercise 18.7 (Limits involving exponential functions)

Sketch the following curves (questions 1 to 6). You may have found the stationary points on many of these curves in Exercise 18.3.

1 $y = xe^{-x}$　　**2** $y = x^2 e^{3x}$　　**3** $y = (x + 2)e^{-x}$　　**4** $y = \dfrac{e^x}{x + 2}$

5 $y = \dfrac{e^x}{x^2}$　　**6** $y = xe^{-2x^2}$

7 Determine the behaviour of $e^{1/x}$,

(i) as $x \to 0$ $(x > 0)$;   (ii) as $x \to 0$ $(x < 0)$;   (iii) as $x \to \pm \infty$.
Sketch the curve $y = e^{1/x}$.

8 By putting $x = \dfrac{1}{z}$, determine the behaviour of $xe^{1/x}$:

(i) as $x \to 0$ $(x > 0)$;   (ii) as $x \to 0$ $(x < 0)$.
Sketch the curve $y = xe^{1/x}$.

9 Sketch the curve $y = e^{-1/x^2}$.

10 Sketch the curve $y = x^3 e^{-1/x^2}$.

Evaluate the following limits (questions 11 to 15).

11 $\lim\limits_{x \to 0} \left(\dfrac{e^{3x} - 1}{x} \right)$ **12** $\lim\limits_{x \to 0} \left(\dfrac{1 - e^{-x}}{x} \right)$ **13** $\lim\limits_{x \to 0} \left(\dfrac{e^x - 1 - x}{x^2} \right)$

14 $\lim\limits_{x \to 0} \left(\dfrac{e^{3x} - e^x}{x} \right)$ **15** $\lim\limits_{x \to 0} \left(\dfrac{e^x + e^{-x} - 2}{x^2} \right)$

19

Logarithmic functions

19.1 The logarithmic function

The function $y = e^x$ is an increasing function, and therefore it has an inverse function. The equation of the inverse is $x = e^y$; y is the power to which e must be raised to obtain x. By the definition of logarithms, y is the logarithm of x to base e, i.e.

$$y = \log_e x$$

Thus the inverse function of e^x is $\log_e x$ (see Fig. 19.1).

Logarithms to base e are called natural logarithms (or Napierian logarithms). We shall write $\ln x$ for $\log_e x$,

i.e. $y = \ln x$ means $e^y = x$

Notice that:

1) $\ln x$ is not defined when $x \leqslant 0$;
2) $\ln x < 0$ when $0 < x < 1$;
3) $\ln 1 = 0$;
4) $\ln x > 0$ when $x > 1$;
5) $\ln e = 1$;
6) $\ln x$ is an increasing function:
$$\ln x \to +\infty \quad \text{as } x \to +\infty$$
$$\ln x \to -\infty \quad \text{as } x \to 0 \quad (x > 0).$$

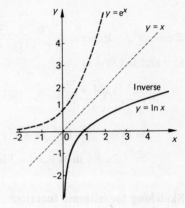

Fig. 19.1

From the properties of logarithms, we have

$$\ln (uv) = \ln u + \ln v$$
$$\ln \left(\frac{u}{v} \right) = \ln u - \ln v$$
$$\ln (u^n) = n \ln u$$

Values of $\ln x$ can be obtained directly on a calculator, using the 'ln' key. Use your calculator to evaluate $\ln (3.2)$ and $\ln (0.43)$ (the answers should be 1.1632 and -0.8440).

Example 1 Show that

(i) $e^{\ln x} = x$ (if $x > 0$) (ii) $\ln (e^x) = x$ (iii) $\ln \left(\dfrac{1}{x} \right) = -\ln x$

(i) If $y = \ln x$, then $e^y = x$, i.e. $e^{\ln x} = x$.

(ii) $\ln (e^x) = x \ln e = x$

(iii) $\ln \left(\dfrac{1}{x} \right) = \ln 1 - \ln x = -\ln x$

343

Example 2   Write as a single logarithm
(i) $3 \ln 2 + \ln 5 - 2 \ln 7$ (ii) $2 \ln (x+1) - \ln (2x+3)$

(i) $3 \ln 2 + \ln 5 - 2 \ln 7 = \ln (2^3) + \ln 5 - \ln (7^2)$

$$= \ln \left(\frac{2^3 \times 5}{7^2} \right) = \ln \left(\frac{40}{49} \right) = -\ln \left(\frac{49}{40} \right)$$

(ii) $2 \ln (x+1) - \ln (2x+3) = \ln \left[(x+1)^2 \right] - \ln (2x+3)$

$$= \ln \left(\frac{(x+1)^2}{2x+3} \right)$$

Example 3   Express $\ln \left(\dfrac{e^{-2x} \sqrt{4+x^2}}{x^3 (3x+1)^5} \right)$ in terms of $\ln (4+x^2)$,
$\ln x$ and $\ln (3x+1)$.

$\ln \left(\dfrac{e^{-2x} \sqrt{4+x^2}}{x^3 (3x+1)^5} \right)$

$= \ln (e^{-2x}) + \ln (\sqrt{4+x^2}) - \ln (x^3) - \ln \left[(3x+1)^5 \right]$

$= -2x + \frac{1}{2} \ln (4+x^2) - 3 \ln x - 5 \ln (3x+1)$

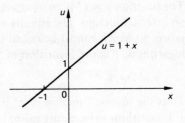

Fig. 19.2

Sketching logarithmic functions

Example 1   Sketch the curve $y = \ln (1+x)$.

We first sketch $u = 1+x$ (Fig. 19.2). Then $y = \ln u$.
  $y = 0$ when $u = 1$, i.e. when $x = 0$.
  When $x \leqslant -1$, y is not defined (since $u \leqslant 0$).
  When $-1 < x < 0$, $y < 0$ (since $0 < u < 1$).
  When $x > 0$, $y > 0$ (since $u > 1$).
  Finally,

$\quad y \to +\infty$ as $x \to +\infty$
$\quad y \to -\infty$ as $x \to -1$ $(x > -1)$.
The curve is sketched in Fig. 19.3.

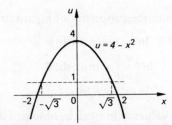

Fig. 19.3

Example 2   Sketch the curve $y = \ln (4 - x^2)$.

We first sketch $u = 4 - x^2$ (Fig. 19.4). Then $y = \ln u$
  $y = 0$ when $u = 1$, i.e. when $x = \pm \sqrt{3}$.
  y is not defined when $u \leqslant 0$, i.e. when $x \leqslant -2$ or $x \geqslant 2$.
  $y > 0$ when $u > 1$, i.e. when $-\sqrt{3} < x < \sqrt{3}$.
  $y < 0$ when $0 < u < 1$, i.e. when $-2 < x < -\sqrt{3}$ or $\sqrt{3} < x < 2$.
  Finally,

$\quad y \to -\infty$ as $x \to 2$ $(x < 2)$
$\quad y \to -\infty$ as $x \to -2$ $(x > -2)$
The curve is sketched in Fig. 19.5.

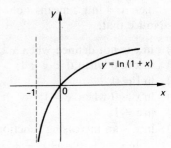

Fig. 19.4

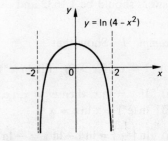

Fig. 19.5

Example 3 Sketch the curve $y = \ln\left(\dfrac{x+1}{x-3}\right)$

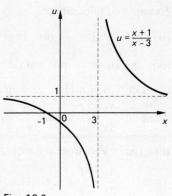

We first sketch $u = \dfrac{x+1}{x-3}$ (Fig. 19.6). Then $y = \ln u$.

y is not defined when $-1 \leqslant x \leqslant 3$.
As $x \to -1$ $(x < -1)$, $u \to 0$, so $y \to -\infty$.
As $x \to 3$ $(x > 3)$, $u \to +\infty$, so $y \to +\infty$.
As $x \to \pm\infty$, $u \to 1$, so $y \to 0$.
The curve is sketched in Fig. 19.7.

Fig. 19.6

Exercise 19.1 (Logarithmic functions: properties and sketching)

1 Simplify
(i) $e^{-\ln x}$ (ii) $e^{2\ln x}$ (iii) $\ln(e^{-\cos x})$.

2 Use $a = e^{\ln a}$ to show that, if $a > 0$, $a^x = e^{x\ln a}$.

Express the following as a single logarithm (questions 3 to 5).

3 $4\ln 3 + 2\ln 6 - 3\ln 2$ **4** $\ln(x+3) + 2\ln x$

5 $3\ln(2x+1) + \frac{1}{2}\ln x - 2\ln(4-x)$

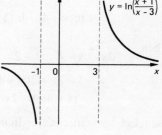

6 Express $\ln\sqrt{\dfrac{2+x}{2-x}}$ in terms of $\ln(2+x)$ and $\ln(2-x)$.

7 Express $\ln\left(\dfrac{(1+x)^2 e^{3x}}{\sqrt{(6x-5)^3}}\right)$ in terms of $\ln(1+x)$ and $\ln(6x-5)$.

Fig. 19.7

From the following equations (questions 8 to 12) express y in terms of x.

8 $\ln y = x^2 - 1$ **9** $e^{2y} = \dfrac{x+1}{x-1}$ **10** $\ln y = \ln(x+3) - 2\ln x$

11 $e^{-y} = (2x+3)e^{4x}$ **12** $\ln(y+2) - \ln(y-2) = \frac{1}{2}\ln x$

Sketch the following curves (questions 13 to 20).

13 $y = \ln 2x$ **14** $y = 2\ln x$ **15** $y = \ln(-\frac{1}{3}x)$ **16** $y = \ln\sqrt{x}$

17 $y = -\ln(x+2)$ **18** $y = \ln(1-3x)$ **19** $y = \ln\left(\dfrac{x-2}{x+4}\right)$ **20** $y = \ln(\sin x)$

19.2 Differentiation of logarithmic functions

If $y = \ln x$ then $x = e^y$.
Differentiating with respect to y,

$$\frac{dx}{dy} = e^y \quad \text{and so} \quad \frac{dy}{dx} = \frac{1}{e^y} = \frac{1}{x}$$

Hence

$$\boxed{\frac{d}{dx}(\ln x) = \frac{1}{x}}$$

Example　Differentiate

(i)　$\ln(5x-7)$　　　(ii)　$\ln(x^2+x+1)$　　　(iii)　$\ln\left(\dfrac{x+4}{3-2x}\right)$.

(iv)　$\ln(\cos x)$　　(v)　$(\ln x)^2$　　(vi)　$x^2\ln x$　　(vii)　$\log_{10}x$

(i)　Let　$y=\ln(5x-7)=\ln u$　where $u=5x-7$

Then　$\dfrac{dy}{dx}=\dfrac{1}{u}\times 5=\dfrac{5}{5x-7}$

(ii)　Let　$y=\ln(x^2+x+1)=\ln u$　where $u=x^2+x+1$

Then　$\dfrac{dy}{dx}=\dfrac{1}{u}\times(2x+1)=\dfrac{2x+1}{x^2+x+1}$

(iii)　Let　$y=\ln\left(\dfrac{x+4}{3-2x}\right)$

$\qquad\qquad=\ln(x+4)-\ln(3-2x)$

Then　$\dfrac{dy}{dx}=\dfrac{1}{x+4}-\dfrac{-2}{3-2x}$

$\qquad\quad=\dfrac{3-2x+2(x+4)}{(x+4)(3-2x)}=\dfrac{11}{(x+4)(3-2x)}$

(iv)　Let　$y=\ln(\cos x)=\ln u$　where $u=\cos x$

Then　$\dfrac{dy}{dx}=\dfrac{1}{u}\times(-\sin x)$

$\qquad\quad=-\dfrac{\sin x}{\cos x}=-\tan x$

(v)　Let　$y=(\ln x)^2=u^2$　where $u=\ln x$

Then　$\dfrac{dy}{dx}=2u\times\dfrac{1}{x}$

$\qquad\quad=\dfrac{2\ln x}{x}$

(vi)　Let　$y=x^2\ln x$　This is the product of x^2 and $\ln x$.

Thus　$\dfrac{dy}{dx}=x^2\left(\dfrac{1}{x}\right)+(2x)\ln x$

$\qquad\quad=x+2x\ln x$

(vii)　Let　$y=\log_{10}x=\dfrac{\log_e x}{\log_e 10}$　$\left(\text{using the general result }\log_b x=\dfrac{\log_a x}{\log_a b}.\right)$

so　　　$y=\dfrac{\ln x}{\ln 10}$

Thus　$\dfrac{dy}{dx}=\dfrac{1}{\ln 10}\times\dfrac{1}{x}$

Logarithmic differentiation

The differentiation of some complicated functions (especially products and quotients) can be made easier by taking logarithms (to base e) first.

Example 1 Differentiate $\dfrac{x^2\sqrt{3-x}}{x^3+1}$.

Let $\qquad y = \dfrac{x^2\sqrt{3-x}}{x^3+1}$

Taking logarithms of both sides,

$$\ln y = \ln\left(\frac{x^2\sqrt{3-x}}{x^3+1}\right)$$

$$= 2\ln x + \tfrac{1}{2}\ln(3-x) - \ln(x^3+1)$$

We now differentiate each term with respect to x. Note that

$$\frac{d}{dx}(\ln y) = \frac{d}{dy}(\ln y) \times \frac{dy}{dx}$$

$$= \frac{1}{y}\frac{dy}{dx}$$

Thus $\qquad \dfrac{1}{y}\dfrac{dy}{dx} = \dfrac{2}{x} + \dfrac{1}{2}\left(\dfrac{-1}{3-x}\right) - \dfrac{3x^2}{x^3+1}$

So $\qquad \dfrac{dy}{dx} = y\left(\dfrac{2}{x} - \dfrac{1}{2(3-x)} - \dfrac{3x^2}{x^3+1}\right)$

$$= \frac{x^2\sqrt{3-x}}{x^3+1}\left(\frac{2}{x} - \frac{1}{2(3-x)} - \frac{3x^2}{x^3+1}\right)$$

Example 2 Differentiate $x^3 e^{-x}\sin x$.

Let $\qquad y = x^3 e^{-x}\sin x$

Then $\quad \ln y = 3\ln x + \ln(e^{-x}) + \ln(\sin x)$

$$= 3\ln x - x + \ln(\sin x)$$

Differentiating,

$$\frac{1}{y}\frac{dy}{dx} = \frac{3}{x} - 1 + \frac{\cos x}{\sin x}$$

So $\qquad \dfrac{dy}{dx} = x^3 e^{-x}\sin x\left(\dfrac{3}{x} - 1 + \cot x\right)$

This method can also be used to differentiate powers.

Example 3   Differentiate a^x (where a is a positive constant).

Let $y = a^x$
Then $\ln y = \ln(a^x) = \ln a$

Differentiating,

$$\frac{1}{y}\frac{dy}{dx} = \ln a$$

and so

$$\frac{dy}{dx} = a^x \ln a$$

Thus, for example,

$$\frac{d}{dx}(2^x) = 2^x \ln 2$$

$$\approx 0.693 \times 2^x$$

and $\dfrac{d}{dx}(3^x) = 3^x \ln 3$

$$\approx 1.099 \times 3^x$$

confirming our discoveries at the beginning of Chapter 18.

Example 4   Differentiate x^x.

Let $y = x^x$
Then $\ln y = \ln(x^x) = x \ln x$

Differentiating,

$$\frac{1}{y}\frac{dy}{dx} = x\left(\frac{1}{x}\right) + (1)\ln x$$

and so

$$\frac{dy}{dx} = x^x(1 + \ln x)$$

Exercise 19.2   (Differentiation of logarithmic functions)

Differentiate the following functions with respect to x (questions 1 to 16).

1 $\ln(3x+2)$ **2** $\ln(4-7x)$ **3** $\ln(2x^2-x+7)$ **4** $\ln(x^3)$

5 $\ln\sqrt{x^2+1}$ **6** $\ln\left(\dfrac{2-5x}{3x+4}\right)$ **7** $\ln\left(\dfrac{x}{1+x^3}\right)$ **8** $\sqrt{\ln x}$

9 $\ln(\sin x)$ **10** $\ln(\sec x+\tan x)$ **11** $x \ln x$ **12** $\dfrac{x^2}{\ln x}$

13 $e^{-x}\ln x$ **14** $\ln(\ln x)$ **15** $\log_{10}(2x+3)$ **16** $x\log_{10}(x^2+1)$

By first taking logarithms, differentiate the following functions (questions 17 to 24).

17 $\dfrac{x^5}{\sqrt{3x+5}}$ 18 $\dfrac{x^3(2x-1)^5}{(x+1)^2}$ 19 $e^{5x}x^4 \cos 3x$ 20 10^x

21 2^{x^2} 22 x^{-x} 23 $x^{1/x}$ 24 $x^{\sin x}$

Find and classify the stationary points on the following curves (questions 25 to 30).

25 $y = x^2 \ln x$ 26 $y = \dfrac{\ln x}{x}$ 27 $y = \dfrac{x}{\ln x - 2}$

28 $y = \dfrac{1}{x} + \ln x$ 29 $y = x^x$ 30 $y = x^{1/x}$

31 Find the point of inflexion on the curve $y = \dfrac{\ln x}{x}$.

32 For the curve $\ln(xy) = x^2 - y^2$, find $\dfrac{dy}{dx}$ in terms of x and y.

33 For the curve
$$\begin{cases} x = \ln(1-t) \\ y = \ln(1+t) \end{cases}$$
find $\dfrac{dy}{dx}$ and $\dfrac{d^2y}{dx^2}$ in terms of t.
   Find the equation of the tangent at the origin.
34 Find the value of c if $y = 4x + c$ is a tangent to the curve $y = \ln x$
35 By sketching the graphs of $y = \ln x$ and $y = 1 - \frac{1}{2}x$, show that the equation $\ln x = 1 - \frac{1}{2}x$ has just one real root. Use Newton's method to find the root correct to 2 decimal places.
36 Find, correct to 3 significant figures, the gradient of the curve $y = x^2 - 25 \log_{10} x$ at the point $(10, 75)$.

19.3 Integration using logarithmic functions

Up till now we have had to exclude $n = -1$ from the general rule

$$\int x^n dx = \frac{x^{n+1}}{n+1} + C$$

Since

$$\frac{d}{dx}(\ln x) = \frac{1}{x}$$

we can now fill in this gap:

$$\int \frac{1}{x} dx = \ln x + C$$

This integration is valid only when $x > 0$ (otherwise $\ln x$ is not defined).

When $x < 0$,

$$\frac{d}{dx}\left[\ln(-x)\right] = \frac{-1}{-x} = \frac{1}{x}$$

and so $\int \frac{1}{x} dx = \ln(-x) + C$

Thus $\int \frac{1}{x} dx = \begin{cases} \ln x + C & \text{if } x > 0 \\ \ln(-x) + C & \text{if } x < 0 \end{cases}$

and we may combine these two cases by writing

$$\boxed{\int \frac{1}{x} dx = \ln|x| + C}$$

For example,

$$\int_2^3 \frac{1}{x} dx = \left[\ln|x|\right]_2^3$$

$$= \ln 3 - \ln 2 = \ln \tfrac{3}{2} \approx 0.405$$

and $\int_{-4}^{-2} \frac{1}{x} dx = \left[\ln|x|\right]_{-4}^{-2} = \ln 2 - \ln 4 = \ln \tfrac{2}{4} = \ln \tfrac{1}{2}$

$$= -\ln 2 \approx -0.693$$

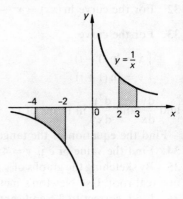

Fig. 19.8

$\int_{-4}^{-2} \frac{1}{x} dx$ is expected to be negative, since the curve $y = \frac{1}{x}$ is
below the x-axis between $x = -4$ and $x = -2$ (see Fig. 19.8).

Note that, for example, $\int_{-2}^3 \frac{1}{x} dx$ cannot be evaluated, because
the range of integration includes $x = 0$, for which the integral,
$\ln|x|$, is not defined.

Example 1

$$\int \frac{1}{4x + 5} dx$$

Let $\frac{1}{4x + 5} = \frac{1}{u}$ where $u = 4x + 5$ is a linear function
So we have

$$\int \frac{1}{4x + 5} dx = \tfrac{1}{4}\ln|4x + 5| + C$$

Example 2

$$\int \frac{1}{3-2x}dx = -\tfrac{1}{2}\ln|3-2x|+C$$

Example 3

$$\int_2^3 \frac{1}{1-x}dx = \left[-\ln|1-x|\right]_2^3$$

$$= -\ln 2 - (-\ln 1) = -\ln 2$$

In general, we have

$$\boxed{\int \frac{1}{ax+b}dx = \frac{1}{a}\ln|ax+b|+C}$$

Example 4

$$\int \frac{x^2}{1+x^3}dx$$

Since x^2 is (apart from a constant factor) the derivative of $1+x^3$, we put

$$1+x^3 = t$$

then $\qquad 3x^2 dx = dt$

Hence

$$\int \frac{x^2}{1+x^3}dx = \int \frac{1}{t}(\tfrac{1}{3}dt)$$

$$= \tfrac{1}{3}\ln|t|+C$$

$$= \tfrac{1}{3}\ln|1+x^3|+C$$

Example 5

$$\int \frac{\cos x}{3+\sin x}dx$$

Put $\qquad 3+\sin x = t$

Then $\qquad \cos x\,dx = dt$

Hence

$$\int \frac{\cos x}{3+\sin x}dx = \int \frac{1}{t}dt = \ln|t|+C$$

$$= \ln(3+\sin x)+C$$

(There is no need for a modulus sign, since $3+\sin x$ is always positive.)

Example 6

$$\int \tan x \, dx = \int \frac{\sin x}{\cos x} \, dx$$

Put $\cos x = u$

Then $-\sin x \, dx = du$

Hence

$$\int \tan x \, dx = \int \frac{1}{u}(-du)$$

$$= -\ln|u| + C = -\ln|\cos x| + C$$

$$= \ln\left|\frac{1}{\cos x}\right| + C = \ln|\sec x| + C$$

Example 7

$$\int_0^{\ln 9} \frac{e^x}{1+e^x} \, dx$$

Put $1 + e^x = u$

Then $e^x \, dx = du$

When $x = 0$, $u = 2$, and when $x = \ln 9$, $u = 1 + e^{\ln 9} = 1 + 9 = 10$.

Hence

$$\int_0^{\ln 9} \frac{e^x}{1+e^x} \, dx = \int_2^{10} \frac{1}{u} \, du$$

$$= \left[\ln u\right]_2^{10} = \ln 10 - \ln 2$$

$$= \ln(\tfrac{10}{2}) = \ln 5 \quad (\approx 1.609)$$

We can see how to integrate a quotient when the numerator is the derivative of the denominator. The integral is the logarithm of the denominator, i.e.

$$\boxed{\int \frac{f'(x)}{f(x)} \, dx = \ln|f(x)| + C}$$

To prove this, if

$$y = \ln|f(x)| = \ln|u| \quad \text{where } u = f(x)$$

then $$\frac{dy}{dx} = \frac{dy}{du} \times \frac{du}{dx} = \frac{1}{u} \times f'(x)$$

$$= \frac{f'(x)}{f(x)}$$

Example 8

$$\int \frac{3x+5}{x^2+1}\,dx = \int \frac{3x}{x^2+1}\,dx + \int \frac{5}{x^2+1}\,dx$$

$$= \frac{3}{2}\int \frac{2x}{x^2+1}\,dx + 5\int \frac{1}{x^2+1}\,dx$$

$$= \tfrac{3}{2}\ln(x^2+1) + 5\tan^{-1}x + C$$

Example 9

$$\int_2^3 \frac{1}{2x^2+x-3}\,dx$$

We first write

$$\frac{1}{2x^2+x-3} = \frac{1}{(x-1)(2x+3)}$$

in partial fractions

$$= \frac{A}{x-1} + \frac{B}{2x+3}$$

We find that $A = \tfrac{1}{5}$ and $B = -\tfrac{2}{5}$.

$$\int_2^3 \frac{1}{2x^2+x-3}\,dx = \int_2^3 \left(\frac{\frac{1}{5}}{x-1} - \frac{\frac{2}{5}}{2x+3}\right)dx$$

$$= \left[\frac{1}{5}\ln|x-1| - \frac{2}{5}\times\frac{1}{2}\ln|2x+3|\right]_2^3$$

$$= \left[\frac{1}{5}\ln\left|\frac{x-1}{2x+3}\right|\right]_2^3 = \tfrac{1}{5}\ln\tfrac{2}{9} - \tfrac{1}{5}\ln\tfrac{1}{7}$$

$$= \frac{1}{5}\ln\left(\frac{14}{9}\right) \quad (\approx 0.0884)$$

Integration of logarithmic functions

To integrate $\ln x$, we use integration by parts. Consider

$$\int \ln x\,dx$$

Put $u = \ln x$ and $\dfrac{dv}{dx} = 1$

Then $\dfrac{du}{dx} = \dfrac{1}{x}$ $v = x$

Hence $\displaystyle\int \ln x\,dx = (\ln x)\,(x) - \int\left(\frac{1}{x}\right)x\,dx$

$$= x\ln x - \int 1\,dx$$

$$= x\ln x - x + C$$

A similar method can be used to integrate $x^n \ln x$, provided that $n \neq -1$.

Example 1

$$\int_1^e x^3 \ln x \, dx$$

Put $u = \ln x$ and $\dfrac{dv}{dx} = x^3$

$$\dfrac{du}{dx} = \dfrac{1}{x} \qquad\qquad v = \dfrac{1}{4}x^4$$

Hence $\displaystyle\int_1^e x^3 \ln x \, dx = \left[\dfrac{1}{4}x^4 \ln x\right]_1^e - \int_1^e \left(\dfrac{1}{x}\right)\left(\dfrac{1}{4}x^4\right) dx$

$$= \left[\dfrac{1}{4}x^4 \ln x\right]_1^e - \int_1^e \dfrac{1}{4}x^3 \, dx$$

$$= \left[\dfrac{1}{4}x^4 \ln x - \dfrac{1}{16}x^4\right]_1^e$$

$$= (\tfrac{1}{4}e^4 \ln e - \tfrac{1}{16}e^4) - (0 - \tfrac{1}{16})$$

$$= \tfrac{1}{4}e^4 - \tfrac{1}{16}e^4 + \tfrac{1}{16}$$

$$= \tfrac{1}{16}(3e^4 + 1) \quad (\approx 10.3)$$

Example 2

$$\int \dfrac{\ln x}{x} \, dx$$

Since $\dfrac{1}{x}$ is the derivative of $\ln x$, we put

$$\ln x = t$$

Then $\qquad \dfrac{1}{x} dx = dt$

Hence $\displaystyle\int \dfrac{\ln x}{x} dx = \int t \, dt = \tfrac{1}{2}t^2 + C$

$$= \tfrac{1}{2}(\ln x)^2 + C$$

Example 3

$$\int \dfrac{1}{x \ln x} \, dx$$

Put $\qquad\qquad \ln x = t$

Then $\qquad \dfrac{1}{x} dx = dt$

Hence $\displaystyle\int \dfrac{1}{x \ln x} dx = \int \dfrac{1}{t} dt = \ln|t| + C$

$$= \ln|\ln x| + C$$

Exercise 19.3   (Integration using logarithmic functions)

Integrate the following functions with respect to x (questions 1 to 25)

1 $\dfrac{1}{3x}$

2 $\dfrac{1}{x+5}$

3 $\dfrac{1}{2x-3}$

4 $\dfrac{1}{7-4x}$

5 $\dfrac{2x}{x^2-4}$

6 $\dfrac{x^3}{x^4-1}$

7 $\dfrac{\cos x - \sin x}{\sin x + \cos x}$

8 $\cot x$

9 $\dfrac{e^{2x}}{3+4e^{2x}}$

10 $\dfrac{4x-3}{1+x^2}$

11 $\dfrac{x-1}{1+4x^2}$

12 $\dfrac{x}{x+2}$ [Put $x+2=u$.]

13 $\dfrac{x}{(x-4)^2}$

14 $\dfrac{x^2-2}{(x+1)^2}$

15 $\dfrac{1}{x(x-1)}$ $\left[\text{Express in partial fractions } \dfrac{A}{x}+\dfrac{B}{x-1}.\right]$

16 $\dfrac{3}{(x-2)(x-5)}$

17 $\dfrac{1}{3x^2-5x-2}$

18 $\dfrac{x+8}{x^2-4}$

19 $\dfrac{1}{x+\sqrt{x}}$ [Put $x=t^2$]

20 $\dfrac{\sin(\ln x)}{x}$

21 $x^2 \ln x$

22 $\dfrac{\ln x}{x^4}$

23 $\dfrac{(\ln x)^2}{x}$

24 $(\ln x)^2$ [Put $x=e^t$.]

25 $\dfrac{1}{x(\ln x+1)}$

26 If $I = \displaystyle\int \dfrac{\ln x}{x} dx$ use integration by parts to show that $I = (\ln x)^2 - I$.
Hence find I.

Evaluate the following definite integrals (questions 27 to 31).

27 $\displaystyle\int_{-1}^{3} \dfrac{1}{2x-9} dx$

28 $\displaystyle\int_{0}^{1} \dfrac{2x+3}{1+x^2} dx$

29 $\displaystyle\int_{4}^{8} \dfrac{(x-4)^2}{x-3} dx$

30 $\displaystyle\int_{9}^{16} \dfrac{16-x}{(x-2)(x+5)} dx$

31 $\displaystyle\int_{1}^{e} \dfrac{\ln x}{x^3} dx$

32 Criticise
$$\int_{1}^{4} \dfrac{1}{x-3} dx = \Big[\ln|x-3|\Big]_{1}^{4} = \ln 1 - \ln 2 = -\ln 2$$

33 Find the area under the curve $y = \dfrac{x}{1+x^2}$ between $x=0$ and $x=2$.

34 Find the area under the curve $y = \dfrac{2x}{x+3}$ between $x=0$ and $x=3$.
Find also the volume of the solid generated when this area is rotated about the x-axis.

35 Find the average value of $\dfrac{1}{x+1}$ over the interval $0 \leqslant x \leqslant 4$.

36 Obtain an approximate value for $\displaystyle\int_{0}^{1} \ln(1+x^2)dx$ by using Simpson's Rule with 5 ordinates.

19.4 Maclaurin's series for $\ln(1+x)$

We cannot expand $\ln x$ as a Maclaurin series, since $\ln x$ is not defined when $x = 0$. However, we can expand $\ln(1+x)$.

Let

$$f(x) = \ln(1+x) \quad \text{so} \quad f(0) = \ln 1 = 0$$

$$f'(x) = \frac{1}{1+x} \quad \text{so} \quad f'(0) = 1$$

$$f''(x) = \frac{-1}{(1+x)^2} \quad \text{so} \quad f''(0) = -1$$

$$f'''(x) = \frac{2}{(1+x)^3} \quad \text{so} \quad f'''(0) = 2$$

$$f^{iv}(x) = \frac{-6}{(1+x)^4} \quad \text{so} \quad f^{iv}(0) = -6$$

Thus $\ln(1+x) = 0 + x(1) + \dfrac{x^2}{2!}(-1) + \dfrac{x^3}{3!}(2) + \dfrac{x^4}{4!}(-6) + \ldots$

i.e.
$$\ln(1+x) = x - \tfrac{1}{2}x^2 + \tfrac{1}{3}x^3 - \tfrac{1}{4}x^4 + \ldots$$

This series is valid when $-1 < x \leqslant 1$.

Replacing x by $-x$, we obtain

$$\ln(1-x) = -x - \tfrac{1}{2}x^2 - \tfrac{1}{3}x^3 - \tfrac{1}{4}x^4 - \ldots$$

which is valid when $-1 \leqslant x < 1$.

If we subtract these two series, we obtain
$$\ln(1+x) - \ln(1-x) = 2x + \tfrac{2}{3}x^3 + \tfrac{2}{5}x^5 + \ldots$$

i.e.
$$\ln\left(\frac{1+x}{1-x}\right) = 2x + \tfrac{2}{3}x^3 + \tfrac{2}{5}x^5 + \ldots$$

This series is valid when $-1 < x < 1$, and it can be used to evaluate the natural logarithm of any positive number, as shown in the following example.

Example 1 Use the series for $\ln\left(\dfrac{1+x}{1-x}\right)$ to evaluate $\ln 3$.

We require $\dfrac{1+x}{1-x} = 3$, i.e. $1 + x = 3 - 3x$, i.e. $x = \tfrac{1}{2}$. Thus

$$\ln 3 = 2(\tfrac{1}{2}) + \tfrac{2}{3}(\tfrac{1}{2})^3 + \tfrac{2}{5}(\tfrac{1}{2})^5 + \tfrac{2}{7}(\tfrac{1}{2})^7 + \tfrac{2}{9}(\tfrac{1}{2})^9 + \ldots$$
$$= 1 + 0.0833 + 0.0125 + 0.0022 + 0.0004 + \ldots \approx 1.098$$

Example 2   Expand $\ln(3+2x)$ as a power series.

$$\ln(3+2x) = \ln\left[3(1+\tfrac{2}{3}x)\right]$$
$$= \ln 3 + \ln(1+\tfrac{2}{3}x)$$
$$= \ln 3 + (\tfrac{2}{3}x) - \tfrac{1}{2}(\tfrac{2}{3}x)^2 + \tfrac{1}{3}(\tfrac{2}{3}x)^3 - \tfrac{1}{4}(\tfrac{2}{3}x)^4 + \dots$$
$$= \ln 3 + \tfrac{2}{3}x - \tfrac{2}{9}x^2 + \tfrac{8}{81}x^3 - \tfrac{4}{81}x^4 + \dots$$

This is valid when $-1 < \tfrac{2}{3}x \leqslant 1$, i.e. when $-\tfrac{3}{2} < x \leqslant \tfrac{3}{2}$.

Example 3   Expand $\ln(\sin x + \cos x)$ as a power series, up to the term in x^3.

We shall find the Maclaurin series directly. Let

$$y = \ln(\sin x + \cos x)$$

When $x = 0$,

$$y = \ln 1 = 0$$

$$\frac{dy}{dx} = \frac{\cos x - \sin x}{\sin x + \cos x}$$

When $x = 0$,

$$\frac{dy}{dx} = 1$$

Writing this as

$$(\sin x + \cos x)\frac{dy}{dx} = \cos x - \sin x$$

and differentiating,

$$(\sin x + \cos x)\frac{d^2y}{dx^2} + (\cos x - \sin x)\frac{dy}{dx} = -\sin x - \cos x$$

When $x = 0$,

$$1 \times \frac{d^2y}{dx^2} + 1 \times 1 = 0 - 1 \quad \text{so} \quad \frac{d^2y}{dx^2} = -2$$

Differentiating again,

$$(\sin x + \cos x)\frac{d^3y}{dx^3} + (\cos x - \sin x)\frac{d^2y}{dx^2} + (\cos x - \sin x)\frac{d^2y}{dx^2} + (-\sin x - \cos x)\frac{dy}{dx}$$
$$= -\cos x + \sin x$$

When $x = 0$,

$$\frac{d^3y}{dx^3} - 2 - 2 - 1 = -1 \quad \text{so} \quad \frac{d^3y}{dx^3} = 4$$

Hence

$$\ln(\sin x + \cos x) = 0 + x(1) + \frac{x^2}{2!}(-2) + \frac{x^3}{3!}(4) + \dots$$

$$= x - x^2 + \tfrac{2}{3}x^3 + \dots$$

Exercise 19.4   (Series expansions for logarithmic functions)

1   Use the series for $\ln(1-x)$ to evaluate $\ln(0.95)$, correct to 6 decimal places.

2   Use the series for $\ln\left(\dfrac{1+x}{1-x}\right)$ to evaluate $\ln 5$, correct to 2 decimal places.

3   Find the sum to infinity of the series

$$1 - \tfrac{1}{2} + \tfrac{1}{3} - \tfrac{1}{4} + \tfrac{1}{5} - \tfrac{1}{6} + \cdots$$

by comparing it with the series for $\ln(1+x)$ with a suitable value of x.

4   Find the sum to infinity of the series

$$1 + \frac{1}{3\times 9} + \frac{1}{5\times 9^2} + \frac{1}{7\times 9^3} + \cdots$$

by comparing it with the series for $k\ln\left(\dfrac{1+x}{1-x}\right)$ with suitable values of k and x.

By using 'standard series', obtain power series expansions for the following functions (questions 5 to 12). In each case state the values of x for which the expansion is valid, and give the first three non-zero terms.

5   $\ln(1-2x)$ 6 $\ln(1+x^2)$ 7 $\ln\left(\dfrac{1+3x}{1-2x}\right)$ 8 $\ln\left(\dfrac{\sqrt{1+x}}{1-x}\right)$

9   $\ln(4-5x)$ 10 $e^x\ln(1-x)$ 11 $(\sin x)\ln(1-3x)$ 12 $\dfrac{\ln(1+x)}{\sqrt{1-x^2}}$

13   Find the Maclaurin series for $\ln(\cos x)$, up to the term in x^4.

14   If $y = \ln(1+e^x)$, show that $(1+e^x)\dfrac{dy}{dx} = e^x$.

Find the Maclaurin series for y (up to x^4).

15   If $y = \tan\ln(1+x)$, show that $(1+x)\dfrac{dy}{dx} = 1+y^2$.

Find the Maclaurin series for y (up to x^3).

16   If $y = \ln\left(\dfrac{e^x + e^{-x}}{2}\right)$, show that $(e^{2x}+1)\dfrac{dy}{dx} = e^{2x} - 1$.

Find the Maclaurin series for y (up to x^4).

19.5   Limits involving ln x

Behaviour as $x \to +\infty$

We know that $\ln x \to +\infty$ as $x \to +\infty$. We now consider the behaviour of $\dfrac{\ln x}{x^n}$ (where n is positive). Put

$$y = \ln x$$

so that

$$x = e^y$$

As $x \to +\infty$, $y \to +\infty$, and

$$\frac{\ln x}{x^n} = \frac{y}{e^{ny}} = ye^{-ny} \to 0$$

Hence, if $n > 0$,

$$\boxed{\frac{\ln x}{x^n} \to 0 \text{ as } x \to +\infty}$$

For example, as $x \to +\infty$, $\dfrac{\ln x}{x^2} \to 0$ and $\dfrac{\ln x}{\sqrt{x}} \to 0$.

We say that $\ln x$ tends to infinity more 'slowly' than any (positive) power of x.

Taking the reciprocal, we have, if $n > 0$,

$$\boxed{\frac{x^n}{\ln x} \to +\infty \text{ as } x \to +\infty}$$

If $n \leqslant 0$, say $n = -m$, then

$$\frac{\ln x}{x^n} = x^m \ln x \to +\infty \text{ as } x \to +\infty$$

and $\quad \dfrac{x^n}{\ln x} = \dfrac{1}{x^m \ln x} \to 0 \text{ as } x \to +\infty$

Behaviour as $x \to 0$

We know that $\ln x \to -\infty$ as $x \to 0$ $(x > 0)$. We now consider the behaviour of $x^n \ln x$ (where n is positive). Put $z = \dfrac{1}{x}$. As $x \to 0$ $(x > 0)$, $z \to +\infty$, and

$$x^n \ln x = \frac{1}{z^n} \ln\left(\frac{1}{z}\right) = \frac{-\ln z}{z^n} \to 0$$

Hence, if $n > 0$,

$$\boxed{x^n \ln x \to 0 \text{ as } x \to 0 \ (x > 0)}$$

and

$$\boxed{\frac{1}{x^n \ln x} \to -\infty \text{ as } x \to 0 \ (x > 0)}$$

If $n \leqslant 0$, say $n = -m$, then

$$x^n \ln x = \frac{\ln x}{x^m} \to -\infty \quad \text{as} \quad x \to 0$$

and $\quad \dfrac{1}{x^n \ln x} = \dfrac{x^m}{\ln x} \to 0 \quad \text{as} \quad x \to 0$

Sketching curves

Example Sketch the curve $y = \dfrac{x^2}{\ln x + 1}$.

y is not defined when $x \leqslant 0$ (since then $\ln x$ is not defined). The denominator is zero when $\ln x = -1$, i.e. $x = e^{-1}$. The curve has a vertical asymptote at $x = e^{-1}$.

y is positive when $\ln x > -1$, i.e. when $x > e^{-1}$, and negative when $\ln x < -1$, i.e. when $0 < x < e^{-1}$.

$$\text{As } x \to +\infty, \; y \approx \frac{x^2}{\ln x} \to +\infty.$$

$$\text{As } x \to 0 \; (x > 0), \quad y \approx \frac{x^2}{\ln x} \to 0.$$

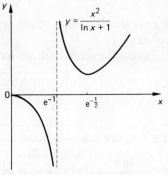

Fig. 19.9

Also

$$\frac{dy}{dx} = \frac{(\ln x + 1)(2x) - (x^2)\left(\dfrac{1}{x}\right)}{(\ln x + 1)^2} = \frac{x(2\ln x + 1)}{(\ln x + 1)^2}$$

so there is a stationary point when $\ln x = -\frac{1}{2}$, i.e. $x = e^{-\frac{1}{2}}$, and this is clearly a minimum.

The curve is sketched in Fig. 19.9.

Other limits

Some other limits may be evaluated by using the series for $\ln(1 + x)$.

Example Evaluate $\displaystyle\lim_{x \to 1}\left(\frac{\ln x}{x - 1}\right)$.

Let $x = 1 + h$. Then

$$\frac{\ln x}{x - 1} = \frac{\ln(1 + h)}{h} = \frac{h - \frac{1}{2}h^2 + \cdots}{h}$$

$$= 1 - \tfrac{1}{2}h + \cdots \to 1 \text{ as } h \to 0 \text{ (i.e. as } x \to 1)$$

Hence $\displaystyle\lim_{x \to 1}\left(\frac{\ln x}{x - 1}\right) = 1$.

The limit of $\left(1 + \dfrac{x}{n}\right)^n$ as $n \to \infty$

We have

$$\ln\left(1 + \frac{x}{n}\right)^n = n \ln\left(1 + \frac{x}{n}\right) = n\left[\frac{x}{n} - \frac{1}{2}\left(\frac{x}{n}\right)^2 + \frac{1}{3}\left(\frac{x}{n}\right)^3 - \cdots\right]$$

$$= x - \frac{x^2}{2n} + \frac{x^3}{3n^2} - \cdots$$

If x is fixed, and $n \to +\infty$, then $\dfrac{x^2}{2n}, \dfrac{x^3}{3n^2}, \ldots$ all tend to zero, thus

$$\ln\left(1 + \frac{x}{n}\right)^n \to x$$

and so $\left(1 + \dfrac{x}{n}\right)^n \to e^x$

Hence

$$\lim_{n \to +\infty} \left(1 + \frac{x}{n}\right)^n = e^x$$

[Note that this is the limit as n (and not x) tends to infinity.]
For example, we can find the value of e by putting $x = 1$. Then

$$e = \lim_{n \to +\infty} \left(1 + \frac{1}{n}\right)^n$$

and we obtain the following values:

n	10	100	1000	10000	100000
$\left(1 + \dfrac{1}{n}\right)^n$	2.594	2.705	2.717	2.718	2.718

Thus $e \approx 2.718$.

Exercise 19.5   (Limits involving $\ln x$)

1   State the limiting values of the following functions, as $x \to +\infty$.

(i) $\ln x$ (ii) $\dfrac{\ln x}{x}$ (iii) $x^{-\frac{1}{3}} \ln x$ (iv) $\dfrac{x^2}{\ln x}$ (v) $\dfrac{1}{x \ln x}$

(vi) $x^2 \ln x$ (vii) $\dfrac{(\ln x + 1)^2}{x}$ (viii) $\dfrac{x}{\ln x - 2}$ (ix) $\dfrac{1}{x} + \ln x$ (x) $\sqrt{x} \ln x$

2   State the limiting values of the functions in question 1, as $x \to 0$ $(x > 0)$.

Sketch the following curves (questions 3 to 6). You may already have found the stationary points in Exercise 19.2.

3 $y = x^2 \ln x$ **4** $y = \dfrac{\ln x}{x}$ **5** $y = \dfrac{x}{\ln x - 2}$ **6** $y = \dfrac{1}{x} + \ln x$

7   If $y = x^x$, find the limiting values of $\ln y$ as $x \to 0$ and as $x \to +\infty$, and deduce the corresponding limiting values of y. Sketch the curve $y = x^x$ (for $x > 0$).
8   If $y = x^{1/x}$, find the limiting values of $\ln y$ as $x \to 0$ and as $x \to +\infty$, and deduce the corresponding limiting values of y. Sketch the curve $y = x^{1/x}$ (for $x > 0$).

9   By writing $x = 2 + h$, and expanding $\dfrac{\ln(x-1)}{x-2}$ as a power series in h, evaluate $\lim\limits_{x \to 2} \left(\dfrac{\ln(x-1)}{x-2}\right)$.

10   Evaluate $\lim\limits_{x \to 1} \left(\dfrac{\ln[\frac{1}{2}(1+x)]}{x-1}\right)$ by first writing $x = 1 + h$.

Use the result $\lim\limits_{n \to \infty} \left(1 + \dfrac{x}{n}\right)^n = e^x$ to evaluate the following: **11** $e^{3.4}$ **12** $e^{-0.8}$

20

Differential equations

20.1 Introduction

An equation which contains derivatives is called a differential equation. For example,

$$\frac{dy}{dx} = 4x - 5$$

is a differential equation of the **first order**, since it only contains the first derivative $\frac{dy}{dx}$. An equation like

$$\frac{d^2y}{dx^2} + 3\frac{dy}{dx} + 2y = e^{-x}$$

is a **second order** differential equation, since it contains $\frac{d^2y}{dx^2}$.

We have seen (in Chapter 12) how differential equations occur in practical situations involving rates of change. Differential equations are to be found in mechanics, physics, chemistry, biology, geography and economics, and we shall look at some of these applications in Chapter 21.

A 'solution' of a differential equation is a function which satisfies the equation; to verify that a given function is a solution, we differentiate it, then substitute it into the equation, and check that the two sides of the equation are identical. Usually, of course, we will not be given a solution; the process of finding the solutions of a given differential equation is called 'solving' the equation. We have already seen how to do this in the following four cases.

(i) $\frac{dy}{dx}$ given in terms of x (see p. 69). We simply integrate with respect to x, to obtain y in terms of x.

Example 1

If $\frac{dy}{dx} = 4x - 5$

then

$$y = 2x^2 - 5x + C$$

Each value of the arbitrary constant C gives a different solution; for example,

$$y = 2x^2 - 5x + 1 \quad \text{and} \quad y = 2x^2 - 5x - 2$$

are 'individual' solutions. ['Individual' solutions are sometimes called 'particular' solutions, but we shall use the term 'particular solution' in a more specialised way; see p. 368.]

When plotted, the solutions give a family of curves, each one of which may be obtained from any other by a translation parallel to the y-axis (see Fig. 20.1).

If we are given the value of y for some value of x, for example, $y = 0$ when $x = 1$, then we know that the solution curve passes through the point $(1, 0)$ and we can pick out the individual curve from the family which satisfies this condition. Substituting $y = 0$ and $x = 1$ into $y = 2x^2 - 5x + C$ gives $0 = 2 - 5 + C$; thus $C = 3$ and the individual solution is $y = 2x^2 - 5x + 3$.

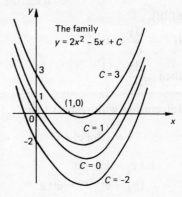

Fig. 20.1

(ii) $\dfrac{dy}{dx}$ given in terms of y (see p. 183). We rearrange to find $\dfrac{dx}{dy}$, and then integrate with respect to y.

Example 2

If
$$\frac{dy}{dx} = \sec y$$

then
$$\frac{dx}{dy} = \frac{1}{\sec y} = \cos y$$

so, integrating with respect to y,

$$x = \sin y + C$$

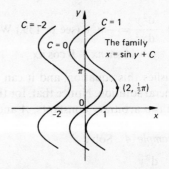

Fig. 20.2

This gives a family of curves which are related by translations parallel to the x-axis (see Fig. 20.2).

If, for example, $y = \frac{1}{2}\pi$ when $x = 2$, then, substituting, $2 = 1 + C$; thus $C = 1$ and the individual solution is $x = \sin y + 1$.

(iii) $\dfrac{dy}{dx} = ky$ (see p. 335). We discovered that $y = Ae^{kx}$ is a solution of this equation.

Example 3

If
$$\frac{dy}{dx} = 2y$$

then
$$y = Ae^{2x}$$

The solution curves are all the multiples of e^{2x} (see Fig. 20.3). If, for example, $y = 4$ when $x = 0$, then $4 = Ae^0$; thus $A = 4$, and the individual solution is $y = 4e^{2x}$.

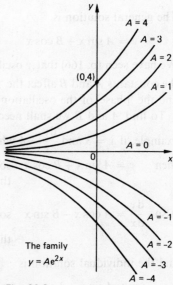

Fig. 20.3

We can show that $y = Ae^{kx}$ is indeed the most general solution of the differential equation $\dfrac{dy}{dx} = ky$, by regarding it as a special case of (ii).

If $\quad \dfrac{dy}{dx} = ky$

then $\quad \dfrac{dx}{dy} = \dfrac{1}{ky}$

Integrating with respect to y,

$$x = \frac{1}{k}\ln|y| + C$$

$$\ln|y| = kx - kC$$

$$|y| = e^{kx-kC} = e^{kx}e^{-kC}$$

Hence $\quad y = Ae^{kx} \quad$ (where $A = \pm e^{-kC}$)

Notice that the general solution of a first order differential equation always contains *one* arbitrary constant.

(iv) $\dfrac{d^2 y}{dx^2} = -n^2 y$ (see p. 139.) We discovered that

$$y = A \sin nx + B \cos nx$$

satisfies this equation, and it can be shown that this is the most general solution. Notice that, for this second order equation, there are *two* arbitrary constants, A and B.

Example 4   Solve

$$\frac{d^2 y}{dx^2} = -y$$

The general solution is

$$y = A \sin x + B \cos x$$

We have seen (p. 166) that y oscillates about zero with period 2π; the constants A and B affect the 'amplitude' (which is $\sqrt{A^2 + B^2}$) and the 'phase' of the oscillations.

To find A and B we shall need two items of information. For example, if $y = 3$ and $\dfrac{dy}{dx} = 4$ when $x = 0$,

then $\quad y = A \sin x + B \cos x \quad$ so $\quad 3 = 0 + B$

$\qquad\qquad\qquad\qquad\qquad$ thus $\quad B = 3$

Also $\dfrac{dy}{dx} = A \cos x - B \sin x \quad$ so $\quad 4 = A - 0$

$\qquad\qquad\qquad\qquad\qquad$ thus $\quad A = 4$

and the individual solution is

$$y = 4 \sin x + 3 \cos x \quad \text{(see Fig. 20.4)}$$

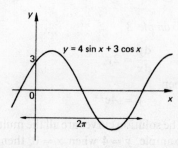

Fig. 20.4

In this chapter we study some further methods for solving differential equations, to obtain a relationship between x and y (not involving derivatives). However, for many differential equations this is not possible; we then use numerical methods, and obtain an approximate solution in the form of a table of values (see Chapter 21).

Differential equation from a family of curves

We have seen that the solutions of a differential equation form a family of curves. Conversely, if we are given a family of curves, we can find a differential equation of which all the members of the family are solutions.

Example Find the differential equation which is satisfied by all members of the family of curves

$$y = \frac{A}{1+x^2}$$

We have $y = \dfrac{A}{1+x^2}$

and, differentiating,

$$\frac{dy}{dx} = \frac{-2Ax}{(1+x^2)^2}$$

We obtain the differential equation by eliminating A between these two equations. Thus

$$A = (1+x^2)y$$

and so $\dfrac{dy}{dx} = \dfrac{-2(1+x^2)yx}{(1+x^2)^2} = \dfrac{-2xy}{1+x^2}$

Hence the differential equation is

$$(1+x^2)\frac{dy}{dx} + 2xy = 0$$

Exercise 20.1 (Differential equations; revision)

In questions 1 to 5, verify that the given function is a solution of the given differential equation.

1 $y = e^{3x} + 2$; $\dfrac{dy}{dx} = 3y - 6$ **2** $y = \dfrac{3}{x}$; $x\dfrac{dy}{dx} + y = 0$ **3** $y = x \sin x$; $\dfrac{d^2y}{dx^2} + y = 2\cos x$

4 $y = 4e^{-x} - 2e^{3x}$; $\dfrac{d^2y}{dx^2} - 2\dfrac{dy}{dx} - 3y = 0$ **5** $y = e^{-x} \sin 2x$; $\dfrac{d^2y}{dx^2} + 2\dfrac{dy}{dx} + 5y = 0$

In questions 6 to 10, find the general solution of the differential equation.

6 $\dfrac{dy}{dx} = x(2 - x)$ **7** $\dfrac{dy}{dx} = 3y^2$ **8** $\dfrac{dy}{dx} = -5y$ **9** $\dfrac{d^2y}{dx^2} = -16y$

10 $\dfrac{dy}{dx} = y(2 - y)$ $\left[\text{Use } \dfrac{1}{y(2-y)} = \dfrac{1}{2}\left(\dfrac{1}{y} + \dfrac{1}{2-y}\right) \right]$

In questions 11 to 14, find the individual solution of the differential equation, which satisfies the given condition.

11   $\dfrac{dy}{dx} = \cos^2 2x;\ y = 3$ when $x = 0$.

12   $\dfrac{dy}{dx} = e^y;\ y = 0$ when $x = 4$.

13   $\dfrac{d^2 y}{dx^2} = -4y;\ y = 2$ and $\dfrac{dy}{dx} = -6$ when $x = 0$.

14   $\dfrac{dy}{dx} = \sin^2 y;\ y = \frac{1}{2}\pi$ when $x = 3$.

For each of the following differential equations (questions 15 and 16), sketch the family of curves which are solutions. Pick out in colour the individual solution indicated.

15   $\dfrac{dy}{dx} = \sin x$; pick out that for which $y = 0$ when $x = 0$.

16   $\dfrac{dy}{dx} = -y$; pick out that for which $y = -1$ when $x = 0$.

In questions 17 to 20, find the differential equation which is satisfied by all members of the given family of curves.

17   $y = Ax$

18   $y = \dfrac{C}{x^2}$

19   $y = \dfrac{1}{x+C}$

20   $y = x^3 + Cx$

20.2   Differential equations of the form

$a\dfrac{dy}{dx} + by = f(x)$ (where a and b are constants)

This is called a first order 'linear' differential equation with constant coefficients.

If $f(x) = 0$, the equation

$$a\frac{dy}{dx} + by = 0$$

may be written

$$\frac{dy}{dx} = -\frac{b}{a}y$$

and we can write down the general solution as

$$y = Ae^{-\frac{b}{a}x}$$

We now consider some examples in which $f(x)$ is not zero.

Example 1

$$\frac{dy}{dx} - 2y = 5$$

We can solve this as follows:

$$\frac{dy}{dx} = 2y + 5$$

so $\dfrac{dx}{dy} = \dfrac{1}{2y+5}$

Integrating with respect to y,

$$x = \tfrac{1}{2}\ln|2y+5| + C$$
$$\ln|2y+5| = 2x - 2C$$
$$|2y+5| = e^{2x-2C} = e^{2x}e^{-2C}$$
$$2y+5 = Be^{2x} \quad (\text{where } B = \pm e^{-2C})$$
$$y = Ae^{2x} - \tfrac{5}{2} \quad (\text{where } A = \tfrac{1}{2}B)$$

Notice that

$$y = Ae^{2x}$$

is the general solution of

$$\frac{dy}{dx} - 2y = 0$$

and $y = -\tfrac{5}{2}$ is a solution of the given equation

$$\frac{dy}{dx} - 2y = 5$$

$$\left(\text{if } y = -\tfrac{5}{2} \text{ then } \frac{dy}{dx} = 0, \text{ and so } \frac{dy}{dx} - 2y = 0 - 2(-\tfrac{5}{2}) = 5\right).$$

Example 2

$$\frac{dy}{dx} + y = 2x + x^2$$

$y = x^2$ is an obvious solution of this equation. The general solution of

$$\frac{dy}{dx} + y = 0$$

is $y = Ae^{-x}$

We therefore try

$$y = Ae^{-x} + x^2$$

as the general solution of the given equation. If

$$y = Ae^{-x} + x^2$$

then $\dfrac{dy}{dx} = -Ae^{-x} + 2x$

and $\dfrac{dy}{dx} + y = -Ae^{-x} + 2x + Ae^{-x} + x^2$

$$= 2x + x^2$$

Hence $y = Ae^{-x} + x^2$

satisfies the given equation.

These two examples suggest the following method for solving

$$a \frac{dy}{dx} + by = f(x)$$

(i) Write down the general solution of

$$a \frac{dy}{dx} + by = 0$$

This is $\quad y = Ae^{-\frac{b}{a}x}$

and it is called the **complementary function** (CF).

(ii) Find one '**particular solution**' (PS; sometimes called a 'particular integral') of the given equation

$$a \frac{dy}{dx} + by = f(x)$$

This is usually done by trying a function of similar form to $f(x)$.

(iii) The general solution is obtained by adding the complementary function and the particular solution, i.e.

$$y = CF + PS$$

(iv) If initial conditions are given, substitute into the general solution found in (iii), to find the arbitrary constant (A), and hence obtain the individual solution.

We call this the 'CF and PS method', and we shall give a justification for it later (see p. 382).

Example 3

$$\frac{dy}{dx} + y = 2x + 4$$

given that $y = 4$ when $x = 0$.

The CF $\left(\text{i.e. the general solution of } \frac{dy}{dx} + y = 0\right)$ is

$$y = Ae^{-x}$$

Since the RHS, $2x + 4$, is a linear function, we try to find a linear function satisfying the equation. We try

$$y = px + q$$

Then $\quad \dfrac{dy}{dx} = p$

and, substituting into the equation,

$$p + px + q = 2x + 4$$

Comparing coefficients of x,

$$p = 2$$

Comparing the constant term,

$$p + q = 4$$

Thus $p = 2$, $q = 2$, and

$$y = 2x + 2$$

is a particular solution of the equation.
   The general solution is therefore

$$y = Ae^{-x} + 2x + 2$$

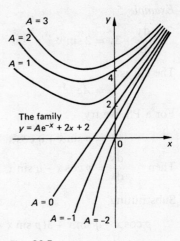

The family
$y = Ae^{-x} + 2x + 2$

Fig. 20.5 shows this family of curves. Note that the particular
solution $y = 2x + 2$ corresponds to $A = 0$.
   Substituting $y = 4$ when $x = 0$,

Fig. 20.5

$$4 = Ae^0 + 0 + 2$$

Thus $A = 2$ and the individual solution is

$$y = 2e^{-x} + 2x + 2$$

To find a PS we try a function of the same type as $f(x)$, the RHS of
the given equation. The following examples may be helpful.

If	try
$f(x) = 5$	$y = p$
$f(x) = 3x$	$y = px + q$
$f(x) = 7 - 4x$	$y = px + q$
$f(x) = x^2$	$y = px^2 + qx + r$
$f(x) = 4\cos x - 2\sin x$	$y = p\sin x + q\cos x$
$f(x) = 3\sin 2x$	$y = p\sin 2x + q\cos 2x$
$f(x) = 5e^{2x}$	$y = pe^{2x}$
$f(x) = 7e^{-3x}$	$y = pe^{-3x}$

and so on.

Example 4

$$\frac{dy}{dx} - 4y = 7$$

The CF is

$$y = Ae^{4x}$$

For a PS, we try $y = p$, then $\dfrac{dy}{dx} = 0$, and substituting into the

equation, $0 - 4p = 7$. Thus $p = -\frac{7}{4}$, and a PS is $y = -\frac{7}{4}$.
   Hence the general solution is

$$y = Ae^{4x} - \frac{7}{4}$$

Example 5

$$\frac{dy}{dx} + 3y = 2 \sin x$$

The CF is

$$y = Ae^{-3x}$$

For a PS, we try

$$y = p \sin x + q \cos x$$

Then $\quad \dfrac{dy}{dx} = p \cos x - q \sin x$

Substituting,

$$p \cos x - q \sin x + 3(p \sin x + q \cos x) = 2 \sin x$$

Comparing coefficients of sin x:

$$-q + 3p = 2$$

Comparing coefficients of cos x:

$$p + 3q = 0$$

which give $p = \frac{3}{5}$, $q = -\frac{1}{5}$. Thus a PS is

$$y = \tfrac{3}{5} \sin x - \tfrac{1}{5} \cos x$$

Hence the general solution is

$$y = Ae^{-3x} + \tfrac{3}{5} \sin x - \tfrac{1}{5} \cos x$$

Example 6

$$\frac{dy}{dx} + 5y = e^{-2x}$$

given that $y = 1$ when $x = 0$.
  The CF is

$$y = Ae^{-5x}$$

For a PS, we try

$$y = pe^{-2x}$$

Then $\quad \dfrac{dy}{dx} = -2pe^{-2x}$

and, substituting,

$$-2pe^{-2x} + 5pe^{-2x} = e^{-2x}$$

Thus $-2p + 5p = 1$, so $p = \dfrac{1}{3}$, and a PS is

$$y = \frac{1}{3}e^{-2x}$$

The general solution is

$$y = Ae^{-5x} + \frac{1}{3}e^{-2x}$$

Substituting $y = 1$ when $x = 0$, $1 = A + \frac{1}{3}$. Thus $A = \frac{2}{3}$, and the individual solution is

$$y = \frac{2}{3}e^{-5x} + \frac{1}{3}e^{-2x}$$

Example 7

$$\frac{dy}{dx} + y = 3e^{-x}$$

The CF is

$$y = Ae^{-x}$$

For a PS, if we try

$$y = pe^{-x}$$

then

$$\frac{dy}{dx} = -pe^{-x}$$

and we require

$$-pe^{-x} + pe^{-x} = 3e^{-x}$$

which is impossible. This means that we cannot find a PS of the form

$$y = pe^{-x}$$

The reason is that

$$y = pe^{-x}$$

is of the same form as the CF, and therefore satisfies

$$\frac{dy}{dx} + y = 0$$

so it cannot also satisfy

$$\frac{dy}{dx} + y = 3e^{-x}$$

In such cases we multiply our previous attempt by x, and try again (but see p. 379 for an alternative approach, which explains why this works).

We now try

$$y = pxe^{-x}$$

Then

$$\frac{dy}{dx} = -pxe^{-x} + pe^{-x}$$

Substituting,

$$-pxe^{-x} + pe^{-x} + pxe^{-x} = 3e^{-x}$$
$$pe^{-x} = 3e^{-x}$$

Thus $p = 3$, and a PS is

$$y = 3xe^{-x}$$

Hence the general solution is

$$y = Ae^{-x} + 3xe^{-x}$$

Exercise 20.2 $\left(\text{Differential equations: type } a\dfrac{dy}{dx} + by = f(x)\right)$

Find the general solution of the following differential equations (questions 1 to 12).

1 $\dfrac{dy}{dx} - 2y = 0$ **2** $3\dfrac{dy}{dx} + 5y = 0$ **3** $\dfrac{dy}{dx} + y = 2x - 5$ **4** $\dfrac{dy}{dx} - 2y = 6$

5 $\dfrac{dy}{dx} + 4y = x^2 + 3x - 2$ **6** $\dfrac{dy}{dx} - 3y = x^2$ **7** $2\dfrac{dy}{dx} + y = x - 1$ **8** $\dfrac{dy}{dx} - y = 2\cos x$

9 $\dfrac{dy}{dx} + 2y = \sin 3x$ **10** $\dfrac{dy}{dx} + 3y = e^x$ **11** $4\dfrac{dy}{dx} - 3y = 2e^{-3x}$

12 $\dfrac{dy}{dx} - y = 2e^x$ $\left[\text{Try } y = pxe^x \text{ for the PS.}\right]$

For the following differential equations (questions 13 to 20) find the solution which satisfies the given condition.

13 $8\dfrac{dy}{dx} - 3y = 0; \; y = 2$ when $x = 0$.   **14** $\dfrac{dy}{dx} - 4y = 0; \; y = 0$ when $x = 3$.

15 $\dfrac{dy}{dx} + y = 0; \; y = 1$ when $x = 2$.   **16** $\dfrac{dy}{dx} + 2y = 5; \; y = 3$ when $x = 0$.

17 $\dfrac{dy}{dx} - y = 2x + 7; \; y = 0$ when $x = 0$.   **18** $\dfrac{dy}{dx} + 3y = \sin 2x; \; y = 1$ when $x = 0$.

19 $\dfrac{dy}{dx} - 2y = 3e^{-x}; \; y = -1$ when $x = 0$.   **20** $\dfrac{dy}{dx} + 4y = 2e^{-4x}; \; y = 2$ when $x = 0$.

20.3 Differential equations with separable variables

This method, which can be used to solve a wide variety of first order equations, follows from the differentiation of implicit relations.

Example 1

$$\frac{dy}{dx} = \frac{4 - 2x}{2y + 6}$$

This may be written

$$(2y + 6)\frac{dy}{dx} = 4 - 2x$$

Now $(2y+6)\dfrac{dy}{dx}$ is the result of differentiating $y^2 + 6y$ with respect to x; hence the solution of the differential equation is given by the implicit relation

$$y^2 + 6y = 4x - x^2 + C$$

Effectively, we have integrated $2y + 6$ with respect to y (to obtain $y^2 + 6y$) and $4 - 2x$ with respect to x (to obtain $4x - x^2$). The solution may conveniently be set out as follows:

$$\frac{dy}{dx} = \frac{4 - 2x}{2y + 6}$$

$$\int (2y + 6)\, dy = \int (4 - 2x)\, dx$$

(Here we have 'multiplied by dx', and rearranged so that all the terms containing y are on the LHS, with 'dy', and all the terms containing x are on the RHS, with 'dx'. The variables, x and y, have been 'separated'.)

Then

$$y^2 + 6y = 4x - x^2 + C$$

(It is only necessary to include an arbitrary constant on one side of this equation. Otherwise,

$$y^2 + 6y + A = 4x - x^2 + B$$

which is the same as

$$y^2 + 6y = 4x - x^2 + C \quad \text{with } C = B - A.)$$

In general, this method will work if we can write the differential equation in the form

$$\frac{dy}{dx} = \frac{f(x)}{g(y)}$$

Then, separating the variables,

$$\int g(y)\, dy = \int f(x)\, dx$$

The solution will be in the form of an implicit relation. If it is reasonable to do so, we should rearrange the solution to give y in terms of x.

Example 2

$$\frac{dy}{dx} = \frac{4y}{2x+3}$$

Separating the variables,

$$\int \frac{1}{y} dy = \int \frac{4}{2x+3} dx$$

$\ln|y| = 2\ln|2x+3| + C$

$\ln|y| = 2\ln|2x+3| + \ln B$ (writing $C = \ln B$)

$\ln|y| = \ln[B(2x+3)^2]$

Hence $y = A(2x+3)^2$ (where $A = \pm B$)

Example 3

$$\frac{dy}{dx} = \frac{1}{y\sqrt{1+4x}}$$

given that $y = 1$ when $x = 2$.

Separating the variables,

$$\int y \, dy = \int \frac{1}{\sqrt{1+4x}} dx$$

$\frac{1}{2}y^2 = \frac{1}{4} \times 2\sqrt{1+4x} + C$

$y^2 = \sqrt{1+4x} + A$ (where $A = 2C$)

Substituting $y = 1$ and $x = 2$, $1 = 3 + A$. Thus $A = -2$, and the individual solution is

$$y^2 = \sqrt{1+4x} - 2$$

Example 4

$$(1+x^2)\frac{dy}{dx} - \cos^2 y = 0$$

$$(1+x^2)\frac{dy}{dx} = \cos^2 y$$

Separating the variables,

$$\int \frac{1}{\cos^2 y} dy = \int \frac{1}{1+x^2} dx$$

$$\int \sec^2 y \, dy = \int \frac{1}{1+x^2} dx$$

Hence $\tan y = \tan^{-1} x + C$

Exercise 20.3 (Differential equations; separable variables)

In questions 1 to 15 give the general solutions of the differential equations.

1 $\dfrac{dy}{dx} + \dfrac{3}{x^2} = 0$ **2** $\dfrac{dy}{dx} = \dfrac{y^2}{4}$ **3** $\dfrac{dy}{dx} = \dfrac{2x}{y}$

4 $\dfrac{dy}{dx} = \dfrac{y}{x}$ **5** $(1-y)\dfrac{dy}{dx} = 1+x$ **6** $(1-x)\dfrac{dy}{dx} = 1+y$

7 $x(y-1)\dfrac{dy}{dx} = 2y$ **8** $\dfrac{dy}{dx} = x(y+2)$ **9** $\dfrac{dy}{dx} = x^2 y^2$

10 $xy\dfrac{dy}{dx} = \dfrac{y^2+1}{x^2+1}$ **11** $xy\dfrac{dy}{dx} = \dfrac{1+x^2}{1+y}$ **12** $x\sqrt{y^2-1} - y\sqrt{x^2-1}\dfrac{dy}{dx} = 0$

13 $\dfrac{dy}{dx} = y^2 \sin x$ **14** $\sin x \cos y = \sin y \cos x \dfrac{dy}{dx}$ **15** $\cos^2 x \dfrac{dy}{dx} = 2y$

In questions 16 to 20 find the solution which satisfies the given condition.

16 $\dfrac{dy}{dx} = 2xy$; $y = 1$ when $x = 1$.   **17** $\dfrac{dy}{dx} = -\dfrac{x}{y}$; $y = 5$ when $x = 0$.

18 $(1+x^2)\dfrac{dy}{dx} = 4e^y$; $y = 0$ when $x = 0$.   **19** $x\dfrac{dy}{dx} = y+1$; $y = 1$ when $x = 2$.

20 $\dfrac{dy}{dx} = y \cos x$; $y = 0$ when $x = \dfrac{\pi}{4}$.

20.4 The integrating factor method

This method enables us to solve differential equations of the form

$$\frac{dy}{dx} + f(x)\,y = g(x)$$

which is the general first order 'linear' equation.

Example 1

$$\frac{dy}{dx} + \frac{2}{x}y = 4x+3$$

If we multiply this equation by x^2, we have

$$x^2 \frac{dy}{dx} + 2xy = 4x^3 + 3x^2$$

Now the LHS, $x^2\dfrac{dy}{dx} + 2xy$, is the result of differentiating the product $x^2 y$. Thus

$$\frac{d}{dx}(x^2 y) = 4x^3 + 3x^2$$

Integrating,

$$x^2 y = x^4 + x^3 + C$$

Hence

$$y = x^2 + x + \frac{C}{x^2}$$

Now consider the general equation

$$\frac{dy}{dx} + f(x)y = g(x)$$

The idea is to multiply the equation by a suitable function, as in Example 1, so that the LHS becomes the derivative of a product. Suppose we multiply by $m(x)$. Then

$$m(x)\frac{dy}{dx} + m(x)f(x)y = m(x)g(x)$$

Now $\qquad \frac{d}{dx}(m(x)y) = m(x)\frac{dy}{dx} + m'(x)y$

and so we need

$$m'(x) = m(x)f(x)$$

i.e. m satisfies the differential equation

$$\frac{dm}{dx} = mf(x)$$

In this equation, the variables are separable,

$$\int \frac{1}{m}dm = \int f(x)dx$$

$$\ln m = \int f(x)dx$$

$$m(x) = e^{\int f(x)dx}$$

We then have

$$\frac{d}{dx}[m(x)y] = m(x)g(x)$$

which we can integrate, obtaining

$$m(x)y = \int m(x)g(x)dx$$

$m(x)$ is called the **integrating factor.**

For the differential equation $\dfrac{dy}{dx} + f(x)y = g(x)$
the integrating factor is $m(x) = e^{\int f(x)dx}$

and then the equation may be written as $\dfrac{d}{dx}[m(x)y] = m(x)g(x)$

Example 2

$$(1+x^2)\frac{dy}{dx} - xy = x(1+x^2)$$

We first divide through by $(1+x^2)$, so that the equation is in the standard form,

$$\frac{dy}{dx} - \frac{x}{1+x^2}y = x$$

It is essential that we work with *this* equation (so that $f(x) = -\frac{x}{1+x^2}$, and $g(x) = x$), and not the original equation. The integrating factor is*

$$e^{\int \frac{-x}{1+x^2}dx} = e^{-\frac{1}{2}\ln(1+x^2)} = e^{-\ln\sqrt{1+x^2}} = \frac{1}{\sqrt{1+x^2}}$$

Multiplying by $\dfrac{1}{\sqrt{1+x^2}}$ gives

$$\frac{1}{\sqrt{1+x^2}}\frac{dy}{dx} - \frac{x}{\sqrt{(1+x^2)^3}}y = \frac{x}{\sqrt{1+x^2}}$$

i.e.

$$\frac{d}{dx}\left(\frac{1}{\sqrt{1+x^2}}y\right) = \frac{x}{\sqrt{1+x^2}}$$

Hence

$$\frac{1}{\sqrt{1+x^2}}y = \int \frac{x}{\sqrt{1+x^2}}dx = \sqrt{1+x^2} + C$$

Thus

$$y = 1 + x^2 + C\sqrt{1+x^2}$$

We can also use an integrating factor for equations of the form

$$a\frac{dy}{dx} + by = f(x)$$

which we solved previously using the complementary function and particular solution method.

Example 3

$$\frac{dy}{dx} + y = 2x + 4$$

The integrating factor is

$$e^{\int 1\,dx} = e^x$$

---

* There is no need to include an arbitrary constant at this stage. If we do, the integrating factor is $e^{-\ln\sqrt{1+x^2}+C} = e^C e^{-\ln\sqrt{1+x^2}} = \dfrac{A}{\sqrt{1+x^2}}$, and we can divide through by the constant A.

Mulplying by e^x gives

$$e^x \frac{dy}{dx} + e^x y = e^x(2x+4)$$

i.e. $\quad \frac{d}{dx}(e^x y) = e^x(2x+4)$

Integrating,

$$e^x y = \int e^x(2x+4)dx$$

which we integrate by parts.

Put $\quad u = 2x+4 \quad$ and $\quad \frac{dv}{dx} = e^x$

Then $\quad \frac{du}{dx} = 2 \quad\quad$ and $\quad v = e^x$

So $\quad e^x y = (2x+4)e^x - \int 2e^x dx$

$$= (2x+4)e^x - 2e^x - C$$

$$= (2x+2)e^x + C$$

Thus $\quad y = 2x+2+Ce^{-x}$

(This is the same result as before; see p. 368, Example 3.)

Example 4

$$\frac{dy}{dx} + 5y = e^{-2x}$$

The integrating factor is

$$e^{\int 5dx} = e^{5x}$$

Multiplying by e^{5x} gives

$$e^{5x} \frac{dy}{dx} + 5e^{5x} y = e^{3x}$$

i.e. $\quad \frac{d}{dx}(e^{5x} y) = e^{3x}$

Integrating

$$e^{5x} y = \int e^{3x} dx = \tfrac{1}{3}e^{3x} + C$$

Thus $\quad\quad y = \tfrac{1}{3}e^{-2x} + Ce^{-5x}$

(This is the same result as before; see p. 370, Example 6.)

Example 5

$$\frac{dy}{dx} + y = 3e^{-x}$$

(This is the example in which we encountered some difficulty in finding the particular solution; see p. 371, Example 7.)

The integrating factor is

$$e^{\int 1\,dx} = e^x$$

Multiplying by e^x gives

$$e^x \frac{dy}{dx} + e^x y = 3$$

i.e. $$\frac{d}{dx}(e^x y) = 3$$

Integrating,

$$e^x y = \int 3\,dx = 3x + C$$

Thus $$y = 3xe^{-x} + Ce^{-x} \quad \text{(as before)}$$

Exercise 20.4   (Differential equations, using an integrating factor)

In questions 1 to 10 give the general solutions of the differential equations.

1 $\dfrac{dy}{dx} - 2xy = 2x$ 2 $x\dfrac{dy}{dx} + y + x = 0$ 3 $\dfrac{dy}{dx} + y = e^{-x}$ 4 $x\dfrac{dy}{dx} - y = x^4$

5 $(1 - x^2)\dfrac{dy}{dx} - xy = 1$ 6 $\cos x \dfrac{dy}{dx} + y \sin x = 1$ 7 $\cos x \dfrac{dy}{dx} - y \sin x = 1$

8 $(1 + x^2)\dfrac{dy}{dx} + 4xy = 3$ 9 $\tan x \dfrac{dy}{dx} - y = \sin^2 x$ 10 $\dfrac{dy}{dx} - y = x^3 e^x$

In questions 11 to 14 find the solution which satisfies the given condition.

11 $x\dfrac{dy}{dx} + 2y = x^3$; $y = 2$ when $x = 1$.   12 $(1 + x)\dfrac{dy}{dx} + 2y = x^2$; $y = 0$ when $x = 0$.

13 $\sin x \dfrac{dy}{dx} - y \cos x = 1$; $y = 3$ when $x = \dfrac{\pi}{2}$.

14 $x(x + 1)\dfrac{dy}{dx} + y = (x + 1)^2 e^x$; $y = 0$ when $x = 1$.

15 Solve the differential equation

$$\frac{dy}{dx} - 2y = 3$$

(i) by CF and PS method;   (ii) by separating variables;   (iii) using an integrating factor.

20.5   Use of a substitution

Some differential equations can be reduced to a form in which the variables are separable, or to a form suitable for the integrating factor method, by means of a substitution.

We introduce a new variable, say v, related to x and y, and we express $\dfrac{dy}{dx}$ in terms of $\dfrac{dv}{dx}$. By substituting for y and $\dfrac{dy}{dx}$ in the original equation, we obtain a differential equation for v and x. The substitution is successful if we can solve this new equation to give a relation between v and x. Finally, we rewrite this solution in terms of y and x.

Example 1   Use the substitution $y = vx$ to solve the differential equation

$$x^2 \frac{dy}{dx} = xy + y^2$$

Let

$$y = vx$$

$\left(\text{We are thus defining a new variable } v \text{ by } v = \dfrac{y}{x}.\right)$ Then

$$\frac{dy}{dx} = (v)(1) + \frac{dv}{dx} x$$

$$= v + x \frac{dv}{dx}$$

Substituting these into the equation

$$x^2 \frac{dy}{dx} = xy + y^2$$

$$x^2 \left(v + x \frac{dv}{dx}\right) = x(vx) + (vx)^2$$

Dividing by x^2,

$$v + x \frac{dv}{dx} = v + v^2$$

$$x \frac{dv}{dx} = v^2$$

The variables are now separable.

$$\int \frac{1}{v^2} dv = \int \frac{1}{x} dx$$

$$-\frac{1}{v} = \ln x + C$$

Since $\qquad v = \dfrac{y}{x},$

$$-\dfrac{x}{y} = \ln x + C$$

Hence $\qquad y = \dfrac{-x}{\ln x + C}$

Note that the substitution $y = vx$ can be used whenever $\dfrac{dy}{dx}$ is a function of $\dfrac{y}{x}$ only. This equation can be written $\dfrac{dy}{dx} = \dfrac{y}{x} + \left(\dfrac{y}{x}\right)^2$.

Such equations are called homogeneous differential equations. Another example is

$$(2y - x)\dfrac{dy}{dx} = (3x + y)$$

which can be written

$$\left(2\dfrac{y}{x} - 1\right)\dfrac{dy}{dx} = 3 + \dfrac{y}{x}$$

Example 2   Use the substitution $\dfrac{1}{y^3} = u$ to solve the differential equation

$$x\dfrac{dy}{dx} + 5y = xy^4$$

Let $\dfrac{1}{y^3} = u$. Differentiating this with respect to x,

$$-\dfrac{3}{y^4}\dfrac{dy}{dx} = \dfrac{du}{dx}$$

Dividing the original equation by y^4,

$$\dfrac{x}{y^4}\dfrac{dy}{dx} + \dfrac{5}{y^3} = x$$

and then, substituting,

$$x\left(-\dfrac{1}{3}\dfrac{du}{dx}\right) + 5u = x$$

$$\dfrac{du}{dx} - \dfrac{15}{x}u = -3$$

This can now be solved using the integrating factor

$$e^{\int -\frac{15}{x}dx} = e^{-15\ln x} = \dfrac{1}{x^{15}}$$

Multiplying by $\dfrac{1}{x^{15}}$,

$$\frac{1}{x^{15}}\frac{du}{dx} - \frac{15}{x^{16}}u = -\frac{3}{x^{15}}$$

i.e. $$\frac{d}{dx}\left(\frac{1}{x^{15}}u\right) = -\frac{3}{x^{15}}$$

Hence $$\frac{1}{x^{15}}u = \int -\frac{3}{x^{15}}dx = \frac{3}{14x^{14}}+C$$

$$u = \frac{3}{14}x + Cx^{15}$$

Thus $$\frac{1}{y^3} = \frac{3}{14}x + Cx^{15}$$

Application to the CF and PS method

We can now give a justification for the CF and PS method, which we used earlier (see p. 368) to solve equations of the form

$$a\frac{dy}{dx} + by = f(x)$$

Example   Consider again the equation

$$\frac{dy}{dx} + y = 2x + x^2$$

(see p. 367, Example 2).

We observe that $y = x^2$ is a particular solution. To find the general solution, put

$$y = u + x^2$$

(We are defining a new variable u by $u = y - x^2$.) Then

$$\frac{dy}{dx} = \frac{du}{dx} + 2x$$

Substituting,

$$\left(\frac{du}{dx} + 2x\right) + (u + x^2) = 2x + x^2$$

$$\frac{du}{dx} + u = 0$$

$$u = Ae^{-x}$$

Hence the general solution, $y = u + x^2$, is indeed

$$y = Ae^{-x} + x^2$$

In general, suppose we wish to solve the equation

$$a\frac{dy}{dx} + by = f(x)$$

and we have found one function, w say (the PS), which satisfies the equation, i.e.

$$a\frac{dw}{dx} + bw = f(x)$$

Now put

$$y = u + w \quad \text{where } u = y - w \text{ is a new variable}$$

Then
$$\frac{dy}{dx} = \frac{du}{dx} + \frac{dw}{dx}$$

and, substituting into

$$a\frac{dy}{dx} + by = f(x)$$

gives $\quad a\left(\dfrac{du}{dx} + \dfrac{dw}{dx}\right) + b(u + w) = f(x)$

Then $\qquad a\dfrac{du}{dx} + bu = 0 \quad$ since $a\dfrac{dw}{dx} + bw = f(x)$

so $\qquad\qquad\qquad u = Ae^{-\frac{b}{a}x}$

Hence $\qquad\qquad\quad y = u + w$
$$= Ae^{-\frac{b}{a}x} + w$$

and we are justified in adding the CF (which is the general solution of $a\dfrac{du}{dx} + bu = 0$) to a PS (which is w) in order to obtain the general solution of

$$a\frac{dy}{dx} + by = f(x)$$

Exercise 20.5   (Differential equations, using a substitution)

Use the substitution $y = vx$ to solve the following differential equations (questions 1 to 7).

1 $(y - x)\dfrac{dy}{dx} + (x + y) = 0$ 2 $y\dfrac{dy}{dx} + (x - 2y) = 0$ 3 $(2y - x)\dfrac{dy}{dx} = x$ 4 $x^2 + y^2 = 2xy\dfrac{dy}{dx}$

5 $y^2 - 2xy = (x^2 - 2xy)\dfrac{dy}{dx}$ 6 $(xy + x^2)\dfrac{dy}{dx} + y^2 = 0$, given that $y = 1$ when $x = 1$.

7 $y\dfrac{dy}{dx} = 2x + y$, given that $y = 2$ when $x = 2$.

8 Use the substitution $y = e^x u$ to solve the differential equation

$$\frac{dy}{dx} - y = e^x$$

9 Using the substitution $y = u - x$, solve the differential equation

$$(x + y)\frac{dy}{dx} = x^2 + xy + x + 1$$

10 Solve the differential equation

$$x\frac{dy}{dx} + y = x$$

(i) using an integrating factor;   (ii) using the substitution $y = vx$.

20.6 Second order differential equations of the form

$$a\frac{d^2y}{dx^2} + b\frac{dy}{dx} + cy = 0 \text{ (where } a, b \text{ and } c \text{ are constants)}$$

We now consider some second order differential equations.

Example 1

$$\frac{d^2y}{dx^2} - 3\frac{dy}{dx} + 2y = 0$$

We write this as

$$\frac{d^2y}{dx^2} - \frac{dy}{dx} - 2\frac{dy}{dx} + 2y = 0$$

$$\frac{d}{dx}\left(\frac{dy}{dx} - y\right) - 2\left(\frac{dy}{dx} - y\right) = 0$$

If we substitute

$$\frac{dy}{dx} - y = z$$

then $\dfrac{dz}{dx} - 2z = 0$

and so $z = Ae^{2x}$

Now $\dfrac{dy}{dx} - y = Ae^{2x}$

We solve this using the integrating factor $e^{\int -1 dx} = e^{-x}$. We have

$$e^{-x}\frac{dy}{dx} - e^{-x}y = Ae^x$$

i.e. $\dfrac{d}{dx}(e^{-x}y) = Ae^x$

Hence $e^{-x}y = \int Ae^x dx = Ae^x + B$

Thus $y = Ae^{2x} + Be^x$

Notice that the general solution contains two arbitrary constants.

The form of this solution suggests an alternative method of approach, in which we try to find solutions of the form $y = Ae^{kx}$.

Suppose we try $y = Ae^{kx}$ as a solution of the given equation. Then

$$\frac{dy}{dx} = kAe^{kx}$$

$$\frac{d^2y}{dx^2} = k^2 Ae^{kx}$$

Substituting into the equation

$$\frac{d^2y}{dx^2} - 3\frac{dy}{dx} + 2y = 0$$

$$k^2 Ae^{kx} - 3kAe^{kx} + 2Ae^{kx} = 0$$

Thus $y = Ae^{kx}$ is a solution provided that

$$k^2 - 3k + 2 = 0$$
$$(k-2)(k-1) = 0$$
$$k = 2 \quad \text{or} \quad 1$$

Hence $y = Ae^{2x}$ and $y = Be^x$ are solutions, and the general solution is obtained by adding these;

$$y = Ae^{2x} + Be^x$$

In general, $y = Ae^{kx}$ is a solution of the differential equation

$$a\frac{d^2y}{dx^2} + b\frac{dy}{dx} + cy = 0$$

if
$$a(k^2 Ae^{kx}) + b(kAe^{kx}) + c(Ae^{kx}) = 0$$

i.e.
$$ak^2 + bk + c = 0$$

This is called the 'auxiliary equation'. The roots of this quadratic equation determine the solution of the differential equation

$$a\frac{d^2y}{dx^2} + b\frac{dy}{dx} + cy = 0$$

When $ak^2 + bk + c = 0$ has 2 distinct real roots   (i.e. $b^2 > 4ac$)

If the roots are k_1 and k_2, then the general solution of the differential equation is

$$y = Ae^{k_1 x} + Be^{k_2 x}$$

Example 2

$$\frac{d^2y}{dx^2} - 2\frac{dy}{dx} - 3y = 0$$

given $y = 3$ and $\frac{dy}{dx} = 1$ when $x = 0$.

The auxiliary equation is

$$k^2 - 2k - 3 = 0$$
$$(k + 1)(k - 3) = 0$$
$$k = -1 \quad \text{or} \quad 3$$

Thus the general solution is

$$y = Ae^{-x} + Be^{3x}$$

Substituting $y = 3$ when $x = 0$,

$$3 = A + B$$

Also, $$\frac{dy}{dx} = -Ae^{-x} + 3Be^{3x}$$

Substituting $\frac{dy}{dx} = 1$ when $x = 0$,

$$1 = -A + 3B$$

These simultaneous equations give $A = 2$, $B = 1$; hence the individual solution (see Fig. 20.6) is

$$y = 2e^{-x} + e^{3x}$$

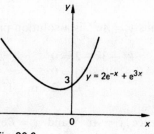

Fig. 20.6

Example 3

$$\frac{d^2y}{dx^2} + 4\frac{dy}{dx} = 0$$

The auxiliary equation is

$$k^2 + 4k = 0$$
$$k(k + 4) = 0$$
$$k = 0 \quad \text{or} \quad -4$$

Hence the general solution is

$$y = Ae^{0x} + Be^{-4x}$$

or $$y = A + Be^{-4x}$$

Example 4

$$\frac{d^2y}{dx^2} - 4y = 0$$

given that $\frac{dy}{dx} = 1$ when $x = 0$ and $y \to 0$ as $x \to +\infty$.

The auxiliary equation is

$$k^2 - 4 = 0$$
$$(k-2)(k+2) = 0$$
$$k = 2 \quad \text{or} \quad -2$$

Thus the general solution is

$$y = Ae^{2x} + Be^{-2x}$$

As $x \to +\infty$, $Be^{-2x} \to 0$, but $e^{2x} \to +\infty$. Since $y \to 0$, we must have $A = 0$. So

$$y = Be^{-2x}$$

and then

$$\frac{dy}{dx} = -2Be^{-2x}$$

Substituting $\dfrac{dy}{dx} = 1$ when $x = 0$, $1 = -2B$, and thus $B = -\tfrac{1}{2}$.

Hence the individual solution is

$$y = -\tfrac{1}{2}e^{-2x}$$

When $ak^2 + bk + c = 0$ has complex roots   (i.e. $b^2 < 4ac$)

The roots are

$$k = \frac{-b \pm \sqrt{b^2 - 4ac}}{2a}$$

$$= \frac{-b \pm j\sqrt{4ac - b^2}}{2a}$$

$$= m \pm jn$$

where $m = -\dfrac{b}{2a}$ and $n = \dfrac{\sqrt{4ac - b^2}}{2a}$ are real numbers (assuming that a, b and c are real), and the general solution of the differential equation is

$$y = Ce^{(m+jn)x} + De^{(m-jn)x}$$

where C and D are arbitrary (complex) constants. Then

$$y = e^{mx}(Ce^{jnx} + De^{-jnx})$$

$$= e^{mx}[C(\cos nx + j\sin nx) + D(\cos nx - j\sin nx)]$$

(using $e^{j\theta} = \cos\theta + j\sin\theta$ with $\theta = nx$, and then with $\theta = -nx$).

Thus

$$y = e^{mx}(A\sin nx + B\cos nx)$$

where

$$A = j(C - D) \quad \text{and} \quad B = C + D$$

Example 5

$$\frac{d^2y}{dx^2} - 4\frac{dy}{dx} + 13y = 0$$

The auxiliary equation

$$k^2 - 4k + 13 = 0$$

has roots

$$k = \frac{4 \pm \sqrt{16 - 52}}{2} = \frac{4 \pm 6j}{2}$$

$$= 2 \pm 3j$$

Hence the general solution is

$$y = e^{2x}(A \sin 3x + B \cos 3x)$$

Example 6

$$\frac{d^2y}{dx^2} + 4\frac{dy}{dx} + 5y = 0$$

given $y = 0$ and $\dfrac{dy}{dx} = 3$ when $x = 0$.

The auxiliary equation

$$k^2 + 4k + 5 = 0$$

has roots

$$k = \frac{-4 \pm \sqrt{16 - 20}}{2} = \frac{-4 \pm 2j}{2}$$

$$= -2 \pm j$$

Thus the general solution is

$$y = e^{-2x}(A \sin x + B \cos x)$$

Substituting $y = 0$ when $x = 0$, $0 = 1 \times (0 + B)$, and so $B = 0$.
Then

$$y = Ae^{-2x} \sin x$$

and

$$\frac{dy}{dx} = Ae^{-2x} \cos x - 2Ae^{-2x} \sin x$$

Substituting $\dfrac{dy}{dx} = 3$ when $x = 0$, $3 = A - 0$, and so $A = 3$.

Hence the individual solution is

$$y = 3e^{-2x} \sin x$$

This represents 'damped' oscillations, for which the amplitude decreases as x increases (Fig. 20.7: note that y returns to zero at constant intervals of π).

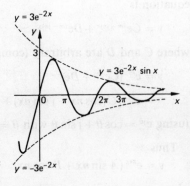

Fig. 20.7

Note that, for the special case

$$\frac{d^2y}{dx^2}+n^2y=0$$

the auxiliary equation $k^2+n^2=0$ has roots $k=\pm jn=0\pm jn$, and the general solution is

$$y=e^{0x}(A\sin nx+B\cos nx)$$

or
$$y=A\sin nx+B\cos nx$$

This agrees with our previous work.

When $ak^2+bk+c=0$ has equal roots   (i.e. $b^2=4ac$)

Example 7

$$\frac{d^2y}{dx^2}-4\frac{dy}{dx}+4y=0$$

The auxiliary equation is

$$k^2-4k+4=0$$
$$(k-2)^2=0$$
$$k=2 \quad \text{(twice)}$$

Thus $y=Ae^{2x}$ is a solution. However this contains only one arbitrary constant, and so it is not the most general solution.

To solve the equation completely, we make the substitution $y=e^{2x}u$ (i.e. we define a new variable $u=e^{-2x}y$). Then

$$\frac{dy}{dx}=e^{2x}\frac{du}{dx}+2e^{2x}u$$

and $\dfrac{d^2y}{dx^2}=e^{2x}\dfrac{d^2u}{dx^2}+4e^{2x}\dfrac{du}{dx}+4e^{2x}u$

Substituting into the equation

$$\frac{d^2y}{dx^2}-4\frac{dy}{dx}+4y=0$$

gives $e^{2x}\dfrac{d^2u}{dx^2}+4e^{2x}\dfrac{du}{dx}+4e^{2x}u-4\left(e^{2x}\dfrac{du}{dx}+2e^{2x}u\right)+4e^{2x}u=0$

which gives $\dfrac{d^2u}{dx^2}=0$

Integrating, $\dfrac{du}{dx}=B$

and, integrating again,

$$u=Bx+A$$

Hence the general solution is

$$y = e^{2x}u = e^{2x}(Bx + A)$$
or
$$y = Ae^{2x} + Bxe^{2x}$$

In general, if the auxiliary equation has a double root $k = m$, say, then the solution of the differential equation is

$$y = Ae^{mx} + Bxe^{mx}$$

Example 8

$$\frac{d^2y}{dx^2} + 2\frac{dy}{dx} + y = 0$$

given $y = 3$ and $\dfrac{dy}{dx} = -1$ when $x = 0$.

The auxiliary equation is

$$k^2 + 2k + 1 = 0$$
$$(k + 1)^2 = 0$$
$$k = -1 \quad \text{(twice)}$$

Thus the general solution is

$$y = Ae^{-x} + Bxe^{-x}$$

Substituting $y = 3$ when $x = 0$, $3 = A$. Also

$$\frac{dy}{dx} = -Ae^{-x} - Bxe^{-x} + Be^{-x}$$

Substituting $\dfrac{dy}{dx} = -1$ when $x = 0$,

$$-1 = -A + B$$

Thus $A = 3$ and $B = 2$, and the individual solution (see Fig. 20.8) is

$$y = 3e^{-x} + 2xe^{-x}$$
or $\quad y = e^{-x}(2x + 3)$

Summarising, we have the following results.

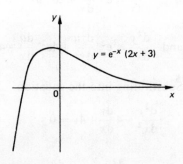

$y = e^{-x}(2x + 3)$

Fig. 20.8

When the roots of the auxiliary equation $ak^2 + bk + c = 0$ are:	The general solution of $a\dfrac{d^2y}{dx^2} + b\dfrac{dy}{dx} + cy = 0$ is:
real and distinct, $k = k_1$ or k_2	$y = Ae^{k_1 x} + Be^{k_2 x}$
complex, $k = m \pm jn$	$y = e^{mx}(A \sin nx + B \cos nx)$
equal, $k = m$ (twice)	$y = Ae^{mx} + Bxe^{mx}$

Exercise 20.6 $\left(\text{Differential equations: type } a\dfrac{d^2y}{dx^2} + b\dfrac{dy}{dx} + cy = 0\right)$

In questions 1 to 15 give the general solutions of the differential equations.

1 $\dfrac{d^2y}{dx^2} - 4\dfrac{dy}{dx} + 3y = 0$ 2 $\dfrac{d^2y}{dx^2} + 3\dfrac{dy}{dx} - 4y = 0$ 3 $\dfrac{d^2y}{dx^2} + 6\dfrac{dy}{dx} + 8y = 0$

4 $\dfrac{d^2y}{dx^2} - 9y = 0$ 5 $3\dfrac{d^2y}{dx^2} - 4\dfrac{dy}{dx} - 4y = 0$ 6 $\dfrac{d^2y}{dx^2} + \dfrac{dy}{dx} - 3y = 0$

7 $\dfrac{d^2y}{dx^2} + y = 0$ 8 $\dfrac{d^2y}{dx^2} + 25y = 0$ 9 $\dfrac{d^2y}{dx^2} - 4\dfrac{dy}{dx} + 20y = 0$

10 $\dfrac{d^2y}{dx^2} + 6\dfrac{dy}{dx} + 10y = 0$ 11 $4\dfrac{d^2y}{dx^2} - 4\dfrac{dy}{dx} + 5y = 0$ 12 $\dfrac{d^2y}{dx^2} + 3\dfrac{dy}{dx} + 3y = 0$

13 $\dfrac{d^2y}{dx^2} - 2\dfrac{dy}{dx} + y = 0$ 14 $\dfrac{d^2y}{dx^2} + 6\dfrac{dy}{dx} + 9y = 0$ 15 $\dfrac{d^2y}{dx^2} - 2\dfrac{dy}{dx} = 0$

In questions 16 to 20 find the solution which satisfies the given initial conditions.

16 $\dfrac{d^2y}{dx^2} - \dfrac{dy}{dx} - 2y = 0;$ $y = 0, \dfrac{dy}{dx} = 2$ when $x = 0$.

17 $\dfrac{d^2y}{dx^2} - 6\dfrac{dy}{dx} + 5y = 0;$ $y = 3, \dfrac{dy}{dx} = -1$ when $x = 0$.

18 $\dfrac{d^2y}{dx^2} - 16y = 0;$ $y = 1, \dfrac{dy}{dx} = 0$ when $x = 0$.

19 $\dfrac{d^2y}{dx^2} + 4y = 0;$ $y = 2, \dfrac{dy}{dx} = -1$ when $x = 0$.

20 $\dfrac{d^2y}{dx^2} + 2\dfrac{dy}{dx} + 5y = 0;$ $y = 0, \dfrac{dy}{dx} = 2$ when $x = 0$.

20.7 Second order differential equations of the form

$a\dfrac{d^2y}{dx^2} + b\dfrac{dy}{dx} + cy = f(x)$ (where a, b and c are constants)

This is a second order 'linear' differential equation with constant coefficients. To solve such an equation we use a complementary function and particular solution method very similar to that for the corresponding first order equations. Consider the equation

$$a\dfrac{d^2y}{dx^2} + b\dfrac{dy}{dx} + cy = f(x)$$

(i) Solve the auxiliary equation $ak^2 + bk + c = 0$, and hence write down (using the summary on p. 390) the general solution of $a\dfrac{d^2y}{dx^2} + b\dfrac{dy}{dx} + cy = 0$. This is the CF.

(ii)   Find one particular solution (PS) of the given equation $a\dfrac{d^2y}{dx^2}+b\dfrac{dy}{dx}+cy=f(x)$, by trying a function of the same type as $f(x)$ (the examples on p. 369 may be helpful). Difficulties will arise here if $f(x)$ is of the same form as part of the CF; in such cases we usually multiply our first attempt by x, and try again.
(iii)   We obtain the general solution by adding the CF to the PS.
(iv)   If initial conditions are given, we substitute these into the general solution found in (iii) (if a value of $\dfrac{dy}{dx}$ is given it will be necessary to differentiate the general solution). We find the arbitrary constants and hence obtain the individual solution.

Example 1

$$\frac{d^2y}{dx^2}+4\frac{dy}{dx}+3y=2e^{-2x}$$

given that $y=0$ and $\dfrac{dy}{dx}=0$ when $x=0$. The auxiliary equation $k^2+4k+3=0$, or $(k+1)(k+3)=0$, has roots $k=-1$ or -3; thus the CF is

$$y=Ae^{-x}+Be^{-3x}$$

For a PS, try $y=pe^{-2x}$; then

$$\frac{dy}{dx}=-2pe^{-2x}$$

and $$\frac{d^2y}{dx^2}=4pe^{-2x}$$

and we require

$$4pe^{-2x}+4(-2pe^{-2x})+3pe^{-2x}=2e^{-2x}$$
$$4p-8p+3p=2$$

Thus $p=-2$, and a PS is $y=-2e^{-2x}$.
   The general solution is therefore

$$y=Ae^{-x}+Be^{-3x}-2e^{-2x}$$

and then

$$\frac{dy}{dx}=-Ae^{-x}-3Be^{-3x}+4e^{-2x}$$

Substituting $y=0$ when $x=0$ gives $0=A+B-2$; and $\dfrac{dy}{dx}=0$ when $x=0$ gives $0=-A-3B+4$. Thus $A=1$, $B=1$. Hence the individual solution is

$$y=e^{-x}+e^{-3x}-2e^{-2x}$$

Example 2

$$\frac{d^2y}{dx^2}-6\frac{dy}{dx}+9y=4$$

The auxiliary equation $k^2-6k+9=0$, or $(k-3)^2=0$, has a double root $k=3$; thus the CF is

$$y=Ae^{3x}+Bxe^{3x}$$

For a PS, try $y=p$; then $\dfrac{dy}{dx}=0$, $\dfrac{d^2y}{dx^2}=0$, and we require

$0-0+9p=4$. Thus $p=\frac{4}{9}$, and a PS is $y=\frac{4}{9}$.
Hence the general solution is

$$y=Ae^{3x}+Bxe^{3x}+\tfrac{4}{9}$$

Example 3

$$\frac{d^2y}{dx^2}-7\frac{dy}{dx}+12y=6\cos x$$

given that $y=1$ and $\dfrac{dy}{dx}=1$ when $x=0$.

The auxiliary equation $k^2-7k+12=0$, or $(k-3)(k-4)=0$, has roots $k=3$ or 4, thus the CF is

$$y=Ae^{3x}+Be^{4x}$$

For a PS, try

$$y=p\sin x+q\cos x$$

Then $\quad\dfrac{dy}{dx}=p\cos x-q\sin x$

$$\frac{d^2y}{dx^2}=-p\sin x-q\cos x$$

and we require

$$(-p\sin x-q\cos x)-7(p\cos x-q\sin x)+12(p\sin x+q\cos x)=6\cos x$$

Comparing coefficients of $\sin x$:

$$-p+7q+12p=0$$

Comparing coefficients of $\cos x$:

$$-q-7p+12q=6$$

Thus $\quad\begin{aligned}11p+7q&=0\\-7p+11q&=6\end{aligned}$

This gives $p=-21/85$ and $q=33/85$; and so a PS is

$$y=-\frac{21}{85}\sin x+\frac{33}{85}\cos x$$

The general solution is therefore

$$y = Ae^{3x} + Be^{4x} - \frac{21}{85}\sin x + \frac{33}{85}\cos x$$

and then

$$\frac{dy}{dx} = 3Ae^{3x} + 4Be^{4x} - \frac{21}{85}\cos x - \frac{33}{85}\sin x$$

Substituting $y = 1$ when $x = 0$ gives $1 = A + B + \frac{33}{85}$; and $\frac{dy}{dx} = 1$

when $x = 0$ gives $1 = 3A + 4B - \frac{21}{85}$. Thus

$$A + B = \frac{52}{85}$$

$$3A + 4B = \frac{106}{85}$$

which gives $A = \frac{6}{5}$, $B = -\frac{10}{17}$.

Hence the individual solution is

$$y = \frac{6}{5}e^{3x} - \frac{10}{17}e^{4x} - \frac{21}{85}\sin x + \frac{33}{85}\cos x$$

Example 4

$$\frac{d^2y}{dx^2} - 4\frac{dy}{dx} + 5y = 5x - 9$$

given that $y = 3$ and $\frac{dy}{dx} = 1$ when $x = 0$.

The auxiliary equation $k^2 - 4k + 5 = 0$ has roots

$$k = \frac{4 \pm \sqrt{16 - 20}}{2} = 2 \pm j. \text{ Thus the CF is}$$

$$y = e^{2x}(A\sin x + B\cos x)$$

For a PS, try $y = px + q$; then $\frac{dy}{dx} = p$, $\frac{d^2y}{dx^2} = 0$, and we require

$$0 - 4p + 5(px + q) = 5x - 9$$

Comparing coefficients of x:

$$5p = 5$$

Comparing the constant term:

$$-4p + 5q = -9$$

Thus $p = 1$, $q = -1$, and a PS is $y = x - 1$.

The general solution is therefore

$$y = e^{2x}(A\sin x + B\cos x) + x - 1$$

Then $\dfrac{dy}{dx}=e^{2x}(A\cos x-B\sin x)+2e^{2x}(A\sin x+B\cos x)+1$

Substituting $y=3$ when $x=0$ gives $3=B-1$; and $\dfrac{dy}{dx}=1$ when

$x=0$ gives $1=A+2B+1$. Thus $B=4$, $A=-8$, and the individual solution is

$$y=e^{2x}(-8\sin x+4\cos x)+x-1$$

The following examples show some of the problems which can occur in finding the particular solution.

Example 5

$$\frac{d^2y}{dx^2}-2\frac{dy}{dx}-3y=e^{3x}$$

The auxiliary equation $k^2-2k-3=0$ has roots $k=3$ or -1, and so the CF is

$$y=Ae^{3x}+Be^{-x}$$

$y=pe^{3x}$ will not work as a PS, since this occurs as part of the CF, so instead we try

$$y=pxe^{3x}$$

Then $\dfrac{dy}{dx}=3pxe^{3x}+pe^{3x}$

$$\frac{d^2y}{dx^2}=9pxe^{3x}+6pe^{3x}$$

and we require

$$(9pxe^{3x}+6pe^{3x})-2(3pxe^{3x}+pe^{3x})-3pxe^{3x}=e^{3x}$$
$$4pe^{3x}=e^{3x}$$

Thus $p=\tfrac{1}{4}$, and a PS is $y=\tfrac{1}{4}xe^{3x}$. Hence the general solution is

$$y=Ae^{3x}+Be^{-x}+\tfrac{1}{4}xe^{3x}$$

Alternatively, we may solve the equation by using a substitution.*

Define a new variable u by $y=e^{3x}u$. Then

$$\frac{dy}{dx}=e^{3x}\frac{du}{dx}+3e^{3x}u$$

$$\frac{d^2y}{dx^2}=e^{3x}\frac{d^2u}{dx^2}+6e^{3x}\frac{du}{dx}+9e^{3x}u$$

*The equation $a\dfrac{d^2y}{dx^2}+b\dfrac{dy}{dx}+cy=f(x)$ can always be solved by means of the

substitution $y=gu$, where g is any one solution of $a\dfrac{d^2y}{dx^2}+b\dfrac{dy}{dx}+cy=0$.

Substituting into the differential equation,

$$\left(e^{3x}\frac{d^2u}{dx^2} + 6e^{3x}\frac{du}{dx} + 9e^{3x}u\right) - 2\left(e^{3x}\frac{du}{dx} + 3e^{3x}u\right) - 3e^{3x}u = e^{3x}$$

$$\frac{d^2u}{dx^2} + 4\frac{du}{dx} = 1$$

Now put $w = \dfrac{du}{dx}$; then.

$$\frac{dw}{dx} + 4w = 1$$

Using the integrating factor $e^{\int 4dx} = e^{4x}$, we have

$$e^{4x}\frac{dw}{dx} + 4e^{4x}w = e^{4x}$$

i.e. $$\frac{d}{dx}(e^{4x}w) = e^{4x}$$

Hence $$e^{4x}w = \int e^{4x}dx = \tfrac{1}{4}e^{4x} + C$$

Now $$\frac{du}{dx} = w = \tfrac{1}{4} + Ce^{-4x}$$

Integrating,

$$u = \tfrac{1}{4}x - \tfrac{1}{4}Ce^{-4x} + D$$

Hence $$y = e^{3x}u = \tfrac{1}{4}xe^{3x} - \tfrac{1}{4}Ce^{-x} + De^{3x}$$

(This is the same result as before, with $A = D$ and $B = -\tfrac{1}{4}C$.)

Example 6

$$\frac{d^2y}{dx^2} + 9y = 4\sin 3x$$

The CF is

$$y = A\sin 3x + B\cos 3x$$

$y = p\sin 3x + q\cos 3x$ will not work as a PS, because this is of the same form as the CF, so we try

$$y = x(p\sin 3x + q\cos 3x)$$

Then $$\frac{dy}{dx} = x(3p\cos 3x - 3q\sin 3x) + (p\sin 3x + q\cos 3x)$$

$$\frac{d^2y}{dx^2} = x(-9p\sin 3x - 9q\cos 3x) + 6p\cos 3x - 6q\sin 3x$$

and we require

$$x(-9p\sin 3x - 9q\cos 3x) + 6p\cos 3x - 6q\sin 3x + 9x(p\sin 3x + q\cos 3x) = 4\sin 3x$$
$$6p\cos 3x - 6q\sin 3x = 4\sin 3x$$

Thus $p = 0$, $q = -\frac{2}{3}$, and a PS is

$$y = -\tfrac{2}{3}x\cos 3x$$

Hence the general solution is

$$y = A\sin 3x + B\cos 3x - \tfrac{2}{3}x\cos 3x$$

Example 7

$$\frac{d^2y}{dx^2} + 4\frac{dy}{dx} = 2x$$

The auxiliary equation $k^2 + 4k = 0$ has roots $k = 0, -4$, and so the CF is

$$y = A + Be^{-4x}$$

$y = px + q$ will not work as a PS, since then

$$\frac{d^2y}{dx^2} + 4\frac{dy}{dx} = 0 + 4p$$

which contains no x term. We try instead

$$y = px^2 + qx$$

Then

$$\frac{dy}{dx} = 2px + q$$

$$\frac{d^2y}{dx^2} = 2p$$

and we require

$$2p + 4(2px + q) = 2x$$

Comparing coefficients of x:
$$8p = 2$$

Comparing the constant term:
$$2p + 4q = 0$$

Thus $p = \frac{1}{4}$, $q = -\frac{1}{8}$, and a PS is

$$y = \tfrac{1}{4}x^2 - \tfrac{1}{8}x$$

Hence the general solution is

$$y = A + Be^{-4x} + \tfrac{1}{4}x^2 - \tfrac{1}{8}x$$

Alternatively, we put $w = \dfrac{dy}{dx}$; then

$$\frac{dw}{dx} + 4w = 2x$$

Using the integrating factor e^{4x}, we have

$$e^{4x}\frac{dw}{dx} + 4e^{4x}w = 2xe^{4x}$$

ie. $\dfrac{d}{dx}(e^{4x}w) = 2xe^{4x}$

Hence $e^{4x}w = \int 2xe^{4x}\,dx$

$\qquad\qquad = \tfrac{1}{2}xe^{4x} - \int \tfrac{1}{2}e^{4x}dx$ (integrating by parts)

$\qquad\qquad = \tfrac{1}{2}xe^{4x} - \tfrac{1}{8}e^{4x} + C$

Now $\dfrac{dy}{dx} = w = \tfrac{1}{2}x - \tfrac{1}{8} + Ce^{-4x}$

Integrating, $y = \tfrac{1}{4}x^2 - \tfrac{1}{8}x - \tfrac{1}{4}Ce^{-4x} + D$

(This is the same result as before, with $A = D$ and $B = -\tfrac{1}{4}C$.)

Example 8

$$\frac{d^2y}{dx^2} - 6\frac{dy}{dx} + 9y = e^{3x}$$

The auxiliary equation $k^2 - 6k + 9 = 0$ has a double root $k = 3$, and so the CF is

$$y = Ae^{3x} + Bxe^{3x}$$

$y = pe^{3x}$ will not work as a PS, since it is part of the CF. Neither will $y = pxe^{3x}$, because this is also part of the CF. We try

$$y = px^2e^{3x}$$

Then $\dfrac{dy}{dx} = 3px^2e^{3x} + 2pxe^{3x}$

$\dfrac{d^2y}{dx^2} = 9px^2e^{3x} + 12pxe^{3x} + 2pe^{3x}$

and we require

$$(9px^2e^{3x} + 12pxe^{3x} + 2pe^{3x}) - 6(3px^2e^{3x} + 2pxe^{3x}) + 9px^2e^{3x} = e^{3x}$$
$$2pe^{3x} = e^{3x}$$

Thus $p = \tfrac{1}{2}$, and a PS is

$$y = \tfrac{1}{2}x^2e^{3x}$$

Hence the general solution is

$$y = Ae^{3x} + Bxe^{3x} + \tfrac{1}{2}x^2e^{3x}$$

Exercise 20.7 $\left(\text{Differential equations: type } a\dfrac{d^2y}{dx^2}+b\dfrac{dy}{dx}+cy=f(x)\right)$

In questions 1 to 18, give the general solution of the differential equation.

1 $\dfrac{d^2y}{dx^2}-5\dfrac{dy}{dx}+4y=8$ **2** $\dfrac{d^2y}{dx^2}-4y=2x+10$ **3** $\dfrac{d^2y}{dx^2}-2\dfrac{dy}{dx}-3y=x^2+4x-5$

4 $\dfrac{d^2y}{dx^2}+3\dfrac{dy}{dx}+2y=4\sin x$ **5** $\dfrac{d^2y}{dx^2}-y=2e^{3x}$

6 $\dfrac{d^2y}{dx^2}-6\dfrac{dy}{dx}+8y=3e^{2x}$ [Try $y=pxe^{2x}$ for the PS.]

7 $\dfrac{d^2y}{dx^2}-4\dfrac{dy}{dx}-5y=4e^{-x}$ **8** $\dfrac{d^2y}{dx^2}+4y=3$ **9** $\dfrac{d^2y}{dx^2}+y=3x-1$

10 $\dfrac{d^2y}{dx^2}+9y=\sin x$ **11** $\dfrac{d^2y}{dx^2}+2\dfrac{dy}{dx}+2y=3e^{-2x}$ **12** $\dfrac{d^2y}{dx^2}-4\dfrac{dy}{dx}+4y=x$

13 $\dfrac{d^2y}{dx^2}+y=\cos x$ [Try $y=px\sin x$ for the PS.]

14 $\dfrac{d^2y}{dx^2}+2\dfrac{dy}{dx}+y=3e^{-x}$ [Try $y=px^2e^{-x}$ for the PS.]

15 $\dfrac{d^2y}{dx^2}+\dfrac{dy}{dx}=2x+1$ **16** $\dfrac{d^2y}{dx^2}-4\dfrac{dy}{dx}=\sin 2x$ **17** $\dfrac{d^2y}{dx^2}-\dfrac{dy}{dx}=4e^x$

18 $\dfrac{d^2y}{dx^2}-6\dfrac{dy}{dx}+9y=e^{3x}$ [Use the substitution $y=e^{3x}u$.]

In questions 19 to 25, find the solution which satisfies the given conditions.

19 $\dfrac{d^2y}{dx^2}-4\dfrac{dy}{dx}+3y=4x-2$; $y=2$, $\dfrac{dy}{dx}=1$ when $x=0$.

20 $\dfrac{d^2y}{dx^2}-9y=4\sin x$; $y=0$, $\dfrac{dy}{dx}=-2$ when $x=0$.

21 $4\dfrac{d^2y}{dx^2}+8\dfrac{dy}{dx}-5y=e^{2x}$; $y=0$, $\dfrac{dy}{dx}=0$ when $x=0$.

22 $\dfrac{d^2y}{dx^2}+4\dfrac{dy}{dx}+3y=e^{-x}$; $y=0$, $\dfrac{dy}{dx}=0$ when $x=0$.

23 $\dfrac{d^2y}{dx^2}+y=3$; $y=1$, $\dfrac{dy}{dx}=0$ when $x=\frac{1}{2}\pi$.

24 $\dfrac{d^2y}{dx^2}+4\dfrac{dy}{dx}+4y=12$; $y=3$, $\dfrac{dy}{dx}=0$ when $x=0$.

25 $\dfrac{d^2y}{dx^2}+2\dfrac{dy}{dx}=5e^{-2x}$; $y=0$, $\dfrac{dy}{dx}=0$ when $x=0$.

20.8   Miscellaneous differential equations

The following exercise contains differential equations of all types.

Exercise 20.8   (Miscellaneous differential equations)

Solve the following differential equations.

Straightforward first order equations

1   $\dfrac{dy}{dx} = -3y$, given that $y = 4$ when $x = 0$.

2   $\dfrac{dy}{dx} = 4y^2$, given that $y = \frac{1}{2}$ when $x = -2$.

3   $\cos^2 x \dfrac{dy}{dx} - e^y = 0$, given that $y = 0$ when $x = 0$.

4   $ye^{2x} \dfrac{dy}{dx} = x$, given that $y = 2$ when $x = 0$.

5   $\dfrac{dy}{dx} = (y+1)(2x+1)$, given that $y = 2$ when $x = 0$.

6   $\dfrac{dy}{dx} + 6y = 3x$, given that $y = 0$ when $x = 0$.

7   $\dfrac{dy}{dx} = \dfrac{x(y^2-1)}{y(x^2+1)}$, given that $y = 3$ when $x = 1$.

8   $\cos y \dfrac{dy}{dx} = (\cot x)(1 + \sin y)$.

9   $2\dfrac{dy}{dx} + 3y = e^{-2x}$, given that $y = 1$ when $x = 0$.

10   $\dfrac{dy}{dx} = \dfrac{1}{2-x}$, given that $y = 0$ when $x = 1$.

Harder first order equations

11   $(x+y)\dfrac{dy}{dx} = 3y$, given that $y = 1$ when $x = 1$. [Use the substitution $y = vx$.]

12   $\dfrac{dy}{dx} - 2xy = 2xe^{x^2}$, given that $y = 0$ when $x = 0$.

13   $x\dfrac{dy}{dx} + 2y = xe^x$, given that $y = 2e$ when $x = 1$.

14   $\dfrac{dy}{dx} = \cos x (\sec y - \tan y)$, given that $y = 0$ when $x = \frac{1}{2}\pi$.

15 $(x^2 + 1)\dfrac{dy}{dx} + 4xy = 12x^3$, given that $y = 1$ when $x = 1$.

16 $\dfrac{dy}{dx} + 2y = x$, given that $\dfrac{dy}{dx} = 0$ when $x = 0$.

17 $\dfrac{dy}{dx} = \dfrac{5x + 7y}{x - y}$ [Put $y = vx$.]

18 $\sin x \dfrac{dy}{dx} - y \cos x = \sin^2 x \cos x$, given that $y = 2$ when $x = \frac{1}{2}\pi$.

19 $(x^2 - 1)\dfrac{dy}{dx} - 2xy = 2x(x^2 - 1)^2$, given that $y = -1$ when $x = 0$.

20 $\dfrac{dy}{dx} = \left(\dfrac{x}{y}\right)e^{x+y}$

21 $x(x - 1)\dfrac{dy}{dx} + y = x^2(2x - 1)$

22 $\dfrac{dy}{dx} = xy \ln x$, given that $y = 1$ when $x = 1$.

23 $x\dfrac{dy}{dx} = y + xe^{-y/x}$, given that $y = 1$ when $x = 1$. [Put $y = vx$.]

24 $x \ln x \dfrac{dy}{dx} = \tan y$, given that $y = \frac{1}{6}\pi$ when $x = e$.

25 $x\dfrac{dy}{dx} + (2x + 3)y = 5x^2 e^{-2x}$

Straightforward second order equations

26 $\dfrac{d^2y}{dx^2} - 9y = 18$, given that $y = 0$ and $\dfrac{dy}{dx} = -6$ when $x = 0$.

27 $\dfrac{d^2y}{dx^2} + 4y = 1$, given that $y = 0$ and $\dfrac{dy}{dx} = \frac{1}{2}$ when $x = 0$.

28 $4\dfrac{d^2y}{dx^2} + y + 8 = 0$, given that $y = 0$ when $x = 0$ and $y = 0$ when $x = \pi$.

29 $9\dfrac{d^2y}{dx^2} - y + 1 = 0$, given that $y = -2$ and $\dfrac{dy}{dx} = 4$ when $x = 0$.

30 $\dfrac{d^2y}{dx^2} - 5\dfrac{dy}{dx} + 6y = 6x$, given that $y = -\frac{1}{6}$ and $\dfrac{dy}{dx} = 0$ when $x = 0$.

31 $\dfrac{d^2y}{dx^2} + 2\dfrac{dy}{dx} - 8y = 14e^{3x}$, given that $y = 6$ and $\dfrac{dy}{dx} = 8$ when $x = 0$.

Harder second order equations

32 $\dfrac{d^2y}{dx^2} + 2\dfrac{dy}{dx} + 5y = 8e^{-x}$, given that $y = 0$ when $x = 0$ and when $x = \frac{1}{4}\pi$.

33 $\dfrac{d^2y}{dx^2} - y = e^x$, given that $y = 0$ and $\dfrac{dy}{dx} = 0$ when $x = 0$.

34 $\dfrac{d^2y}{dx^2} - 4\dfrac{dy}{dx} + 5y = 2e^{2x}$, given that $y = 2$ and $\dfrac{dy}{dx} = 3$ when $x = 0$.

35 $3\dfrac{d^2y}{dx^2} + 2\dfrac{dy}{dx} - 5y + 10x = 0$, given that $y = \dfrac{4}{5}$ and $\dfrac{dy}{dx} = \dfrac{38}{3}$ when $x = 0$.

36 $\dfrac{d^2y}{dx^2} - 2\dfrac{dy}{dx} + 2y = 2\sin x$, given that $y = 0$ and $\dfrac{dy}{dx} = 0$ when $x = 0$.

37 Find the solution of the differential equation $\dfrac{d^2y}{dx^2} + 9y = 18$ for which y has a maximum at $(\frac{1}{2}\pi, 6)$. Find the minimum value of y, and the values of x for which $y = 0$.   (O)

38 Find the general solution of the differential equation

$$\frac{d^2y}{dx^2} - 4\frac{dy}{dx} + 4y = 2e^{2x}$$

by putting $y = e^{2x}z$, or otherwise.   (O)

39 By using the substitution $x = e^t$, solve the differential equation

$$x^2\frac{d^2y}{dx^2} - 4x\frac{dy}{dx} + 4y = 8x^3$$

given that $y = 2$ and $\dfrac{dy}{dx} = 0$ when $x = 1$.

$$\left[\text{If } x = e^t, \text{ show that } \frac{dy}{dx} = e^{-t}\frac{dy}{dt} \text{ and } \frac{d^2y}{dx^2} = e^{-2t}\left(\frac{d^2y}{dt^2} - \frac{dy}{dt}\right).\right]$$ (L)

40 Given that $y = Ae^{Bx}$ where A and B are constants, find $\dfrac{dy}{dx}$ and $\dfrac{d^2y}{dx^2}$. Hence form a differential equation which does not contain these constants. Find the possible values of $\dfrac{dy}{dx}$ when $y = 4$ and $\dfrac{d^2y}{dx^2} = 9$.   (L)

21

Applications of differential equations

Introduction

Differential equations arise from practical situations involving rates of change, and we have already studied some simple applications (in Chapter 12). A typical solution involves three stages.

(i) Choosing two suitable variables, and using the given information to form a differential equation involving them (remember that a differential equation containing $\dfrac{dW}{dt}$, say, cannot be solved if it contains any variables other than W and t). We shall normally be working with letters other than x and y, and very often one of the variables will be time, t. Note that a dot may be used to denote differentiation with respect to t; thus $\dot{x}$ means $\dfrac{dx}{dt}$, $\ddot{x}$ means $\dfrac{d^2x}{dt^2}$, and so on.

(ii) Solving the differential equation, and finding the values of any arbitrary constants (if possible).

(iii) Interpreting the solution in a form relevant to the original situation.

Here we give some examples illustrating applications in various fields.

21.1 Applications in mechanics

For a particle of mass m moving along a straight line, we have (from Newton's second law of motion)*, $F = ma$, where F is the net force acting on the particle in some direction, and a is the acceleration of the particle in that direction (provided that the units are suitable; for example, m could be in kg, F in newtons and a in $m\,s^{-2}$).

If, at time t, the displacement of the particle from a fixed point O on the line is x, then its velocity is

$$v = \frac{dx}{dt}$$

---

*Strictly, 'rate of change of momentum is proportional to the applied force', giving $F = \dfrac{d}{dt}(mv) = m\dfrac{dv}{dt} + \dfrac{dm}{dt}v$. This version must be used if the mass of the particle is changing (for example, an accelerating electron).

and its acceleration is

$$a = \frac{dv}{dt} = \frac{d^2x}{dt^2}$$

(see Chapter 11). Another, very useful, form of the acceleration is obtained as follows:

$$a = \frac{dv}{dt} = \frac{dv}{dx} \times \frac{dx}{dt} = \frac{dv}{dx} \times v$$

$$= v\frac{dv}{dx}$$

The acceleration is always measured in the direction of increasing x (i.e. the 'positive direction'). To obtain a differential equation, we first express the net force F acting on the particle in the positive direction in terms of x, v or t (or a combination of these). We then write $F = ma$, using either $\frac{dv}{dt}$ or $\frac{d^2x}{dt^2}$ or $v\frac{dv}{dx}$ for the acceleration, depending on which is most suited to the problem.

Example 1   A particle of mass m moves along a straight line, and experiences a resistive force mkv, where v is the velocity of the particle and k a constant. If, initially, the particle is at O and has velocity U, find an expression for its displacement in terms of time, and show that the displacement never exceeds $\dfrac{U}{k}$.

When the velocity is v, the force is mkv in the opposite direction (since it is a resistive force). Thus the force on the particle is $F = -mkv$. As we require expressions in terms of time, we shall use $\dfrac{dv}{dt}$ for the acceleration.

Hence the differential equation is

$$-mkv = m\frac{dv}{dt}$$

or $$\frac{dv}{dt} = -kv$$

We can solve this immediately:

$$v = Ae^{-kt}$$

When $t = 0$, $v = U$, so $U = A$, and

$$v = Ue^{-kt}$$

Now $$v = \frac{dx}{dt}$$

thus $$\frac{dx}{dt} = Ue^{-kt}$$

$$x = -\frac{U}{k}e^{-kt} + C$$

When $t = 0$, $x = 0$, so $0 = -\dfrac{U}{k} + C$ and $C = \dfrac{U}{k}$. Hence

$$x = -\frac{U}{k}e^{-kt} + \frac{U}{k} = \frac{U}{k}(1 - e^{-kt})$$

Since e^{-kt} is always positive, x is always less than $\dfrac{U}{k}$. In fact, as

$t \to +\infty$, $x \to \dfrac{U}{k}$.

Example 2 A particle of mass m is thrown vertically upwards with initial velocity U. The particle experiences a force mg vertically downwards (the force due to gravity, where g is a constant) and a resistive force mkv, where v is the velocity. Find the time taken for the particle to reach its greatest height.

While the particle is moving upwards with velocity v, it experiences forces mg and mkv vertically downwards. Thus the net force in the positive direction (upwards) is

$$F = -mg - mkv$$

Hence the equation of motion is

$$-mg - mkv = m\frac{dv}{dt}$$

or $\qquad \dfrac{dv}{dt} + kv = -g$

This has CF $v = Ae^{-kt}$ and a PS is $v = -\dfrac{g}{k}$; thus the general

solution is

$$v = Ae^{-kt} - \frac{g}{k}$$

When $t = 0$, $v = U$, so $U = A - \dfrac{g}{k}$ and $A = U + \dfrac{g}{k}$. Hence

$$v = \left(U + \frac{g}{k}\right)e^{-kt} - \frac{g}{k}$$

The particle reaches its greatest height when $v = 0$; then

$$\frac{g}{k} = \left(U + \frac{g}{k}\right)e^{-kt}$$

$$e^{kt} = \frac{k}{g}\left(U + \frac{g}{k}\right) = 1 + \frac{kU}{g}$$

$$kt = \ln\left(1 + \frac{kU}{g}\right)$$

The time taken by the particle to reach its greatest height is

$$t = \frac{1}{k}\ln\left(1 + \frac{kU}{g}\right)$$

Example 3   A particle of mass m is thrown vertically upwards with initial velocity U, as in Example 2, except that the resistive force is now mkv^2, where v is the velocity. Find the greatest height reached by the particle.

Suppose that, at time t, the height is y and the velocity is v. The net force on the particle in the positive direction is

$$F = -mg - mkv^2$$

Since F is given in terms of v, and we wish to find the height y, we use $v\dfrac{dv}{dy}$ for the acceleration. Hence the differential equation is

$$-mg - mkv^2 = mv\frac{dv}{dy}$$

Separating the variables,

$$\int dy = \int \frac{-v}{g + kv^2}\, dv$$

$$\int dy = -\frac{1}{2k} \int \frac{2kv}{g + kv^2}\, dv$$

$$y = -\frac{1}{2k} \ln(g + kv^2) + C$$

When $y = 0$, $v = U$, so

$$0 = -\frac{1}{2k} \ln(g + kU^2) + C$$

and $$C = \frac{1}{2k} \ln(g + kU^2)$$

Thus $$y = -\frac{1}{2k} \ln(g + kv^2) + \frac{1}{2k} \ln(g + kU^2) = \frac{1}{2k} \ln\left(\frac{g + kU^2}{g + kv^2}\right)$$

The particle reaches its greatest height when $v = 0$; then

$$y = \frac{1}{2k} \ln\left(\frac{g + kU^2}{g}\right) = \frac{1}{2k} \ln\left(1 + \frac{kU^2}{g}\right)$$

Hence the greatest height is

$$\frac{1}{2k} \ln\left(1 + \frac{kU^2}{g}\right)$$

Example 4   A particle P of mass m hangs on the end of a spring, and can rest in equilibrium at the point O. If P is moved to a position at a distance x below O, it experiences a net force upwards of kx (where k is a constant, called the 'stiffness' of the spring). [Note that this may be called a 'restoring force', since it tends to restore the particle to its equilibrium position (at O).] Show that, if P is disturbed from its equilibrium position, it executes simple harmonic motion, and find the period of the motion.

The spring is shown in Fig. 21.1. Since the displacement x is measured downwards, the positive direction is now vertically downwards. When the displacement is x the net force downwards is

$$F = -kx$$

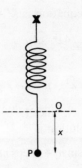

Using $\dfrac{d^2x}{dt^2}$ for the acceleration, the equation of motion is

$$-kx = m\frac{d^2x}{dt^2}$$

Fig. 21.1

or $\quad \dfrac{d^2x}{dt^2} = -\dfrac{k}{m}x$

This is of the form

$$\frac{d^2x}{dt^2} = -\omega^2 x$$

with $\omega = \sqrt{\dfrac{k}{m}}$, and so P executes simple harmonic motion (see p. 165) with centre O and period

$$\frac{2\pi}{\omega} = 2\pi\sqrt{\frac{m}{k}}$$

Example 5 A particle is moving along a straight line. When its displacement from a fixed point O on the line is x, and its velocity is v, it experiences a restoring force $m\mu x$ (towards O) and a resistive force $2m\lambda v$ (where μ and λ are constants, and m is the mass of the particle). Describe the motion in the case when $\lambda^2 < \mu$.

Consider the particle moving with x and v positive. The forces $m\mu x$ and $2m\lambda v$ then act in the negative direction, and so the net force in the positive direction is

$$F = -m\mu x - 2m\lambda v$$

We therefore have

$$-m\mu x - 2m\lambda v = m\frac{d^2x}{dt^2}$$

This contains three variables, x, v and t. However, $v = \dfrac{dx}{dt}$ and so we obtain the differential equation

$$\frac{d^2x}{dt^2} + 2\lambda\frac{dx}{dt} + \mu x = 0$$

The auxiliary equation is $k^2 + 2\lambda k + \mu = 0$, which has roots

$$k = \frac{-2\lambda \pm \sqrt{4\lambda^2 - 4\mu}}{2}$$

$$= -\lambda \pm \sqrt{\lambda^2 - \mu} = -\lambda \pm jn$$

where $n = \sqrt{\mu - \lambda^2}$, since $(\lambda^2 - \mu)$ is negative.
The general solution is

$$x = e^{-\lambda t}(A \sin nt + B \cos nt)$$

The particle is oscillating about O, and it returns to O at regular intervals of time. However, the amplitude of the oscillations decreases exponentially (see Fig. 21.2).

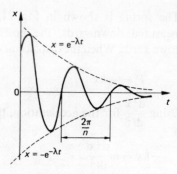

Fig. 21.2

Exercise 21.1   (Differential equations in mechanics)

1   A particle is released from rest and falls under gravity in a resisting medium so that its downward acceleration at time t after release is $g - \dfrac{gv}{c}$ where g, c are positive constants and v is the velocity of the particle at time t. Show that $v = c(1 - e^{-gt/c})$.

Deduce that $v < c$. Find the distance fallen by the particle at time t.   (O and C)

2   The motion of a particle is such that its speed v at time t is given by

$$\frac{dv}{dt} = \tfrac{1}{2}(v - v^2)$$

and $v = 0.2$ when $t = 0$. By solving the differential equation (i) find the value of t when $v = 0.5$, giving your answer correct to two decimal places; (ii) express v in terms of t.

By considering the differential equation in the form

$$v\frac{dv}{dx} = \tfrac{1}{2}(v - v^2)$$

or otherwise, where x is the distance travelled when the speed is v, and $x = 0$ when $v = 0.2$, show that the value of x when $v = 0.8$ is double its value when $v = 0.6$.   (JMB)

3   A vehicle of mass m moves in a straight line subject to a resistance $(P + Qv^2)$, where v is the speed and P, Q are constants. Form an equation of motion, using the expression $v(dv/dx)$ for the acceleration. Hence show (a) that if $P = 0$ the distance required to slow down from speed $\tfrac{3}{2}U$ to speed U is $(m/Q)\ln\tfrac{3}{2}$; (b) that if $P > 0$ the distance D required to stop from speed U is given by $D = \lambda \ln(1 + \mu U^2)$ where λ, μ are constants. Express these constants in terms of the data.

Use the above results to estimate the landing run of an aircraft of mass 10^5 kg assuming that the speed falls from $90 \, \text{m s}^{-1}$ to $60 \, \text{m s}^{-1}$ under air resistance only, given by $125v^2$ N, and that subsequently the air resistance is supplemented by a constant braking force of 7.5×10^5 N.

(MEI)

4   A particle of mass m moves along the positive x-axis, and when it is at a distance x from O it experiences a force $\dfrac{72m}{x^2}$ directed away from O. Initially the particle is at a distance 4 from O and is moving with speed 8 towards O. Find the shortest distance of the particle from O during the subsequent motion, and show that its speed is always less than 10.

5   A particle of mass m is thrown vertically upwards with initial velocity u. It experiences a constant force mg vertically downwards, together with a resistive force mkv^2, where v is the velocity and g and k are constants. Show that the time taken for the particle to reach its greatest height is

$$\frac{1}{\sqrt{gk}} \tan^{-1}\left(u\sqrt{\frac{k}{g}}\right)$$

6   A particle of mass m moves along the x-axis. When its displacement from O is x it experiences a restoring force $m\lambda^2 x$ directed towards O. It is also subject to a constant force mQ in the positive direction. If the particle is initially at rest at O, find x in terms of time, and describe the motion.

7   A particle is moving on the end of a spring. When it is a distance x below its equilibrium position O, it experiences a restoring force (upwards) of $64mx$. There is a frictional resistance to motion of $\frac{1}{4}mv$, where v is the velocity. An additional force $m(\cos 4t + 48 \sin 4t)$ is applied vertically downwards. Show that

$$\frac{d^2x}{dt^2} + \frac{1}{4}\frac{dx}{dt} + 64x = \cos 4t + 48 \sin 4t$$

and prove that after a sufficient lapse of time the displacement is approximately $x = \sin 4t$.

8   A rope is wrapped round a rough cylinder, leaving the surface at points A and B. The tension in the rope at A remains constant at 20 newtons, and the tension at B is just sufficient to cause the rope to slip. If T is the tension in the rope at the point making an angle θ measured round from A, then $\frac{dT}{d\theta} = \mu T$, where μ is the coefficient of friction. When the rope makes one and a half turns round the cylinder, the tension at B is 50 newtons. What is the tension at B if the rope is wrapped two and a half times round the cylinder?

21.2   Applications in physics

Example 1   According to Newton's law of cooling, the rate at which a hot object cools is proportional to the difference between the temperature of the object and the temperature of the surrounding air (assumed to be constant). If an object cools from 100° to 80° in 10 minutes, and from 80° to 65° in another 10 minutes, find the temperature of the surrounding air and the temperature of the object after a further 10 minutes.   (O)

Suppose the temperature of the surrounding air is θ_0. If the temperature of the object at time t is θ, then the rate of change of θ, $\frac{d\theta}{dt}$ is negative (since the object is cooling), and is proportional to $(\theta - \theta_0)$.

Thus

$$\frac{d\theta}{dt} = -k(\theta - \theta_0)$$

where k is a constant. We are given values of θ and t, so we need to solve this equation before attempting to find the constants k and θ_0.

The differential equation is

$$\frac{d\theta}{dt} + k\theta = k\theta_0$$

which has CF $\theta = Ae^{-kt}$ and PS $\theta = \theta_0$. Hence

$$\theta = Ae^{-kt} + \theta_0$$

When $t = 0$, $\theta = 100$, so $100 = A + \theta_0$,

and
$$A = 100 - \theta_0 \qquad (1)$$

When $t = 10$, $\theta = 80$, so $80 = Ae^{-10k} + \theta_0$,

and
$$Ae^{-10k} = 80 - \theta_0 \qquad (2)$$

When $t = 20$, $\theta = 65$, so $65 = Ae^{-20k} + \theta_0$,

and
$$Ae^{-20k} = 65 - \theta_0 \qquad (3)$$

Dividing,

$(2) \div (1)$ gives $e^{-10k} = \dfrac{80 - \theta_0}{100 - \theta_0}$

$(3) \div (2)$ gives $e^{-10k} = \dfrac{65 - \theta_0}{80 - \theta_0}$

Thus

$$\frac{80 - \theta_0}{100 - \theta_0} = \frac{65 - \theta_0}{80 - \theta_0}$$

$$6400 - 160\theta_0 + \theta_0^2 = 6500 - 165\theta_0 + \theta_0^2$$
$$\theta_0 = 20$$

Then $e^{-10k} = \dfrac{80 - \theta_0}{100 - \theta_0} = \dfrac{60}{80} = \dfrac{3}{4}$

and $A = 100 - \theta_0 = 80$

Hence

$$\theta = 80e^{-kt} + 20$$

When $t = 30$.

$$\theta = 80e^{-30k} + 20 = 80(e^{-10k})^3 + 20$$
$$= 80 \times (\tfrac{3}{4})^3 + 20 = 53.75$$

Hence the temperature of the surrounding air is 20°, and the temperature of the object after a further 10 minutes is 53.75°.

Example 2 A radioactive element decays at a rate proportional to the mass of the element remaining at any one instant. If two-thirds of the initial mass remains after 100 days, find the time after which one half of the initial mass remains.

Suppose that the mass is initially M_0 and that, after t days, the mass remaining is M. The rate of change of M, $\dfrac{dM}{dt}$, is negative and proportional to M, thus

$$\frac{dM}{dt} = -kM$$

where k is a constant. Hence

$$M = A e^{-kt}$$

When $t = 0$, $M = M_0$, so

$$M_0 = A$$

When $t = 100$, $M = \tfrac{2}{3}M_0$, so

$$\tfrac{2}{3}M_0 = A e^{-100k}$$

Dividing,

$$\tfrac{3}{2} = e^{100k}$$

Thus $k = \dfrac{1}{100}\ln\tfrac{3}{2}$

and we have

$$M = M_0 e^{-kt}$$

When $M = \tfrac{1}{2}M_0$,

$$e^{-kt} = \tfrac{1}{2}$$
$$kt = \ln 2$$

and the time taken is

$$t = \frac{\ln 2}{k} = \frac{100 \ln 2}{\ln \tfrac{3}{2}}$$

$$\approx 171 \text{ days}$$

Example 3 When an alternating emf $V \cos \omega t$ is applied across an electrical circuit consisting of an inductance L in series with a resistance R (where V, ω, L and R are constants), the current i flowing in the circuit satisfies the differential equation

$$L\frac{di}{dt} + Ri = V \cos \omega t$$

Find i in terms of t when t is large.

The differential equation

$$L\frac{di}{dt} + Ri = V\cos\omega t$$

has CF $i = Ae^{-Rt/L}$

For a PS, try

$$i = p\sin\omega t + q\cos\omega t$$

Then $\dfrac{di}{dt} = p\omega\cos\omega t - q\omega\sin\omega t$

We require

$$L(p\omega\cos\omega t - q\omega\sin\omega t) + R(p\sin\omega t + q\cos\omega t) = V\cos\omega t$$

Comparing coefficients of $\sin\omega t$:

$$-Lq\omega + Rp = 0$$

Comparing coefficients of $\cos\omega t$:

$$Lp\omega + Rq = V$$

This gives

$$p = \frac{V\omega L}{R^2 + \omega^2 L^2} \quad \text{and} \quad q = \frac{VR}{R^2 + \omega^2 L^2}$$

Hence the general solution is

$$i = Ae^{-Rt/L} + \frac{V}{R^2 + \omega^2 L^2}(\omega L\sin\omega t + R\cos\omega t)$$

As $t \to +\infty$, $Ae^{-Rt/L} \to 0$; thus whatever the initial conditions, when t is large,

$$i \approx \frac{V}{R^2 + \omega^2 L^2}(\omega L\sin\omega t + R\cos\omega t)$$

Exercise 21.2   (Differential equations in physics)

1   By Newton's law of cooling the rate of change at time t of the surface temperature T of a sphere in an atmosphere of constant temperature T_0 is proportional to the difference of these temperatures. Form a differential equation for T and show that, if $T = 3T_0$ when $t = 0$ and $T = 2T_0$ when $t = 1$,

$$T = T_0(1 + 2^{1-t}) \tag{O and C}$$

2   The temperature y degrees of a body, t minutes after being placed in a certain room, satisfies the differential equation

$$6\frac{d^2y}{dt^2} + \frac{dy}{dt} = 0$$

By using the substitution $z = \dfrac{dy}{dt}$, or otherwise, find y in terms of t, given that $y = 63$ when $t = 0$ and $y = 36$ when $t = 6\ln 4$.

Find after how many minutes the rate of cooling of the body will have fallen below one degree per minute, giving your answer correct to the nearest minute.

How cool does the body get? (O and C)

3   Heat is supplied to an electric kettle at a constant rate of 2000 watts, but heat is lost to the surroundings at a rate of 20 watts for every °C difference between the temperature of the kettle and that of the surroundings. One watt causes the temperature of the kettle to rise at a rate of $\frac{1}{50}$°C per minute. If the temperature of the surroundings is 15°C, and θ°C is the temperature of the kettle after t minutes, show that

$$\frac{d\theta}{dt} = 40 - \tfrac{2}{5}(\theta - 15)$$

How long will it take for the temperature to rise from 15°C to 100°C?

4   In a certain electrical circuit, the current i satisfies the differential equation

$$L\frac{di}{dt} + Ri = E$$

Show that, if the current initially is zero, at time t

$$i = \frac{E}{R}(1 - e^{-Rt/L})$$

Sketch the graph of i against t.

5   When an alternating emf $E \sin pt$ is applied across an electrical circuit consisting of an inductance L in series with a capacitance C, the current i flowing at time t satisfies the equation

$$L\frac{d^2i}{dt^2} + \frac{i}{C} = Ep \cos pt$$

Define $\omega^2 = \dfrac{1}{LC}$ and $F = \dfrac{Ep}{L}$.

(i)   If $p \neq \omega$, find i in terms of t, given that $i = 0$ and $\dfrac{di}{dt} = 0$ when $t = 0$. What happens when p is close to ω?

(ii)   If $p = \omega$, find i in terms of t, given that $i = 0$ and $\dfrac{di}{dt} = 0$ when $t = 0$. Sketch the graph of i against t in this case.

6   A chemical substance X decays, at a rate equal to twice the quantity of X present, so that

$$\frac{dx}{dt} = -2x$$

where x is the quantity of X present at time t. Given that initially $x = a$, find an expression for x in terms of a and t. The quantity, y, of another substance Y changes so that its rate of increase is equal to $2ae^{-2t} - \dfrac{y}{2}$. Given that initially $y = 0$, find an expression for y at time t and determine the time at which y is a maximum, leaving your answer in terms of natural logarithms.   (L)

21.3   Applications in chemistry and biology

Example 1   In a chemical reaction there are present, at time t, x kg of substance X and y kg of substance Y, and initially there is 1 kg of X and 2 kg of Y. The variables x and y satisfy the equations

$$\frac{dx}{dt} = -x^2y \quad \text{and} \quad \frac{dy}{dt} = -xy^2$$

Find $\dfrac{dy}{dx}$ in terms of x and y, and express y in terms of x. Hence obtain a differential equation in x and t only, and so find an expression for x in terms of t.

We have

$$\frac{dy}{dx} = \frac{\dfrac{dy}{dt}}{\dfrac{dx}{dt}} = \frac{-xy^2}{-x^2y} = \frac{y}{x}$$

Separating the variables,

$$\int \frac{1}{y}dy = \int \frac{1}{x}dx$$

$$\ln y = \ln x + C$$

When $x = 1$, $y = 2$, so $\ln 2 = 0 + C$, and thus

$$\ln y = \ln x + \ln 2$$

$$= \ln (2x)$$

Hence $y = 2x$

Then $\dfrac{dx}{dt} = -x^2y = -2x^3$

$$\int \frac{1}{x^3}dx = \int(-2)dt$$

$$-\frac{1}{2x^2} = -2t + B$$

When $t = 0$, $x = 1$, so $-\frac{1}{2} = B$, and $-\dfrac{1}{2x^2} = -2t - \frac{1}{2}$.

Hence

$$x = \frac{1}{\sqrt{1 + 4t}}$$

Example 2 An infectious disease spreads at a rate which is proportional to the product of the number infected and the number uninfected. Initially one half of the population is infected and the rate of spread is such that, were it to remain constant, the whole population would become infected in 24 days. Calculate the proportion of the population which is infected after 12 days.

If the infected *proportion* is x, then the uninfected proportion is $(1 - x)$. $\dfrac{dx}{dt}$ is the rate at which the infected proportion is increasing; this is proportional to the product of x and $(1 - x)$. Thus

$$\frac{dx}{dt} = kx(1 - x) \quad \text{where } k \text{ is a constant.}$$

Initially, $x = \frac{1}{2}$ and $\frac{dx}{dt}$ is equal to the constant rate at which the

remaining $\frac{1}{2}$ would become infected in 24 days,

i.e. $\dfrac{dx}{dt} = \dfrac{\frac{1}{2}}{24} = \dfrac{1}{48}$

Substituting $x = \frac{1}{2}$ into the differential equation,

$\dfrac{1}{48} = k \times \frac{1}{2} \times \frac{1}{2}$

Thus $k = \dfrac{1}{12}$, and we have

$$\frac{dx}{dt} = \frac{1}{12}x(1-x)$$

$$\int \frac{1}{x(1-x)}dx = \int \frac{1}{12}dt$$

$$\int \left(\frac{1}{x} + \frac{1}{1-x}\right)dx = \int \frac{1}{12}dt$$

$$\ln x - \ln(1-x) = \frac{1}{12}t + C$$

When $t = 0$, $x = \frac{1}{2}$, so $\ln\frac{1}{2} - \ln\frac{1}{2} = 0 + C$, and $C = 0$. Thus

$$\ln\left(\frac{x}{1-x}\right) = \frac{1}{12}t$$

$$\frac{x}{1-x} = e^{\frac{1}{12}t}$$

When $t = 12$, $\dfrac{x}{1-x} = e$, giving

$$x = \frac{e}{1+e} \approx 0.73$$

Hence about 73% of the population is infected after 12 days.

Exercise 21.3 (Differential equations in chemistry and biology)

1 In appropriate units, the relation between the concentration c and the time t in a chemical reaction is given by

$$\frac{dc}{dt} = -kc^2$$

where k is a constant. Prove that

$$kt = \frac{1}{c} - A$$

where A is a constant.

It is given that when $t = 0$, $c = 1$ and when $t = 30$, $c = 0.5$. Find c when $t = 120$.   (L)

2   It is found that when quantities of two liquids A and B are boiling in the same container the ratio of the rates of vaporisation of the two liquids being vaporised at any instant is a fixed multiple of the ratio of the amounts remaining. If x and y are the respective amounts of A and B remaining at any instant show that $b^k x = a y^k$, where k is a constant and a and b are the initial values of x and y. (O and C)

3   The size S of a population at time t satisfies approximately the differential equation $\dfrac{dS}{dt} = kS$, where k is a constant. Integrate this equation to find S as a function of t.

The population numbered 32 000 in the year 1900 and had increased to 48 000 by 1970. Estimate what its size will be (correct to the nearest 1000) in the year 2000. (SMP)

4   At time $t = 0$ the number of bacteria in a certain culture is N_0. At time t the birth-rate of the bacteria is numerically equal to half of the number N of living bacteria and there is a constant death-rate of $\frac{1}{4} N_0$. Assuming N to be a continuous variable, form a differential equation for N and obtain N as a function of t. Find the time that elapses before the number of living bacteria doubles. (S)

5   In a colony of organisms, it is known that the natural growth rate of the colony is λ organisms per organism per minute. Express this fact as a differential equation connecting n and t, where n is the number of organisms present at time t minutes. If, in addition, organisms die at the rate of μ organisms per minute, show that $\dfrac{dn}{dt} = \lambda n - \mu$. In this latter case, if it is also known that at time t_0 there were n_0 organisms present: (a) find the time that elapses before the colony is wiped out, given that $n_0 = 200$, $\lambda = 2$ and $\mu = 500$; (b) find the number of organisms in the colony after $\frac{1}{2}$ minute, given that $n_0 = 200$, $\lambda = 4$ and $\mu = 500$. (C)

6   The food calories taken in by a human body go partly to increase the mass and partly to fulfil the requirements of the body; these daily requirements are taken to be proportional to the mass M. The rate of increase of mass is proportional to the number of calories available for this. Write down a differential equation connecting M, the time t and the daily intake of calories $f(t)$.

A man's mass is 100 kg; if he took in no calories he would reduce his weight by 10 per cent in 10 days. How long would it take him to reduce by this amount if, instead, he took in exactly half the number of calories needed to keep his mass constant at 100 kg. (MEI)

7   A plant grows in a pot which contains a volume V of soil. At time t the mass of the plant is m and the volume of soil utilised by the roots is αm, where α is a constant. The rate of increase of the mass of the plant is proportional to the mass of the plant times the volume of soil not yet utilised by the roots. Obtain a differential equation for m, and verify that it can be written in the form

$$V \beta \frac{dt}{dm} = \frac{1}{m} + \frac{\alpha}{V - \alpha m}$$

where β is a constant.

The mass of the plant is initially $V/4\alpha$. Find, in terms of V and β, the time taken for the plant to double its mass. Find also the mass of the plant at time t. (JMB)

8   In a certain process, the rate of production of yeast is kx grams per minute, where x grams is the amount already produced and k is a constant. Write down a differential equation relating x and the time t measured in minutes. Show that, if $k = 0.003$, then the amount of yeast is doubled in about 230 minutes.

If, in addition, yeast is removed at a constant rate of m grams per minute, so that the differential equation becomes $\dfrac{dx}{dt} = kx - m$, find the amount of yeast at time t minutes, given that at $t = 0$ there were p grams.

Deduce that if $m \leqslant kp$ the supply of yeast is never exhausted. Find the value of m to three significant figures if $k = 0.003$, $p = 20\,000$ and the supply is exhausted in 100 minutes. (C)

21.4 Numerical methods

There are many differential equations which cannot be solved analytically (i.e. in the form of a relation involving algebraic, circular, exponential and logarithmic functions.). We now consider some methods by which approximate values of the solution may be obtained.

The solution of a differential equation can usually be expressed as an infinite Taylor series, and we have seen how this may be found by differentiating the equation (see p. 315). The series can then be used to obtain values, to any desired degree of accuracy (as long as the series is valid). However, the series may only be valid in a small neighbourhood of the starting point. Also, the repeated differentiation of the equation may involve considerable work.

We have also seen how the formulae

$$f'(x) \approx \frac{f(x+h) - f(x-h)}{2h}$$

and $f''(x) \approx \dfrac{f(x+h) - 2f(x) + f(x-h)}{h^2}$

provide 'step by step' methods for building up a table of values (see p. 285).

First order equations

The step by step method, based on

$$f'(x) \approx \frac{f(x+h) - f(x-h)}{2h}$$

requires two adjacent starting values, whereas we shall usually be given just one value of the solution. One way of overcoming this is to use a Taylor series for the first step. We then have two adjacent values, and we can continue by using the formula above.

Example If $\dfrac{dy}{dx} = 2xy$, and $y = 1$ when $x = 1$, obtain approximate values of y when $x = 1.1, 1.2, 1.3$.

If the solution is $y = f(x)$, the Taylor series is

$$f(1+h) = 1 + 2h + 3h^2 + \frac{10}{3}h^3 + \frac{19}{6}h^4 + \cdots$$

(see p. 317, Example 2).
Putting $h = 0.1$,

$$f(1.1) = 1 + 0.2 + 0.03 + 0.0033 + 0.0003 + \cdots$$
$$\approx 1.234$$

We have $f'(x) \approx \dfrac{f(x+h) - f(x-h)}{2h}$

and $f'(x) = 2xf(x)$

Thus $\dfrac{f(x+h)-f(x-h)}{2h} \approx 2xf(x)$

We shall use $h = 0.1$.

Putting $x = 1.1$,

$$\dfrac{f(1.2)-f(1.0)}{0.2} \approx 2 \times 1.1 \times f(1.1)$$

$$\dfrac{f(1.2)-1}{0.2} \approx 2 \times 1.1 \times 1.234$$

giving $\qquad f(1.2) \approx 1.543$

Putting $x = 1.2$,

$$\dfrac{f(1.3)-f(1.1)}{0.2} \approx 2 \times 1.2 \times f(1.2)$$

$$\dfrac{f(1.3)-1.234}{0.2} \approx 2 \times 1.2 \times 1.543$$

giving $\qquad f(1.3) \approx 1.975$

Hence $f(1.1) \approx 1.234$, $f(1.2) \approx 1.543$, $f(1.3) \approx 1.975$.

We can solve this equation (the variables are separable) to obtain the exact solution $y = e^{x^2-1}$. The true values are thus $f(1.1) = 1.234$, $f(1.2) = 1.553$, $f(1.3) = 1.994$.

Euler's method

We now describe a simpler (but less accurate) method, which uses the approximation

$$f(a+h) \approx f(a) + hf'(a)$$

obtained by taking just the first two terms of the Taylor series for $f(a+h)$.

Given a first order differential equation, we can build up a table of values of the solution by using the above approximation, where $f'(a)$ is calculated from the differential equation.

Example If $\dfrac{dy}{dx} = 2xy$, and $y = 1$ when $x = 1$, use Euler's method to obtain approximate values of y when $x = 1.1, 1.2, 1.3$.

If $y = f(x)$ is the solution, we are given that $f(1) = 1$, and thus, since $\dfrac{dy}{dx} = 2xy$, $f'(1) = 2 \times 1 \times 1 = 2$.

Applying

$$f(a+h) \approx f(a) + hf'(a)$$

with $h = 0.1$, we obtain

$$f(1.1) \approx f(1) + 0.1 + f'(1)$$
$$= 1 + 0.1 \times 2 = 1.2$$

and thus

$$f'(1.1) = 2 \times 1.1 \times 1.2 = 2.64$$

Also

$$f(1.2) \approx f(1.1) + 0.1 \times f'(1.1)$$
$$= 1.2 + 0.1 \times 2.64 = 1.464$$

and thus

$$f'(1.2) = 2 \times 1.2 \times 1.464 = 3.5136$$

Finally,

$$f(1.3) \approx f(1.2) + 0.1 \times f'(1.2)$$
$$= 1.464 + 0.1 \times 3.5136 = 1.815$$

Hence $f(1.1) \approx 1.2$, $f(1.2) \approx 1.464$, $f(1.3) \approx 1.815$.

We see that this method is less accurate than the previous one.

Suppose A is a point on a solution curve $y = f(x)$. When we use Euler's method, we effectively replace the curve, between $x = a$ and $x = a + h$, by its tangent AB at the point A (which has gradient $f'(a)$).

The method predicts the point B as the next point on the solution curve (see Fig. 21.3), whereas the true point is P. We would not expect this to give very accurate results.

However, the accuracy could be improved by using a smaller value of h.

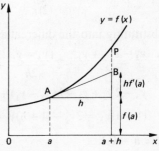

Fig. 21.3

Second order equations

Suppose the following is part of a table of values of a function satisfying a second order differential equation.

x	$x_2 - h$	x_2	$x_2 + h$
y	y_1	y_2	y_3

When $x = x_2$, we have

$$y = y_2,$$
$$\frac{dy}{dx} \approx \frac{y_3 - y_1}{2h}$$

and $\dfrac{d^2 y}{dx^2} \approx \dfrac{y_3 - 2y_2 + y_1}{h^2}$

If we substitute these into the differential equation, we obtain a relation between y_1, y_2 and y_3. If we know y_1 and y_2, we can calculate y_3; we can then use y_2 and y_3 to calculate the next value, and so on.

Example If

$$\frac{d^2y}{dx^2} - 2\frac{dy}{dx} + xy = e^x$$

and $y = 2$, $\frac{dy}{dx} = 1$ when $x = 0$, obtain approximate values of y when $x = 0.1, 0.2, 0.3$.

Suppose the table of values is

x	x_0	x_1	x_2	x_3	x_4
	(-0.1)	(0)	(0.1)	(0.2)	(0.3)
y	y_0	y_1	y_2	y_3	y_4
		(2)			

Substituting into the differential equation: when $x = x_2$,

$$\frac{y_3 - 2y_2 + y_1}{h^2} - 2\frac{(y_3 - y_1)}{2h} + x_2 y_2 = e^{x_2}$$

$$y_3 - 2y_2 + y_1 - h(y_3 - y_1) + h^2 x_2 y_2 = h^2 e^{x_2}$$

$$(1 - h)y_3 = 2y_2 - (1 + h)y_1 + h^2(e^{x_2} - x_2 y_2)$$

Since $h = 0.1$,

$$0.9y_3 = 2y_2 - 1.1y_1 + 0.01(e^{x_2} - x_2 y_2)$$

This is the recurrence relation for calculating the next value. However, we cannot use it immediately, because we only know one value of y.

We are given $\frac{dy}{dx} = 1$ when $x = 0$; thus

$$\frac{y_2 - y_0}{0.2} \approx 1$$

$$y_0 = y_2 - 0.2$$

Also, we have

$$0.9y_2 = 2y_1 - 1.1y_0 + 0.01(e^{x_1} - x_1 y_1)$$
$$0.9y_2 = 2 \times 2 - 1.1(y_2 - 0.2) + 0.01(1 - 0)$$

giving $y_2 = 2.115$. We now use

$$0.9y_3 = 2y_2 - 1.1y_1 + 0.01 (e^{x_2} - x_2 y_2)$$
$$0.9y_3 = 2 \times 2.115 - 1.1 \times 2 + 0.01 (e^{0.1} - 0.1 \times 2.115)$$

giving $y_3 = 2.265$. Then

$$0.9y_4 = 2y_3 - 1.1y_2 + 0.01(e^{x_3} - x_3 y_3)$$
$$0.9y_4 = 2 \times 2.265 - 1.1 \times 2.115 + 0.01(e^{0.2} - 0.2 \times 2.265)$$

giving $y_4 = 2.457$. Hence the approximate table of values is

x	0	0.1	0.2	0.3
y	2	2.115	2.265	2.457

Exercise 21.4   (Numerical solution of differential equations)

1   If $\dfrac{dy}{dx} = 2x + y$, and $y = 2$ when $x = 1$, use Euler's method to obtain approximate values of y for

$x = 1.2, 1.4, 1.6, 1.8, 2$.

2   If $\dfrac{dy}{dx} + y \sin x = x^2$, and $y = 0$ when $x = 0$, use Euler's method to obtain approximate values of y

for $x = 0.2, 0.4, 0.6$.

3   If $\dfrac{dy}{dx} + \dfrac{1}{x} y = \dfrac{1}{x} e^{x^2}$, and $y = 0$ when $x = 1$, use Euler's method to obtain approximate values of y

when $x = 1.1$ and $x = 1.2$. Show that $xy = \displaystyle\int_1^x e^{t^2} dt$, and hence find an approximate value for

$\displaystyle\int_1^{1.2} e^{t^2} dt$.

4   If $\dfrac{dy}{dx} = 2x + y$, and $y = 2$ when $x = 1$, use a Taylor series to calculate the value of y when $x = 1.2$.

Then use the approximation $f'(x) \approx \dfrac{f(x+h) - f(x-h)}{2h}$

to calculate approximate values of y when $x = 1.4, 1.6, 1.8, 2$. Compare these values with those
obtained in question 1, and with the true values (given by $y = 6e^{x-1} - 2x - 2$).

5   If $\dfrac{dy}{dx} = x - y^2$, and $y = -0.5$ when $x = 0$, use a Maclaurin series to calculate the value of y when

$x = 0.1$. Use this to obtain approximate values of y for $x = 0.2, 0.3, 0.4$.

6   Given that $\dfrac{dy}{dx} = \cos x + y \sin x$, subject to $y = 0$ when $x = 0$, find the Maclaurin expansion for y in

terms of x as far as the term involving x^5. Hence find the value of y when $x = 0.1$, correct to 6
significant figures.

Justify by means of a sketch the approximation $\left(\dfrac{dy}{dx}\right)_0 \approx \dfrac{1}{2h}(y_1 - y_{-1})$

and apply it to find the approximate value of y when $x = 0.2$, correct to 4 significant figures.
Check the accuracy of this value by using the Maclaurin expansion to evaluate y when $x = 0.2$.

(MEI)

7   If $\dfrac{d^2y}{dx^2} + 3\dfrac{dy}{dx} - \sin y = 8x^2$, obtain an approximate relation between three successive values,

y_1, y_2, y_3, of y corresponding to values $x_2 - h$, x_2, $x_2 + h$, of x.

If $y = 0.3$ and $\dfrac{dy}{dx} = 0$ when $x = 0$, use your relation, with $h = 0.1$, to obtain approximate values of y

when $x = 0.1, 0.2, 0.3$.

8   The function $f(x)$ satisfies the differential equation $\dfrac{d^2f}{dx^2} = 16f - 12$, and $f(0) = 0$, $f(1) = 0$.

(i)   Apply the result $f''(x) \approx \dfrac{f(x+h) - 2f(x) + f(x-h)}{h^2}$

with $x = 0.5$ and $h = 0.5$ to find an approximation to $f(0.5)$.
(ii)   Writing $f(0.25) = y_1$, $f(0.5) = y_2$, $f(0.75) = y_3$, apply the result above with $h = 0.25$ and
$x = 0.25, 0.5$ and 0.75. Hence obtain another approximation to $f(0.5)$.

Revision questions E

Revision paper E1

1 (i) Differentiate with respect to x:

(a) $2^x \tan x$ (b) $\dfrac{x}{\cos^2 2x}$

(ii) Find the values of the first three derivatives of $\sec x$ with respect to x at $x = 0$, and hence obtain the expansion of $\sec x$ in ascending powers of x if x is so small that powers of x greater than the third may be neglected.

Using your expansion, find an approximate root of the equation $\sec x = 4x$. (O and C)

2 Find the values of x for which the function

$$f(x) = e^x(2x^2 - 3x + 2)$$

has (i) a maximum, (ii) a minimum, (iii) an inflexion.

Draw a rough sketch of the graph of the function. (O)

3 Sketch the curve whose equation is

$$y = \frac{x+2}{x}$$

and state the equations of its asymptotes.

By considering this sketch, or otherwise, show that the curve whose equation is

$$y = \ln\left(\frac{x+2}{x}\right)$$

has no points whose x-coordinates lie in the interval $-2 \leqslant x \leqslant 0$. Sketch this curve on a separate diagram.

Prove that the area bounded by the *second* curve, the x-axis and the lines $x = 1$ and $x = 2$ is $3 \ln\frac{4}{3}$.
(JMB)

4 (i) (a) Evaluate

$$\int_1^e \ln x \, dx$$

(b) Using the substitution $3x = 2 + 2\tan\theta$, or otherwise, find

$$\int \frac{dx}{9x^2 - 12x + 8}$$
(L)

(ii) Find the value of

$$\int_0^1 (1+x)e^{1+x} dx$$
(L)

5   Taking 1.55 as a first approximation to a root of the equation $x - 2 + \ln x = 0$, use one application of the Newton–Raphson method to obtain a second approximation.   (L)

6   Evaluate

$$\int_0^{0.4} x \ln (1 + x) \, dx$$

using Simpson's rule with 5 ordinates and correcting your answer to two significant figures. Show clearly how your answer has been obtained.   (L)

7   The number per unit area of the species *Daphnia* (a water-flea) is n. Observations begin when $n = 100$ at $t = 0$, where t is measured in days. It is observed that n continually increases, but that it does so eventually at a decreasing rate. It is noted that n appears to approach 300.

We can construct a mathematical model of these observations by approximating n by a continuous variable which satisfies a differential equation. One, and only one, of the following differential equations is consistent with the observations above:

$$\frac{dn}{dt} = n - \frac{1}{300} n^2; \quad \frac{dn}{dt} = n + \frac{1}{300} n^2; \quad \frac{d^2 n}{dt^2} = 1 - \frac{1}{300} n$$

Choose the consistent equation, and solve it to find n in terms of t. [Note that you are asked to solve only *one* differential equation.]

From your solution estimate to 1 decimal place the value of t when $n = 200$.

Give one possible interpretation of each of the observations that (a) n increases continually, and (b) n increases eventually at a decreasing rate.   (MEI)

Revision paper E2

1   Show that the expansions in ascending powers of x of $\ln (1 + a \sin x)$ and $\sin \ln (1 + ax)$ $(a \neq 0)$ agree as far as the terms in x^2 inclusive.

For what values of a do they agree as far as the terms in x^3 inclusive?   (O)

2   Find the maximum and minimum values of the expression

$$y = x \exp (- 2x^2)$$

and justify your assertion as to which is which.

Find also the values of x which give inflexions.

[Give all your answers to 3 decimal places.]   (O and C)

3   The curve whose equation is

$$y = (1 - x)e^x$$

meets the x-axis at A and the y-axis at B. The region bounded by the arc AB of the curve and the line segments OA and OB, where O is the origin, is rotated through a complete revolution about the x-axis. Show that the volume swept out is

$$\tfrac{1}{4}\pi(e^2 - 5)$$ (JMB)

4   (a) Integrate the following functions with respect to x:

(i) $(x^3 + 1)^2$ (ii) $x(1 - x^2)^{\frac{1}{2}}$ (iii) $x \ln x$

(b) Find the area enclosed between the curve $y = \dfrac{1}{x - 2}$, the ordinates at $x = -2$ and $x = 1$, and the x-axis.   (S)

5   It is required to tabulate the root, lying between 1 and 10, of the equation $1 + \ln x = kx$ for $k = 0.5(0.1)0.9$. Using any method or combination of methods, form such a table. More credit will be given for a complete table with moderate accuracy than for fewer results with higher accuracy. The accuracy claimed should in any case be stated and substantiated.   (MEI)

6   By applying the trapezium rule to the integral $\int_{n-1}^{n+1} \dfrac{dx}{x}$, obtain the approximation

$$\ln\left(\frac{n+1}{n-1}\right) \approx \frac{1}{2}\left(\frac{1}{n-1} + \frac{2}{n} + \frac{1}{n+1}\right)$$

where $n > 1$.

Show that, if n is sufficiently large for powers of $\dfrac{1}{n}$ greater than the third to be neglected, the error in the approximation is approximately $\dfrac{1}{3n^3}$.

Use the approximation to estimate $\ln\left(\frac{12}{10}\right)$ in decimal form and give a numerical estimate of the accuracy of your value.   (MEI)

7   A point P is travelling in the positive direction on the x-axis with acceleration proportional to the square of its speed v. At time $t = 0$ it passes through the origin with speed gT and with acceleration g. Show that

$$\frac{dv}{dx} = \frac{v}{gT^2}$$

and hence obtain an expression for v in terms of x, g and T. Prove that, at time t,

$$x = gT^2 \ln\left(\frac{T}{T-t}\right)$$

Sketch the graph of x against t for $0 \leqslant t < T$.   (L)

Miscellaneous revision questions: paper E3

1   Find the Taylor expansion of $f(x) = (1 + x + 2x^2)^{-\frac{1}{2}}$ about $x = 1$ up to and including the term in $(x - 1)^2$.

Using these terms find an approximation to the value of $f(1.04)$, giving your answer correct to three places of decimals. [Do *not* use mathematical tables or calculators.]   (O)

2   (a)   (i)   Find $\dfrac{dy}{dx}$ when $y = e^{\tan x}$.

(ii)   Evaluate $\dfrac{dy}{dx}$ when $x = 2$ if

$$y = \ln\sqrt{\frac{1+x^2}{2x-1}}$$

(b)   Find the turning points on the curve $y = \cos x + 2\cos\frac{1}{2}x$ for $0 \leqslant x \leqslant 2\pi$. Hence sketch the curve over this range.

3   Sketch the curve $y = \ln(x - 2)$.

The inner surface of a bowl is of the shape formed by rotating completely about the y-axis that part of the x-axis between $x = 0$ and $x = 3$ and that part of the curve $y = \ln(x - 2)$ between $y = 0$ and

$y = 2$. The bowl is placed with its axis vertical and water is poured in. Calculate the volume of water in the bowl when the bowl is filled to a depth $h (< 2)$.

If water is poured into the bowl at a rate of 50 cubic units per second, find the rate at which the water level is rising when the depth of the water is 1.5 units. (AEB)

4 A right circular cone with semi-vertical angle θ is inscribed in a sphere of radius a, with its vertex and the rim of its base on the surface of the sphere. Prove that its volume is $\frac{8}{3}\pi a^3 \cos^4 \theta \sin^2 \theta$. If a is fixed and θ varies, find the limits within which this volume must lie. (O and C)

5 (i) Differentiate with respect to x

(a) $\sqrt{\tan x}$ (b) $x^2 \ln\left(\dfrac{1}{x}\right)$

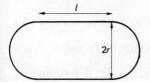

Fig. E.1

(ii) Fig. E1 shows a playing field in the form of a rectangle of length l with a semicircle of radius r at each end. A 400 m race track is to form the perimeter of the field. Find l and r if the rectangular part of the field is to have as large an area as possible. (L)

6 The function f is defined by

$$f(x) = \frac{e^{ax}}{1 + x^2} \quad \text{for all real } x$$

Find conditions upon a for f to have both a maximum and a minimum. *Sketch* (do *not* plot) the graphs of f for the following values of a: (a) $-\frac{1}{3}$; (b) 1; (c) 3. (O and C)

7 A curve C has the parametric form

$$x = \frac{at^2}{1 + t^2}, \quad y = \frac{at^3}{1 + t^2}$$

The part of the first quadrant enclosed by C, the x-axis and the line $x = \frac{1}{2}a$ is rotated about the x-axis to give a solid of revolution. Find its volume. (O)

8 By means of the expansions of e^x and $\ln(1 + x)$, or otherwise, prove that, when n is large,

$$\left(1 + \frac{1}{n}\right)^n = e\left(1 - \frac{1}{2n} + \frac{11}{24n^2} - \frac{7}{16n^3} + \cdots\right)$$

Hence show that e is given by the formula

$$2e = (1.1)^{10} + (0.9)^{-10}$$

with an error of approximately 0.46%. (MEI)

9 The following pairs of values of x and y satisfy approximately a relation of the form $y = ax^n$, where a and n are integers. By plotting the graph of lg y against lg x find the values of the integers a and n. [Note that lg N denotes $\log_{10} N$.]

x	0.7	0.9	1.1	1.3	1.5
y	1.37	2.92	5.32	8.80	13.50

Estimate the value of the integral

$$\int_{0.7}^{1.5} y \, dx$$

(a)   by Simpson's rule, using five ordinates and clearly indicating your method;
(b)   by using the relation $y = ax^n$ with the values found for a and n.   (L)

10   Given that $y = \dfrac{(x-4)^2}{x-3}$, where x and y are real, find the stationary values of y, distinguishing

between maximum and minimum values.

Show, graphically or otherwise, that as x varies from $-\infty$ to $+\infty$, y takes each value from $-\infty$ to $+\infty$ twice except for values in an interval of length 4 units.

Show that the area of the finite region bounded by the portion of the x-axis between $x = 4$ and $x = 8$, the line $x = 8$ and the arc of the curve $y = \dfrac{(x-4)^2}{x-3}$ between the points $(4, 0)$ and $(8, \frac{16}{5})$ is

$4 + \ln 5$ square units. Hence, by using Simpson's rule with four equal intervals, i.e. five ordinates, deduce a first approximation to the value of $\ln 5$.   (MEI)

11   Show graphically that the equation

$$x^2 = 7 \log_{10} x + 2.347$$

has two real positive roots.

Taking $x = 2.2$ as an initial approximation to the larger of these roots, obtain a second approximation (i) by Newton's method; (ii) by writing the equation in the form

$$x = \sqrt{7 \log_{10} x + 2.347}$$

and using an iterative method.

Work to three decimal places and give your answers to two decimal places.   (MEI)

12   Obtain the general solution of the differential equation

$$\frac{d^2 x}{dt^2} + \frac{dx}{dt} - 6x = e^{-t}$$

A particle moves in a plane in such a manner that its coordinates (x, y) at time t satisfy the equations

$$2\frac{dy}{dt} + x = e^{-t} \qquad \frac{dx}{dt} + x + 12y = -7e^{-t}$$

Prove that

$$\frac{d^2 x}{dt^2} + \frac{dx}{dt} - 6x = e^{-t}$$

and hence determine x and y as functions of t, given that x remains finite as $t \to \infty$ and $y = 1$ when $t = 0$.   (MEI)

13   A radioactive substance P decays and changes (without loss of mass) into a substance Q, which itself similarly changes into a third substance R. R suffers no further change. The masses of P, Q and R present at time t are given by p, q and r grams respectively. The rates of change are such that

$$\frac{dp}{dt} = -2p \quad \text{and} \quad \frac{dr}{dt} = q \qquad (1)$$

Show that

$$\frac{dq}{dt} = 2p - q$$

Initially (at time $t = 0$) there is one gram of the substance P and none of Q. Integrate equation (1) and hence show that q satisfies the differential equation

$$\frac{dq}{dt} + q = 2e^{-2t} \qquad (2)$$

Show that (2) may be written in the form

$$\frac{d}{dt}(qe^t) = 2e^{-t}$$

and integrate to find q as a function of t. Hence prove that at any subsequent time there is never more than $\frac{1}{2}$ gram of Q present. **(SMP)**

14 (a)   If $y = \sin x + 5c \sin 5x$, where c is a constant, and if M and N denote the mean values of y and y^2, respectively, over the interval $0 \leqslant x \leqslant \pi$, prove that

$$M = 2(1 + c)/\pi \quad \text{and} \quad N = \tfrac{1}{2}(1 + 25c^2)$$

Verify that $N > M^2$ for all values of c.

(b)   Evaluate

$$\int_0^\pi |a - x| \sin x \, dx$$

distinguishing the cases $a < 0$, $0 \leqslant a \leqslant \pi$ and $\pi < a$. **(MEI)**

15   A particle of mass m is in motion along a straight path through O. It is subject to a force of repulsion from O of magnitude $\dfrac{k}{x^2}$ (where x is its displacement from O at time t and k is a positive constant). Show that x satisfies the differential equation

$$v\frac{dv}{dx} = \frac{k}{mx^2}$$

Show that, if P is initially projected towards O with speed u from a point with displacement a (> 0) from 0, then x will decrease to a value

$$\frac{2ka}{2k + mau^2}$$

What can you say about the sign of $\dfrac{dv}{dt}$? Deduce that in the subsequent motion x will increase without limit. Show also that v will increase, but to a limiting value, and find this value. **(SMP)**

22

Integration III

22.1 Systematic integration: standard integrals, and algebraic functions

Standard integrals

We studied methods of integration in Chapter 15. Since then we have introduced some new functions, e^x and $\ln x$, so we now review our 'standard' integrals.

$$\int x^n \, dx = \frac{x^{n+1}}{n+1} + C \qquad \text{(provided } n \neq -1)$$

$$\int \frac{1}{x} \, dx = \ln|x| + C$$

$$\int \sin x \, dx = -\cos x + C$$

$$\int \cos x \, dx = \sin x + C$$

$$\int \tan x \, dx = \ln|\sec x| + C$$

$$\int \sec x \, dx = \ln|\sec x + \tan x| + C$$

$$\int e^x \, dx = e^x + C$$

$$\int a^x \, dx = \frac{a^x}{\ln a} + C \quad \text{(where } a \text{ is a positive constant)}$$

$$\int \frac{1}{\sqrt{a^2 - x^2}} \, dx = \sin^{-1} \frac{x}{a} + C$$

$$\int \frac{1}{a^2 + x^2} \, dx = \frac{1}{a} \tan^{-1} \frac{x}{a} + C$$

$$\int \frac{1}{\sqrt{a^2 + x^2}} \, dx = \ln(x + \sqrt{a^2 + x^2}) + C$$

$$\int \frac{1}{\sqrt{x^2 - a^2}} \, dx = \ln(x + \sqrt{x^2 - a^2}) + C$$

$$\int \frac{f'(x)}{f(x)} dx \qquad = \ln|f(x)| + C$$

We can check these by differentiation. For example,

$$\frac{d}{dx}(\ln|\sec x + \tan x| + C) = \frac{\sec x \tan x + \sec^2 x}{\sec x + \tan x} = \frac{\sec x(\tan x + \sec x)}{\sec x + \tan x}$$

$$= \sec x$$

$$\frac{d}{dx}\left(\frac{a^x}{\ln a} + C\right) = \frac{1}{\ln a}\frac{d}{dx}(a^x) = \frac{1}{\ln a}(a^x \ln a) \text{ (see p. 348)}$$

$$= a^x$$

$$\frac{d}{dx}[\ln(x + \sqrt{a^2 + x^2})] = \frac{1}{x + \sqrt{a^2 + x^2}}\left(1 + \frac{x}{\sqrt{a^2 + x^2}}\right)$$

$$= \frac{1}{x + \sqrt{a^2 + x^2}} \times \frac{\sqrt{a^2 + x^2} + x}{\sqrt{a^2 + x^2}}$$

$$= \frac{1}{\sqrt{a^2 + x^2}}$$

Related functions can be integrated by applying the 'linear function rule'.

Example 1

$$\int 3^{2x+1} dx$$

This is of the form $\int a^x dx$, with $a = 3$ and x replaced by $(2x + 1)$.
Thus

$$\int 3^{2x+1} dx = \frac{1}{2} \times \frac{3^{2x+1}}{\ln 3} + C = \frac{3^{2x+1}}{2 \ln 3} + C$$

Example 2

$$\int \sec 5x \, dx = \frac{1}{5} \ln|\sec 5x + \tan 5x| + C$$

Example 3

$$\int \frac{1}{\sqrt{25 - 9x^2}} dx = \int \frac{1}{\sqrt{5^2 - (3x)^2}} dx$$

This is of the form $\int \frac{1}{\sqrt{a^2 - x^2}} dx$, with $a = 5$ and x replaced by $3x$.

Thus

$$\int \frac{1}{\sqrt{25 - 9x^2}} dx = \frac{1}{3} \sin^{-1}\left(\frac{3x}{5}\right) + C$$

Integrals involving $(ax+b)^n$

Simple algebraic functions with denominator $(ax+b)$, or containing any power of $(ax+b)$, may be integrated by means of the substitution $ax+b=t$.

Example 1

$$\int \frac{x^2}{x+3}\,dx$$

Put $\qquad x+3=t$

Then $\qquad dx=dt$

and $\displaystyle\int \frac{x^2}{x+3}\,dx = \int \frac{(t-3)^2}{t}\,dt$

$$= \int \frac{t^2-6t+9}{t}\,dt$$

$$= \int \left(t-6+\frac{9}{t}\right)dt$$

$$= \tfrac{1}{2}t^2-6t+9\ln|t|+C$$

$$= \tfrac{1}{2}(x+3)^2-6(x+3)+9\ln|x+3|+C$$

$$= \tfrac{1}{2}x^2-3x+9\ln|x+3|+C'$$

(Here, all the constant terms have been absorbed in the arbitrary constant C'.)

For integrals involving $\sqrt{ax+b}$, the substitution $ax+b=u^2$ gives a neater solution.

Example 2

$$\int x\sqrt{1-4x}\,dx$$

Put $\qquad 1-4x=u^2$

Then $\qquad -4\,dx=2u\,du$

and $\displaystyle\int x\sqrt{1-4x}\,dx = \int \tfrac{1}{4}(1-u^2)u(-\tfrac{1}{2}u\,du)$

$$= \int \tfrac{1}{8}(u^4-u^2)du$$

$$= \tfrac{1}{40}u^5-\tfrac{1}{24}u^3+C$$

$$= \tfrac{1}{40}\sqrt{(1-4x)^5}-\tfrac{1}{24}\sqrt{(1-4x)^3}+C$$

Integrals of the type $\displaystyle\int \frac{px+q}{ax^2+bx+c}\,dx$

If the denominator is a quadratic expression which factorises, we first write the function in its partial fractions.

Example 1

$$\int \frac{2x-5}{x^2+x-12}\,dx$$

Writing $\displaystyle\frac{2x-5}{x^2+x-12} = \frac{2x-5}{(x-3)(x+4)} = \frac{A}{x-3} + \frac{B}{x+4}$

we find that $A = \frac{1}{7}$ and $B = \frac{13}{7}$. Thus

$$\int \frac{2x-5}{x^2+x-12}\,dx = \int\left(\frac{\frac{1}{7}}{x-3} + \frac{\frac{13}{7}}{x+4}\right)dx$$

$$= \tfrac{1}{7}\ln|x-3| + \tfrac{13}{7}\ln|x+4| + C$$
$$= \tfrac{1}{7}\ln|(x-3)(x+4)^{13}| + C$$

Example 2

$$\int \frac{1}{9-x^2}\,dx$$

Writing $\displaystyle\frac{1}{9-x^2} = \frac{1}{(3+x)(3-x)} = \frac{A}{3+x} + \frac{B}{3-x}$

we find that $A = \frac{1}{6}$ and $B = \frac{1}{6}$. Thus

$$\int \frac{1}{9-x^2}\,dx = \int\left(\frac{\frac{1}{6}}{3+x} + \frac{\frac{1}{6}}{3-x}\right)dx$$

$$= \tfrac{1}{6}\ln|3+x| - \tfrac{1}{6}\ln|3-x| + C$$

$$= \tfrac{1}{6}\ln\left|\frac{3+x}{3-x}\right| + C$$

Example 3

$$\int \frac{x}{4x^2-9}\,dx$$

This could be done using partial fractions, but, since the derivative of $(4x^2-9)$ is $8x$, it is simpler to proceed as follows.

$$\int \frac{x}{4x^2-9}\,dx = \tfrac{1}{8}\int \frac{8x}{4x^2-9}\,dx$$

$$= \tfrac{1}{8}\ln|4x^2-9| + C$$

If the denominator cannot be factorised, then it may be necessary
to complete the square.

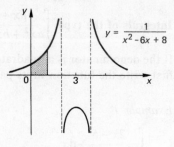

Fig. 22.1

Example 4

$$\int \frac{1}{x^2 + 4x + 13} dx = \int \frac{1}{(x+2)^2 + 9} dx$$

$$= \int \frac{1}{3^2 + (x+2)^2} dx$$

$$= \tfrac{1}{3} \tan^{-1}\left(\frac{x+2}{3}\right) + C$$

Example 5

$$\int \frac{1}{9x^2 + 6x + 4} dx = \int \frac{1}{(3x+1)^2 + 3} dx$$

$$= \int \frac{1}{(\sqrt{3})^2 + (3x+1)^2} dx$$

$$= \tfrac{1}{3} \times \frac{1}{\sqrt{3}} \tan^{-1}\left(\frac{3x+1}{\sqrt{3}}\right) + C$$

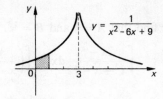

Fig. 22.2

Example 6

$$\int \frac{x+5}{x^2+4} dx = \int \left(\frac{x}{x^2+4} + \frac{5}{x^2+4}\right) dx$$

$$= \tfrac{1}{2} \int \frac{2x}{x^2+4} dx + 5 \int \frac{1}{2^2 + x^2} dx$$

$$= \tfrac{1}{2} \ln(x^2+4) + 5 \times \tfrac{1}{2} \tan^{-1}\frac{x}{2} + C$$

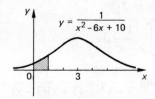

Fig. 22.3

Example 7

$$\int \frac{3x-4}{x^2+x+1} dx$$

The derivative of $(x^2 + x + 1)$ is $(2x + 1)$; we therefore rearrange the
numerator to include a multiple of this.

$$\int \frac{3x-4}{x^2+x+1} dx = \int \frac{\tfrac{3}{2}(2x+1) - \tfrac{11}{2}}{x^2+x+1} dx = \tfrac{3}{2}\int \frac{2x+1}{x^2+x+1} dx - \tfrac{11}{2}\int \frac{1}{(x+\tfrac{1}{2})^2 + \tfrac{3}{4}} dx$$

$$= \tfrac{3}{2}\ln(x^2+x+1) - \tfrac{11}{2}\int \frac{1}{(\frac{\sqrt{3}}{2})^2 + (x+\tfrac{1}{2})^2} dx$$

$$= \tfrac{3}{2}\ln(x^2+x+1) - \tfrac{11}{2} \times \frac{2}{\sqrt{3}} \tan^{-1}\left(\frac{x+\tfrac{1}{2}}{\frac{\sqrt{3}}{2}}\right) + C$$

$$= \tfrac{3}{2}\ln(x^2+x+1) - \frac{11}{\sqrt{3}} \tan^{-1}\left(\frac{2x+1}{\sqrt{3}}\right) + C$$

Example 8   Find

$$\int \frac{1}{x^2 - 6x + 8}\,dx, \quad \int \frac{1}{x^2 - 6x + 9}\,dx \quad \text{and} \quad \int \frac{1}{x^2 - 6x + 10}\,dx$$

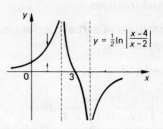

Fig. 22.4

Illustrate the results graphically.

We have

$$\int \frac{1}{x^2 - 6x + 8}\,dx = \int \frac{1}{(x-4)(x-2)}\,dx$$

$$= \int \tfrac{1}{2}\left(\frac{1}{x-4} - \frac{1}{x-2}\right)dx$$

$$= \tfrac{1}{2}\ln\left|\frac{x-4}{x-2}\right| + C$$

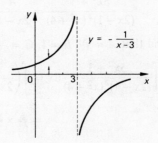

Fig. 22.5

$$\int \frac{1}{x^2 - 6x + 9}\,dx = \int \frac{1}{(x-3)^2}\,dx$$

$$= -\frac{1}{x-3} + C$$

$$\int \frac{1}{x^2 - 6x + 10}\,dx = \int \frac{1}{(x-3)^2 + 1}\,dx$$

$$= \tan^{-1}(x-3) + C$$

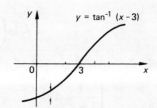

Fig. 22.6

These integrals have quite different forms. Although the three original functions have similar formulae, their graphs are very different (Figs 22.1, 22.2 and 22.3). The integral functions (Figs 22.4, 22.5 and 22.6) represent the area under these curves; for example, the area between $x = 0$ and $x = 1$ (shown shaded) is equal to the difference in the values of the integral function at $x = 0$ and $x = 1$ (shown arrowed).

The curve $y = \dfrac{1}{x^2 - 6x + 8}$ has two discontinuities (at $x = 2$ and $x = 4$), and so has its integral. The curve $y = \dfrac{1}{x^2 - 6x + 9}$ has a single discontinuity (at $x = 3$), and so has its integral. The curve $y = \dfrac{1}{x^2 + 6x + 10}$ is continuous for all x, and so is its integral.

Partial fractions

The method of partial fractions can be used for more complicated denominators.

Example 1

$$\int \frac{5x^2+3}{(2x-1)^2(x^2+4)} dx$$

Writing

$$\frac{5x^2+3}{(2x-1)^2(x^2+4)} = \frac{A}{2x-1} + \frac{B}{(2x-1)^2} + \frac{Cx+D}{x^2+4}$$

we find that $A = \frac{8}{17}$, $B = 1$, $C = -\frac{4}{17}$ and $D = \frac{15}{17}$. Thus

$$\int \frac{5x^2+3}{(2x-1)^2(x^2+4)} dx = \int \left(\frac{\frac{8}{17}}{2x-1} + \frac{1}{(2x-1)^2} + \frac{-\frac{4}{17}x+\frac{15}{17}}{x^2+4} \right) dx$$

$$= \tfrac{8}{17} \times \tfrac{1}{2}\ln|2x-1| + \tfrac{1}{2}\left(\frac{-1}{2x-1} \right) - \tfrac{2}{17}\int \frac{2x}{x^2+4} dx + \tfrac{15}{17}\int \frac{1}{2^2+x^2} dx$$

$$= \tfrac{4}{17}\ln|2x-1| - \frac{1}{2(2x-1)} - \tfrac{2}{17}\ln(x^2+4) + \tfrac{15}{17} \times \tfrac{1}{2}\tan^{-1}\left(\frac{x}{2}\right) + C$$

$$= \tfrac{2}{17}\ln\left(\frac{(2x-1)^2}{x^2+4} \right) - \frac{1}{2(2x-1)} + \frac{15}{34}\tan^{-1}\frac{x}{2} + C$$

Example 2

$$\int \frac{1}{1+x^3} dx$$

Writing

$$\frac{1}{1+x^3} = \frac{1}{(1+x)(1-x+x^2)}$$

$$= \frac{A}{1+x} + \frac{Bx+C}{1-x+x^2}$$

we find that $A = \frac{1}{3}$, $B = -\frac{1}{3}$ and $C = \frac{2}{3}$. Thus

$$\int \frac{1}{1+x^3} dx = \int \left(\frac{\frac{1}{3}}{1+x} + \frac{-\frac{1}{3}x+\frac{2}{3}}{1-x+x^2} \right) dx$$

$$= \tfrac{1}{3}\ln|1+x| + \int \frac{-\frac{1}{6}(2x-1)+\frac{1}{2}}{1-x+x^2} dx$$

$$= \tfrac{1}{3}\ln|1+x| - \tfrac{1}{6}\int \frac{2x-1}{1-x+x^2} dx + \tfrac{1}{2}\int \frac{1}{(\frac{\sqrt{3}}{2})^2+(x-\frac{1}{2})^2} dx$$

$$= \tfrac{1}{3}\ln|1+x| - \tfrac{1}{6}\ln(1-x+x^2) + \frac{1}{2} \times \frac{2}{\sqrt{3}}\tan^{-1}\frac{x-\frac{1}{2}}{\frac{\sqrt{3}}{2}} + C$$

$$= \tfrac{1}{6}\ln\left(\frac{(1+x)^2}{1-x+x^2} \right) + \frac{1}{\sqrt{3}}\tan^{-1}\left(\frac{2x-1}{\sqrt{3}} \right) + C$$

Exercise 22.1   (Algebraic functions)

Find the following integrals.

1 $\displaystyle\int \frac{1}{\sqrt{2x+1}}\,dx$ 2 $\displaystyle\int \frac{x}{2x+1}\,dx$ 3 $\displaystyle\int \frac{x}{2x^2+1}\,dx$ 4 $\displaystyle\int \frac{x}{\sqrt{3x+2}}\,dx$

5 $\displaystyle\int \left(x-\frac{2}{x}\right)^2 dx$ 6 $\displaystyle\int \frac{1}{9+4x^2}\,dx$ 7 $\displaystyle\int \frac{1}{x(x+2)}\,dx$ 8 $\displaystyle\int \frac{x}{(x+2)^2}\,dx$

9 $\displaystyle\int x(1+x^2)^5 dx$ 10 $\displaystyle\int \frac{1}{(x+1)(2x+1)}\,dx$ 11 $\displaystyle\int \frac{1}{2x^2-2x+5}\,dx$

12 $\displaystyle\int \frac{1}{x(x^2-1)}\,dx$ 13 $\displaystyle\int \frac{x^2+3x-2}{x^2+3x+2}\,dx$ 14 $\displaystyle\int \frac{25}{(x+2)(2x-1)^2}\,dx$

15 $\displaystyle\int \frac{2}{(1-x)(1+x^2)}\,dx$ 16 $\displaystyle\int_0^1 x\sqrt{1-x}\,dx$ 17 $\displaystyle\int_0^1 \frac{x}{x^2-5x+6}\,dx$

18 $\displaystyle\int_1^{\sqrt{3}} \frac{1+x}{1+x^2}\,dx$ 19 $\displaystyle\int_0^2 \frac{3x+4}{(x^2+4)(x-3)}\,dx$ 20 $\displaystyle\int_1^3 \frac{3(x+1)}{x^2(x^2+3)}\,dx$

22.2 Systematic integration: circular functions and parts

Circular functions

We recall (from Chapter 10) that we can integrate $\sin x$, $\cos x$, $\sec^2 x$, $\csc^2 x$, $\sec x \tan x$ and $\csc x \cot x$. Also, the integrals of $\tan x$ and $\sec x$ are listed as 'standard' integrals (see p. 428).

We can sometimes use trigonometrical identities to express the required function in terms of these.

Example 1

$$\int \sqrt{1+\cos x}\,dx = \int \sqrt{1+(2\cos^2(\tfrac{1}{2}x)-1)}\,dx$$
$$= \int \sqrt{2\cos^2(\tfrac{1}{2}x)}\,dx$$
$$= \int \sqrt{2}\cos(\tfrac{1}{2}x)\,dx = \sqrt{2}\times 2\sin(\tfrac{1}{2}x)+C$$
$$= 2\sqrt{2}\sin(\tfrac{1}{2}x)+C$$

Again, we should look for a function occurring with its derivative.

Example 2

$$\int \cot x\,dx = \int \frac{\cos x}{\sin x}\,dx$$
$$= \ln|\sin x|+C$$

Example 3

$\int \sec^3 x \tan x \, dx$

Since the derivative of $\sec x$ is $\sec x \tan x$, we put $\sec x = u$; then $\sec x \tan x \, dx = du$, and

$$\int \sec^3 x \tan x \, dx = \int \sec^2 x \times \sec x \tan x \, dx$$
$$= \int u^2 \, du = \tfrac{1}{3}u^3 + C$$
$$= \tfrac{1}{3}\sec^3 x + C$$

Example 4

$$\int \frac{\cos x}{\sin x + 2\cos x} \, dx$$

We first express the numerator as a multiple of the denominator $(\sin x + 2\cos x)$ plus a multiple of its derivative $(\cos x - 2\sin x)$. Write

$$\cos x = p(\sin x + 2\cos x) + q(\cos x - 2\sin x)$$

Equating coefficients of $\sin x$:

$$0 = p - 2q$$

Equating coefficients of $\cos x$:

$$1 = 2p + q$$

This gives $p = \tfrac{2}{5}$, $q = \tfrac{1}{5}$. Thus

$$\int \frac{\cos x}{\sin x + 2\cos x} \, dx = \int \frac{\tfrac{2}{5}(\sin x + 2\cos x) + \tfrac{1}{5}(\cos x - 2\sin x)}{\sin x + 2\cos x} \, dx$$

$$= \int \tfrac{2}{5} \, dx + \tfrac{1}{5} \int \frac{\cos x - 2\sin x}{\sin x + 2\cos x} \, dx$$

$$= \tfrac{2}{5}x + \tfrac{1}{5} \ln|\sin x + 2\cos x| + C$$

The integral of $\sin^m x \cos^n x$

If m is odd, we use the substitution $\cos x = u$; if n is odd, we use the substitution $\sin x = u$. If m and n are both even we use trigonometrical identities such as $\sin^2 x = \tfrac{1}{2}(1 - \cos 2x)$ (see p. 273).

Example

$\int \sin^5 x \cos^4 x \, dx$

Put $\qquad \cos x = u$

Then $-\sin x \, dx = du$

and $\int \sin^5 x \cos^4 x \, dx = \int \sin^4 x \cos^4 x \sin x \, dx = \int (1 - u^2)^2 u^4 (-du)$

$$= \int (-u^4 + 2u^6 - u^8) du$$
$$= -\tfrac{1}{5} u^5 + \tfrac{2}{7} u^7 - \tfrac{1}{9} u^9 + C$$
$$= -\tfrac{1}{5} \cos^5 x + \tfrac{2}{7} \cos^7 x - \tfrac{1}{9} \cos^9 x + C$$

The substitution $\tan (\tfrac{1}{2}x) = t$

Any integral involving circular functions only may be transformed to an algebraic integral by means of the substitution $\tan (\tfrac{1}{2}x) = t$.
Then

$$\sin x = \frac{2t}{1+t^2}, \quad \cos x = \frac{1-t^2}{1+t^2} \quad \text{and} \quad \tan x = \frac{2t}{1-t^2}$$

Also $\tfrac{1}{2} \sec^2 (\tfrac{1}{2}x) dx = dt$

i.e. $\tfrac{1}{2}(1 + t^2) dx = dt$

and so $dx = \dfrac{2}{1+t^2} dt$

Example 1

$\int \operatorname{cosec} x \, dx$

Put $\tan (\tfrac{1}{2}x) = t$

Then $dx = \dfrac{2}{1+t^2} dt$

and $\int \operatorname{cosec} x \, dx = \int \left(\dfrac{1+t^2}{2t} \right) \left(\dfrac{2}{1+t^2} dt \right)$

$$= \int \frac{1}{t} dt = \ln |t| + C = \ln |\tan (\tfrac{1}{2}x)| + C$$

Example 2

$\int \sec x \, dx$

Put $\tan(\tfrac{1}{2}x) = t$

Then $dx = \dfrac{2}{1+t^2} dt$

and $\int \sec x \, dx = \int \left(\dfrac{1+t^2}{1-t^2} \right) \left(\dfrac{2}{1+t^2} dt \right) = \int \dfrac{2}{1-t^2} dt$

$$= \int \left(\frac{1}{1+t} + \frac{1}{1-t} \right) dt$$

$$= \ln |1+t| - \ln |1-t| + C = \ln \left| \frac{1+t}{1-t} \right| + C$$

$$= \ln \left| \frac{1 + \tan (\tfrac{1}{2}x)}{1 - \tan (\tfrac{1}{2}x)} \right| + C$$

Note that this result appears to be different from the standard integral

$$\int \sec x \, dx = \ln |\sec x + \tan x| + C$$

However,

$$\sec x + \tan x = \frac{1+t^2}{1-t^2} + \frac{2t}{1-t^2} = \frac{(1+t)^2}{1-t^2} = \frac{1+t}{1-t}$$

Example 3

$$\int \frac{1}{4+5 \sin x} \, dx$$

Put $\tan(\tfrac{1}{2}x) = t$

Then $dx = \dfrac{2}{1+t^2} \, dt$

and $\displaystyle \int \frac{1}{4+5 \sin x} \, dx = \int \frac{1}{4 + \dfrac{10t}{1+t^2}} \left(\frac{2}{1+t^2} \right) dt$

$$= \int \frac{2}{4(1+t^2)+10t} \, dt = \int \frac{1}{2+2t^2+5t} \, dt$$

$$= \int \frac{1}{(1+2t)(2+t)} \, dt = \int \left(\frac{\tfrac{2}{3}}{1+2t} + \frac{-\tfrac{1}{3}}{2+t} \right) dt$$

$$= \tfrac{2}{3} \times \tfrac{1}{2} \ln|1+2t| - \tfrac{1}{3} \ln|2+t| + C$$

$$= \tfrac{1}{3} \ln \left| \frac{1+2t}{2+t} \right| + C$$

$$= \tfrac{1}{3} \ln \left| \frac{1+2 \tan (\tfrac{1}{2}x)}{2 + \tan (\tfrac{1}{2}x)} \right| + C$$

Integration by parts

When we have to integrate a product of two different types of function, we usually integrate 'by parts'.

Example 1

$$\int x e^{-5x} \, dx$$

Put $u = x$ and $\dfrac{dv}{dx} = e^{-5x}$

Then $\dfrac{du}{dx} = 1$ and $v = -\tfrac{1}{5} e^{-5x}$

Then $\displaystyle \int x e^{-5x} \, dx = x(-\tfrac{1}{5} e^{-5x}) - \int (1)(-\tfrac{1}{5} e^{-5x}) dx$

$$= -\tfrac{1}{5} x e^{-5x} - \tfrac{1}{25} e^{-5x} + C$$

Example 2

$\int xe^{-x^2}dx$

The derivative of $(-x^2)$ is $(-2x)$; in this case there is no need to use parts. We put $-x^2 = t$; then $-2x\,dx = dt$, and

$$\int xe^{-x^2}\,dx = \int e^t(-\tfrac{1}{2}dt)$$
$$= -\tfrac{1}{2}e^t + C$$
$$= -\tfrac{1}{2}e^{-x^2} + C$$

Remember that functions such as $\sin^{-1}x$ and $\ln x$ can be integrated by parts.

Example 3

$\int \tan^{-1}x\,dx$

Put $u = \tan^{-1}x$ and $\dfrac{dv}{dx} = 1$

Then $\dfrac{du}{dx} = \dfrac{1}{1+x^2}$ and $v = x$

Hence $\displaystyle\int \tan^{-1}x\,dx = (\tan^{-1}x)(x) - \int\left(\frac{1}{1+x^2}\right)(x)dx$

$$= x\tan^{-1}x - \tfrac{1}{2}\int\frac{2x}{1+x^2}dx$$
$$= x\tan^{-1}x - \tfrac{1}{2}\ln(1+x^2) + C$$

Exercise 22.2 (Circular functions, and integration by parts)

Find the following integrals.

1 $\int \sin^2 3x\,dx$ 2 $\int \sin 2x\cos 3x\,dx$ 3 $\int \cot x\,dx$ 4 $\int \sin^3 x\,dx$

5 $\displaystyle\int \frac{\cos x}{1+\sin x}dx$ 6 $\int x\cos x\,dx$ 7 $\int xe^{-2x}\,dx$ 8 $\int \sin^2 x\cos^2 x\,dx$

9 $\int x^2 \sin x\,dx$ 10 $\displaystyle\int \frac{\ln x}{x^3}dx$ 11 $\int x\ln 2x\,dx$ 12 $\int \sin^2 x\cos^3 x\,dx$

13 $\displaystyle\int \frac{1-\sin x}{1+\sin x}dx$ 14 $\displaystyle\int \frac{1}{1+\sin x + \cos x}dx$ 15 $\int x\sec^2 x\,dx$

16 $\displaystyle\int_{-1}^{1} xe^x\,dx$ 17 $\displaystyle\int_{0}^{\frac{1}{2}} \sin^{-1}x\,dx$ 18 $\displaystyle\int_{0}^{1} \tan^{-1}x\,dx$

19 $\displaystyle\int_{0}^{\pi} x\sin x\sin 2x\,dx$ 20 $\displaystyle\int_{0}^{\frac{1}{4}\pi} \frac{\sin x}{\sin x + \cos x}dx$

22.3   Systematic integration: further algebraic functions

Integrals of the type $\displaystyle\int \frac{px+q}{\sqrt{ax^2+bx+c}}\,dx$

For integrals involving the square root of a quadratic expression, it is usually necessary to 'complete the square'.

Example 1

$$\int \frac{1}{\sqrt{3+2x-x^2}}\,dx = \int \frac{1}{\sqrt{4-(x-1)^2}}\,dx$$

This is now of the form $\displaystyle\int \frac{1}{\sqrt{a^2-x^2}}\,dx$, with $a = 2$ and x replaced by $(x-1)$. Then

$$\int \frac{1}{\sqrt{3+2x-x^2}}\,dx = \sin^{-1}\left(\frac{x-1}{2}\right)+C$$

Example 2

$$\int \frac{1}{\sqrt{1-12x-4x^2}}\,dx = \int \frac{1}{\sqrt{10-(2x+3)^2}}\,dx$$

$$= \int \frac{1}{\sqrt{(\sqrt{10})^2-(2x+3)^2}}\,dx$$

$$= \tfrac{1}{2}\sin^{-1}\left(\frac{2x+3}{\sqrt{10}}\right)+C$$

Always look for a function occurring with its derivative. In such cases we may use a simple substitution (or integrate by inspection).

Example 3

$$\int \frac{5-2x}{\sqrt{1+5x-x^2}}\,dx$$

$(5-2x)$ is the derivative of $(1+5x-x^2)$, so it is not necessary to complete the square. We put $1+5x-x^2 = u^2$; then $(5-2x)dx = 2u\,du$ and

$$\int \frac{5-2x}{\sqrt{1+5x-x^2}}\,dx = \int \frac{1}{u}(2u\,du)$$

$$= \int 2\,du = 2u+C$$

$$= 2\sqrt{1+5x-x^2}+C$$

Example 4

$$\int \frac{x}{\sqrt{5+4x-x^2}} \, dx$$

The derivative of $(5+4x-x^2)$ is $(4-2x)$. We rearrange the numerator to include a multiple of $(4-2x)$. Then

$$\int \frac{x}{\sqrt{5+4x-x^2}} \, dx = \int \frac{-\frac{1}{2}(4-2x)+2}{\sqrt{5+4x-x^2}} \, dx$$

$$= -\frac{1}{2} \int \frac{4-2x}{\sqrt{5+4x-x^2}} \, dx + 2 \int \frac{1}{\sqrt{5+4x-x^2}} \, dx$$

$$= -\frac{1}{2} \times 2\sqrt{5+4x-x^2} + 2 \int \frac{1}{\sqrt{9-(x-2)^2}} \, dx$$

$$= -\sqrt{5+4x-x^2} + 2 \sin^{-1}\left(\frac{x-2}{3}\right) + C$$

Integrals of the type $\int \sqrt{ax^2+bx+c} \, dx$

For $\int \sqrt{a^2-x^2} \, dx$, we use the substitution $x = a \sin \theta$

Example 1

$$\int \sqrt{9-x^2} \, dx$$

Put $\qquad\qquad x = 3 \sin \theta$

Then $\qquad\qquad dx = 3 \cos \theta \, d\theta$

and $\quad \int \sqrt{9-x^2} \, dx = \int \sqrt{9 - 9 \sin^2 \theta} \; (3 \cos \theta \, d\theta)$

$$= \int (3 \cos \theta)(3 \cos \theta \, d\theta)$$

$$= \int 9 \cos^2 \theta \, d\theta = \int \tfrac{9}{2}(1 + \cos 2\theta) \, d\theta$$

$$= \tfrac{9}{2}\theta + \tfrac{9}{4} \sin 2\theta + C$$

$$= \tfrac{9}{2}\theta + \tfrac{9}{2} \sin \theta \cos \theta + C$$

$$= \tfrac{9}{2}\theta + \tfrac{9}{2} \sin \theta \sqrt{1 - \sin^2 \theta} + C$$

$$= \tfrac{9}{2} \sin^{-1}(\tfrac{1}{3}x) + \tfrac{9}{2}(\tfrac{1}{3}x) \sqrt{1 - \tfrac{1}{9}x^2} + C$$

$$= \tfrac{9}{2} \sin^{-1}(\tfrac{1}{3}x) + \tfrac{1}{2}x \sqrt{9-x^2} + C$$

It may be necessary to complete the square first, as in the next example.

Example 2

$$\int \sqrt{6x - 9x^2}\, dx = \int \sqrt{1 - (3x - 1)^2}\, dx$$

Put $\qquad\qquad 3x - 1 = \sin\theta$
Then $\qquad\qquad 3\, dx = \cos\theta\, d\theta$

and $\displaystyle\int \sqrt{6x - 9x^2}\, dx = \int \cos\theta(\tfrac{1}{3}\cos\theta\, d\theta)$

$$= \int \tfrac{1}{6}(1 + \cos 2\theta)\, d\theta$$
$$= \tfrac{1}{6}\theta + \tfrac{1}{12}\sin 2\theta + C$$
$$= \tfrac{1}{6}\theta + \tfrac{1}{6}\sin\theta\cos\theta + C$$
$$= \tfrac{1}{6}\sin^{-1}(3x - 1) + \tfrac{1}{6}(3x - 1)\sqrt{1 - (3x - 1)^2} + C$$
$$= \tfrac{1}{6}\sin^{-1}(3x - 1) + \tfrac{1}{6}(3x - 1)\sqrt{6x - 9x^2} + C$$

Exercise 22.3   (Further algebraic functions)

Find the following integrals.

1 $\displaystyle\int \frac{1}{\sqrt{1 + 6x - 3x^2}}\, dx$ 2 $\displaystyle\int \frac{1}{\sqrt{8 + 2x - x^2}}\, dx$ 3 $\displaystyle\int \sqrt{\frac{x - 2}{4 - x}}\, dx$ $\left[\text{Multiply by } \sqrt{\dfrac{x - 2}{x - 2}}.\right]$

4 $\displaystyle\int \frac{2x^2 - 1}{\sqrt{1 - x^2}}\, dx$ 5 $\displaystyle\int \frac{x^4}{\sqrt{1 - x^2}}\, dx$ 6 $\displaystyle\int x^2\sqrt{1 - x^2}\, dx$ 7 $\displaystyle\int \frac{x - 2}{\sqrt{x^2 + 4}}\, dx$

8 $\displaystyle\int_0^1 \frac{1}{(1 + x^2)^{\frac{3}{2}}}\, dx$ [Put $x = \tan\theta$.]   9 $\displaystyle\int_0^1 \frac{1}{(1 + x)\sqrt{2 + x - x^2}}\, dx$ $\left[\text{Put } 1 + x = \dfrac{1}{y}.\right]$

10 $\displaystyle\int_{-1}^1 \frac{1}{(2 - x)\sqrt{1 - x^2}}\, dx$

22.4   Miscellaneous integration

The following exercise contains integrals of all types.

Exercise 22.4   (Miscellaneous integration)

Find the following integrals (questions 1 to 30).

1 $\displaystyle\int \frac{x}{\sqrt{25 - x^2}}\, dx$ 2 $\displaystyle\int \sin 2x \sin x\, dx$ 3 $\displaystyle\int \frac{1}{\sqrt{1 - 2x}}\, dx$

4 $\displaystyle\int \frac{e^x}{e^x + 1}\, dx$ 5 $\displaystyle\int \frac{9x}{3x - 8}\, dx$ 6 $\displaystyle\int \sin^3 x \cos^2 x\, dx$

7 $\displaystyle\int x \sin 2x\, dx$ 8 $\displaystyle\int \frac{3 + x}{(1 - x)(1 + 3x)}\, dx$ 9 $\displaystyle\int \frac{1}{\sqrt{4 - 3x^2}}\, dx$

10 $\displaystyle\int \frac{1}{1+\cos x}\,dx$

11 $\displaystyle\int (\ln x)^2\,dx$

12 $\displaystyle\int \sin 3x \cos 2x\,dx$

13 $\displaystyle\int \frac{x^3+2}{x^2+2x+2}\,dx$

14 $\displaystyle\int \ln(x+\sqrt{x})\,dx$
[Put $x = u^2$.]

15 $\displaystyle\int \frac{x^2+x+2}{(x^2+1)(x-1)}\,dx$

16 $\displaystyle\int xe^{2x}\,dx$

17 $\displaystyle\int \frac{1}{(x+2)(x+3)}\,dx$

18 $\displaystyle\int xe^{x^2}\,dx$

19 $\displaystyle\int \frac{1}{x^2-2x+10}\,dx$

20 $\displaystyle\int \frac{e^{2x}}{(1+e^{2x})^2}\,dx$

21 $\displaystyle\int \frac{e^x}{4-e^{2x}}\,dx$ [Put $e^x = t$.]

22 $\displaystyle\int \frac{x^2-2x-1}{x^2-3x+2}\,dx$

23 $\displaystyle\int x^2 \ln x\,dx$

24 $\displaystyle\int \sec^3 x\,dx$

25 $\displaystyle\int_8^{24} \frac{1}{3x-8}\,dx$

26 $\displaystyle\int_0^{\frac{1}{2}} \frac{\sin^{-1} x}{\sqrt{1-x^2}}\,dx$

27 $\displaystyle\int_0^1 (2x^2+1)\sqrt{2x^3+3x+4}\,dx$

28 $\displaystyle\int_0^{\frac{1}{4}\pi} \sin^5 x\,dx$

29 $\displaystyle\int_e^{e^3} \frac{1}{x(\ln x)^2}\,dx$
[Put $\ln x = t$.]

30 $\displaystyle\int_0^1 \frac{1}{x+\sqrt{1-x^2}}\,dx$ [Put $x = \sin\theta$.]

31 Evaluate $\displaystyle\int_0^\pi |1-2\sin x|\,dx$.

32 Sketch the graph of the function
$$y = x \sin 2x$$
for values of x from $-\pi$ to $+\pi$.
Evaluate

(a) $\displaystyle\int_{-\pi}^\pi x \sin 2x\,dx$

(b) $\displaystyle\int_{-\pi}^\pi |x \sin 2x|\,dx$

(c) $\displaystyle\int_{-\pi}^\pi |x| \sin 2x\,dx$

(d) $\displaystyle\int_{-\pi}^\pi x|\sin 2x|\,dx$

(O and C)

22.5 Properties of definite integrals

The following results can help us to evaluate certain definite integrals. We shall just explain why we expect the results to be true; it is possible to prove them formally, working from the definition of a definite integral as the limit of a sum.

For the first two properties, consider evaluating the definite integrals by the usual method (i.e. integrating, putting x equal to the limits of integration, and subtracting).

(I) $\displaystyle\int_b^a f(x)\,dx = -\int_a^b f(x)\,dx$

(II) $\displaystyle\int_a^a f(x)\,dx = 0$

The remaining properties will become clearer if we interpret the definite integrals in terms of areas.

Using symmetry, we have already seen that

(III)   If $f(x)$ is an odd function (i.e. $f(-x) = -f(x)$, see p. 83), then

$$\int_{-a}^{a} f(x)\,dx = 0$$

(IV)   If $f(x)$ is an even function (i.e. $f(-x) = f(x)$), then

$$\int_{-a}^{a} f(x)\,dx = 2\int_{0}^{a} f(x)\,dx$$

We also have the following:

(V) $\displaystyle\int_{a}^{b} f(x)\,dx = \int_{a}^{c} f(x)\,dx + \int_{c}^{b} f(x)\,dx$

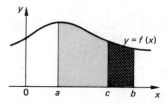

For simplicity we shall assume that $f(x) > 0$ and $a < c < b$. $\int_{a}^{c} f(x)\,dx$ is the area under the curve $y = f(x)$ from $x = a$ to $x = c$.

$\int_{c}^{b} f(x)\,dx$ is the area under the curve from $x = c$ to $x = b$.

If we add these, we obtain the area under the curve from $x = a$ to $x = b$, i.e. $\int_{a}^{b} f(x)\,dx$ (see Fig. 22.7).

Fig. 22.7

Example 1   Evaluate $\int_{1}^{6} f(x)\,dx$, where

$$f(x) = \begin{cases} x^2 & \text{if } x \leqslant 3 \\ 15 - 2x & \text{if } x > 3 \end{cases}$$

The curve is sketched in Fig. 22.8.

We have

$$\int_{1}^{6} f(x)\,dx = \int_{1}^{3} f(x)\,dx + \int_{3}^{6} f(x)\,dx$$

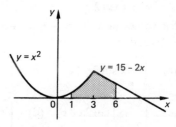

Fig. 22.8

Between $x = 1$ and $x = 3$, $f(x) = x^2$, and so

$$\int_{1}^{3} f(x)\,dx = \int_{1}^{3} x^2\,dx$$

Between $x = 3$ and $x = 6$, $f(x) = 15 - 2x$, and so

$$\int_{3}^{6} f(x)\,dx = \int_{3}^{6} (15 - 2x)\,dx$$

Thus $\displaystyle\int_{1}^{6} f(x)\,dx = \int_{1}^{3} x^2\,dx + \int_{3}^{6} (15 - 2x)\,dx$

$$= \left[\frac{1}{3}x^3\right]_{1}^{3} + \left[15x - x^2\right]_{3}^{6}$$

$$= 9 - \frac{1}{3} + (90 - 36) - (45 - 9) = \frac{80}{3}$$

(VI)   If $f(x) < g(x)$ in the interval $a < x < b$, then

$$\int_a^b f(x)\,dx < \int_a^b g(x)\,dx$$

The curve $y = f(x)$ is underneath the curve $y = g(x)$, and so the area under the first curve is less than the area under the second (see Fig. 22.9).

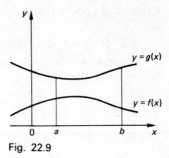

Fig. 22.9

Example 2   Show that

$$\frac{\sqrt{2}-1}{2} < \int_0^{\frac{1}{4}\pi} \frac{\sin x}{\sqrt{1+\cos^2 x}}\,dx < \frac{\sqrt{2}-1}{\sqrt{3}}$$

If $0 < x < \frac{1}{4}\pi$, then

$$\frac{1}{\sqrt{2}} < \cos x < 1$$

so $$\tfrac{3}{2} < 1 + \cos^2 x \quad < 2$$

and $$\sqrt{\tfrac{3}{2}} < \sqrt{1+\cos^2 x} < \sqrt{2}$$

Thus

$$\frac{\sin x}{\sqrt{2}} < \frac{\sin x}{\sqrt{1+\cos^2 x}} < \frac{\sin x}{\sqrt{\tfrac{3}{2}}}$$

and hence

$$\int_0^{\frac{1}{4}\pi} \frac{\sin x}{\sqrt{2}}\,dx < \int_0^{\frac{1}{4}\pi} \frac{\sin x}{\sqrt{1+\cos^2 x}}\,dx < \int_0^{\frac{1}{4}\pi} \frac{\sin x}{\sqrt{\tfrac{3}{2}}}\,dx$$

Now

$$\int_0^{\frac{1}{4}\pi} \sin x\,dx = \left[-\cos x \right]_0^{\frac{1}{4}\pi}$$

$$= \left(-\frac{1}{\sqrt{2}} \right) - (-1) = 1 - \frac{1}{\sqrt{2}}$$

so, if

$$I = \int_0^{\frac{1}{4}\pi} \frac{\sin x}{\sqrt{1+\cos^2 x}}\,dx$$

then

$$\frac{1}{\sqrt{2}}\left(1 - \frac{1}{\sqrt{2}} \right) < I < \sqrt{\frac{2}{3}}\left(1 - \frac{1}{\sqrt{2}} \right)$$

i.e. $$\frac{\sqrt{2}-1}{2} < I < \frac{\sqrt{2}-1}{\sqrt{3}}$$

Example 3   Show that

$$\tfrac{1}{5} < \int_1^4 \frac{1}{2x+7} dx < \tfrac{1}{3}$$

If $1 < x < 4$, then

$$\tfrac{1}{15} < \frac{1}{2x+7} < \tfrac{1}{9}$$

and so

$$\int_1^4 \tfrac{1}{15} dx < \int_1^4 \frac{1}{2x+7} dx < \int_1^4 \tfrac{1}{9} dx$$

i.e. $$\left[\frac{1}{15}x\right]_1^4 < \int_1^4 \frac{1}{2x+7} dx < \left[\frac{1}{9}x\right]_1^4$$

i.e. $$\tfrac{1}{5} < \int_1^4 \frac{1}{2x+7} dx < \tfrac{1}{3}$$

(VII) $$\int_a^b f(x) dx = \int_a^b f(t) dt$$

Using t (or any other letter) instead of x does not affect the value of a definite integral. We can verify this by substitution. Put $x = t$; then $dx = dt$, and hence

$$\int_a^b f(x) dx = \int_a^b f(t) dt$$

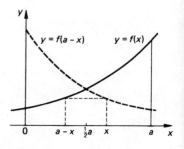

Fig. 22.10

(VIII) $$\int_0^a f(x) dx = \int_0^a f(a-x) dx$$

The curve $y = f(a-x)$ is the reflection of the curve $y = f(x)$ in the line $x = \tfrac{1}{2}a$ (see Fig. 22.10). So the areas under the two curves, from $x = 0$ to $x = a$, are the same.

We can also show this using substitution. Put $x = a - t$; then $dx = -dt$. When $x = 0$, $t = a$, and when $x = a$, $t = 0$; hence

$$\int_0^a f(x) dx = \int_a^0 f(a-t)(-dt) = \int_0^a f(a-t) dt$$

$$= \int_0^a f(a-x) dx$$

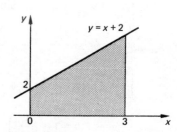

Fig. 22.11

Example 4   Suppose $f(x) = x+2$, and $a = 3$. Then

$$f(a-x) = f(3-x)$$

$$= (3-x)+2 = 5-x$$

(see Figs 22.11 and 22.12) and therefore we have

$$\int_0^3 (x+2) dx = \int_0^3 (5-x) dx$$

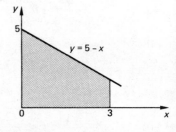

Fig. 22.12

We can check this by direct evaluation:

$$\int_0^3 (x+2)\,dx = \left[\frac{1}{2}x^2 + 2x\right]_0^3 = (\tfrac{9}{2} + 6) - 0 = \tfrac{21}{2}$$

$$\int_0^3 (5-x)\,dx = \left[5x - \frac{1}{2}x^2\right]_0^3 = (15 - \tfrac{9}{2}) - 0 = \tfrac{21}{2}$$

Exercise 22.5 (Properties of definite integrals)

1 The function $f(x)$ is defined by

$$f(x) = \begin{cases} x^2 & \text{if } x < 1 \\ 1 & \text{if } 1 \leqslant x < 3 \\ 4 - x & \text{if } x \geqslant 3 \end{cases}$$

Sketch the graph of $y = f(x)$.

Use $\displaystyle\int_0^4 f(x)\,dx = \int_0^1 f(x)\,dx + \int_1^3 f(x)\,dx + \int_3^4 f(x)\,dx$

to evaluate $\displaystyle\int_0^4 f(x)\,dx$.

2 Evaluate $\displaystyle\int_0^5 f(x)\,dx$, where

$$f(x) = \begin{cases} 2x & \text{if } x < 3 \\ (5-x)^2 & \text{if } x \geqslant 3 \end{cases}$$

Sketch the graph of $y = f(x)$, and indicate the area given by this definite integral.

3 By considering the range of values taken by $\sqrt{1+x^2}$, show that

$$\frac{1}{3\sqrt{2}} < \int_0^1 \frac{x^2}{\sqrt{1+x^2}}\,dx < \tfrac{1}{3}$$

and $\displaystyle\frac{26}{3\sqrt{10}} < \int_1^3 \frac{x^2}{\sqrt{1+x^2}}\,dx < \frac{26}{3\sqrt{2}}$

4 Show that

$$\tfrac{1}{6} < \int_0^{\frac{1}{6}\pi} \frac{\cos x}{3 - \sin x}\,dx < \tfrac{1}{5}$$

5 If $0 < x < 2$, show that $\dfrac{1}{6} < \dfrac{1}{4+x} < \dfrac{1}{4}$, and deduce that

$$\tfrac{1}{3} < \int_0^2 \frac{1}{4+x}\,dx < \tfrac{1}{2}$$

Show similarly that

$$\tfrac{3}{8} < \int_6^{12} \frac{1}{4+x}\,dx < \tfrac{3}{5}$$

6   By considering the range of values taken by $\dfrac{1}{\sqrt{9-2\sin^2 x}}$ in the interval $0 < x < \frac{1}{4}\pi$,

show that

$$\frac{\pi}{12} < \int_0^{\frac{1}{4}\pi} \frac{1}{\sqrt{9-2\sin^2 x}}\,dx < \frac{\pi}{8\sqrt{2}}$$

7   What does the result

$$\int_0^a f(x)\,dx = \int_0^a f(a-x)\,dx$$

give when $f(x) = 2x + 1$ and $a = 4$? Verify this by direct evaluation of both integrals.

8   Use the substitution $x = a + b - t$ to show that

$$\int_a^b f(x)\,dx = \int_a^b f(a+b-x)\,dx$$

Verify this result directly when $a = 2$, $b = 3$, and $f(x) = \dfrac{1}{(4-x)^2}$.

22.6   Evaluating definite integrals

The properties of the previous section can often be used to simplify the evaluation of a definite integral $\int_a^b f(x)\,dx$. Even when we cannot integrate $f(x)$ it is sometimes possible to evaluate the definite integral.

Example 1   Evaluate

$$\int_{-\pi}^{\pi} \frac{\sin x}{\sqrt{1+x^2}}\,dx$$

It may be seen that

$$f(x) = \frac{\sin x}{\sqrt{1+x^2}} \quad \text{is an odd function}$$

since $f(-x) = \dfrac{\sin(-x)}{\sqrt{1+(-x)^2}} = \dfrac{-\sin x}{\sqrt{1+x^2}}$

$$= -f(x)$$

and hence

$$\int_{-\pi}^{\pi} \frac{\sin x}{\sqrt{1+x^2}}\,dx = 0$$

The property

$$\int_0^a f(x)\,dx = \int_0^a f(a-x)\,dx$$

is often useful when circular functions are involved. We take a to be

$\frac{1}{2}\pi$ or π, and use results such as

$$\sin\left(\tfrac{1}{2}\pi - x\right) = \cos x$$

$$\cos\left(\tfrac{1}{2}\pi - x\right) = \sin x$$

$$\cos(\pi - x) = -\cos x$$

and so on.

Example 2   Evaluate

$$\int_0^{\frac{1}{2}\pi} \frac{\sin^2 x \cos x}{\sin x + \cos x}\, dx$$

We put $x = \tfrac{1}{2}\pi - y$; then $dx = -dy$. When $x = 0$, $y = \tfrac{1}{2}\pi$; and when $x = \tfrac{1}{2}\pi$, $y = 0$.

Let $I = \displaystyle\int_0^{\frac{1}{2}\pi} \frac{\sin^2 x \cos x}{\sin x + \cos x}\, dx$

$$= \int_{\frac{1}{2}\pi}^0 \frac{\sin^2\left(\tfrac{1}{2}\pi - y\right)\cos\left(\tfrac{1}{2}\pi - y\right)}{\sin\left(\tfrac{1}{2}\pi - y\right) + \cos\left(\tfrac{1}{2}\pi - y\right)}(-dy)$$

$$= \int_0^{\frac{1}{2}\pi} \frac{\cos^2 y \sin y}{\cos y + \sin y}\, dy$$

$$= \int_0^{\frac{1}{2}\pi} \frac{\cos^2 x \sin x}{\cos x + \sin x}\, dx$$

Thus $2I = \displaystyle\int_0^{\frac{1}{2}\pi} \frac{\sin^2 x \cos x}{\sin x + \cos x}\, dx + \int_0^{\frac{1}{2}\pi} \frac{\cos^2 x \sin x}{\cos x + \sin x}\, dx$

$$= \int_0^{\frac{1}{2}\pi} \frac{\sin x \cos x\,(\sin x + \cos x)}{\sin x + \cos x}\, dx$$

$$= \int_0^{\frac{1}{2}\pi} \sin x \cos x\, dx$$

$$= \left[\tfrac{1}{2}\sin^2 x\right]_0^{\frac{1}{2}\pi} = \left(\tfrac{1}{2}\right) - 0 = \tfrac{1}{2}$$

Hence $I = \tfrac{1}{4}$, i.e.

$$\int_0^{\frac{1}{2}\pi} \frac{\sin^2 x \cos x}{\sin x + \cos x}\, dx = \tfrac{1}{4}$$

Exercise 22.6   (Evaluating definite integrals)

In questions 1 to 5, evaluate the definite integrals.

1 $\displaystyle\int_{-\frac{1}{3}\pi}^{\frac{1}{3}\pi} \sin^3 x\, dx$ **2** $\displaystyle\int_{-4}^{4} \frac{x^5}{1+x^2}\, dx$ **3** $\displaystyle\int_{-3}^{3} (x^9 - 7x^5 + 2x^2 - 3x + 5)\, dx$

4 $\displaystyle\int_{-\frac{1}{4}\pi}^{\frac{1}{4}\pi} \frac{x\cos x - 2\sin x + 1}{\cos^2 x}\,dx$ **5** $\displaystyle\int_{-\frac{1}{2}}^{\frac{1}{2}} \frac{x^3 + 1}{\sqrt{1-x^2}}\,dx$

6 Use the substitution $x = \pi - t$ to show that

$$\int_0^{\frac{1}{2}\pi} \cos^3 x\,dx = -\int_{\frac{1}{2}\pi}^{\pi} \cos^3 x\,dx$$

Deduce that $\displaystyle\int_0^{\pi} \cos^3 x\,dx = 0$. Evaluate

$$\int_0^{\pi} \cos^5 x\,dx \quad \text{and} \quad \int_0^{\pi} \cos^7 x\,dx$$

7 If $I = \displaystyle\int_0^{\frac{1}{2}\pi} \frac{\cos^4 x \sin x}{\sin^3 x + \cos^3 x}\,dx$

show that

$$I = \int_0^{\frac{1}{2}\pi} \frac{\sin^4 x \cos x}{\cos^3 x + \sin^3 x}\,dx$$

and deduce the value of I.

8 If $I = \displaystyle\int_0^{\pi} x \sin x \cos^6 x\,dx$

use the substitution $x = \pi - y$ to show that

$$I = \int_0^{\pi} (\pi - x)\sin x \cos^6 x\,dx$$

Deduce that

$$2I = \int_0^{\pi} \pi \sin x \cos^6 x\,dx$$

and find the value of I.

9 Show that

$$I = \int_0^{\frac{1}{2}\pi} \frac{\sin\theta\,d\theta}{\sin\theta + \cos\theta} = \int_0^{\frac{1}{2}\pi} \frac{\cos\phi\,d\phi}{\sin\phi + \cos\phi}$$

and hence find the value of I. (MEI)

10 Show that

$$I = \int_0^{\pi} \frac{x\sin x}{1+\cos^2 x}\,dx = \int_0^{\pi} \frac{(\pi - x)\sin x}{1+\cos^2 x}\,dx$$

Deduce that

$$I = \pi \int_0^{\pi/2} \frac{\sin x}{1+\cos^2 x}\,dx$$

and hence, or otherwise, show that $I = \dfrac{\pi^2}{4}$.

(L)

22.7   Improper integrals

Consider the definite integral

$$\int_2^b \frac{1}{x^2}\,dx = \left[-\frac{1}{x}\right]_2^b$$

$$= -\frac{1}{b} - \left(-\frac{1}{2}\right) = \frac{1}{2} - \frac{1}{b}$$

As $b \to +\infty$,

$$\int_2^b \frac{1}{x^2}\,dx \to \frac{1}{2}$$

We write

$$\int_2^\infty \frac{1}{x^2}\,dx = \frac{1}{2}$$

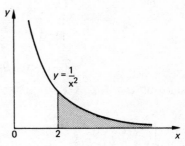

The area under the curve $y = \dfrac{1}{x^2}$, from $x = 2$ onwards, is therefore
$\frac{1}{2}$ (see Fig. 22.13).

Fig. 22.13

In practice, the evaluation would be set out as follows:

$$\int_2^\infty \frac{1}{x^2}\,dx = \left[-\frac{1}{x}\right]_2^\infty$$

$$= (0) - \left(-\frac{1}{2}\right) = \frac{1}{2}$$

Example 1

$$\int_0^\infty e^{-4x}\,dx = \left[-\frac{1}{4}e^{-4x}\right]_0^\infty$$

$$= (0) - \left(-\frac{1}{4}\right) = \frac{1}{4}$$

Example 2

$$\int_{-\infty}^\infty \frac{1}{1+x^2}\,dx = \left[\tan^{-1} x\right]_{-\infty}^\infty$$

$$= \left(\frac{1}{2}\pi\right) - \left(-\frac{1}{2}\pi\right) = \pi$$

Example 3

$$\int_0^\infty x^2 e^{-3x}\,dx$$

Put $\quad u = x^2 \quad$ and $\quad \dfrac{dv}{dx} = e^{-3x}$

then $\quad \dfrac{du}{dx} = 2x \quad$ and $\quad v = -\frac{1}{3}e^{-3x}$

So $\displaystyle\int_0^\infty x^2 e^{-3x}\,dx = \left[-\tfrac{1}{3}x^2 e^{-3x}\right]_0^\infty + \int_0^\infty \tfrac{2}{3}x e^{-3x}\,dx$

Now put $u = \tfrac{2}{3}x$ and $\dfrac{dv}{dx} = e^{-3x}$

then $\dfrac{du}{dx} = \tfrac{2}{3}$ and $v = -\tfrac{1}{3}e^{-3x}$

So $\displaystyle\int_0^\infty x^2 e^{-3x}\,dx = \left[-\tfrac{1}{3}x^2 e^{-3x} - \tfrac{2}{9}x e^{-3x}\right]_0^\infty + \int_0^\infty \tfrac{2}{9}e^{-3x}\,dx$

$$= \left[-\tfrac{1}{3}x^2 e^{-3x} - \tfrac{2}{9}x e^{-3x} - \tfrac{2}{27}e^{-3x}\right]_0^\infty$$

$$= (-0-0-0) - (-0-0-\tfrac{2}{27})$$

since, as $x \to +\infty$, $x^2 e^{-3x}$, $x e^{-3x}$ and e^{-3x} all tend to 0. Thus

$$\int_0^\infty x^2 e^{-3x}\,dx = \tfrac{2}{27}$$

Now consider the definite integral $\displaystyle\int_0^1 \dfrac{1}{\sqrt{x}}\,dx$.

The function $\dfrac{1}{\sqrt{x}}$ is not defined when $x = 0$; it has a vertical asymptote there.

If $a > 0$, we have

$$\int_a^1 \dfrac{1}{\sqrt{x}}\,dx = \left[2\sqrt{x}\,\right]_a^1 = 2 - 2\sqrt{a}$$

and as $a \to 0$,

$$\int_a^1 \dfrac{1}{\sqrt{x}}\,dx \to 2$$

We write $\displaystyle\int_0^1 \dfrac{1}{\sqrt{x}}\,dx = 2$

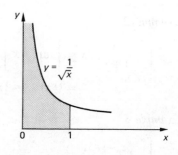

Fig. 22.14

The area under the curve $y = \dfrac{1}{\sqrt{x}}$, from $x = 0$ to $x = 1$ (see Fig. 22.14) is equal to 2, even though $\dfrac{1}{\sqrt{x}} \to +\infty$ as $x \to 0\ (x > 0)$.

The evaluation would normally be set out as:

$$\int_0^1 \dfrac{1}{\sqrt{x}}\,dx = \left[2\sqrt{x}\,\right]_0^1$$

$$= 2 - 0 = 2$$

Example 4

$$\int_0^1 \frac{1}{\sqrt{1-x^2}} dx = \left[\sin^{-1} x \right]_0^1$$

$$= \tfrac{1}{2}\pi - 0 = \tfrac{1}{2}\pi$$

The definite integrals studied in this section are called 'improper integrals', because they cannot be defined directly as the limit of a sum. (They are limiting values of proper integrals.)

Exercise 22.7 (Improper integrals)

Evaluate the following 'improper' integrals (questions 1 to 11).

1 $\displaystyle\int_1^\infty \frac{1}{x^3} dx$ 2 $\displaystyle\int_{-\infty}^\infty \frac{1}{4+9x^2} dx$ 3 $\displaystyle\int_0^\infty \frac{1}{x^2+2x+2} dx$

4 $\displaystyle\int_0^\infty xe^{-4x} dx$ 5 $\displaystyle\int_0^\infty x^2 e^{-2x} dx$ 6 $\displaystyle\int_0^\infty e^{-x} \sin x\, dx$

7 $\displaystyle\int_1^\infty \frac{\ln x}{x^2} dx$ 8 $\displaystyle\int_0^8 x^{-\frac{1}{3}} dx$ 9 $\displaystyle\int_{\frac{3}{2}}^6 \frac{1}{\sqrt{2x-3}} dx$

10 $\displaystyle\int_0^1 \ln x\, dx$ [Remember that $x \ln x \to 0$ as $x \to 0$.]   11 $\displaystyle\int_0^1 \frac{1}{\sqrt{x(1-x)}} dx$

12 Let A be the area under the curve $y = \dfrac{1}{x}$ between $x = 1$ and $x = a$, and let V be the volume of the solid of revolution formed when this area is rotated about the x-axis. Find A and V in terms of a, and show that, as $a \to +\infty$, the area A becomes arbitrarily large, but the volume V approaches a finite limit. State the limiting value of V.

22.8 Reduction formulae

It is sometimes possible to express an integral in terms of a simpler one of the same form. This is usually done by integrating the function as a product, i.e. 'by parts'.

Example If

$$I_n = \int_0^{\frac{1}{2}\pi} \sin^n x\, dx$$

where $n \geqslant 2$, express I_n in terms of I_{n-2}. Hence evaluate

$\displaystyle\int_0^{\frac{1}{2}\pi} \sin^6 x\, dx$ and $\displaystyle\int_0^{\frac{1}{2}\pi} \sin^7 x\, dx$.

$$I_n = \int_0^{\frac{1}{2}\pi} \sin^n x\, dx$$

Put $u = \sin^{n-1} x$ and $\dfrac{dv}{dx} = \sin x$

Then $\dfrac{du}{dx} = (n-1)\sin^{n-2} x \cos x$ and $v = -\cos x$

Then $\quad I_n = \left[-\sin^{n-1} x(\cos x) \right]_0^{\frac{1}{2}\pi} + (n-1) \int_0^{\frac{1}{2}\pi} \sin^{n-2} x \cos^2 x \, dx$

$$= 0 - 0 + (n-1) \int_0^{\frac{1}{2}\pi} \sin^{n-2} x (1 - \sin^2 x) dx$$

$$= (n-1) \int_0^{\frac{1}{2}\pi} \sin^{n-2} x \, dx - (n-1) \int_0^{\frac{1}{2}\pi} \sin^n x \, dx$$

$$= (n-1) I_{n-2} - (n-1) I_n$$

Thus $\quad nI_n = (n-1)I_{n-2}$

i.e. $\quad I_n = \dfrac{n-1}{n} I_{n-2}$

This is a **reduction formula** since it reduces the problem of finding I_n to one of finding I_{n-2}.

To evaluate I_6, we have $I_6 = \frac{5}{6} I_4$. Also $I_4 = \frac{3}{4} I_2$ and $I_2 = \frac{1}{2} I_0$. We know that

$$I_0 = \int_0^{\frac{1}{2}\pi} 1 \, dx = \left[x \right]_0^{\frac{1}{2}\pi} = \frac{1}{2}\pi$$

Thus $\quad I_6 = \frac{5}{6} \times \frac{3}{4} \times \frac{1}{2} \times \frac{1}{2}\pi = \frac{5}{32}\pi$, i.e.

$$\int_0^{\frac{1}{2}\pi} \sin^6 x \, dx = \frac{5}{32}\pi$$

Similarly, $I_7 = \frac{6}{7} I_5 = \frac{6}{7} \times \frac{4}{5} I_3 = \frac{6}{7} \times \frac{4}{5} \times \frac{2}{3} I_1$, and

$$I_1 = \int_0^{\frac{1}{2}\pi} \sin x \, dx = \left[-\cos x \right]_0^{\frac{1}{2}\pi}$$

$$= 0 - (-1) = 1$$

Thus $\quad I_7 = \frac{6}{7} \times \frac{4}{5} \times \frac{2}{3} = \frac{16}{35}$, i.e.

$$\int_0^{\frac{1}{2}\pi} \sin^7 x \, dx = \frac{16}{35}$$

Exercise 22.8 (Reduction formulae)

1 If $I_n = \displaystyle\int_0^t \sec^n x \, dx$, show that

$$(n-1)I_n = \sec^{n-2} t \tan t + (n-2)I_{n-2}$$

Evaluate

$$\int_0^{\frac{1}{4}\pi} \sec^6 x \, dx$$

(L)

2 Show that $\int_0^\pi e^x \sin x \, dx = \dfrac{e^\pi + 1}{2}$.

Let n be a non-negative integer, and let $I_n = \int_0^\pi e^x \sin^n x \, dx$. Show that if $n \geqslant 2$ then

$$(n^2 + 1)I_n = n(n-1)I_{n-2}$$

Hence, or otherwise, evaluate I_4 and I_5. (W)

3 If $I_n = \int_1^e x(\ln x)^n \, dx$, where n is a positive integer, prove that

$$I_n = \tfrac{1}{2}e^2 - \tfrac{1}{2}nI_{n-1}$$

Evaluate I_3. (O)

4 Find $\int x(4 - x^2)^{\frac{1}{2}} \, dx$.

Given that $I_n = \int_0^2 x^n(4 - x^2)^{\frac{1}{2}} \, dx$, prove that, for $n \geqslant 2$,

$$(n+2)I_n = 4(n-1)I_{n-2}$$

Evaluate I_4. (JMB)

5 If $I_n = \int_0^1 x^n e^{-x} \, dx$ where $n \geqslant 0$, find the relation between I_n and I_{n-1} where $n \geqslant 1$.

Express $\int_0^1 x^4 e^{-x} \, dx$ in terms of e. (L)

6 If $I_n = \int_{-1}^0 \dfrac{dx}{(x^2 + 2x + 2)^n}$ prove that

$$2nI_{n+1} = (2n - 1)I_n + 2^{-n}$$

Find the value of $\int_{-1}^0 \dfrac{dx}{(x^2 + 2x + 2)^3}$. (L)

7 I_n is defined by the formula $I_n = \int_0^{\frac{1}{4}\pi} \tan^{2n} x \, dx$ provided that $n \geqslant 0$; prove that

$$I_n + I_{n+1} = \dfrac{1}{2n + 1}$$

and hence evaluate $\int_0^{\frac{1}{4}\pi} \tan^6 x \, dx$. (O and C)

8 (i) Prove that, if $I_n = \int_0^{\frac{1}{2}\pi} \sin^n \theta \cos^2 \theta \, d\theta$ and $n \geqslant 2$,

$$I_n = \dfrac{n-1}{n+2}I_{n-2}$$

(ii) Prove that (a) $\int_0^{\frac{1}{2}\pi} \sin^4 \theta \cos^2 \theta \, d\theta = \dfrac{\pi}{32}$; (b) $\int_0^{\frac{1}{2}\pi} \sin^7 \theta \cos^2 \theta \, d\theta = \dfrac{16}{315}$. (O and C)

22.9 Definite integrals with variable limits

The definite integral $\displaystyle\int_a^b f(x)dx$ is defined as the limit of the sum

$\displaystyle\sum_{x=a}^{x=b} f(x)\delta x$, as $\delta x \to 0$. It is a number which depends on the function $f(x)$ and on the values of a and b.

In this section it will be convenient to use a different letter, say t, in the integral. Thus the integral $\displaystyle\int_a^b f(t)dt$ is, of course, the same as

$$\int_a^b f(x)dx.$$

$\displaystyle\int_a^x f(t)\,dt$ as a function of x

If a is constant, and x varies, the value of the definite integral $\displaystyle\int_a^x f(t)dt$ will depend on x, and so it may be regarded as a function of x. The integral $\displaystyle\int_a^x f(t)dt$ may be interpreted as the area under the curve $y = f(t)$ up to $t = x$, which is the area function A discussed in Chapter 5. There we discovered that $\dfrac{dA}{dx} = f(x)$. We now state this formally in terms of integrals.

The fundamental theorem

If $f(t)$ is a continuous function, then

$$\frac{d}{dx}\left(\int_a^x f(t)dt\right) = f(x)$$

Example 1

$$\frac{d}{dx}\left(\int_0^x \frac{t^2}{1+t^4}dt\right) = \frac{x^2}{1+x^4}$$

Example 2 Find

$$\frac{d}{dx}\left(\int_0^{x^2} \frac{t^2}{1+t^4}dt\right)$$

Let $u = x^2$; then, by the chain rule,

$$\frac{d}{dx}\left(\int_0^{x^2} \frac{t^2}{1+t^4}dt\right) = \frac{d}{du}\left(\int_0^u \frac{t^2}{1+t^4}dt\right) \times \frac{du}{dx} = \frac{u^2}{1+u^4} \times 2x$$

$$= \frac{(x^2)^2}{1+(x^2)^4} \times 2x = \frac{2x^5}{1+x^8}$$

Example 3 Find

$$\frac{d}{dx}\left(\int_{-x}^{x} \frac{t^2}{1+t^4}\,dt\right)$$

$\dfrac{t^2}{1+t^4}$ is an even function, so

$$\int_{-x}^{x} \frac{t^2}{1+t^4}\,dt = 2\int_{0}^{x} \frac{t^2}{1+t^4}\,dt$$

Thus

$$\frac{d}{dx}\left(\int_{-x}^{x} \frac{t^2}{1+t^4}\,dt\right) = \frac{2x^2}{1+x^4}$$

Defining functions as integrals

By the fundamental theorem, the function $\displaystyle\int_{a}^{x} f(t)\,dt$ differentiates to give $f(x)$, and so it is an (indefinite) integral of $f(x)$. It is in fact the integral which is zero when $x = a$. For example, taking $f(t) = t^2$, and $a = 2$,

$$\int_{2}^{x} t^2\,dt = \left[\tfrac{1}{3}t^3\right]_{2}^{x} = \tfrac{1}{3}x^3 - \tfrac{8}{3}$$

This is an integral of x^2, and it is zero when $x = 2$.

We cannot integrate $\sqrt{1+x^3}$ using functions which we have met so far. However, we may define a new function, say $Q(x)$, by

$$Q(x) = \int_{0}^{x} \sqrt{1+t^3}\,dt$$

Then $Q(x)$ is an integral of $\sqrt{1+x^3}$, and

$$\int \sqrt{1+x^3}\,dx = Q(x) + C$$

For any given value of x, we can evaluate $Q(x) = \displaystyle\int_{0}^{x} \sqrt{1+t^3}\,dt$ to any required degree of accuracy, using Simpson's rule (see p. 281) or infinite series (see p. 314).

The theory of exponential and logarithmic functions can be developed starting from the definition of $\ln x$ as

$$\ln x = \int_{1}^{x} \frac{1}{t}\,dt$$

The theory of circular functions may be developed from definitions such as

$$\tan^{-1} x = \int_{0}^{x} \frac{1}{1+t^2}\,dt$$

Exercise 22.9 (Definite integrals with variable limits)

1 Differentiate with respect to x:

(i) $\displaystyle\int_0^x \frac{1}{1+t}\,dt$ (ii) $\displaystyle\int_0^x \sec\theta\,d\theta$ (iii) $\displaystyle\int_1^x \frac{t^3}{2+t^4}\,dt$ (iv) $\displaystyle\int_0^{2x} \frac{1-t}{1+t}\,dt$

(v) $\displaystyle\int_0^{x^3} \tan\theta\,d\theta$ (vi) $\displaystyle\int_0^{\sin x} t\sin^{-1}t\,dt$ (vii) $\displaystyle\int_{-x}^x \frac{\cos\theta}{\sqrt{1-\theta^2}}\,d\theta$ (viii) $\displaystyle\int_{-x}^x \frac{t^5}{1+t^8}\,dt$

2 By writing $\displaystyle\int_{-x}^0 f(t)\,dt = -\int_0^{-x} f(t)\,dt$ show that

$$\frac{d}{dx}\left(\int_{-x}^0 f(t)\,dt \right) = f(-x)$$

Deduce that

$$\frac{d}{dx}\left(\int_{-x}^x f(t)\,dt \right) = f(x)+f(-x)$$

3 Find $\displaystyle\frac{d}{dx}\left(\int_{-x}^x \frac{dt}{1+t^3} \right).$

4 If $\displaystyle F(x) = \int_0^x \sqrt{1+t^3}\,dt$ differentiate $F(x)$ and $xF(x)$ with respect to x.

5 Write down $G'(x)$ when $\displaystyle G(x) = \int_0^x f(t)\,dt.$

Find $F'(x)$ if $\displaystyle F(x) = \int_0^x xf(t)\,dt.$ (MEI)

6 Let $\displaystyle F(x) = \int_0^{x^2} \sin^3 t\,dt.$ Without obtaining the integral, find $F'(x)$, where the dash denotes

differentiation with respect to x. Check your answer by performing the integration.   (MEI)

7 Given that $\displaystyle \ln x = \int_1^x \frac{dt}{t}$, prove, by substituting $t = u^n$ in the integral corresponding to $\ln(x^n)$, that

$\ln(x^n) = n\ln x$ (L)

8 Show graphically that $e^{-x} < 1$ when $x > 0$.
By considering $\int_0^x e^{-t}\,dt$ and $\int_0^x 1\,dt$, show that, when $x > 0$,

$$e^{-x} > 1-x$$

By considering $\int_0^x e^{-t}\,dt$ and $\int_0^x (1-t)\,dt$, show that, when $x > 0$,

$$e^{-x} < 1-x+\tfrac{1}{2}x^2$$

Show similarly that, when $x > 0$,

$$e^{-x} > 1 - x + \tfrac{1}{2}x^2 - \tfrac{1}{6}x^3 \quad \text{and} \quad e^{-x} < 1 - x + \tfrac{1}{2}x^2 - \tfrac{1}{6}x^3 + \tfrac{1}{24}x^4$$

By putting $x = \tfrac{1}{2}$, deduce that $\left(\dfrac{384}{233}\right)^2 < e < \left(\dfrac{48}{29}\right)^2$

22.10 Integrals, series and inequalities

Since a definite integral is defined as the limit of a sum of a series it is clear that integrals and series are often linked. The integral is also the area 'under the curve' and if we can find upper and lower bounds for this area then we can see how integrals, series and inequalities may arise together.

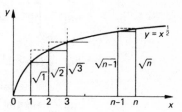

Fig. 22.15

Example 1 By considering the area under the curve $y = x^{\frac{1}{2}}$, between $x = 1$ and $x = n$ (a positive integer), show that

$$2n\sqrt{n} + 1 \leqslant 3(\sqrt{1} + \sqrt{2} + \ldots + \sqrt{n}) \leqslant (2n+3)\sqrt{n} - 2$$

The curve is shown in Fig. 22.15. Since $f(x) = x^{\frac{1}{2}}$ is an increasing function we have

$$\begin{array}{ccc}\text{sum of the rectangles} \\ \text{under the curve}\end{array} \leqslant \text{area under the curve} \leqslant \begin{array}{c}\text{sum of the rectangles} \\ \text{over the curve}\end{array}$$

$$1(\sqrt{1} + \sqrt{2} + \ldots + \sqrt{n-1}) \leqslant \int_1^n x^{\frac{1}{2}} \, dx \leqslant 1(\sqrt{2} + \sqrt{3} + \ldots + \sqrt{n})$$

so $\quad \sqrt{1} + \sqrt{2} + \ldots + \sqrt{n-1} \leqslant \tfrac{2}{3}(n\sqrt{n} - 1) \leqslant \sqrt{2} + \sqrt{3} + \ldots + \sqrt{n}$

The inequality is now considered in two parts.

(a) $\quad \sqrt{1} + \sqrt{2} + \ldots + \sqrt{n-1} \leqslant \tfrac{2}{3}(n\sqrt{n} - 1)$

$\quad 3(\sqrt{1} + \sqrt{2} + \ldots + \sqrt{n-1}) \leqslant 2n\sqrt{n} - 2$

So $\quad 3(\sqrt{1} + \sqrt{2} + \ldots + \sqrt{n}) \leqslant 2n\sqrt{n} - 2 + 3\sqrt{n}$

$$\leqslant (2n+3)\sqrt{n} - 2$$

(b) $\tfrac{2}{3}(n\sqrt{n} - 1) \leqslant \sqrt{2} + \sqrt{3} + \ldots + \sqrt{n}$

$\quad 2n\sqrt{n} - 2 \leqslant 3(\sqrt{2} + \sqrt{3} + \ldots + \sqrt{n})$

$\quad 2n\sqrt{n} + 1 \leqslant 3(\sqrt{1} + \sqrt{2} + \sqrt{3} + \ldots + \sqrt{n})$

Hence

$$2n\sqrt{n} + 1 \leqslant 3(\sqrt{1} + \sqrt{2} + \ldots + \sqrt{n}) \leqslant (2n+3)\sqrt{n} - 2$$

Example 2   Evaluate $\int_0^2 x^2 dx$ 'from first principles'.

Dividing the area under the curve $y = x^2$ between $x = 0$ and $x = 2$
into n strips of width $\dfrac{2}{n}$ (Fig. 22.16), the total area of the rectangles
'above' the curve is

$$\frac{2}{n}\left[\left(\frac{2}{n}\right)^2 + \left(\frac{4}{n}\right)^2 + \left(\frac{6}{n}\right)^2 + \ldots + \left(\frac{2n}{n}\right)^2\right]$$

The definite integral $\int_0^2 x^2 dx$ is the limit of the sum of this series, i.e.

$$\int_0^2 x^2 dx = \lim_{n \to \infty} \frac{2}{n}\left[\left(\frac{2}{n}\right)^2 + \left(\frac{4}{n}\right)^2 + \ldots + \left(\frac{2n}{n}\right)^2\right]$$

$$= \lim_{n \to \infty} \frac{2}{n^3}\left[2^2 + 4^2 + \ldots + (2n)^2\right]$$

$$= \lim_{n \to \infty} \frac{8}{n^3}(1^2 + 2^2 + \ldots + n^2)$$

$$= \lim_{n \to \infty} \frac{8}{n^3} \times \tfrac{1}{6}n(n+1)(2n+1)$$

$$= \lim_{n \to \infty} \tfrac{8}{6}\left(1 + \frac{1}{n}\right)\left(2 + \frac{1}{n}\right) = \tfrac{8}{6} \times 1 \times 2 = \tfrac{8}{3}$$

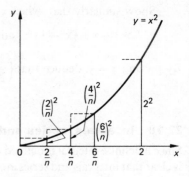

Fig. 22.16

We now reverse the process of the previous example, and use a
definite integral to find the limit of a series.

Example 3   Find the limit of

$$\frac{n}{n^2 + 1^2} + \frac{n}{n^2 + 2^2} + \frac{n}{n^2 + 3^2} + \ldots + \frac{n}{n^2 + n^2} \quad \text{as } n \to \infty.$$

This series may be written

$$S_n = \frac{1}{n}\left[\frac{1}{1 + \left(\dfrac{1}{n}\right)^2} + \frac{1}{1 + \left(\dfrac{2}{n}\right)^2} + \ldots + \frac{1}{1 + \left(\dfrac{n}{n}\right)^2}\right]$$

Consider the area under the curve $y = \dfrac{1}{1 + x^2}$ between $x = 0$ and

$x = 1$, divided into n strips of width $\dfrac{1}{n}$ (Fig. 22.17). S_n is the sum of
the areas of the rectangles 'below' the curve, and so the limit of this
sum is given by a definite integral. Thus

$$\lim_{n \to \infty} S_n = \int_0^1 \frac{1}{1 + x^2} dx$$

$$= \left[\tan^{-1} x\right]_0^1 = \tfrac{1}{4}\pi$$

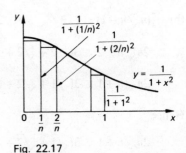

Fig. 22.17

Exercise 22.10   (Definite integrals and series)

1   By considering $\int_1^n x^{\frac{1}{3}}dx$, where n is an integer greater than 1, prove that

$$1 + 3n\sqrt[3]{n} \leqslant 4(\sqrt[3]{1} + \sqrt[3]{2} + \sqrt[3]{3} + \ldots + \sqrt[3]{n}) \leqslant \sqrt[3]{n}(3n+4) - 3 \qquad \text{(MEI)}$$

2   The function f is continuous, positive and strictly decreasing for $x \geqslant 1$. By using a graphical interpretation of the 'area under the curve', show that

$$0 \leqslant f(1) + f(2) + \ldots + f(n) - \int_1^{n+1} f(x)dx \leqslant f(1) \qquad \text{(MEI)}$$

3   By interpreting $A_m = 1 + \frac{1}{2} + \frac{1}{3} + \ldots + \frac{1}{m} - \ln(m+1)$ as an area, show that A_m tends to a (finite) limit as $m \to \infty$.   (MEI)

4   Interpret

$$\frac{\pi}{n} \sum_{r=1}^{n} \sin \frac{r\pi}{n}$$

in terms of the area under the curve $y = \sin x$ between $x = 0$ and $x = \pi$. Given that

$$\sum_{r=1}^{n} \sin \frac{r\pi}{n} = \cot \frac{\pi}{2n}$$

evaluate $\int_0^\pi \sin x \, dx$ from first principles.

5   Interpret

$$S_n = \frac{1^4 + 2^4 + 3^4 + \ldots + n^4}{n^5}$$

in terms of the area under the curve $y = x^4$ between suitable values of x, and hence evaluate $\lim_{n \to \infty} S_n$.

6   Find

$$\lim_{n \to \infty} n\left(\frac{1}{(n+1)^2} + \frac{1}{(n+2)^2} + \ldots + \frac{1}{(n+n)^2}\right)$$

by interpreting it in terms of the area under the curve $y = \dfrac{1}{(1+x)^2}$.

7   By relating

$$S_n = \frac{1}{n}\left[\ln 1 + \ln\left(1 + \frac{1}{n}\right) + \ln\left(1 + \frac{2}{n}\right) + \ldots + \ln\left(1 + \frac{n-1}{n}\right)\right]$$

to an area, evaluate $\lim_{n \to \infty} S_n$, using an appropriate integral.   (MEI)

8   Find the limit of $\displaystyle\sum_{r=1}^{n} \frac{1}{\sqrt{4n^2 - r^2}}$ as $n \to \infty$.

23

Hyperbolic functions

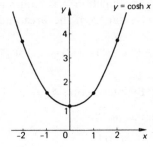

Fig. 23.1

23.1 Definitions and graphs

The hyperbolic cosine, cosh x (sometimes abbreviated to ch x), and the hyperbolic sine, sinh x (pronounced 'shine' x, and sometimes written sh x) are defined for all values of x by

$$\cosh x = \frac{e^x + e^{-x}}{2} \qquad \sinh x = \frac{e^x - e^{-x}}{2}$$

We also define

$$\tanh x = \frac{\sinh x}{\cosh x} \qquad \operatorname{sech} x = \frac{1}{\cosh x}$$

$$\operatorname{cosech} x = \frac{1}{\sinh x} \qquad \coth x = \frac{\cosh x}{\sinh x}$$

Note that tanh x is pronounced 'tanch' x, and is sometimes written th x; coth x is sometimes written cotanh x.

Using a calculator (or a book of tables), we can obtain values and draw the graphs of cosh x (Fig. 23.1), sinh x (Fig. 23.2) and tanh x (Fig. 23.3).

Note that

$$\tanh x = \frac{e^x - e^{-x}}{e^x + e^{-x}}$$

$$= \frac{1 - e^{-2x}}{1 + e^{-2x}} \to 1 \text{ as } x \to +\infty \text{ (since } e^{-2x} \to 0)$$

$$\text{and} \quad \tanh x = \frac{e^{2x} - 1}{e^{2x} + 1} \to -1 \text{ as } x \to -\infty \text{ (since } e^{2x} \to 0)$$

The graphs of their reciprocals, sech x, cosech x and coth x are shown in Figs 23.4, 23.5 and 23.6 respectively.

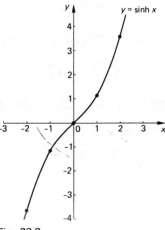

Fig. 23.2

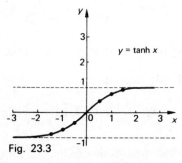

Fig. 23.3

Exercise 23.1 (Graphs of hyperbolic functions)

1 Show that cosh x is an even function, and that sinh x is an odd function. Which of the following functions are odd, and which are even: tanh x, sech x, cosech x, coth x?

2 Show that $\cosh x + \sinh x = e^x$ and $\cosh x - \sinh x = e^{-x}$

Hence express each of the following functions as the sum of an even function and an odd function:

(i) e^{2x} (ii) $2e^{3x} - 4e^{-5x}$ (iii) $e^{3x} \sinh x$

Sketch the following curves:

3 $y = 2 \cosh \tfrac{1}{2} x$ 4 $y = \sinh^2 x$ 5 $y = \operatorname{sech} x \tanh x$ 6 $y = x \cosh x$

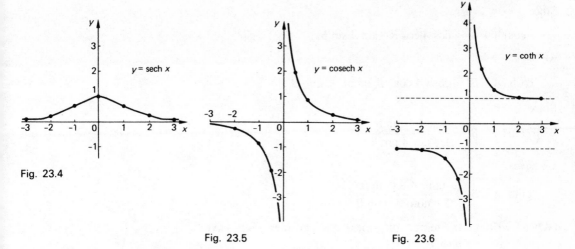

Fig. 23.4

Fig. 23.5

Fig. 23.6

23.2 Formulae and equations

We have

$$\cosh^2\theta - \sinh^2\theta = \left(\frac{e^\theta + e^{-\theta}}{2}\right)^2 - \left(\frac{e^\theta - e^{-\theta}}{2}\right)^2$$

$$= \frac{(e^{2\theta} + 2 + e^{-2\theta}) - (e^{2\theta} - 2 + e^{-2\theta})}{4}$$

$$= 1$$

It follows that $x = a\cosh\theta$, $y = b\sinh\theta$ are parametric equations for the hyperbola

$$\frac{x^2}{a^2} - \frac{y^2}{b^2} = 1$$

Notice that this gives only one branch of the hyperbola. The other branch is given by $x = -a\cosh\theta$, $y = b\sinh\theta$. Remember that $x = a\cos\theta$, $y = b\sin\theta$ are the usual parametric equations for the ellipse

$$\frac{x^2}{a^2} + \frac{y^2}{b^2} = 1$$

The result $\cosh^2\theta - \sinh^2\theta = 1$ reminds us of $\cos^2\theta + \sin^2\theta = 1$. In fact all the formulae relating circular functions give rise to corresponding formulae for hyperbolic functions, according to the following rule.

Osborn's rule   In any formula connecting the circular functions of general angles, replace each circular function by the corresponding hyperbolic function, and change the sign of every product (or implied product) of two sines.

Thus, for example, since

$$\sin 2\theta = 2\sin\theta\cos\theta$$

we have $\sinh 2\theta = 2\sinh\theta\cosh\theta$

Since

$$\cos(A+B) = \cos A \cos B - \sin A \sin B$$

we have

$$\cosh(A+B) = \cosh A \cosh B + \sinh A \sinh B$$

Since

$$\tan(A-B) = \frac{\tan A - \tan B}{1 + \tan A \tan B}$$

we have

$$\tanh(A-B) = \frac{\tanh A - \tanh B}{1 - \tanh A \tanh B}$$

(When written in terms of sines and cosines, we have $\tan A \tan B = \dfrac{\sin A \sin B}{\cos A \cos B}$ which contains a product of two sines.)

Osborn's rule provides a simple method for remembering formulae connecting the hyperbolic functions. However, these formulae must be proved by referring to the original definitions.

Hyperbolic functions appear to behave (so far as formulae are concerned) in a very similar way to their corresponding circular functions. However, their graphs are different; whereas the circular functions are periodic, this is not true for the hyperbolic functions.

Solving equations

We sometimes meet equations containing hyperbolic functions. These are usually easier to solve by reference to the definitions than by the methods used with circular functions.

Example 1   Solve the equations

(i) $9 \cosh x + 6 \sinh x = 23$ (ii) $3 \sinh^2 x - 2 \cosh x = 2$

(i)   Using the definition of $\cosh x$ and $\sinh x$,

$$9\left(\frac{e^x + e^{-x}}{2}\right) + 6\left(\frac{e^x - e^{-x}}{2}\right) = 23$$

$$15e^x + 3e^{-x} = 46$$
$$15e^{2x} - 46e^x + 3 = 0$$

Putting $e^x = y$,

$$15y^2 - 46y + 3 = 0$$
$$(y-3)(15y-1) = 0$$
$$y = 3 \quad \text{or} \quad y = \tfrac{1}{15}$$

i.e. $e^x = 3$ or $e^x = \tfrac{1}{15}$

Hence $x = \ln 3$ or $x = \ln \tfrac{1}{15} = -\ln 15$

(ii) $3 \sinh^2 x - 2 \cosh x = 2$

Using $\cosh^2 x - \sinh^2 x = 1$,

$$3(\cosh^2 x - 1) - 2 \cosh x = 2$$
$$3 \cosh^2 x - 2 \cosh x - 5 = 0$$
$$(3 \cosh x - 5)(\cosh x + 1) = 0$$

So $\cosh x = \frac{5}{3}$ (since $\cosh x \geqslant 1$)

Using the definition of $\cosh x$,

$$\frac{e^x + e^{-x}}{2} = \frac{5}{3}$$

$$3e^x + 3e^{-x} = 10$$
$$3e^{2x} - 10e^x + 3 = 0$$

Putting $e^x = y$,

$$3y^2 - 10y + 3 = 0$$
$$(3y - 1)(y - 3) = 0$$
$$y = \tfrac{1}{3} \quad \text{or} \quad y = 3$$

i.e. $e^x = \tfrac{1}{3}$ or $e^x = 3$

$$x = \ln \tfrac{1}{3} = -\ln 3 \quad \text{or} \quad x = \ln 3$$

Hence $x = \pm \ln 3$.

Example 2   If x and y satisfy the equations

$$\cosh x \cosh y = 2$$
$$\sinh x \sinh y = -1$$

show that $x = -y = \pm \ln(1 + \sqrt{2})$.   (L)

Adding the equations,

$$\cosh x \cosh y + \sinh x \sinh y = 1$$
$$\cosh (x + y) = 1$$
$$x + y = 0$$

Thus $x = -y$. Substituting into the first equation,

$$\cosh x \cosh (-x) = 2$$
$$\cosh^2 x = 2 \quad \text{(since $\cosh(-x) = \cosh x$)}$$
$$\cosh x = \sqrt{2} \quad \text{(since $\cosh x \geqslant 1$)}$$
$$\frac{e^x + e^{-x}}{2} = \sqrt{2}$$

Put $e^x = y$; then

$$y + \frac{1}{y} = 2\sqrt{2}$$

$$y^2 - 2\sqrt{2}y + 1 = 0$$

$$y = \frac{2\sqrt{2} \pm \sqrt{8-4}}{2} = \sqrt{2} \pm 1$$

$$x = \ln(\sqrt{2}+1) \quad \text{or} \quad x = \ln(\sqrt{2}-1)$$

$$= -\ln\left(\frac{1}{\sqrt{2}-1}\right)$$

$$= -\ln\left(\frac{\sqrt{2}+1}{(\sqrt{2}-1)(\sqrt{2}+1)}\right)$$

$$= -\ln(\sqrt{2}+1)$$

Hence $x = -y = \pm\ln(\sqrt{2}+1)$.

Exercise 23.2   (Formulae and equations)

1 From the result $\cosh^2\theta - \sinh^2\theta = 1$, deduce that $\operatorname{sech}^2\theta = 1 - \tanh^2\theta$ and that $\operatorname{cosech}^2\theta = \coth^2\theta - 1$.

In questions 2 to 10, write down the corresponding hyperbolic formulae. In questions 2 to 4, prove that the formula you have written down is true.

2 $\cos 2\theta = \cos^2\theta - \sin^2\theta$ **3** $\cos 2\theta = 2\cos^2\theta - 1$ **4** $\cos 2\theta = 1 - 2\sin^2\theta$

5 $\cos(A - B) = \cos A \cos B + \sin A \sin B$ **6** $\sin(A + B) = \sin A \cos B + \cos A \sin B$

7 $\tan(A + B) = \dfrac{\tan A + \tan B}{1 - \tan A \tan B}$ **8** $2\sin A \cos B = \sin(A + B) + \sin(A - B)$

9 $\sin 3\theta = 3\sin\theta - 4\sin^3\theta$ **10** $\cos C - \cos D = -2\sin\left(\dfrac{C+D}{2}\right)\sin\left(\dfrac{C-D}{2}\right)$

11 Show that if $\theta = \ln\tan\phi$, then $\tanh\theta = -\cos 2\phi$.   (O and C)

12 Find the real values of x which satisfy the equation

$3\cosh x - \sinh x = 3$ (MEI)

13 Solve the equation

$4\tanh x - \operatorname{sech} x = 1$

giving your solution in logarithmic form.   (L)
14 Find, in logarithmic form, the value of x for which $\coth x + 2 = 0$.   (L)
15 Find the real values of x and y which satisfy the equations

$$\sinh x + \sinh y = \frac{25}{12}$$

$$\cosh x - \cosh y = \frac{5}{12}$$ (MEI)

23.3   Differentiation and integration

We have

$$\frac{d}{dx}(\cosh x) = \frac{d}{dx}\left(\frac{e^x + e^{-x}}{2}\right) = \frac{e^x - e^{-x}}{2} = \sinh x$$

$$\frac{d}{dx}(\sinh x) = \frac{d}{dx}\left(\frac{e^x - e^{-x}}{2}\right) = \frac{e^x + e^{-x}}{2} = \cosh x$$

$$\frac{d}{dx}(\tanh x) = \frac{d}{dx}\left(\frac{\sinh x}{\cosh x}\right) = \frac{(\cosh x)(\cosh x) - (\sinh x)(\sinh x)}{\cosh^2 x}$$

$$= \frac{1}{\cosh^2 x} = \operatorname{sech}^2 x$$

$$\frac{d}{dx}(\operatorname{sech} x) = \frac{d}{dx}\left(\frac{1}{\cosh x}\right) = \frac{0 - \sinh x}{\cosh^2 x}$$

$$= -\left(\frac{1}{\cosh x}\right)\left(\frac{\sinh x}{\cosh x}\right) = -\operatorname{sech} x \tanh x$$

$$\frac{d}{dx}(\operatorname{cosech} x) = \frac{d}{dx}\left(\frac{1}{\sinh x}\right) = \frac{0 - \cosh x}{\sinh^2 x}$$

$$= -\left(\frac{1}{\sinh x}\right)\left(\frac{\cosh x}{\sinh x}\right) = -\operatorname{cosech} x \coth x$$

$$\frac{d}{dx}(\coth x) = \frac{d}{dx}\left(\frac{\cosh x}{\sinh x}\right) = \frac{\sinh^2 x - \cosh^2 x}{\sinh^2 x}$$

$$= \frac{-1}{\sinh^2 x} = -\operatorname{cosech}^2 x$$

Example 1   Differentiate $\sinh 2x$

If $y = \sinh 2x = \sinh u$ where $u = 2x$, then

$$\frac{dy}{dx} = \frac{dy}{du} \times \frac{du}{dx} = (\cosh u) \times 2 = 2\cosh 2x$$

Example 2   Differentiate $\tan^{-1}(\tanh x)$.

If $y - \tan^{-1}(\tanh x) = \tan^{-1} u$ where $u = \tanh x$, then

$$\frac{dy}{dx} = \left(\frac{1}{1 + u^2}\right)(\operatorname{sech}^2 x) = \left(\frac{1}{1 + \tanh^2 x}\right)\left(\frac{1}{\cosh^2 x}\right)$$

$$= \frac{1}{\cosh^2 x + \sinh^2 x} = \frac{1}{\cosh 2x} = \operatorname{sech} 2x$$

(We shall refer to this result later; see Example 6.)

We have

$$\int \sinh x \, dx = \cosh x + C$$

$$\int \cosh x \, dx = \sinh x + C$$

$$\int \text{sech}^2 x \, dx = \tanh x + C$$

and so on. Usually, hyperbolic functions are integrated in a similar way to the corresponding circular functions, but sometimes it is easier to return to the definitions of $\cosh x$ and $\sinh x$ in terms of e^x and e^{-x}.

Example 3

$$\int \sinh^2 x \, dx = \int \tfrac{1}{2}(\cosh 2x - 1) \, dx$$

$$\text{(since } \cosh 2x = 1 + 2 \sinh^2 x)$$

$$= \tfrac{1}{4} \sinh 2x - \tfrac{1}{2}x + C$$

Example 4

$$\int \cosh 3x \cosh 2x \, dx = \int \tfrac{1}{2}(\cosh 5x + \cosh x) \, dx$$

$$\text{(using } 2 \cosh A \cosh B = \cosh(A+B) + \cosh(A-B))$$

$$= \tfrac{1}{10} \sinh 5x + \tfrac{1}{2} \sinh x + C$$

Example 5

$$\int \coth 2x \, dx = \int \frac{\cosh 2x}{\sinh 2x} \, dx$$

$$= \tfrac{1}{2} \int \frac{2 \cosh 2x}{\sinh 2x} \, dx$$

$$= \tfrac{1}{2} \ln |\sinh 2x| + C$$

Example 6

$$\int \text{sech } 2x \, dx = \int \frac{2}{e^{2x} + e^{-2x}} \, dx$$

$$= \int \frac{2e^{2x}}{e^{4x} + 1} \, dx$$

Now put $e^{2x} = u$; so $2e^{2x} \, dx = du$ and

$$\int \text{sech } 2x \, dx = \int \frac{1}{u^2 + 1} \, du$$

$$= \tan^{-1} u + C$$

$$= \tan^{-1}(e^{2x}) + C$$

Now this appears to be different from the answer we expect from Example 2, i.e. $\tan^{-1}(\tanh x)$. However,

$$\tanh x = \frac{e^{2x} - 1}{1 + e^{2x}}$$

$$= \tan(\tan^{-1}(e^{2x}) - \tfrac{1}{4}\pi)$$

Hence $\tan^{-1}(\tanh x) = \tan^{-1}(e^{2x}) - \tfrac{1}{4}\pi$

Thus $\tan^{-1}(\tanh x)$ and $\tan^{-1}(e^{2x})$ only differ by a constant.

Exercise 23.3   (Differentiation and integration)

1 Differentiate the following functions with respect to x:

(i) $\cosh 4x$ (ii) $\tanh \tfrac{1}{2}x$ (iii) $2\,\mathrm{sech}\,2x$ (iv) $\cosh^2 2x$

(v) $\ln \cosh x$ (vi) $3\tanh^2 x$ (vii) $\mathrm{cosech}^2\,2x$ (viii) $\coth 3x$

(ix) $\ln \coth \tfrac{1}{2}x$ (x) $\cosh^3 \tfrac{1}{3}x$ (xi) $\sinh(x^3)$ (xii) $\tan^{-1}(\sinh 2x)$

2 Integrate the following functions with respect to x:

(i) $\sinh 2x$ (ii) $\cosh \tfrac{1}{2}x$ (iii) $\tanh x$ (iv) $\mathrm{cosech}^2\,3x$

(v) $\cosh^2 x$ (vi) $\sinh^2 3x$ (vii) $\coth \tfrac{1}{2}x$ (viii) $\mathrm{sech}\,x$

(ix) $\tanh^2 x$ (x) $\cosh 4x \sinh 2x$ (xi) $\sinh^3 x$ (xii) $\cosh^2 x \sinh^3 x$

3 (a) Use the method of integration by parts to find

(i) $\int x \cosh x \, dx$ (ii) $\int x \sinh 3x \, dx$ (iii) $\int x^2 \sinh x \, dx$

(b) Find $\int e^x \sinh x \, dx$.

4 Find the minimum value of $(5\cosh x + 3\sinh x)$.   (L)

5 The parametric equations of a curve are

$$x = a \cosh \theta, \qquad y = a \sinh \theta$$

Obtain the equation of the tangent to this curve at the point with parameter θ.   (L)

6 Show graphically that the equation

$$2 \cosh x = 3 + 2 \cos x$$

has only two real roots.

Taking $x = 1.2$ as a first approximation to one of these roots use Newton's formula once to obtain a better approximation. State the value of the second root to the same degree of accuracy.   (AEB)

7 Let $y = \ln(\sec x + \tan x)$ where $|x| < \tfrac{1}{2}\pi$. Show that $dy/dx = \cosh y$ and hence, or otherwise, find

$$\int \frac{1}{\cosh y} \, dy.$$

Evaluate $\displaystyle\int_1^2 \frac{1}{\cosh x} \, dx$ to three significant figures.   (W)

8 Starting from the inequality $\cosh x \geqslant 1$, deduce, by integration, that for $x > 0$, $\sinh x > x$ and $\cosh x > 1 + \tfrac{1}{2}x^2$.

9 Verify that $y = A \cosh nx + B \sinh nx$ is a solution of the differential equation $\dfrac{d^2 y}{dx^2} - n^2 y = 0$.

10 Find the general solution of the differential equation $\dfrac{d^2 y}{dx^2} - 4y = \sinh x$.

23.4   Inverse hyperbolic functions

In the same way that we have inverse circular functions, we have inverse hyperbolic functions. These are particularly useful in integration.

$\cosh x$ is an increasing function when restricted to $x \geqslant 0$; the corresponding inverse function is written $\cosh^{-1} x$ (see Fig. 23.7).

$\cosh^{-1} x$ is defined for $x \geqslant 1$, and takes all positive values.

$y = \cosh^{-1} x$ is equivalent to $\cosh y = x$ and $y \geqslant 0$.

$\sinh x$ is an increasing function for all x, and so it has an inverse function, $\sinh^{-1} x$ (see Fig. 23.8).

Just as $\sin^{-1} x$ may also be written as arc $\sin x$, $\sinh^{-1} x$ is sometimes written as arg $\sinh x$, or just ar $\sinh x$ and so on.

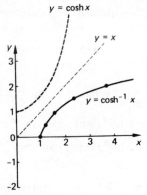

Fig. 23.7

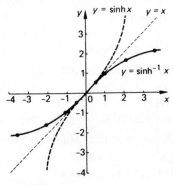

Fig. 23.8

Relationship with the hyperbola

Consider first the point $P(x, y)$ on the circle $x^2 + y^2 = 1$ (see Fig. 23.9). We have $x = \cos \theta$, $y = \sin \theta$, where $A\hat{O}P = \theta$.

The area of the sector OAP (shown shaded) is $\frac{1}{2}\theta = \frac{1}{2}\cos^{-1} x$.
Now consider the rectangular hyperbola $x^2 - y^2 = 1$ (Fig. 23.10), for which parametric equations are $x = \cosh \theta$, $y = \sinh \theta$.

If $Q(\cosh \theta_1, \sinh \theta_1)$ is a point on the hyperbola, the area AQN is

$$\int_0^{\theta_1} y \frac{dx}{d\theta} d\theta = \int_0^{\theta_1} \sinh \theta \sinh \theta \, d\theta$$

$$= \int_0^{\theta_1} \tfrac{1}{2}(\cosh 2\theta - 1) \, d\theta$$

$$= \tfrac{1}{2}\left[\tfrac{1}{2}\sinh 2\theta - \theta \right]_0^{\theta_1}$$

$$= \tfrac{1}{2}(\tfrac{1}{2}\sinh 2\theta_1 - \theta_1)$$

$$= \tfrac{1}{2}\sinh \theta_1 \cosh \theta_1 - \tfrac{1}{2}\theta_1$$

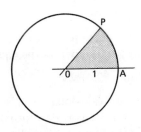

Fig. 23.9

The area OQN is $\frac{1}{2}\sinh \theta_1 \cosh \theta_1$, and so the area OAQ (shown shaded) is

$$OQN - AQN = \tfrac{1}{2}\theta_1$$

i.e. if $Q(x, y)$ is a point on the hyperbola, the shaded area OAQ is $\frac{1}{2}\cosh^{-1} x$.

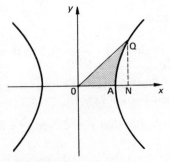

Fig. 23.10

Expression in terms of logarithms

The inverse hyperbolic functions may be written in terms of natural logarithms.

Suppose $y = \cosh^{-1} x$. Then

$$\cosh y = x$$
$$\frac{e^y + e^{-y}}{2} = x$$
$$e^y - 2x + e^{-y} = 0$$
$$(e^y)^2 - 2xe^y + 1 = 0$$

Solving this as a quadratic equation for e^y,

$$e^y = \frac{2x \pm \sqrt{4x^2 - 4}}{2} = x \pm \sqrt{x^2 - 1}$$

Thus $y = \ln(x \pm \sqrt{x^2 - 1})$. Now

$$x - \sqrt{x^2 - 1} = (x - \sqrt{x^2 - 1}) \times \frac{(x + \sqrt{x^2 - 1})}{x + \sqrt{x^2 - 1}}$$

$$= \frac{x^2 - (x^2 - 1)}{x + \sqrt{x^2 - 1}} = \frac{1}{x + \sqrt{x^2 - 1}}$$

and so $\ln(x - \sqrt{x^2 - 1}) = -\ln(x + \sqrt{x^2 - 1})$

Thus $$y = \pm \ln(x + \sqrt{x^2 - 1})$$

But $y = \cosh^{-1} x \geqslant 0$, and so we have

$$\boxed{\cosh^{-1} x = \ln(x + \sqrt{x^2 - 1})}$$

Now suppose $y = \sinh^{-1} x$; so $\sinh y = x$ and

$$\frac{e^y - e^{-y}}{2} = x$$

$$(e^y)^2 - 2xe^y - 1 = 0$$

$$e^y = \frac{2x \pm \sqrt{4x^2 + 4}}{2} = x \pm \sqrt{x^2 + 1}$$

$$e^y = x + \sqrt{x^2 + 1} \quad (\text{since } e^y > 0)$$
$$y = \ln(x + \sqrt{x^2 + 1})$$

Hence

$$\boxed{\sinh^{-1} x = \ln(x + \sqrt{x^2 + 1})}$$

Differentiation

If $y = \cosh^{-1} x$, then $\cosh y = x$. Differentiating with respect to y,

$$\frac{dx}{dy} = \sinh y$$

So $\dfrac{dy}{dx} = \dfrac{1}{\sinh y} = \dfrac{1}{\sqrt{\cosh^2 y - 1}} = \dfrac{1}{\sqrt{x^2 - 1}}$

If $y = \sinh^{-1} x$, then $\sinh y = x$ and $\dfrac{dx}{dy} = \cosh y$. So

$$\frac{dy}{dx} = \frac{1}{\cosh y} = \frac{1}{\sqrt{\sinh^2 y + 1}} = \frac{1}{\sqrt{x^2 + 1}}$$

If $y = \tanh^{-1} x$, then $\tanh y = x$, and $\dfrac{dx}{dy} = \operatorname{sech}^2 y$. So

$$\frac{dy}{dx} = \frac{1}{\operatorname{sech}^2 y} = \frac{1}{1 - \tanh^2 y} = \frac{1}{1 - x^2}$$

Integration using inverse hyperbolic functions

For an integral involving $\sqrt{x^2 - a^2}$, we use the substitution $x = a \cosh \theta$ (since $\cosh^2 \theta - 1 = \sinh^2 \theta$).

For an integral involving $\sqrt{x^2 + a^2}$, we put $x = a \sinh \theta$ (since $\sinh^2 \theta + 1 = \cosh^2 \theta$).

Example 1

$$\int \frac{1}{\sqrt{x^2 - 9}} \, dx$$

Put $x = 3 \cosh \theta$; then $dx = 3 \sinh \theta \, d\theta$ and

$$\int \frac{1}{\sqrt{x^2 - 9}} \, dx = \int \frac{3 \sinh \theta}{\sqrt{9 \cosh^2 \theta - 9}} \, d\theta$$

$$= \int \frac{3 \sinh \theta}{3 \sinh \theta} \, d\theta = \int 1 \, d\theta = \theta + C = \cosh^{-1} \frac{x}{3} + C$$

Alternatively, since $\sec^2 \theta - 1 = \tan^2 \theta$, we may use the substitution $x = 3 \sec \theta$; then $dx = 3 \sec \theta \tan \theta \, d\theta$ and

$$\int \frac{1}{\sqrt{x^2 - 9}} \, dx = \int \frac{3 \sec \theta \tan \theta}{\sqrt{9 \sec^2 \theta - 9}} \, d\theta$$

$$= \int \frac{3 \sec \theta \tan \theta}{3 \tan \theta} \, d\theta = \int \sec \theta \, d\theta$$

$$= \ln (\sec \theta + \tan \theta) + C$$

$$= \ln \left(\frac{x}{3} + \sqrt{\frac{x^2}{9} - 1} \right) + C = \cosh^{-1} \frac{x}{3} + C$$

Example 2

$$\int \sqrt{x^2 + 4}\, dx$$

Put $x = 2 \sinh \theta$; then $dx = 2 \cosh \theta\, d\theta$ and

$$\int \sqrt{x^2 + 4}\, dx = \int \sqrt{4 \sinh^2 \theta + 4}\,(2 \cosh \theta)\, d\theta$$

$$= \int (2 \cosh \theta)\,(2 \cosh \theta)\, d\theta$$

$$= \int 2(\cosh 2\theta + 1)\, d\theta = \sinh 2\theta + 2\theta + C$$

$$= 2 \sinh \theta \cosh \theta + 2\theta + C$$

$$= \tfrac{1}{2}(2 \sinh \theta)\,(2 \cosh \theta) + 2\theta + C$$

$$= \tfrac{1}{2} x \sqrt{x^2 + 4} + 2 \sinh^{-1} \frac{x}{2} + C$$

The following are 'standard integrals'.

$$\int \frac{1}{\sqrt{x^2 - a^2}}\, dx = \cosh^{-1} \frac{x}{a} + C$$

$$\text{or } \ln (x + \sqrt{x^2 - a^2}) + C$$

$$\int \frac{1}{\sqrt{x^2 + a^2}}\, dx = \sinh^{-1} \frac{x}{a} + C$$

$$\text{or } \ln (x + \sqrt{x^2 + a^2}) + C$$

Remember that

$$\int \frac{1}{\sqrt{a^2 - x^2}}\, dx = \sin^{-1} \frac{x}{a} + C$$

Note that

$$\cosh^{-1} \frac{x}{a} = \ln \left(\frac{x}{a} + \sqrt{\left(\frac{x}{a}\right)^2 - 1} \right)$$

$$= \ln \left(\frac{x + \sqrt{x^2 - a^2}}{a} \right)$$

$$= \ln (x + \sqrt{x^2 - a^2}) - \ln a$$

Similarly,

$$\sinh^{-1} \frac{x}{a} = \ln (x + \sqrt{x^2 + a^2}) - \ln a$$

Example 3

$$\int \frac{1}{\sqrt{3+5x^2}}\,dx = \int \frac{1}{\sqrt{(\sqrt{5}x)^2+(\sqrt{3})^2}}\,dx$$

$$= \frac{1}{\sqrt{5}}\sinh^{-1}\left(\frac{\sqrt{5}x}{\sqrt{3}}\right)+C$$

Example 4

$$\int_{-3}^{-2} \frac{1}{\sqrt{x^2-1}}\,dx$$

We cannot say $\displaystyle\int_{-3}^{-2} \frac{1}{\sqrt{x^2-1}}\,dx = \left[\cosh^{-1}x\right]_{-3}^{-2}$

since $\cosh^{-1}x$ is not defined when $x<1$.
  We first put $x=-u$; so $dx=-du$ and then

$$\int_{-3}^{-2} \frac{1}{\sqrt{x^2-1}}\,dx = \int_{3}^{2} \frac{1}{\sqrt{u^2-1}}(-du) = \int_{2}^{3} \frac{1}{\sqrt{u^2-1}}\,du$$

$$= \left[\cosh^{-1}u\right]_{2}^{3} = \cosh^{-1}3-\cosh^{-1}2$$

Exercise 23.4   (Inverse hyperbolic functions)

1   Sketch the graph of $y=\tanh^{-1}x$, stating the values of x for which y is defined.
2   Express $\tanh^{-1}x$ in terms of natural logarithms.
3   Express the following as logarithms, and hence evaluate them correct to 6 significant figures:

(i) $\cosh^{-1}2$ (ii) $\sinh^{-1}(-\tfrac{1}{2})$ (iii) $\tanh^{-1}(0.9)$

4   Differentiate with respect to x:

(i) $\cosh^{-1}2x$ (ii) $\sinh^{-1}\dfrac{4x}{5}$ (iii) $\tanh^{-1}3x$

(iv) $\sinh^{-1}(x-1)$ (v) $\cosh^{-1}(2x^2)$ (vi) $\tanh^{-1}(2x+3)$

Find the following integrals (questions 5 to 16).

5   $\displaystyle\int \frac{1}{\sqrt{x^2-4}}\,dx$ 6 $\displaystyle\int \frac{1}{\sqrt{4+x^2}}\,dx$ 7 $\displaystyle\int \frac{1}{\sqrt{9x^2-4}}\,dx$ 8 $\displaystyle\int \frac{1}{\sqrt{4x^2+3}}\,dx$

9   $\displaystyle\int \frac{1}{\sqrt{2x^2-2x+5}}\,dx$ 10 $\displaystyle\int \frac{1}{\sqrt{x^2+8x+7}}\,dx$ 11 $\displaystyle\int \sqrt{x^2-25}\,dx$ 12 $\displaystyle\int \sqrt{4x^2+9}\,dx$

13   $\displaystyle\int \sinh^{-1}x\,dx$ 14 $\displaystyle\int_{6}^{12} \frac{1}{\sqrt{x^2-9}}\,dx$ 15 $\displaystyle\int_{0}^{2} \sqrt{1+x^2}\,dx$ 16 $\displaystyle\int_{-1}^{3} \frac{1}{\sqrt{x^2-2x+5}}\,dx$

17 Evaluate   (i) $\displaystyle\int_{1}^{2} \cosh^{-1} x\, dx$ (ii) $\displaystyle\int_{\frac{1}{2}}^{1} \cosh^{-1}\left(\frac{1}{x}\right) dx$ (MEI)

18 (i) Explain why $\displaystyle\int_{3}^{6} \frac{1}{\sqrt{x^2-9}}\, dx$ is an 'improper' integral, and evaluate it.

(ii) Explain why $\displaystyle\int_{-9}^{9} \frac{1}{\sqrt{x^2-9}}\, dx$ cannot be evaluated.

(iii) Evaluate $\displaystyle\int_{-9}^{-6} \frac{1}{\sqrt{x^2-9}}\, dx$.

19 Given that $y = (\sinh^{-1} 2x)^2$ show that

$$(4x^2 + 1)\left(\frac{dy}{dx}\right)^2 = 16y$$ (L)

20 Prove that

$$\sinh^{-1}\tfrac{3}{4} + \sinh^{-1}\tfrac{5}{12} = \sinh^{-1}\tfrac{4}{3}$$ (O)

23.5 Maclaurin series

We can easily find the Maclaurin series for $\cosh x$. Let

$$
\begin{aligned}
f(x) &= \cosh x, &\text{so} \quad f(0) &= 1 \\
\text{Then} \quad f'(x) &= \sinh x, &\text{so} \quad f'(0) &= 0 \\
f''(x) &= \cosh x, &\text{so} \quad f''(0) &= 1 \\
f'''(x) &= \sinh x, &\text{so} \quad f'''(0) &= 0
\end{aligned}
$$

and so on. Hence

$$\cosh x = 1 + \frac{x^2}{2!} + \frac{x^4}{4!} + \frac{x^6}{6!} + \cdots$$

Similarly,

$$\sinh x = x + \frac{x^3}{3!} + \frac{x^5}{5!} + \frac{x^7}{7!} + \cdots$$

These series are true for all values of x.

The similarity of these series to

$$\cos x = 1 - \frac{x^2}{2!} + \frac{x^4}{4!} - \frac{x^6}{6!} + \cdots$$

and $\sin x = x - \dfrac{x^3}{3!} + \dfrac{x^5}{5!} - \dfrac{x^7}{7!} + \cdots$

explains why the hyperbolic functions behave in so many ways like their corresponding circular functions.

We can link the two sets of functions concisely by using the imaginary number j (where $j^2 = -1$). We have

$$\cos jx = 1 - \frac{(jx)^2}{2!} + \frac{(jx)^4}{4!} - \frac{(jx)^6}{6!} + \ldots$$

$$= 1 + \frac{x^2}{2!} + \frac{x^4}{4!} + \frac{x^6}{6!} + \ldots = \cosh x$$

$$\sin jx = jx - \frac{(jx)^3}{3!} + \frac{(jx)^5}{5!} - \frac{(jx)^7}{7!} + \ldots$$

$$= j\left(x + \frac{x^3}{3!} + \frac{x^5}{5!} + \frac{x^7}{7!} + \ldots\right) = j \sinh x$$

This also explains why, in Osborn's rule (see p. 463), a product of two sines leads to a change of sign, since

$$\sin jx \sin jy = (j \sinh x)(j \sinh y)$$

$$= - \sinh x \sinh y$$

Exercise 23.5 (Maclaurin series)

1 Obtain the series expansions for $\cosh x$ and $\sinh x$ from the definitions $\cosh x = \frac{1}{2}(e^x + e^{-x})$, $\sinh x = \frac{1}{2}(e^x - e^{-x})$, using the series for e^x and e^{-x}.

Find series expansions for the following functions (questions 2 to 5). Give the first three non-zero terms.

2 $\cosh 2x$ 3 $\sinh(\frac{1}{2}x^2)$ 4 $\tanh x$ 5 $\operatorname{sech} x$

6 By first expanding $\dfrac{1}{\sqrt{1+x^2}}$ as a power series, obtain the Maclaurin series for $\sinh^{-1} x$, up to the term in x^7.

7 Find the Maclaurin series for $\tanh^{-1} x$, up to the term in x^7.

8 We have seen that $\cos jx = \cosh x$ and $\sin jx = j \sinh x$. Find similar formulae for

(i) $\cosh jx$ (ii) $\sinh jx$ (iii) $\tan jx$

24

Curves, areas and surfaces

In this chapter we shall look briefly at more advanced ideas.

24.1 Length of a curve

Suppose we wish to find the length of a curve $y = f(x)$ between $x = a$ and $x = b$ (see Fig. 24.1). If P and Q are two points, close together, on the curve, the length of the curve between P and Q, δs, is approximately equal to the length of the straight line PQ, i.e.

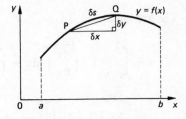

$$\delta s \approx \sqrt{(\delta x)^2 + (\delta y)^2}$$

The total length of the curve is

$$\sum_{x=a}^{x=b} \delta s \approx \sum_{x=a}^{x=b} \sqrt{(\delta x)^2 + (\delta y)^2}$$

$$= \sum_{x=a}^{x=b} \sqrt{1 + \left(\frac{\delta y}{\delta x}\right)^2}\, \delta x$$

$$\approx \sum_{x=a}^{x=b} \sqrt{1 + \left(\frac{dy}{dx}\right)^2}\, \delta x$$

Fig. 24.1

Taking the limit as $\delta x \to 0$, the required arc length is given by

$$s = \int_a^b \sqrt{1 + \left(\frac{dy}{dx}\right)^2}\, dx$$

Note that we are really *defining* the arc length to be the value of this definite integral, provided that it exists.

In parameters
If x and y are given in terms of a parameter t, then we have

$$\delta s \approx \sqrt{(\delta x)^2 + (\delta y)^2}$$

$$= \sqrt{\left(\frac{\delta x}{\delta t}\right)^2 + \left(\frac{\delta y}{\delta t}\right)^2}\, \delta t$$

and the arc length, between the points $t = t_1$ and $t = t_2$, is

$$s = \int_{t_1}^{t_2} \sqrt{\left(\frac{dx}{dt}\right)^2 + \left(\frac{dy}{dt}\right)^2}\, dt$$

We have met this idea before (see p. 249).

Example 1 Find the length of the arc of $y^2 = x^3$ from the origin to the point $(5, 5^{\frac{3}{2}})$.

Now

$$2y\frac{dy}{dx} = 3x^2$$

$$\frac{dy}{dx} = \frac{3x^2}{2y}$$

$$\left(\frac{dy}{dx}\right)^2 = \frac{9x^4}{4y^2} = \frac{9x^4}{4x^3} = \frac{9}{4}x$$

The arc length is

$$s = \int_0^5 \sqrt{1 + \left(\frac{dy}{dx}\right)^2}\,dx = \int_0^5 \sqrt{1 + \tfrac{9}{4}x}\,dx$$

$$= \left[\tfrac{4}{9} \times \tfrac{2}{3}(1 + \tfrac{9}{4}x)^{\frac{3}{2}}\right]_0^5 = \frac{8}{27}\left(\frac{7}{2}\right)^3 - \frac{8}{27} = \frac{335}{27}$$

$$\approx 12.41$$

Example 2 Find the length of arc of the parabola $y^2 = 4ax$ (given parametrically as $x = at^2$, $y = 2at$) from the origin to the point where $t = T$.

We have $\dfrac{dx}{dt} = 2at$ and $\dfrac{dy}{dt} = 2a$, and so the arc length is

$$s = \int_0^T \sqrt{\left(\frac{dx}{dt}\right)^2 + \left(\frac{dy}{dt}\right)^2}\,dt = \int_0^T \sqrt{4a^2t^2 + 4a^2}\,dt$$

$$= 2a\int_0^T \sqrt{t^2 + 1}\,dt$$

Putting $t = \sinh\theta$, then $dt = \cosh\theta\,d\theta$ and

$$s = 2a\int_0^{\sinh^{-1}T} \cosh\theta \cosh\theta\,d\theta$$

$$= a\int_0^{\sinh^{-1}T} (1 + \cosh 2\theta)\,d\theta$$

$$= a\left[\theta + \tfrac{1}{2}\sinh 2\theta\right]_0^{\sinh^{-1}T}$$

$$= a\left[\theta + \sinh\theta\cosh\theta\right]_0^{\sinh^{-1}T}$$

$$= a\left(\sinh^{-1}T + T\sqrt{1 + T^2}\right)$$

Exercise 24.1 (Arc length)

1 Show that the length of the arc of the parabola $x^2 = 4y$ from the point $(0, 0)$ to the point $(4, 4)$ is $2\sqrt{5} + \sinh^{-1} 2$.   (MEI)

2 Prove that the length of the arc $y = a \cosh \dfrac{x}{a}$ $(0 \leqslant x \leqslant a)$ is $\frac{1}{2}a(e - e^{-1})$.   (MEI)

3 A curve is given by the parametric equations $x = a(t - \sin t)$, $y = a(1 - \cos t)$ where $0 \leqslant t \leqslant 2\pi$.

Show that the arc length s satisfies $\dfrac{ds}{dt} = 2a \sin \frac{1}{2}t$, and hence find the total arc length of the curve.   (W)

4 Sketch the curve with parametric equations $x = a \cos^3 t$, $y = a \sin^3 t$ where $a > 0$ and $0 \leqslant t < 2\pi$. Show that the total length of the curve is $6a$.   (L)

5 If s is the arc length, measured from the point $t = 0$ of the curve $x = a\left(t - \dfrac{\sinh t}{\cosh t}\right)$, $y = \dfrac{a}{\cosh t}$

where a is a positive constant, prove that

$$\frac{ds}{dt} = a \tanh t \quad \text{and} \quad \frac{ds}{dy} = -\frac{a}{y}$$

Find y in terms of s.   (MEI)

6 A curve is defined in terms of a positive parameter t by $x = at \cos t$, $y = at \sin t$.

(a) Give a rough sketch of the curve for $0 \leqslant t \leqslant 4\pi$.
(b) Find the arc length along the curve measured from the origin to the point with parameter T.   (O and C)

7 The parametric coordinates of a point on a curve are given by $x = a(\tan t - t)$, $y = a \ln \sec t$. Prove that the arc length s of the curve measured from a certain point O is $a \sec t - a$ and give the coordinates of O.   (O and C)

8 Sketch the curve whose parametric equations are $x = a(2 \cos t - \cos 2t)$, $y = a(2 \sin t - \sin 2t)$ where a is a positive constant. Prove that the length of the curve is $16a$.   (MEI)

24.2   The area of a sector

We now consider the problem of finding the area of a *sector* OAB, where A and B are two points on a curve (see Fig. 24.2). This is the region (shown shaded) which is bounded by the lines OA and OB and the curve between A and B. We suppose that the curve is given parametrically, where A and B are the points $t = t_1$ and $t = t_2$.

Consider two points $P(x, y)$ and $Q(x + \delta x, y + \delta y)$ on the curve (Fig. 24.3). The area of the triangle OPQ is

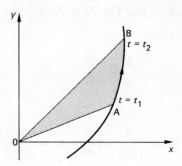

Fig. 24.2

$$\text{triangle OQR} - \text{triangle OPS} - \text{trapezium PQRS}$$
$$= \tfrac{1}{2}(x + \delta x)(y + \delta y) - \tfrac{1}{2}xy - \tfrac{1}{2}(y + y + \delta y)\delta x$$
$$= \tfrac{1}{2}(x\, \delta y - y\, \delta x)$$

The total area of the sector is approximately

$$\sum_{t = t_1}^{t = t_2} \tfrac{1}{2}\left(x\, \delta y - y\, \delta x\right) = \tfrac{1}{2} \sum_{t = t_1}^{t = t_2} \left(x \frac{\delta y}{\delta t} - y \frac{\delta x}{\delta t}\right) \delta t$$

and hence the area of the sector OAB is

$$\tfrac{1}{2} \int_{t_1}^{t_2} \left(x \frac{dy}{dt} - y \frac{dx}{dt}\right) dt$$

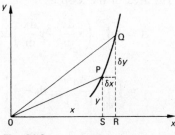

Fig. 24.3

Note that this should not be confused with the problem of finding the area between a curve and the x-axis, which is given by

$$\int_{t_1}^{t_2} y \frac{dx}{dt} \, dt$$ (see p. 248), although in certain circumstances the areas

may be the same.

Example 1   Find the area enclosed by the ellipse
$x = a \cos t, \; y = b \sin t$.

The ellipse is shown in Fig. 24.4. We have

$$\frac{dx}{dt} = -a \sin t, \quad \frac{dy}{dt} = b \cos t$$

and so the area is

$$A = \frac{1}{2} \int_0^{2\pi} \left(x \frac{dy}{dt} - y \frac{dx}{dt} \right) dt$$

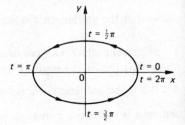

Fig. 24.4

$$= \frac{1}{2} \int_0^{2\pi} \left[(a \cos t)(b \cos t) - (b \sin t)(-a \sin t) \right] dt$$

$$= \frac{1}{2} \int_0^{2\pi} ab \, dt = \frac{1}{2} \left[abt \right]_0^{2\pi} = \pi ab$$

Example 2   Find the area of the loop of the curve

$$x = \frac{3at}{1+t^3}, \; y = \frac{3at^2}{1+t^3}.$$

The loop is from $t = 0$ to $t = +\infty$, and this may be regarded as a sector (see Fig. 24.5). Now $\dfrac{y}{x} = t$, so, differentiating with respect to t,

$$\frac{x \dfrac{dy}{dt} - y \dfrac{dx}{dt}}{x^2} = 1$$

Thus $$x \frac{dy}{dt} - y \frac{dx}{dt} = x^2 = \frac{9a^2 t^2}{(1+t^3)^2}$$

The area of the loop is

$$A = \frac{1}{2} \int_0^{\infty} \left(x \frac{dy}{dt} - y \frac{dx}{dt} \right) dt$$

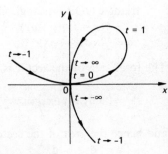

Fig. 24.5

$$= \frac{1}{2} \int_0^{\infty} \frac{9a^2 t^2}{(1+t^3)^2} \, dt = \frac{1}{2} \left[\frac{-3a^2}{1+t^3} \right]_0^{\infty}$$

$$= 0 - (-\tfrac{3}{2}a^2) = \tfrac{3}{2}a^2$$

Exercise 24.2 (Area of a sector in parameters)

1 Find the area cut off between the x-axis and that part of the curve $x = at \cos t$, $y = at \sin t$ for which $0 \leqslant t \leqslant \pi$.

2 O is the origin, and A and B are points on the hyperbola $x = a \cosh \theta$, $y = b \sinh \theta$ with parameters $\theta = -2$ and $\theta = 2$. Find the area enclosed by the lines OA, OB, and the arc of the hyperbola between A and B.

24.3 Curvature

We have an intuitive idea of curvature. A curve like that shown in Fig. 24.6 is curving slowly at P while one like that shown in Fig. 24.7 is curving more quickly at Q.

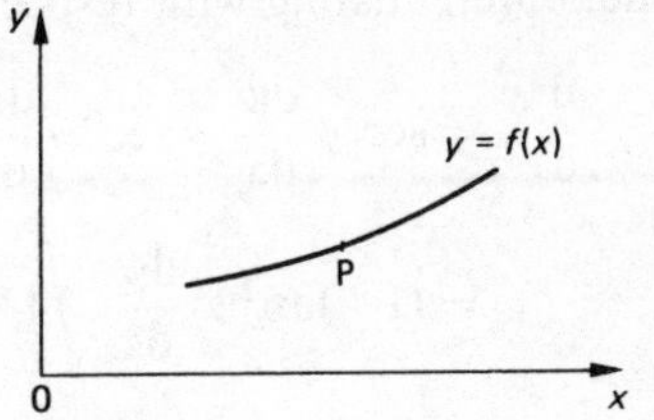

We can imagine drawing in a circle which just 'matches' the curve at P (Fig. 24.8), and at Q (Fig. 24.9). (We can draw a circle passing through any three points on the curve. To obtain the matching circle at, say, P, we consider the limit as the three points are taken closer and closer to P.) Clearly the first circle will have a much larger radius than the second. Hence slow curvature corresponds to a large radius of the matching circle and fast curvature corresponds to a small radius of matching circle.

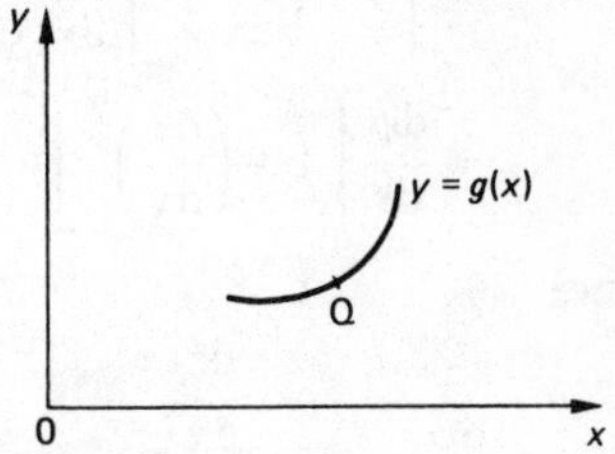

This matching circle is called the **circle of curvature**; its radius is the **radius of curvature**, and its centre is the **centre of curvature**. All these refer to a single point on a curve.

Consider a point A on a curve (see Fig. 24.10). Let s be the arc length (measured from some fixed point), and ψ be the angle which the tangent at A makes with the positive x-axis.

We define the **curvature**, κ, at A to be the rate of change of direction with respect to distance along the curve; thus

$$\kappa = \frac{d\psi}{ds}$$

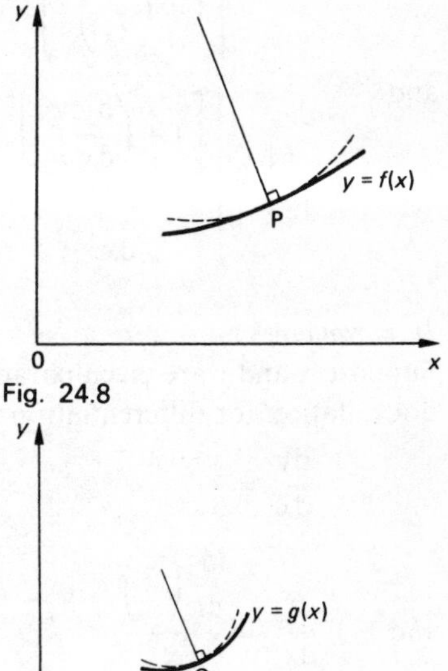

The curve shown in Fig. 24.10 has a positive curvature. A curve like the one shown in Fig. 24.11 has a negative curvature, since ψ decreases as s increases.

Consider two points, A and B, close together on a curve (see Fig. 24.12). The centre of curvature, C, corresponding to the point A lies on the normal at A and (approximately) also on the normal at B. We have

$$\delta s \approx \rho \, \delta \psi$$

and thus

$$\rho \approx \frac{\delta s}{\delta \psi}$$

The radius of curvature is therefore

$$\rho = \left| \frac{ds}{d\psi} \right| = \left| \frac{1}{\kappa} \right|$$

(the radius must be positive).

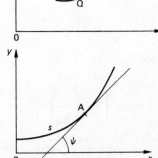

Fig. 24.6

Fig. 24.7

Fig. 24.8

Fig. 24.9

Fig. 24.10

Formulae for κ and ρ

We have

$$\frac{dy}{dx} = \tan \psi$$

and, differentiating with respect to x,

$$\frac{d^2 y}{dx^2} = \sec^2 \psi \frac{d\psi}{dx} = \sec^2 \psi \frac{d\psi}{ds} \frac{ds}{dx}$$

$$= (1 + \tan^2 \psi) \frac{d\psi}{ds} \sqrt{1 + \left(\frac{dy}{dx}\right)^2}$$

$$= \left[1 + \left(\frac{dy}{dx}\right)^2\right] \frac{d\psi}{ds} \sqrt{1 + \left(\frac{dy}{dx}\right)^2}$$

$$= \frac{d\psi}{ds} \left[1 + \left(\frac{dy}{dx}\right)^2\right]^{\frac{3}{2}}$$

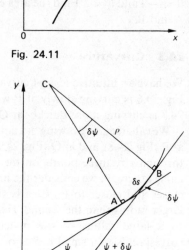

Fig. 24.11

Fig. 24.12

Hence

$$\kappa = \frac{d\psi}{ds} = \frac{\dfrac{d^2 y}{dx^2}}{\left[1 + \left(\dfrac{dy}{dx}\right)^2\right]^{\frac{3}{2}}}$$

and

$$\rho = \left|\frac{1}{\kappa}\right| = \left|\frac{\left[1 + \left(\dfrac{dy}{dx}\right)^2\right]^{\frac{3}{2}}}{\dfrac{d^2 y}{dx^2}}\right|$$

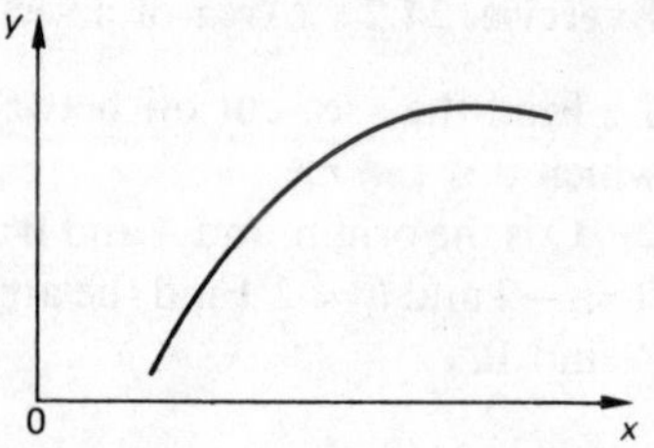

In parameters

Suppose x and y are given parametrically in terms of t. Using the dot notation for differentiation with respect to t, we have

$$\frac{dy}{dx} = \frac{\dot{y}}{\dot{x}}$$

and

$$\frac{d^2 y}{dx^2} = \frac{\dfrac{d}{dt}\left(\dfrac{\dot{y}}{\dot{x}}\right)}{\dot{x}}$$

$$= \frac{\dot{x}\ddot{y} - \dot{y}\ddot{x}}{\dot{x}^3}$$

Hence

$$\kappa = \frac{\dot{x}\ddot{y} - \dot{y}\ddot{x}}{\dot{x}^3 \left[1 + \left(\dfrac{\dot{y}}{\dot{x}}\right)^2\right]^{\frac{3}{2}}}$$

or $\kappa = \dfrac{\dot{x}\ddot{y} - \dot{y}\ddot{x}}{(\dot{x}^2 + \dot{y}^2)^{\frac{3}{2}}}$ and $\rho = \left|\dfrac{1}{\kappa}\right|$

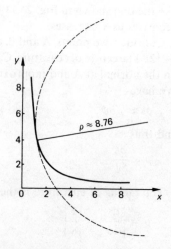

Fig. 24.13

Example Find the radius of curvature at $(1, 4)$ to the curve $xy = 4$.

Now $y = \dfrac{4}{x}, \quad \dfrac{dy}{dx} = -\dfrac{4}{x^2}, \quad \dfrac{d^2y}{dx^2} = \dfrac{8}{x^3}$

(see Fig. 24.13). Thus

$$\rho = \left| \frac{\left(1 + \dfrac{16}{x^4}\right)^{\frac{3}{2}}}{\dfrac{8}{x^3}} \right|$$

When $x = 1$, $\rho \approx 8.76$.

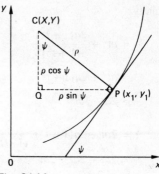

Fig. 24.14

Finding the centre of curvature

Method 1 Suppose $P(x_1, y_1)$ is the point on the curve (Fig. 24.14) and we wish to find the corresponding centre of curvature $C(X, Y)$. Then

$$X = x_1 - \rho \sin \psi$$
$$Y = y_1 + \rho \cos \psi$$

Since $\tan \psi = \dfrac{dy}{dx} = \dfrac{\dot{y}}{\dot{x}}$

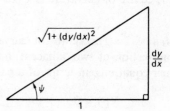

Fig. 24.15

we can find $\sin \psi$ and $\cos \psi$ by considering the triangles in Figs 24.15 or 24.16.

Example Find the centre of curvature of the parabola $y^2 = 4ax$ at the point $(at^2, 2at)$.

With reference to Fig. 24.17 we have

$$X = at^2 + \rho \sin \psi$$
$$Y = 2at - \rho \cos \psi$$

We first find ρ.

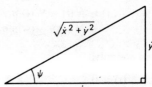

Fig. 24.16

$$\begin{aligned} x &= at^2 & y &= 2at \\ \dot{x} &= 2at & \dot{y} &= 2a \\ \ddot{x} &= 2a & \ddot{y} &= 0 \end{aligned}$$

Thus $\rho = \left| \dfrac{(4a^2t^2 + 4a^2)^{\frac{3}{2}}}{0 - 4a^2} \right| = 2a(1 + t^2)^{\frac{3}{2}}$

Also $\tan \psi = \dfrac{dy}{dx} = \dfrac{\dot{y}}{\dot{x}} = \dfrac{1}{t}$

and thus (from Fig. 24.18)

$$\sin \psi = \frac{1}{\sqrt{1 + t^2}} \quad \text{and} \quad \cos \psi = \frac{t}{\sqrt{1 + t^2}}$$

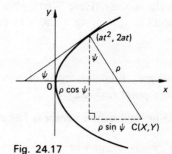

Fig. 24.17

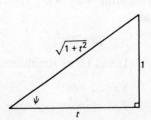

Fig. 24.18

Finally,

$$X = at^2 + \frac{2a(1+t^2)^{\frac{3}{2}}}{(1+t^2)^{\frac{1}{2}}}$$

$$= at^2 + 2a(1+t^2)$$

$$= 2a + 3at^2$$

$$Y = 2at - 2a(1+t^2)^{\frac{3}{2}}\frac{t}{(1+t^2)^{\frac{1}{2}}}$$

$$= 2at - 2at - 2at^3$$

$$= -2at^3$$

The centre of curvature is $C(2a + 3at^2, -2at^3)$.

Method 2   The centre of curvature can be thought of as the intersection of two 'adjacent' normals (see Fig. 24.19). Suppose their equations, in terms of a parameter t are given by

$$N(t) = 0$$

and $N(t + \delta t) = 0$

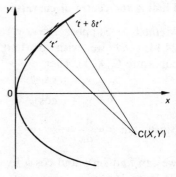

To solve these simultaneously we use

$$N(t + \delta t) \approx N(t) + N'(t)\delta t$$

$$= N'(t)\delta t \quad \text{(when } N(t) = 0)$$

and so we solve

$$\left.\begin{aligned} N(t) &= 0 \\ \text{and} \quad N'(t) &= 0 \end{aligned}\right\}$$

Fig. 24.19

simultaneously. (Here $N'(t)$ is obtained by differentiating $N(t)$ with respect to t, treating x and y as constants.)

Example   Find the centre of curvature of the parabola $y^2 = 4ax$ at the point $(at^2, 2at)$.

The normal to the parabola has equation

$$y + tx = 2at + at^3 \tag{1}$$

Differentiating this with respect to t, treating x and y as constants,

$$x = 2a + 3at^2 \tag{2}$$

Solving (1) and (2) as simultaneous equations for x and y gives

$$x = 2a + 3at^2$$

$$y = -2at^3$$

as before.

Exercise 24.3　(Curvature)

1　Find the radius of curvature at the point where $x = \dfrac{\sqrt{3}c}{2}$ on the hyperbola $xy = c^2$ and show that

the centre of curvature is $\left(\dfrac{43\sqrt{3}c}{36}, \dfrac{19\sqrt{3}c}{16}\right)$.　(O and C)

2　Sketch the curve whose equation is $xy^2 = a^2(a-x)$. Find the radius of curvature at the point $(\tfrac{1}{2}a, a)$ and the coordinates of the corresponding centre of curvature.　(O and C)

3　Find the radius of curvature of the curve $y = 2\cosh\tfrac{1}{2}x$ at the point where $x = 2$. Find also the length of the curve from the point where $x = -2$ to the point where $x = 2$.　(AEB)

4　A cycloid is given in the parametric form

$$x = a(\theta - \sin\theta), \quad y = a(1 - \cos\theta)$$

Prove that the radius of curvature (assumed positive) at the point with parameter θ, where $0 \leqslant \theta \leqslant 2\pi$, is $4a\sin\tfrac{1}{2}\theta$ and that the centre of curvature at this point is $[a(\theta + \sin\theta), -a(1 - \cos\theta)]$.

Describe (either verbally or diagrammatically) the relation between one arch of the given cycloid and the locus of its centre of curvature.　(O and C)

5　Prove that in the parabola defined by the equations

$$x = a\cot^2\psi, \quad y = 2a\cot\psi$$

the angle ψ is the angle between the tangent and the axis of the parabola. Prove also that the radius of curvature is $2a\,\mathrm{cosec}^3\psi$　(O and C)

6　Find the point on the curve $y = e^x$ at which the curvature is a maximum and determine the curvature and the coordinates of the centre of curvature at this point.　(O and C)

24.4　Curved surface area of revolution

When an arc of a curve is rotated about the x-axis, this generates a **surface of revolution**.

Consider first a straight line segment PQ (Fig. 24.20). The surface area generated is the difference between the curved surface areas of the two cones, which is

$$\pi r_2 l_2 - \pi r_1 l_1$$

Now $\dfrac{l_1}{r_1} = \dfrac{l_2}{r_2} = k$, say, so this area is

$$k\pi r_2{}^2 - k\pi r_1{}^2 = k\pi(r_2 + r_1)(r_2 - r_1)$$

$$= \pi(r_2 + r_1)(l_2 - l_1)$$

$$= 2\pi\left(\frac{r_2 + r_1}{2}\right)(l_2 - l_1)$$

$$= 2\pi y\,\delta s$$

where y is the distance of the centre of the segment from the axis, and δs is the length of the segment (see Fig. 24.21).

Alternatively, we may think of the area generated as a band, with approximate circumference $2\pi y$, and width δs.

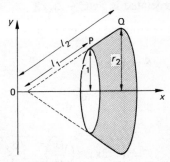

Fig. 24.20

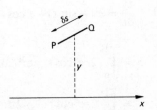

Fig. 24.21

Now consider an arc of a curve (Fig. 24.22). If a small element has length δs, *measured along the curve*, then (treating it as a small line segment) this generates a band whose area is approximately $2\pi y \delta s$, and hence the curved surface area of revolution is given by

$$S = \int_\alpha^\beta 2\pi y \, ds$$

in which we may replace ds by

$$\sqrt{1 + \left(\frac{dy}{dx}\right)^2} \, dx$$

with the limits of integration being limits of x; or

$$\sqrt{\left(\frac{dx}{dt}\right)^2 + \left(\frac{dy}{dt}\right)^2} \, dt$$

with the limits of integration being limits of t.

Similarly, when an arc is rotated about the y-axis, the area of the surface of revolution generated is $\displaystyle\int_\alpha^\beta 2\pi x \, ds$.

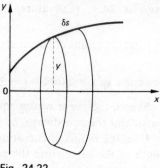

Fig. 24.22

Example A curve is given parametrically by the equations $x = e^t \sin t$, $y = e^t \cos t$. The part of the curve for which $0 \leqslant t \leqslant \frac{1}{2}\pi$ is rotated about the x-axis. Show that the area of the surface generated is $\frac{2}{5}\pi(e^\pi - 2)\sqrt{2}$.

The required surface area is

$$S = \int 2\pi y \, ds$$

$$= \int_0^{\frac{1}{2}\pi} 2\pi y \frac{ds}{dt} \, dt$$

$$= \int_0^{\frac{1}{2}\pi} 2\pi y \sqrt{\left(\frac{dx}{dt}\right)^2 + \left(\frac{dy}{dt}\right)^2} \, dt$$

Now $\dfrac{dx}{dt} = e^t \cos t + e^t \sin t = e^t(\cos t + \sin t)$

and $\dfrac{dy}{dt} = e^t(-\sin t) + e^t \cos t = e^t(\cos t - \sin t)$

$$S = \int_0^{\frac{1}{2}\pi} 2\pi e^t \cos t \sqrt{e^{2t}(\cos t + \sin t)^2 + e^{2t}(\cos t - \sin t)^2} \, dt$$

$$= \int_0^{\frac{1}{2}\pi} 2\pi e^t \cos t \, e^t \sqrt{2} \, dt$$

$$= 2\sqrt{2}\pi \int_0^{\frac{1}{2}\pi} e^{2t} \cos t \, dt$$

Let $\quad I = \int_0^{\frac{1}{2}\pi} e^{2t} \cos t \, dt$

$\qquad = \left[e^{2t} \sin t \right]_0^{\frac{1}{2}\pi} - \int_0^{\frac{1}{2}\pi} 2e^{2t} \sin t \, dt$

$\qquad = e^{\pi} - \left[-2e^{2t} \cos t \right]_0^{\frac{1}{2}\pi} + \int_0^{\frac{1}{2}\pi} 4e^{2t} (-\cos t) \, dt$

$\qquad = e^{\pi} - 2 - 4I$

and so $\quad I = \dfrac{1}{5}(e^{\pi} - 2)$

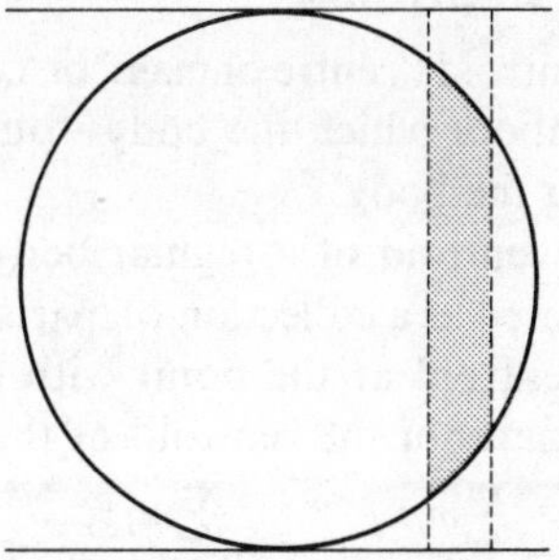

Hence $\quad S = 2\sqrt{2}\pi \times \dfrac{1}{5}(e^{\pi} - 2)$

$\qquad = \dfrac{2}{5}\pi(e^{\pi} - 2)\sqrt{2}$

Fig. 24.23

Exercise 24.4   (Area of surface of revolution)

1   Find the area of surface generated by revolution about Ox of $y = \sqrt{4ax}$ from $x = 0$ to $x = \alpha a$.

2   Prove that the area of the surface generated by the arc $y = a \cosh \dfrac{x}{a}$ (for $0 \leqslant x \leqslant a$) when it is rotated through 2π radians:

(i)   about the x-axis is $\pi a^2 \dfrac{(e^2 + 4 - e^{-2})}{4}$;

(ii)   about the y-axis is $2\pi a^2 (1 - e^{-1})$.   (MEI)

3   A curve has parametric equations

$\qquad x = 4t^2, \quad y = t^4 - 4 \ln t$

Find the length of the arc of the curve from $t = 1$ to $t = 2$.
   Find also the area of the surface formed when this arc is rotated completely about the y-axis.
(AEB)

4   Sketch the curve whose parametric equations are

$\qquad x = a \cos^3 t, \quad y = a \sin^3 t, \quad 0 \leqslant t \leqslant \frac{1}{2}\pi, \quad a > 0$

The points A and B on the curve correspond to the values $t = 0$ and $t = \frac{1}{2}\pi$ respectively. Calculate (a) the length of the arc AB of the curve; (b) the area of the curved surface generated when the arc AB is rotated through 2π radians about the x-axis.   (L)

5   Find the area of surface generated by revolution about Ox of the cycloid

$\qquad x = a(t - \sin t), \quad y = a(1 - \cos t)$

from $t = 0$ to $t = 2\pi$.

6   The arc of the circle $y^2 = a^2 - x^2$, between $x = x_1$ and $x = x_2$, is rotated about the x-axis. Show that the area of the surface generated is $2\pi a(x_2 - x_1)$. Deduce Archimedes' theorem, i.e. if a sphere is placed inside a cylinder of the same radius, then between any two planes drawn perpendicular to the axis of the cylinder, the surface area of the section of the sphere is equal to the surface area of the corresponding section of the cylinder (see Fig. 24.23 above).

24.5   Centroids

The centroid ('centre of mass' or 'centre of gravity') of a body is the point about which the body would balance. It may lie inside or outside the body.

The centroid of a regular body lies at its centre. If a body is considered as a collection of 'parts', and a typical part has mass m_i with centroid at the point with coordinates (x_i, y_i, z_i) then the coordinates of the centroid of the complete body are given by

$$\bar{x} = \frac{\sum m_i x_i}{\sum m_i}, \quad \bar{y} = \frac{\sum m_i y_i}{\sum m_i}, \quad \bar{z} = \frac{\sum m_i z_i}{\sum m_i}$$

where the summations are taken over all the parts of the body (and thus $\sum m_i$ is the total mass of the body).

The centroid of a region

Consider the area between a curve and the x-axis, and divide this area into strips parallel to the y-axis (see Fig. 24.24).

Suppose the mass per unit area is m. A typical strip has mass $m y \delta x$, and its centroid is at its centre, i.e. at the point $(x, \tfrac{1}{2} y)$.

Hence, if the centroid of the region is at $(\bar{x}, \bar{y})$, then

$$\bar{x} = \frac{\sum (m y \delta x) x}{\sum (m y \delta x)} = \frac{\displaystyle\int_a^b m x y \, dx}{\displaystyle\int_a^b m y \, dx}$$

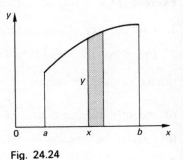

and $\quad \bar{y} = \dfrac{\sum (m y \delta x)(\tfrac{1}{2} y)}{\sum (m y \delta x)} = \dfrac{\tfrac{1}{2}\displaystyle\int_a^b m y^2 \, dx}{\displaystyle\int_a^b m y \, dx}$

Fig. 24.24

The centroid of an arc

Suppose the mass per unit length is n. A typical element has mass $n \delta s$ and its centroid is at (x, y) (see Fig. 24.25). Hence

$$\bar{x} = \frac{\sum (n \delta s) x}{\sum n \delta s} = \frac{\displaystyle\int_\alpha^\beta n x \, ds}{\displaystyle\int_\alpha^\beta n \, ds}$$

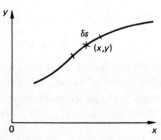

and $\quad \bar{y} = \dfrac{\sum (n \delta s) y}{\sum n \delta s} = \dfrac{\displaystyle\int_\alpha^\beta n y \, ds}{\displaystyle\int_\alpha^\beta n \, ds}$

Fig. 24.25

(Here, for example, ds may be replaced by $\sqrt{1 + \left(\dfrac{dy}{dx}\right)^2} \, dx$, and then α and β are limits of x.)

The centroid of a solid of revolution

We consider the solid to be composed of discs. Suppose that the mass per unit volume is p.

A typical disc (Fig. 24.26) has mass $p\pi y^2 \delta x$, and its centroid is at $(x, 0)$.

The centroid of the solid of revolution is at $(\bar{x}, 0)$ (by symmetry), where

$$\bar{x} = \frac{\sum (p\pi y^2 \delta x) x}{\sum p\pi y^2 \delta x} = \frac{\displaystyle\int_a^b p\pi x y^2 \, dx}{\displaystyle\int_a^b p\pi y^2 \, dx}$$

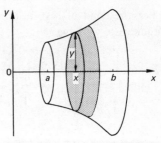

Fig. 24.26

The centroid of a curved surface of revolution

Suppose the mass per unit area is q. Consider a typical 'band' (Fig. 24.27). This has mass $q(2\pi y \delta s)$ and its centroid is at $(x, 0)$.

The centroid of the surface is at $(\bar{x}, 0)$, where

$$\bar{x} = \frac{\sum q(2\pi y \delta s) x}{\sum q(2\pi y \delta s)} = \frac{\displaystyle\int_\alpha^\beta 2\pi q x y \, ds}{\displaystyle\int_\alpha^\beta 2\pi q y \, ds}$$

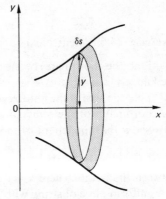

Example Find the coordinates of the centroid of the finite region R bounded by the curve $y = e^x$, the coordinate axes and the line $x = 1$.

This region R is rotated about the x-axis to form a solid of revolution. Find the coordinates of the centroid of this solid. (L)

Fig. 24.27

We first find the centroid of R (Fig. 24.28). Suppose the mass per unit area is m (which we assume to be constant). Then

$$\bar{x} = \frac{\displaystyle\int_0^1 mxy \, dx}{\displaystyle\int_0^1 my \, dx} = \frac{\displaystyle\int_0^1 xe^x \, dx}{\displaystyle\int_0^1 e^x \, dx} = \frac{\Big[xe^x - e^x\Big]_0^1}{\Big[e^x\Big]_0^1} = \frac{1}{e-1}$$

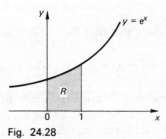

Fig. 24.28

$$\text{and} \quad \bar{y} = \frac{\frac{1}{2}\displaystyle\int_0^1 my^2 \, dx}{\displaystyle\int_0^1 my \, dx} = \frac{\frac{1}{2}\displaystyle\int_0^1 e^{2x} \, dx}{\displaystyle\int_0^1 e^x \, dx} = \frac{\frac{1}{2}\Big[\frac{1}{2}e^{2x}\Big]_0^1}{e-1} = \frac{\frac{1}{4}(e^2 - 1)}{e-1}$$

$$= \tfrac{1}{4}(e+1)$$

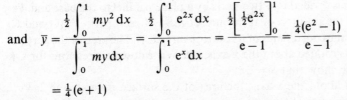

Hence the centroid of R is at $\left(\dfrac{1}{e-1}, \tfrac{1}{4}(e+1)\right)$.

For the solid of revolution (Fig. 24.29), if the mass per unit volume is p, we have

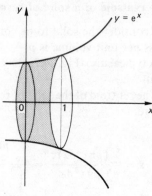

$$\bar{x} = \frac{\int_0^1 p\pi x y^2 \, dx}{\int_0^1 p\pi y^2 \, dx} = \frac{\int_0^1 x e^{2x} \, dx}{\int_0^1 e^{2x} \, dx}$$

$$= \frac{\left[\frac{1}{2} x e^{2x} - \frac{1}{4} e^{2x}\right]_0^1}{\left[\frac{1}{2} e^{2x}\right]_0^1} = \frac{\frac{1}{4}(e^2 + 1)}{\frac{1}{2}(e^2 - 1)} = \frac{1}{2}\left(\frac{e^2 + 1}{e^2 - 1}\right)$$

Fig. 24.29

Thus the centroid is at $\left(\dfrac{1}{2}\left(\dfrac{e^2 + 1}{e^2 - 1}\right), 0\right)$.

Exercise 24.5   (Centroids)

1   Find the coordinates of the centroid of the region bounded by the curve $y = x^2$, the x-axis, and the line $x = 2$.

2   Find the area of the finite region enclosed by the curve $xy = 6$ and the line $x + y = 5$ and calculate the coordinates of the centroid of this region, giving your answers to two significant figures.   (L)

3   Show that the length of the arc of the curve

$$y = \frac{1}{3}\left[(x^2 + 2)^{\frac{3}{2}} - 2^{\frac{3}{2}}\right]$$

between the points where $x = 0$ and $x = a(> 0)$ is $a + \frac{1}{3}a^3$.

A uniform piece of string with length 12 m lies along the above curve in the region $x \geqslant 0$ with one end at $x = 0$. Find the x-coordinate of its centre of mass.   (MEI)

4   A surface of revolution is formed by rotating completely about the x-axis the arc of the parabola $x = at^2$, $y = 2at$ from $t = 0$ to $t = \sqrt{3}$. Show that its surface area, S, is $\dfrac{56\pi a^2}{3}$.

Show also that the x-coordinate, $\bar{x}$, of the centroid of this surface is given by

$$S\bar{x} = 8\pi a^3 \int_0^{\sqrt{3}} t^3 \sqrt{(1 + t^2)} \, dt$$

Hence find $\bar{x}$, using the substitution $u^2 = 1 + t^2$ or otherwise.   (L)

5   Find the length of the curve $x = at^2$, $y = a(t - \frac{1}{3}t^3)$ from the origin to the point $(a, \frac{2}{3}a)$.

Find also the coordinates of the centroid of this part of the curve.   (O and C)

6   A uniform hemisphere of radius a is divided into two parts by a plane parallel to the base and distant $\frac{1}{2}a$ from it. Find the volume and the position of the centre of mass of each part.   (O and C)

7   An arc of a curve has length s and the coordinates of its centroid are $(\bar{x}, \bar{y})$. The area of the surface of revolution generated when this arc is rotated about the x-axis, is S. Write down expressions for s, $\bar{y}$ and S in terms of integrals, and hence show that $S = 2\pi \bar{y}s$.

Show also that if the arc is rotated about the y-axis, the area of the surface generated is $2\pi \bar{x}s$.

8   The circumference of a circle, with centre $(0, a)$ and radius b (where $b < a$) is rotated about the x-axis to form a 'torus'. Using the result of question 7, show that the surface area of the torus is $4\pi^2 ab$.

9   A region (which does not cross the x-axis) has area A, and the coordinates of its centroid are $(\bar{x}, \bar{y})$. The volume of the solid of revolution generated when this region is rotated about the x-axis, is V. Divide the region into strips parallel to the x-axis, and suppose that a typical strip (of width δy) has length z. Show that

$$V = \int 2\pi z y \, dy$$

and write down expressions for A and $\bar{y}$ in terms of integrals with respect to y. Deduce that $V = 2\pi \bar{y} A$. Write down a similar expression for the volume of the solid generated when the region is rotated about the y-axis.

Find the volume enclosed by the torus described in question 8.

10   (a)   Calculate the area of the region R bounded by the curve $y = (1 - x^2)^2$ for $0 \leqslant x \leqslant 1$, the x-axis and the y-axis.

(b)   Find the coordinates of the centroid of R, and the volumes of the solids of revolution when R is rotated about (i) the x-axis, and (ii) the y-axis.

(O and C)

24.6   Polar coordinates

We now consider curves given in terms of polar coordinates (r, θ).

Area of a sector

Suppose we wish to find the area of the sector OAB, where A and B are two points on a curve corresponding to $\theta = \alpha$ and $\theta = \beta$.

The area of a typical sector of small angle $\delta\theta$ is approximately $\frac{1}{2} r^2 \, \delta\theta$, and so the total area of the sector is approximately

$$\sum_{\theta = \alpha}^{\theta = \beta} \tfrac{1}{2} r^2 \, \delta\theta$$

Hence the area of the sector OAB is

$$\tfrac{1}{2} \int_{\alpha}^{\beta} r^2 \, d\theta$$

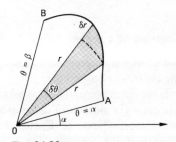

Fig. 24.30

Example   Sketch the curve with polar equation $r = a \cos 3\theta$ ($a > 0$), showing clearly the tangents to the curve at the pole.

Find the area of the finite region enclosed by one loop of the curve.   (L)

The area of one loop (Fig. 24.31) is given by

$$A = \tfrac{1}{2} \int_{-\frac{1}{6}\pi}^{\frac{1}{6}\pi} r^2 \, d\theta$$

$$= \tfrac{1}{2} \int_{-\frac{1}{6}\pi}^{\frac{1}{6}\pi} (a \cos 3\theta)^2 \, d\theta$$

$$= \tfrac{1}{4} a^2 \int_{-\frac{1}{6}\pi}^{\frac{1}{6}\pi} (1 + \cos 6\theta) \, d\theta$$

$$= \tfrac{1}{4} a^2 \left[\theta + \tfrac{1}{6} \sin 6\theta \right]_{-\frac{1}{6}\pi}^{\frac{1}{6}\pi}$$

$$= \tfrac{1}{12} \pi a^2$$

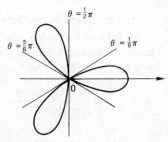

Fig. 24.31

Length of arc

Let δs be the arc length between two neighbouring points P and Q on a curve (Fig. 24.32). $P\hat{R}Q$ is approximately a right angle, and so we have

$$\delta s \approx \sqrt{(r\,\delta\theta)^2 + (\delta r)^2}$$

$$= \sqrt{r^2 + \left(\frac{\delta r}{\delta\theta}\right)^2}\,\delta\theta$$

Hence the arc length between $\theta = \alpha$ and $\theta = \beta$ is

$$\int_{\alpha}^{\beta} \sqrt{r^2 + \left(\frac{dr}{d\theta}\right)^2}\,d\theta$$

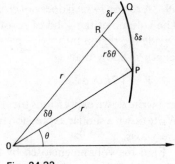

Fig. **24.32**

Curvature

It is possible to find the curvature of a curve given in polar coordinates. We just quote the formula, which is

$$\kappa = \frac{r^2 + 2\left(\dfrac{dr}{d\theta}\right)^2 - r\dfrac{d^2r}{d\theta^2}}{\left[r^2 + \left(\dfrac{dr}{d\theta}\right)^2\right]^{\frac{3}{2}}}$$

Centroid of a region

Sometimes it is convenient to use polar coordinates when finding the centroid of a region.

If the mass per unit area is m, a typical small sector (Fig. 24.33) has mass $(\frac{1}{2}r^2\,\delta\theta)m$ and its centroid is at the point G with polar coordinates $(\frac{2}{3}r, \theta)$, or Cartesian coordinates $(\frac{2}{3}r\cos\theta, \frac{2}{3}r\sin\theta)$. Note that the formulae for $\bar{x}$ and $\bar{y}$ may only be applied in Cartesian coordinates.

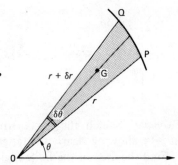

Fig. **24.33**

Example   A sector of a circle of radius a has angle 2α. Find the position of its centroid.

Clearly the centroid lies on the axis of symmetry (see Fig. 24.34), which we take as the initial line for polar coordinates.

The polar equation of the curved edge is then $r = a$, from $\theta = -\alpha$ to $\theta = \alpha$. We have

$$\bar{x} = \frac{\sum (\frac{1}{2}r^2\,\delta\theta)m\,(\frac{2}{3}r\cos\theta)}{\sum (\frac{1}{2}r^2\,\delta\theta)m}$$

$$= \frac{\dfrac{1}{3}\displaystyle\int_{-\alpha}^{\alpha} mr^3\cos\theta\,d\theta}{\dfrac{1}{2}\displaystyle\int_{-\alpha}^{\alpha} mr^2\,d\theta}$$

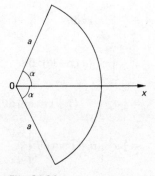

Fig. **24.34**

$$= \frac{\frac{1}{3} \int_{-\alpha}^{\alpha} a^3 \cos\theta \, d\theta}{\frac{1}{2} \int_{-\alpha}^{\alpha} a^2 \, d\theta}$$

$$= \frac{\frac{1}{3} a^3 (2\sin\alpha)}{\frac{1}{2} a^2 (2\alpha)} = \frac{2a\sin\alpha}{3\alpha}$$

Hence the centroid is on the axis of symmetry, at a distance $\dfrac{2a\sin\alpha}{3\alpha}$

from the centre.

Exercise 24.6   (Polar coordinates)

1   Sketch the curve whose polar equation is $r = 3\cos 2\theta$. Calculate the area of one of its loops.
(AEB)

2   Sketch the curve with polar equation $r = a\sin 2\theta$, where $a > 0$, for $0 \leqslant \theta \leqslant \pi/2$, and find the area of the region enclosed by the curve.
(L)

3   Sketch the curve whose polar equation is $r = 3 + 2\cos\theta$. Find the area of the region enclosed by the curve.
(AEB)

4   Sketch the curve $r = a\theta$, where a is a positive constant and $0 \leqslant \theta \leqslant \pi$, and find the area of the region enclosed by this curve and the half-line $\theta = \pi$.
(L)

5   Sketch the curve whose polar equation is $r = \tan\frac{1}{2}\theta$ and find the area enclosed by its loop.
(AEB)

6   Sketch, on the same diagram, the curves whose equations in polar coordinates are $r = 2a\cos 2\theta$, $-\pi/4 \leqslant \theta \leqslant \pi/4$, and $r = a$, where $a > 0$. Show that the area of the finite region lying within both curves is $\dfrac{a^2(4\pi - 3\sqrt{3})}{12}$.
(L)

7   Sketch, on the same diagram, the curves whose polar equations are $r = \sqrt{\cos\frac{1}{2}\theta}$ and $r = 1/\sqrt{2}$. Find the polar coordinates of their points of intersection and the area of that part of the first curve which lies outside the second.
(S)

8   Sketch on the same diagram the curves $r = a$ and $r = a(1 + \cos\theta)$, where $a > 0$.
   Find the area of the region enclosed by the curve $r = a(1 + \cos\theta)$, and show that the area of the part of this region for which $r > a$ is $\dfrac{a^2}{4}(\pi + 8)$.
(L)

25

Partial differentiation

25.1 Surfaces

So far we have only investigated areas and volumes of surfaces formed by the rotation of a curve about an axis. In this chapter we look at surfaces in a more general way.

Building a surface

The locus of a point (x, y, z) such that $z = f(x, y)$ is, in general, a surface. We illustrate this by investigating the surface

$$z = x^2 - xy + y + 1$$

We first make a table of values by allocating values to x and to y (see Fig. 25.1). A model can then be made in several different ways.

y \ x	-3	-2	-1	0	1	2	3
5	30	20	12	6	2	0	0
4	26	17	10	5	2	1	2
3	22	14	8	4	2	2	4
2	18	11	6	3	2	3	6
1	14	8	4	2	2	4	8
0	10	5	2	1	2	5	10
-1	6	2	0	0	2	6	12
-2	2	-1	-2	-1	2	7	14
-3	-2	-4	-4	-2	2	8	16

Fig. 25.1

(A)   We can make a 'block' graph, using centimetre cubes. Since the smallest value of z is -4, by adding 6 to each z we can then build up the graph by placing a vertical pile of 36 cubes on the square, marked 30 above, 32 on the square marked 26 and so on.

This gives a solid model which will only show the surface approximately (see Fig. 25.2).

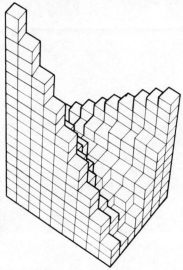

Fig. 25.2

(B)   The more usual method of making the model is to make up the shapes represented by the columns and by the rows from fairly stiff card. These can then be slotted together to make an accurate picture of the surface. (Two typical cards are shown in Figs 25.3 and 25.4.)

An actual model of this surface is shown in Fig. 25.5.

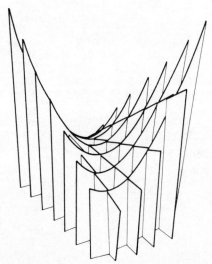

Fig. 25.5

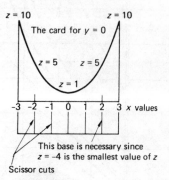

Fig. 25.3

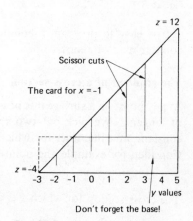

Fig. 25.4

(C)   The model in Fig. 25.5 can be 'boxed in' and filled with 'Polyfilla' to make a solid surface. It is difficult to obtain the right consistency of filler, but if you are lucky you will finish with a solid representing the surface.

Contour lines

We can illustrate the surface by drawing contour lines. This is done by joining points in the xy-plane which have equal z-values (see Fig. 25.6).

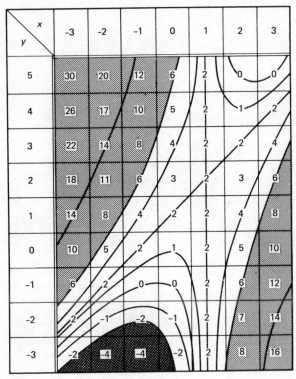

Fig. 25.6

It is easy to program a computer so that it will print out a contour map of a surface.

The gradient of a cross-section

At any point on a surface it is possible to draw an infinite number of tangents. We pick out two special ones: one for which y is constant, and the other for which x is constant.

Consider, for example, the point $P(-2, 2, 11)$ on the surface

$$z = x^2 - xy + y + 1$$

The cross-section for which y is constant (i.e. $y = 2$) is

$$z = x^2 - 2x + 3$$

The gradient of this (given by $2x - 2$) at the point P is -6. The cross-section for which x is constant (i.e. $x = -2$) is

$$z = 4 + 2y + y + 1 = 5 + 3y$$

The gradient of this at the point P is 3.

In general we can find the gradient of a cross-section for which y is constant by differentiating with respect to x, treating y as a constant. This gives the **partial derivative** $\dfrac{\partial z}{\partial x}$ (said 'partial dee z by dee x'). Similarly, we find the gradient of a cross-section for which x is constant by differentiating with respect to y, treating x as a constant. This gives $\dfrac{\partial z}{\partial y}$. Thus, if $z = x^2 - xy + y + 1$, then

$$\frac{\partial z}{\partial x} = 2x - y \quad \text{and} \quad \frac{\partial z}{\partial y} = -x + 1$$

At the point $P(-2, 2, 11)$,

$$\frac{\partial z}{\partial x} = -4 - 2 = -6 \quad \text{and} \quad \frac{\partial z}{\partial y} = 2 + 1 = 3$$

Small changes on a surface

Consider a small element of the surface with points $A(x, y, z)$ and $B(x + \delta x, y + \delta y, z + \delta z)$, close together on it (see Fig. 25.7). We consider the increase δz as $\delta z_1 + \delta z_2$ where AC is in the plane parallel to Ox and CB in the plane parallel to Oy as shown in Fig. 25.8.

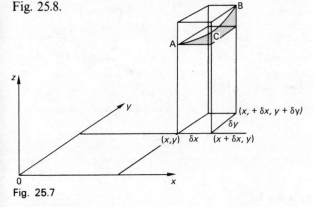

Fig. 25.7

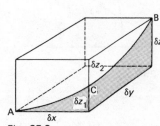

Fig. 25.8

Then

$$\frac{\delta z_1}{\delta x} \approx \frac{\partial z}{\partial x} \quad \text{and} \quad \frac{\delta z_2}{\delta y} \approx \frac{\partial z}{\partial y}$$

Then $\delta z = \delta z_1 + \delta z_2$

so

$$\boxed{\delta z \approx \frac{\partial z}{\partial x} \delta x + \frac{\partial z}{\partial y} \delta y}$$

δz is called the **total differential** of z.

Exercise 25.1   (Surfaces)

1   Make a model of the surface $z = x^2 - y^2 + 10$ taking values of x and y as $-2, -1, 0, 1, 2$.

2   Make tables of values for the following surfaces, taking values of x and y from -3 to $+3$. Hence make contour maps for them.

(i) $z = 3x + 2y$ (ii) $z = x^2 - xy^2$ (iii) $z = x^2 + y^2 - 2xy$ (iv) $z = xy$

(v) $z = 2x^2 + y^2$ (vi) $z = x^2 + y^2 + 1$ (vii) $z = \sqrt{9 - x^2 - y^2}$ (viii) $z = 1 - x^2 - y$

3   By considering the contours, describe the following surfaces

(i) $z = x^2 + y^2$ (ii) $z = x^2 + 4y^2$ (iii) $z = 9x^2 - y^2$

Find $\dfrac{\partial z}{\partial x}$ and $\dfrac{\partial z}{\partial y}$ for the following surfaces (questions 4 to 10).

4 $z = x^2 + y^2$ **5** $z = x^2 - y^2 + 3xy$ **6** $z = \dfrac{x}{y}$ **7** $z = x^3 + y^3 - 3xy + 2y + 1$

8 $z = e^x \cos(x - 3y)$ **9** $z = \tan^{-1}\left(\dfrac{y}{x}\right)$ **10** $\sin z = 2x^2 + 4xy - 2y^2$

25.2   Stationary points

A stationary point on a surface is a point where $\dfrac{\partial z}{\partial x} = 0$ and $\dfrac{\partial z}{\partial y} = 0$.

There are three types of stationary point: a maximum point (Fig. 25.9); a minimum point (Fig. 25.10); and a saddle point (Fig. 25.11). We shall distinguish between these three types by considering the surface in a neighbourhood of a stationary point. It is possible to use the second partial derivatives instead, but we shall not do so here.

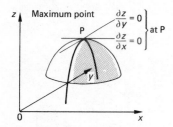

Fig. 25.9

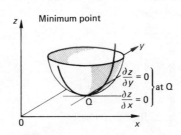

Fig. 25.10

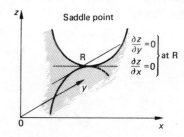

Fig. 25.11

Example 1   Consider the surface $z = x^2 - xy + y + 1$.

We have $\dfrac{\partial z}{\partial x} = 2x - y$ and $\dfrac{\partial z}{\partial y} = -x + 1$

For a stationary point, $2x - y = 0$ and $-x + 1 = 0$, giving $x = 1$, $y = 2$, and hence $z = 2$.

There is only one stationary point, $P(1, 2, 2)$.

Looking at the contour map (see p. 496) we see that some of the cross-sections are maxima, and some are minima (see Fig. 25.12). Hence P is a saddle point.

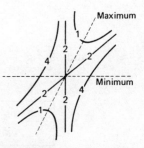

Fig. 25.12

Example 2 Consider the surface $z = x^3 + 2xy - y^2$.

We have $\dfrac{\partial z}{\partial x} = 3x^2 + 2y$ and $\dfrac{\partial z}{\partial y} = 2x - 2y$

For stationary points, $3x^2 + 2y = 0$ and $2x - 2y = 0$, giving $y = x$ and $3x^2 + 2x = 0$, so $x = 0$ or $-\frac{2}{3}$. Hence there are two stationary points $P(0, 0, 0)$ and $Q(-\frac{2}{3}, -\frac{2}{3}, \frac{4}{27})$.

First consider $P(0, 0, 0)$. The cross-section with y constant (i.e. $y = 0$) is

$$z = x^3$$

which has a point of inflexion, and hence P is a saddle point.

Now consider $Q(-\frac{2}{3}, -\frac{2}{3}, \frac{4}{27})$. The cross-section with y constant (i.e. $y = -\frac{2}{3}$) is

$$z = x^3 - \tfrac{4}{3}x - \tfrac{4}{9}$$

which has a maximum at Q.

The cross-section with x constant (i.e. $x = -\frac{2}{3}$) is

$$z = -\tfrac{8}{27} - \tfrac{4}{3}y - y^2$$

which also has a maximum at Q.

This suggests that Q may be a maximum, but it does *not* establish it (some other cross-section could give a minimum). We put $x = -\frac{2}{3} + \lambda$, $y = -\frac{2}{3} + \mu$, where λ and μ are small. Then

$$z = (-\tfrac{2}{3} + \lambda)^3 + 2(-\tfrac{2}{3} + \lambda)(-\tfrac{2}{3} + \mu) - (-\tfrac{2}{3} + \mu)^2$$

$$= \tfrac{4}{27} + \lambda^3 - 2\lambda^2 + 2\lambda\mu - \mu^2 = \tfrac{4}{27} + \lambda^3 - \lambda^2 - (\lambda - \mu)^2$$

$$= \tfrac{4}{27} - \lambda^2(1 - \lambda) - (\lambda - \mu)^2$$

Thus $z < \frac{4}{27}$ whenever (x, y) is close to $(-\frac{2}{3}, -\frac{2}{3})$, and hence Q is indeed a maximum point.

Exercise 25.2 (Stationary points)

1   Show that $z = x^2 + xy + y^2$ has a minimum value and $z = x^2 + xy - y^2$ a saddle point at $(0, 0, 0)$.
2   Sketch the form of the contour map for

$$z = x^3 + y^3 - 3xy$$

Deduce that z has a minimum value at $(1, 1, -1)$ and a saddle point at $(0, 0, 0)$.
3   Show that $z = ax^2 + by^2 + c$ always has a single stationary value which is a minimum, a maximum or a saddle point according to the values of a and b (whether they are both positive, or both negative, or differ in sign).
4   Find the maximum value of

$$z = xy^2(6 - x - y)^3$$

5   Find the stationary points of the surface

$$z = x^3 + xy + y^2$$

and determine their nature.

6   Show that

$$z = x^3 + y^3 - 2(x^2 + y^2) + 3xy$$

has stationary values when $x = 0$, $y = 0$ and when $x = \frac{1}{3}$, $y = \frac{1}{3}$, and investigate their nature.   (L)

7   Find the stationary points of the surface

$$z = x^3 + y^3 - 12x - 3y + 20$$

and determine their nature.

8   Find the stationary points of the surface

$$z = x^2 + xy + y^3$$

and hence find the minimum value of z.

9   Find the stationary points of the function

$$z = x + \frac{4x}{x^2 - 3y^2}$$

i.e. those points at which $\dfrac{\partial z}{\partial x}$ and $\dfrac{\partial z}{\partial y}$ are both zero.

   Sketch the curve of intersection of the given surface with the xy plane. By considering the signs of x, $x^2 - 3y^2$ and $x^2 - 3y^2 + 4$, determine the regions in which z is positive and indicate them in your sketch.

   Determine whether the stationary points of the function that lie in the xy plane correspond to maximum or minimum values of z, or neither.   (O and C)

10   Find the values (x_0, y_0) for which the partial derivatives of the function $z = \dfrac{x(x^2 - a^2)}{(b^2 + y^2)}$ with respect to x and y are both zero.

   By writing $x = x_0 + \varepsilon \cos\theta$, $y = y_0 + \varepsilon \sin\theta$ derive an expression for the value of the function at points on a small circle of radius ε with centre at (x_0, y_0), and hence determine if the function has a maximum or a minimum value there.   (O and C)

25.3   Applications to inaccuracies in measurement

The methods used in partial differentiation are particularly useful in dealing with inaccuracies in measurement. These inaccuracies are due to the instruments or techniques used in the measurements. They are often called errors but should not be equated with mistakes or blunders, for they are unavoidable. All measurements must, by their very nature, be approximate.

Example 1   The volume of a right circular cone is calculated from the formula $v = \frac{1}{3}\pi r^2 \cot\alpha$, where the base radius is $r = 8 \pm 0.04$ cm, and the semi-vertical angle is $\alpha = 45° \pm 0.25°$. Find the greatest possible percentage error in the calculated volume.

We have $v = \frac{1}{3}\pi r^3 \cot \alpha$

so $\delta v \approx \dfrac{\partial v}{\partial r}\delta r + \dfrac{\partial v}{\partial \alpha}\delta \alpha$

Now $\dfrac{\partial v}{\partial r} = \pi r^2 \cot \alpha;\quad \dfrac{\partial v}{\partial \alpha} = -\frac{1}{3}\pi r^3 \operatorname{cosec}^2 \alpha$

Hence $\delta v \approx \pi r^2 \cot \alpha\, \delta r - \frac{1}{3}\pi r^3 \operatorname{cosec}^2 \alpha\, \delta \alpha$

To find the percentage error, we require $\dfrac{\delta v}{v} \times 100$:

$$\frac{\delta v}{v} \times 100 \approx \left(\frac{\pi r^2 \cot \alpha}{\frac{1}{3}\pi r^3 \cot \alpha}\delta r - \frac{\frac{1}{3}\pi r^3 \operatorname{cosec}^2 \alpha}{\frac{1}{3}\pi r^3 \cot \alpha}\delta \alpha \right) \times 100$$

$$= \left(\frac{3}{r}\delta r - \frac{1}{\sin \alpha \cos \alpha}\delta \alpha \right) \times 100$$

When $r = 8$, $\alpha = 45°\,(= \pi/4)$, $\delta r = \pm 0.04$,

$$\delta \alpha = \left(\pm 0.25 \times \frac{\pi}{180} \right) \text{rad}$$

The maximum percentage error occurs when $\delta r > 0$, $\delta \alpha < 0$. In this case

$$\frac{\delta v}{v} \times 100 \approx \left(\frac{3}{8} \times 0.04 + 2 \times 0.25 \times \frac{\pi}{180} \right) \times 100 \approx 2.37$$

The greatest percentage error in the volume is about 2.4%.

Example 2   Find the absolute error in $\dfrac{ab}{a+b}$ due to errors δa in a and δb in b.

Suppose $v = \dfrac{ab}{a+b}$

then $\delta v \approx \dfrac{\partial v}{\partial a}\delta a + \dfrac{\partial v}{\partial b}\delta b$

so $\delta v \approx \dfrac{(a+b)b - ab}{(a+b)^2}\delta a + \dfrac{(a+b)a - ab}{(a+b)^2}\delta b$

$$= \frac{b^2}{(a+b)^2}\delta a + \frac{a^2}{(a+b)^2}\delta b$$

which can be written

$$\frac{ab}{a+b}\left(\frac{\delta a}{a} + \frac{\delta b}{b} - \frac{\delta a + \delta b}{a+b} \right)$$

Example 3   The length of the side a in a triangle is calculated using the 'cosine rule',

$$a^2 = b^2 + c^2 - 2bc \cos A$$

Find the maximum error in a, when $b = 10 \pm 0.5$ cm, $c = 15 \pm 1.0$ cm and $A = 45° \pm 1°$.

Now $a^2 = b^2 + c^2 - 2bc \cos A$

so $\delta(a^2) \approx \dfrac{\partial(a^2)}{\partial b} \delta b + \dfrac{\partial(a^2)}{\partial c} \delta c + \dfrac{\partial(a^2)}{\partial A} \delta A$

so $2a\,\delta a \approx (2b - 2c \cos A)\delta b + (2c - 2b \cos A)\delta c + (2bc \sin A)\delta A$

$$\delta a \approx \frac{1}{a}[(b - c \cos A)\delta b + (c - b \cos A)\delta c + (bc \sin A)\delta A]$$

$$\approx \frac{1}{a}[(10 - 15 \cos 45°)\,\delta b + (15 - 10 \cos 45°)\,\delta c + (10 \times 15 \sin 45°)\delta A]$$

$$\approx \frac{1}{10.62}[-0.607\,\delta b + 7.93\,\delta c + 106\,\delta A]$$

Now, to find the largest error, we must consider $\delta b = -0.5$, $\delta c = +1.0$ and $\delta A = +\dfrac{\pi}{180}$, i.e.

$$\delta a_{max} \approx \frac{1}{10.62}(0.303 + 7.93 + 1.85) \approx 0.95$$

Exercise 25.3   (Application to inaccuracies)

1   If $f(x, y) = x^3 + 3xy^2 + 2y^3$, state the error δf in f due to small errors δx, δy in the independent variables x, y respectively.

If $\delta x = \delta y$, by the method of completing squares, or otherwise, prove that δf has the same sign as δx for all non-zero x, y.

If the percentage errors in x, y are equal, prove that, whatever the values of x, y such that $y = kx$, the error in f is zero if $2k^3 + 3k^2 + 1 = 0$. Show that k takes a value lying between -1 and -2 and find this value of k correct to two decimal places.   (O and C)

2   The side a of a triangle ABC is calculated from the formula

$$a^2 = b^2 + c^2 - 2bc \cos A$$

If small errors β and γ occur in b and c respectively, show that if terms in β^2, γ^2 and $\beta\gamma$ are neglected then the resulting small error α in a is given approximately by

$$a\alpha = (b - c \cos A)\beta + (c - b \cos A)\gamma$$

Find also an expression for the approximate error δ in the area of the triangle calculated from the formula

$$\Delta = \tfrac{1}{2}bc \sin A$$

If the sides b and c are given as 42 mm and 24 mm respectively, each to the nearest mm, and A is $60°$ exactly, calculate values for a and Δ, and estimate in each case the maximum error in the value obtained.   (C)

Revision questions F

Revision paper F1

1 (a) If $y = e^{3x} \sin 2x$, show that

$$\frac{d^2y}{dx^2} = 13e^{3x} \sin(2x + 2\alpha)$$

and state the value of $\tan \alpha$. Write down a formula for $\dfrac{d^n y}{dx^n}$ and prove that it is correct.

(b) Evaluate

(i) $\displaystyle\int_2^3 \frac{dx}{3x^2 - 12x + 13}$ (ii) $\displaystyle\int_0^1 \frac{dx}{\sqrt{2x - x^2}}$ (iii) $\displaystyle\int_1^2 x^2 \ln(x^3)dx$ (MEI)

2 Show, without integrating, that if

$$A = \int_1^{\frac{3}{2}} \frac{dx}{\sqrt{2x - x^2}}, \quad B = \int_1^{\frac{3}{2}} \frac{dx}{x\sqrt{2 - x}}, \quad C = \int_1^{\frac{3}{2}} \frac{dx}{x}$$

then $A > B > C$. Verify this result by evaluating the integrals. [The substitutions $u = 1 - x$ and $v = \sqrt{2 - x}$ will be found suitable.]   (MEI)

3 (i) Differentiate with respect to x (a) $\ln \cosh x$ (b) $\sinh(x^2)$ (c) $\arcsin(2x)$

(ii) Evaluate   (a) $\displaystyle\int_0^1 \frac{1}{\sqrt{4 + x^2}}dx$ (b) $\displaystyle\int_0^1 \frac{1}{\sqrt{4 - x^2}}dx$ (L)

4 Sketch the curve $x = 4\sqrt{2} \sin t$, $y = \sin 2t$. Show that

$$\left(\frac{dx}{dt}\right)^2 + \left(\frac{dy}{dt}\right)^2 = 4(2 + \cos 2t)^2$$

and hence find the total length of the curve.

5 (i) A particular integral of the differential equation

$$\frac{d^2y}{dx^2} + n^2y = \sin nx$$

has the form $x(a \sin nx + b \cos nx)$, where a, b and n are constants. Determine a and b in terms of n. Hence find y in terms of x given that $y = 0$ and $\dfrac{dy}{dx} = 0$ when $x = 0$.

(ii) By writing $z = \dfrac{dy}{dx}$, or otherwise, find the general solution of the differential equation

$$x\frac{d^2y}{dx^2} + 2\frac{dy}{dx} = 0$$ (L)

503

6 Show that the partial derivatives of the function

$$f(x, y) = 2x^2 + 4xy + 3y^2 - 12x - 14y + 20$$

are both zero at the point (2, 1).

Show that $f(x, y) \geqslant f(2, 1)$ for all x and y. 　　　　　　　(O)

7 Express $t^4(1-t)^4$ in the form $(t^2 + 1)Q(t) + R(t)$ where $Q(t)$, $R(t)$ are polynomials and $R(t)$ is linear. Deduce that

$$\int_0^1 \frac{t^4(1-t)^4}{1+t^2} dt = \frac{22}{7} - \pi.$$

Assuming that Simpson's rule with 5 ordinates gives the integral to four decimal places, discuss the accuracy of $\frac{22}{7}$ as an approximation to π. 　　　　　　　(MEI)

Revision paper F2

1 (a) Let p be a positive real number, and let

$$I_p = \int_0^1 x^2(1-x)^p dx$$

Prove that

$$I_p = \int_0^1 x^p(1-x)^2 dx$$

and express I_p in terms of p.

Hence, or otherwise, evaluate

$$\int_0^{\frac{1}{2}\pi} \sin^5 \theta \cos^7 \theta \, d\theta$$

(b) Evaluate $\displaystyle\int_1^{\frac{3}{2}} \sqrt{\frac{2-x}{x}} \, dx$.

　　　　　　　(W)

2 (a) Show that $\displaystyle\int_{-a}^a f(x)dx = \int_0^a [f(x) + f(-x)]dx$.

Hence, or otherwise, show that $\displaystyle\int_{-\frac{1}{4}\pi}^{\frac{1}{4}\pi} \frac{1}{1+\sin x} dx = 2$.

(b) Evaluate, to three significant figures,

$$\int_1^2 \frac{\ln(1+x)}{x^2} dx$$

　　　　　　　(W)

3 Define $\tanh x$ in terms of e^x and e^{-x}, and hence or otherwise show that

$$\operatorname{ar tanh} y = \tfrac{1}{2} \ln\left(\frac{1+y}{1-y}\right) \quad \text{for} \quad |y| < 1.$$

Sketch on one diagram the graphs of $y = \tanh x$ and $y = \operatorname{ar tanh} x$.

Obtain the first two non-zero terms of the series for $\tanh x$ in ascending powers of x, and evaluate

$$\lim_{x \to 0} \frac{\operatorname{ar tanh} x}{x}$$

　　　　　　　(L)

4 Find the coordinates of any turning points and the equations of any asymptotes of the curve $y = \dfrac{x^2 + 1}{x^2 - 4}$. Show that the curve has no point of inflexion. Sketch the curve.

Calculate the radius of curvature of the curve at the point $(3, 2)$. (AEB)

5 A particle moves along the x-axis so that at time t its displacement from the origin O satisfies the differential equation

$$\frac{d^2 x}{dt^2} + 4\frac{dx}{dt} + 13x = 40 \cos 3t$$

Find a particular integral of this differential equation in the form $p \cos 3t + q \sin 3t$, where p and q are constants. Obtain further the complementary function of this differential equation.

Hence express x in terms of t, given that $x = 1$, $\dfrac{dx}{dt} = 12$ when $t = 0$.

Show that, when t is large, the motion approximates to a simple harmonic oscillation about O with period $2\pi/3$ and with amplitude $\sqrt{10}$. (L)

6 (i) Prove that if $\quad u = \dfrac{(x^2 + y^2)}{(x^2 - y^2)^2}$

then $\quad x\dfrac{\partial u}{\partial x} + y\dfrac{\partial u}{\partial y} + 2u = 0$

(ii) If $\quad V = e^{x+y}\cos(x - y)\quad$ and $\quad u = \dfrac{\partial V}{\partial x}, \quad v = \dfrac{\partial V}{\partial y}$

prove that $\quad u + v = 2V$

and $\quad \dfrac{\partial u}{\partial x} + \dfrac{dv}{\partial y} = 0$ (O and C)

7 Use the binomial theorem to expand $(1 + x^3)^{10}$ in ascending powers of x up to and including the term in x^9. Hence estimate I, where

$$I = \int_0^{0.2} (1 + x^3)^{10}\, dx$$

to three decimal places.

Make another estimate of I again to three decimal places by using Simpson's rule with 3 ordinates, showing all your working. (L)

Miscellaneous revision questions: paper F3

1 (i) Evaluate $\displaystyle\int_0^1 \text{arc}\sin x\, dx$.

(ii) Express $\tanh x$ in terms of e^x, and hence evaluate $\displaystyle\int_0^1 e^x \tanh x\, dx$.

(iii) Given that $\quad I_n = \displaystyle\int_0^{\pi/2} \sin^n x\, dx \quad (n \geqslant 0)$

show that $\quad nI_n = (n - 1)I_{n-2} \quad (n \geqslant 2)$

Hence evaluate I_4. (L)

2 (a) Prove that $\displaystyle\int_0^a f(x)\,dx = \int_0^a f(a-x)\,dx$

and hence, or otherwise, show that $\displaystyle\int_0^\pi \frac{x\sin x}{1+\cos^2 x}\,dx = \tfrac14\pi^2$.

(b) Define an *odd function*, and prove that if f is an odd function then $\displaystyle\int_0^\pi f(\cos\theta)\,d\theta = 0$.

(c) Evaluate $\displaystyle\int_0^1 \frac{\sqrt{x}}{\sqrt{4-x}}\,dx$.   (W)

3 *Sketch* on a diagram the region defined by the inequalities

$$0 \leqslant x \leqslant \tfrac12\pi, \quad 0 \leqslant y \leqslant x\cos x$$

Find the area of this region. Find also the coordinates of its centroid.
   Use your results to obtain, without further integration, the mean values of $x\cos x$ and $x^2\cos^2 x$ between $x = 0$ and $x = \tfrac12\pi$.
[In this question give all your answers in terms of π.]   (MEI)

4 Define, in terms of e^x, the functions $\sinh x$ and $\tanh x$ and prove that, for $x > 0$, $\sinh x > \tanh x$.
(a) Sketch, in two separate diagrams, the graphs of $y = \sinh x$, $y = \tanh x$.
(b) Prove that, for small x, the function

$$f(x) = \frac{x}{\sinh x} - \frac{\tanh x}{x}$$

is positive.   (O and C)

5 (a) Show that, for all real numbers $x \geqslant 0$,

$$2\int_1^{\cosh x} \sqrt{u^2 - 1}\,du = \cosh x \sinh x - x$$

(b) Using the substitution $x = \sqrt{\dfrac{1+u}{1-u}}$ or otherwise, show that

$$\int_1^{\sqrt{3}} \frac{x^2-1}{(x^2+1)\sqrt{1+x^4}}\,dx = \frac{1}{\sqrt{2}}\int_0^{\frac12} \frac{u\,du}{\sqrt{1-u^4}}$$

and hence evaluate this integral.   (W)

6 Verify that the function $y = \dfrac{\sin^{-1} x}{\sqrt{1-x^2}}$ satisfies the equation

$$(1-x^2)\frac{d^2y}{dx^2} - 3x\frac{dy}{dx} - y = 0$$

Hence, or otherwise, prove the more general relation

$$(1-x^2)y^{(n+1)} - (2n+1)xy^{(n)} - n^2 y^{(n-1)} = 0 \quad \text{where } y^{(m)} = \frac{d^m y}{dx^m}$$

Determine the value of $y^{(n)}$ at $x = 0$.   (MEI)

7   If $f(x)$ is a steadily decreasing function of x in the closed interval (a, b) and $b - a = nh$, show that

$$h \sum_{r=0}^{n-1} f(a + rh) > \int_a^b f(x)\,dx > h \sum_{r=1}^n f(a + rh)$$

By taking $f(x) = \dfrac{1}{1 + x^2}$, show that if $S_n = \displaystyle\sum_{r=1}^n \dfrac{4n}{n^2 + r^2}$ then $\dfrac{2}{n} > \pi - S_n > 0$ and hence that $\displaystyle\lim_{n \to \infty} S_n$ is π.

(MEI)

8   If $f(x) = f(a - x)$ for all x, where a is a constant, show that

$$\int_0^a x f(x)\,dx = \tfrac{1}{2} a \int_0^a f(x)\,dx = a \int_0^{\frac{1}{2}a} f(x)\,dx$$

Hence, or otherwise, evaluate $\displaystyle\int_0^\pi \dfrac{x \sin^3 x}{1 + \cos^2 x}\,dx$.

(W)

9   (i)   Solve the differential equation

$$\frac{dy}{dx} + y \tan x = \cos^2 x$$

for $-\pi/2 < x < \pi/2$, given that $y = \tfrac{1}{2}$ when $x = \pi/4$.

(ii)   Find the constant k such that ke^{-2x} is a particular integral of the differential equation

$$\frac{d^2 y}{dx^2} + 4\frac{dy}{dx} + 29y = 75e^{-2x}$$

Hence solve the differential equation given that $y = 0$ and $\dfrac{dy}{dx} = 0$ when $x = 0$.

(L)

10   Calculate the length of the arc of the curve whose parametric equation is

$$\begin{cases} x = a(3 \cos t - \cos 3t) \\ y = a(3 \sin t - \sin 3t) \end{cases}$$

between the points corresponding to $t = 0$ and $t = \pi$.

Find the area between this arc and the x-axis.

What is the area of the surface of revolution formed when the arc is rotated through an angle of 2π about the x-axis?

(O and C)

11   (i)   Determine $\displaystyle\int \dfrac{dx}{x^{\frac{1}{2}} + x^{\frac{1}{3}}}$.

(ii)   Determine $\int e^{-x} \sin x\,dx$.

Prove that

$$\int_{n\pi}^{(n+1)\pi} e^{-x} |\sin x|\,dx = \tfrac{1}{2} e^{-n\pi}(1 + e^{-\pi})$$

where n is a positive integer, and hence find

$$\int_0^{(n+1)\pi} e^{-x} |\sin x|\,dx$$

Show that this integral has limit $\dfrac{e^\pi + 1}{2(e^\pi - 1)}$ as $n \to \infty$.

(O and C)

12 (a) Solve the differential equation

$$x\frac{dy}{dx} - (x+1)y = x^4$$

and find the particular solution for which y takes the same value when $x = 0$ as when $x = 1$.
(b) Solve the differential equation

$$\frac{d^2y}{dx^2} - y = \cosh x$$

given that $y = 0$ and $\dfrac{dy}{dx} = 1$ when $x = 0$.   (MEI)

13 Given that for $s > 0$

$$F(s) = \int_0^\infty e^{-st} f(t)\,dt$$

prove that

$$\int_0^\infty e^{-st} f''(t)\,dt = -f'(0) - sf(0) + s^2 F(s)$$

Show that, if $f(0) = f'(0) = 0$, then

$$f''(t) + k^2 f(t) = 1 \quad \text{implies} \quad F(s) = \frac{1}{s(s^2 + k^2)}$$

Given that

$$\int_0^\infty e^{-st} \cos kt\,dt = \frac{s}{s^2 + k^2}$$

use the foregoing to obtain a solution of the differential equation with the given initial conditions.
                                                                                            (MEI)

14 Sketch, on the same diagram, the graphs of $y = \cosh x$, $y = x$ and $y = \cosh^{-1} x$.
Show that the shortest distance between the curve $y = \cosh x$ and the line $y = x$ is
$$\frac{\sqrt{2} - \ln(1 + \sqrt{2})}{\sqrt{2}}.$$
                                                                                            (MEI)

15 Given the differential equation

$$100\frac{d^2y}{dx^2} = 1 + (y - 3)^2$$

with conditions $y(0) = 4$ and $y(1) = 4$, use the approximation

$$y''(x_r) \approx \frac{1}{h^2}[y(x_{r+1}) - 2y(x_r) + y(x_{r-1})] \quad \text{where } y''(x_r) = \left(\frac{d^2y}{dx^2}\right)_{x = x_r}$$

with $h = 1$, to estimate $y(4)$, working to five decimal places.
Given that a similar computation with conditions $y(0) = 4$ and $y(1) = 4.1$ gives $y(4) = 4.545$, use linear interpolation to estimate, to three significant figures, the value of $y(1)$ for which $y(4) = 4.3$.
                                                                                            (MEI)

Answers to exercises

Exercise 1.2 (p. 6)

1 (i) 3 (ii) $-\frac{1}{3}$ (iii) 2 (iv) $-\frac{3}{5}$
2 (i) $y = 2x - 1$ (ii) $y = -\frac{1}{3}x + 3$ (iii) $y = 3x - 1$ (iv) $x + 2y = 5$ (v) $y = 2x - 7$ (vi) $5x + 6y = 32$

Exercise 1.3 (p. 7)

1 3 **2** -2 **3** -1 **4** -1.25

Exercise 1.4 (p. 10)

1 (i) 2 (ii) 3 (iii) 4 (iv) 5 (v) 0.5 (vi) -1 (vii) -2 **2** 32 **3** 0.25 **4** -4 **5** -1

Exercise 1.5 (p. 11)

1 (i) 27 (ii) 48 (iii) $3x^2$ **2** $4x^3$ **3** $5x^4$ **4** $-\dfrac{1}{x^2}$ **5** $-\dfrac{2}{x^3}$ **6** $3x^2, 4x^3, 5x^4, -\dfrac{1}{x^2}, -\dfrac{2}{x^3}$.

Exercise 1.6 (p. 13)

1 $6x$ **2** (i) $10x$ (ii) $18x$ (iii) x (iv) $2kx$ **3** $6x^2$ **4** (i) $9x^2$ (ii) $12x^2$ (iii) $\frac{3}{4}x^2$ (iv) $3kx^2$
5 (i) $8x^3$ (ii) $12x^3$ (iii) $4kx^3$ (iv) $10x^4$ (v) $15x^4$ (vi) $5kx^4$ (vii) $-\dfrac{2}{x^2}$ (viii) $-\dfrac{3}{x^2}$
(ix) $-\dfrac{k}{x^2}$ (x) knx^{n-1}

Exercise 2.1 (p. 16)

1 $5x^4$ **2** $9x^8$ **3** $99x^{98}$ **4** $56x^6$ **5** $\frac{8}{3}x^3$ **6** $3x^2 + 1$ **7** $6x^5 - 4x^3$ **8** $2x - 6$ **9** -4 **10** $6x - 7$
11 $24x^2 + 12x - 37$ **12** $-4x + 4x^3 - 30x^5$ **13** $2x + 2$ **14** $4x - 5$ **15** $-4x^3 + 6x^2 - 6x + 10$
16 $8x + 4$ **17** $3x^2 - 6x + 2$ **18** nkx^{n-1} **19** $(n+1)x^n$ **20** $2ax + b$

Exercise 2.2 (p. 18)

1 2 **2** 8 **3** 3 **4** $\frac{2}{3}$ **5** $\frac{1}{2}$ **6** 1 **7** 32 **8** 16 **9** 27 **10** $\frac{1}{4}$ **11** $\frac{64}{125}$ **12** $\frac{625}{16}$ **13** $\frac{1}{8}$
14 $\frac{1}{20}$ **15** $\frac{1}{3}$ **16** $\frac{49}{16}$ **17** 2 **18** $\frac{32}{243}$ **19** -8 **20** $\frac{1}{16}$ **21** $x^{\frac{1}{2}}$ **22** x^{-1} **23** x^{-4} **24** $x^{-\frac{2}{3}}$
25 $x^{-\frac{3}{4}}$ **26** $\sqrt{x}$ **27** $\sqrt[3]{x^2}$ **28** $\dfrac{1}{\sqrt[5]{x^4}}$ **29** $\dfrac{1}{\sqrt{x^3}}$ **30** $\dfrac{1}{\sqrt[4]{x^3}}$ **31** x^2 **32** x
33 $x + x^{\frac{1}{3}}$; $x + \sqrt[3]{x}$ **34** $x^{\frac{3}{2}} - 3x$; $\sqrt{x^3} - 3x$ **35** $x - x^{\frac{5}{3}}$; $x - \sqrt[3]{x^5}$

Exercise 2.3   (p. 20)

1 $\dfrac{1}{3\sqrt[3]{x^2}}$ **2** $\tfrac{2}{3}x^{-\frac{3}{3}}$ **3** $-\dfrac{1}{x^2}$ **4** $-\dfrac{3}{x^4}$ **5** $-\dfrac{1}{5\sqrt[5]{x^6}}$ **6** $-\tfrac{15}{2}x^{-\frac{7}{2}}$ **7** $\tfrac{8}{3}x^{\frac{1}{3}}-\tfrac{2}{3}x^{-\frac{1}{3}}$ **8** $15x^2-\dfrac{4}{x^2}+\dfrac{21}{x^4}$

9 $\dfrac{11}{x^2}-\dfrac{4}{x^3}$ **10** $\tfrac{1}{2}x^{-\frac{1}{2}}-\tfrac{1}{2}x^{-\frac{3}{2}}$ **11** $6x^2+6x$ **12** $-\dfrac{4}{x^3}-\dfrac{15}{x^4}$ **13** $-\dfrac{8}{3x^3}-\tfrac{4}{3}x$ **14** $-\dfrac{36}{5x^5}+\dfrac{2}{5x^3}$

15 $-\dfrac{1}{2\sqrt{x^3}}+\dfrac{3}{2\sqrt{x}}$ **16** $-\tfrac{8}{3}x^{-\frac{11}{3}}-\tfrac{10}{3}x^{-\frac{13}{3}}$ **17** $-\dfrac{6}{x^3}+\dfrac{21}{x^4}-\dfrac{20}{x^5}$ **18** $\tfrac{3}{2}\sqrt{x}+\dfrac{1}{\sqrt{x}}-\dfrac{1}{2\sqrt{x^3}}$

19 $\dfrac{1}{n\sqrt[n]{x^{n-1}}}$ **20** $-\dfrac{n}{x^{n+1}}$

Exercise 2.4   (p. 21)

1 (i) $3x^2$ (ii) $4-10x$ (iii) $-\dfrac{6}{x^3}$ (iv) nx^{n-1} **2** $2x$ **3** $8x^3$ **4** $2x-7$ **5** $6x^2-2x+5$

6 $\tfrac{3}{2}x^{-\frac{1}{2}}$ **7** $1-\dfrac{9}{x^2}$ **8** $\tfrac{3}{2}\sqrt{x}+\dfrac{2}{\sqrt{x^3}}$ **9** $4x+1$ **10** $\tfrac{7}{3}x^{\frac{4}{3}}+\tfrac{1}{3}x^{-\frac{2}{3}}$

Exercise 2.5   (p. 22)

1 $12,\ -6$ **2** $0,36$ **3** 9 **4** $\tfrac{1}{4},\tfrac{3}{2}$ **5** 17 **6** $5,\ -7$ **7** $2,\ -16$ **8** $\tfrac{1}{2},\ -\tfrac{1}{64}$

Exercise 2.7   (p. 31)

1 $\delta y = 4\delta x + 2(\delta x)^2$; $\dfrac{\delta y}{\delta x} = 4+2\delta x$; gradient is 4.

2 $\delta y = -\delta x + (\delta x)^2$; $\dfrac{\delta y}{\delta x} = -1+\delta x$;   gradient is -1.

3 $6x$ **4** $2x+3$ **5** $-10x$ **6** $8x-9$ **7** 3 **8** -5 **9** $3x^2$ **10** $4x^3$ **11** $1-\dfrac{1}{x^2}$ **12** $-\dfrac{1}{(x+1)^2}$

13 $-\dfrac{2}{(2x-3)^2}$ **14** $-\dfrac{1}{2\sqrt{x^3}}$

Exercise 2.8   (p. 32)

1 2   **2** 4   **3** 0   **4** $-1,0$ **5** 1   **6** 0, 1   **7** (i) 0   (ii) $1+\tfrac{1}{2}\delta x$; no.   **8** (i) 1   (ii) $1+\tfrac{1}{2}\delta x$; yes.

Exercise 3.1   (p. 35)

1 (i) $4x^3,\ 12x^2$ (ii) $21x^2,\ 42x$ (iii) $\dfrac{1}{2\sqrt{x}},\ -\dfrac{1}{4\sqrt{x^3}}$ (iv) $-\dfrac{1}{x^2},\ \dfrac{2}{x^3}$ (v) $-\dfrac{4}{x^5},\ \dfrac{20}{x^6}$ · (vi) 3, 0

(vii) $3x^2+3,\ 6x$ (viii) $2x-7,\ 2$ (ix) $1-\dfrac{1}{x^2},\ \dfrac{2}{x^3}$ (x) $4x-\dfrac{6}{x^3},\ 4+\dfrac{18}{x^4}$ **2** (i) 37, -41, 32

(ii) 2, $\tfrac{1}{12}$, $-\tfrac{1}{144}$ (iii) 1, -16, 128   **3** $4x+3,\ 4$ **7** -3 **8** $a=4,\ b=6$

Exercise 3.3   (p.41)

1 (1, 3) min   **2** $(-1, 3)$ max   **3** $(-3, 23)$ max; $(1, -9)$ min   **4** $(-2, -20)$ min; (4, 88) max
5 $(-2, 1)$ inflexion   **6** (0, 1) inflexion   **7** (0, 3) min   **8** (0, 2) max   **9** $(-2, 16)$ max; $(2, -16)$ min
10 (0, 3) inflexion (3, 30) max   **11** $(-3, -6)$ max; (3, 6) min   **12** $(\tfrac{1}{4}, -\tfrac{1}{4})$ min   **13** $(-3, -\tfrac{5}{27})$ min; (1, 1) max
14 72, -9 **15** 9, 0

Exercise 3.4 (p. 42)

1 $2 - 8q + 9q^2$, $-8 + 18q$ **2** $4t^3 - 2 - \dfrac{1}{t^2}$, $12t^2 + \dfrac{2}{t^3}$ **3** $8\pi r$, 8π **4** $4\pi r^2$, $8\pi r$ **5** (i) $2\pi r + \pi l$ (ii) πr

6 $u + gt$, g

Exercise 3.5 (p. 44)

1 $400\,\text{m}^2$ **2** $48\,\text{m} \times 24\,\text{m}$ **3** (i) 7 (ii) 12 **4** $3\,\text{m}^2$ **5** $4000\,\text{cm}^3$ **6** $8\,\text{cm} \times 8\,\text{cm} \times 8\,\text{cm}$ **7** $200\,\text{cm}^3$
8 $1000\pi\,\text{cm}^3$ **9** $r = 6\,\text{cm}$, $h = 12\,\text{cm}$ **10** $\frac{1}{3}\pi\,\text{m}^3$

Exercise 3.6 (p. 46)

1 $y = 6x - 9$ **2** $y = 8x + 16$ **3** $y = -3$ **4** $y = 18x - 12$ **5** $(2, 11)$ **6** $(4, 4)$; $y = 5x - 16$
7 $y = 3x - 8$, $y = 3x + 8$ **8** $C(-6, 4)$; $\frac{1}{2}$

Exercise 3.7 (p. 47)

1 (i) perpendicular; (ii) neither; (iii) parallel; (iv) perpendicular. **2** $7x - 4y + 34 = 0$
3 $x + 8y = 66$ **4** $16x - 4y = 27$ **5** $x = 2$ **6** $x + 3y = 6$; $(\frac{4}{3}, \frac{14}{9})$ **7** $P(\frac{1}{8}, 8)$
8 $(-1, 4)$, $(\frac{3}{2}, \frac{11}{4})$; normal at $(-1, 4)$. **9** $0, -18$ **10** $(2, \frac{3}{2})$

Revision paper A1 (p. 49)

1 (a) $f(1) = 4$ (b) $f(-2) = 14\frac{1}{2}$ (c) $f'(x) = 6x + \dfrac{1}{x^2}$ (d) $f'(2) = 12\frac{1}{4}$ (e) $f'(a) = 6a + \dfrac{1}{a^2}$

(f) $f''(2) = 5\frac{3}{4}$ (g) $f''(a) = 6 - \dfrac{2}{a^3}$ **2** $x = 0, y = 0$; $x = \frac{4}{3}, y = -\frac{32}{27}$ **3** $x = 4, y = 8$; $y - 8 = -\frac{1}{3}(x - 4)$

4 (i) -9 (ii) $\min x = 0, y = 0$; $\max x = 2, y = 4$ (iii) $x = 1, y = 2$ **5** (i) $200\,\text{m}$ (ii) $177.25\,\text{m}$

Revision paper A2 (p. 49)

1 (i) $f'(x) = 6x^2 - 4x + 1$; $f(1) = 0, f'(1) = 3$ (ii) $\dfrac{dy}{dx} = 3 + \dfrac{1}{x^2}$; 7; $x = -\frac{1}{2}$
2 $x = -3, y = 27$; $x = 1, y = -5$; $x = -4, y = 20$; $x = 0, y = 0$; $x = 4, y = 76$
3 $y - 5 = 2(x - 7)$; $(4\frac{1}{2}, 0)$; $y = -\frac{1}{2}(x - 4\frac{1}{2})$; $x = 2, y = \frac{5}{4}$
4 (i) $\min x = 2, y = -8$; $\max x = -4, y = 100$. (ii) $n = \frac{1}{3}$ or $\frac{2}{3}$ **5** $r = 4\,\text{cm}$

Exercise 4.1 (p. 52)

1 10.5 **2** $83.2\,\text{m}^2$

Exercise 4.2 (p. 53)

1 (i) $0.16, 0.2025$ (iii) $0.36, 0.3025$

Exercise 4.3 (p. 55)

3 (i) $16 \times \frac{1}{4}$ (ii) $81 \times \frac{1}{4}$ (iii) $b^4 \times \frac{1}{4}$ **4** (i) $32 \times \frac{1}{5}$ (ii) $243 \times \frac{1}{5}$ (iii) $b^5 \times \frac{1}{5}$ **5** b **6** $\frac{1}{2}b^2$

Exercise 4.4 (p. 56)

1 8 **2** $\frac{6}{5}$ **3** $29\frac{1}{4}$ **4** $86\frac{2}{3}$ **5** 34 **6** $\frac{176}{105}$

Exercise 5.1   (p. 57)

3 (i) $A = \frac{1}{4}x^4$ (ii) $A = \frac{1}{5}x^5$ (iii) $A = x$ (iv) $A = \frac{1}{2}x^2$ (v) $A = \dfrac{x^{n+1}}{n+1}$

Exercise 5.3   (p. 59)

1 False when $x = 1$; true otherwise.
2 False when $x = 1$ or $x = 3$; true otherwise.
3 True because $f(x)$ is continuous.

Exercise 5.4   (p. 63)

1 (i) 21   (ii) $6\frac{1}{3}$ (iii) 72   2 (i) $3\frac{3}{4}$ (ii) 20   (iii) $42\frac{1}{5}$ (iv) 20   (v) $7\frac{7}{8}$
3 (i) 18   (ii) 26; $y = 4x + 5$ 4 $19; f(x) = \frac{3}{2}\sqrt{x}$

Exercise 6.1   (p. 69)

1 (i) $\frac{1}{4}x^4 + C$ (ii) $7x + C$ (iii) $\frac{1}{2}x^4 + \frac{1}{3}x^3 - 3x^2 + 5x + C$ (iv) $\frac{4}{5}x^{\frac{5}{4}} + C$ (v) $-\dfrac{1}{x^2} + C$ 2 $\frac{1}{6}x^6 + C$

3 $\frac{1}{11}x^{11} + C$ 4 $\frac{2}{3}x^3 - 2x^2 + 3x + C$ 5 $5x + \frac{3}{2}x^2 - \frac{4}{3}x^3 + C$ 6 $\frac{1}{2}x^4 - \frac{7}{2}x^2 + C$ 7 $-\frac{2}{3}x^3 + \frac{9}{2}x^2 - 4x + C$
8 $\frac{4}{3}x^3 - 10x^2 + 25x + C$ 9 $\frac{1}{5}x^5 + \frac{4}{3}x^3 + 4x + C$ 10 $\frac{3}{4}x^{\frac{4}{3}} + C$ 11 $\frac{4}{5}x^{\frac{5}{2}} - 2x^{\frac{3}{2}} + C$ 12 $\frac{5}{18}x^{\frac{9}{5}} + C$

13 $\frac{2}{5}\sqrt{x^5} + 2\sqrt{x^3} + C$ 14 $-\dfrac{1}{3x^3} + C$ 15 $-\dfrac{3}{x} + \dfrac{5}{2x^2} + C$ 16 $2x + \frac{1}{x} + C$ 17 $\frac{1}{2}x^2 + \frac{1}{x} + C$

18 $-\dfrac{4}{x} - \dfrac{2}{x^2} - \dfrac{1}{3x^3} + C$ 19 $\frac{2}{3}\sqrt{x^3} - 6\sqrt{x} + C$ 20 $\frac{4}{5}\sqrt{x^5} + 2\sqrt{x^3} - 8\sqrt{x} + C$

Exercise 6.2   (p. 71)

1 $y = 2x^2 + C$ 2 $y = 4x + C$ 3 $y = x^3 - x + C$ 4 $y = 2x - 2x^2 + C$ 5 $y = -\frac{3}{x} + C$
6 $y = \frac{2}{3}x^3 + \frac{1}{x} + C$ 7 $y = 3x^2 + 4$ 8 $y = \frac{1}{4}x^4 + \frac{1}{2}x^2 + \frac{5}{4}$ 9 $y = x + \frac{1}{x} - \frac{11}{2}$ 10 $y = 4\sqrt{x^3} - 25$
11 $y = \frac{1}{2}x^2 - 2x + 3$ 12 $y = 2x - \frac{1}{x} + 2$ 13 $y = \frac{4}{9}\sqrt{x^3} + \frac{2}{3}\sqrt{x} - \frac{43}{18}$ 14 $x^2 + 2$ 15 $7x - 2x^2 + 38$

Exercise 6.3   (p. 74)

1 $\frac{26}{3}$ 2 12   3 $\frac{41}{6}$ 4 $\frac{3}{2}$ 5 $\frac{51}{4}$ 6 $\frac{43}{30}$ 7 52   8 $\frac{45}{4}$ 9 $-\frac{28}{9}$ 10 -12 11 $\frac{1}{6}$ 12 $-\frac{5}{72}$ 13 $-\frac{11}{24}$
14 $-\frac{5}{12}$ 15 $\frac{154}{5}$

Exercise 6.4   (p. 75)

3 Not valid.   4 Valid.   5 Not valid.   6 Valid.   7 Valid.   8 Not valid.

Exercise 7.1   (p. 78)

1 12   2 10   3 5   4 $\frac{3}{2}$ 5 $6\frac{2}{3}$ 6 $5\frac{1}{6}$ 7 $1\frac{1}{3}$ 8 (i) $1\frac{1}{3}$ (ii) $1\frac{1}{3}$ (iii) $1\frac{1}{3}$; 4; $1\frac{1}{3}$ 9 1; 0   10 $20\frac{5}{6}$
11 $3\frac{1}{12}$ 12 8

Exercise 7.2   (p. 80)

1 6   2 $11\frac{1}{3}$ 3 $18\frac{2}{3}$ 4 $10\frac{1}{2}$ 5 $4\frac{1}{2}$ 6 $20\frac{5}{6}$ 7 $21\frac{1}{3}$ 8 $21\frac{1}{3}$ 9 $40\frac{1}{2}$ 10 8

Exercise 7.3   (p. 81)

1 $8\frac{2}{3}$ 2 16   3 $\frac{2}{5}$ 4 $3\frac{1}{5}$ 5 $11\frac{1}{4}$ 6 $7\frac{1}{2}$ 7 60   8 $7\frac{1}{3}$ 9 2   10 $12\frac{2}{3}$

Exercise 7.4   (p. 83)

1 $5\frac{1}{3}$ **2** 2   **3** $-\frac{2}{3}$ **4** 1   **5** $5\frac{1}{3}$ **6** $2\frac{2}{5}$

Exercise 7.5   (p. 85)

1 Odd   **2** Even   **3** Even   **4** Neither   **5** Odd   **6** Neither   **7** Even   **8** Odd
9 0   **10** 0   **11** $7\frac{7}{15}$ **12** 30   **13** 30   **14** 30   **15** $-4\frac{2}{3}$ **16** 0

Exercise 7.6   (p. 88)

1 $\frac{242}{5}\pi$ **2** $\frac{3}{2}\pi$ **3** $\frac{128}{3}\pi$ **4** $\frac{81}{10}\pi$ **5** $\frac{56}{15}\pi$ **6** $\frac{64}{3}\pi$

7 $\frac{8}{81}\pi r^3$;   (i) $\frac{5}{24}\pi r^3$ (ii) $\frac{1}{3}\pi h^2 (3r - h)$ **8** $\frac{1}{3}\pi m^2 h^3$; $m = \dfrac{r}{h}$

Exercise 7.7   (p. 89)

1 $\frac{2414}{15}\pi$ **2** $\frac{151}{24}\pi$ **3** $\frac{5}{2}\pi$ **4** $\frac{64}{3}\pi$ **5** 4π **6** $\frac{7}{6}\pi$

Exercise 8.1   (p. 94)

1 0.3960, 0.3895, 0.3795, 0.3665, 0.3500, 0.3305, 0.3095, 0.2875, 0.2640
2 1.552, 0.876, 0.148, -0.590, -1.290, -1.912, -2.416, -2.766, -2.948
3 3.01   **4** 0.1   **5** 0.24

Exercise 8.2   (p. 99)

1 4.46, 5.24   **2** 8.38, 9.83   **3** 0.73, 0.83   **4** 1.22, 1.59   **5** 211, 251
6 35.45   **7** 1.87   **8** 6.35   **9** 6.63   **10** 0.472

Exercise 8.3   (p. 100)

1 8.18, 10.49   **2** 1615   **3** 3.15, 5.51

Exercise 8.4   (p. 102)

1 5.60   **2** 1.12   **3** 1.476   **4** 2.26

Revision paper B1   (p. 105)

1 (i) (a) $\dfrac{x^3}{3} + \dfrac{1}{x} + C$ (b) $2x^{\frac{1}{2}} + 2x^{-\frac{1}{2}} + C$ (ii) $\frac{56}{3}$ (iii) 0   **2** (i) $\frac{2}{3}$ (ii) $-2, \frac{5}{3}$; area $24\frac{35}{54}$ sq units.
3 (i) $y = 24x - 4x^3 - 8$; (2, 8); $(-2, -24)$ (ii) $-\frac{3}{2}$ **4** 512π, $496\frac{1}{8}\pi$
5 $\dfrac{100}{v}$ days, $C = \dfrac{15000}{v} + \dfrac{800}{3}v$, $v = 7\frac{1}{2}\,\text{km h}^{-1}$, £4000   **6** (i) 0.83, 0.73   (ii) 0.502

Revision paper B2   (p. 105)

1 (i) (a) $-\frac{9}{8}$ (b) $10\frac{1}{5}$ (ii) $\frac{25}{4}\pi$ **2** $(\frac{3}{2}, \frac{9}{4})$ (a) 36   (b) $\frac{7}{12}$
3 20 sq units; $135\frac{1}{2}\pi$ cu units.   **4** (b) $a = 21$, $b = -4$ Area 36 sq units.
5 Vol per min = $50 \displaystyle\int_0^{10} \sqrt{x - \frac{1}{10}x^3}\ dx$; 687.53 m³ per min [trap rule]   **6** 2.5463

Miscellaneous revision questions: paper B3   (p. 106)

1 (i) $y = \frac{2}{5}x^{\frac{5}{2}} - 2x^{\frac{1}{2}} + \frac{8}{5}$; $y = 10\frac{2}{5}$ when $x = 4$ (ii) $y = 3x - 5$; $(\frac{1}{3}, -1\frac{2}{9})$

2 $x = 0, \dfrac{dy}{dx} = -2$; $x = -3, \dfrac{dy}{dx} = 4$; $x = 1, \dfrac{dy}{dx} = -4$; $10\frac{2}{3}$ sq units.

3 (i) $4y + 3x = 24$ (ii) $y = x^2 + x$ 4 $k = 6$; $108\,\pi$ cu units.

5 (a) Max $(4, 128)$. Grad OP $= 32$; $x = \dfrac{4\sqrt{3}}{3}$ (b) 11 sq units.

6 Max $(-1, 27)$; min $(2, 0)$; min in range $(-3, -25)$; max in range $(4, 52)$.
7 $\frac{4}{3}$m^2 8 6.4 cu units.   9 $x = 10$, $h = 5$ 10 (i) $a = 2, b = -4, c = 4$; $y = 2x + 4$; 9   (ii) $\frac{1}{6}$
11 1.85   12 21.47, 22.80, 7.05

Exercise 9.2   (p. 112)

1 $2x + 2$ 2 $x(8 - 3x)$ 3 $3x^2 - 2x + 1$ 4 $2x(5x^3 + 14)$ 5 $4(2x + 1)$ 6 $\frac{5}{2}x^{\frac{3}{2}} + \frac{3}{2}x^{\frac{1}{2}}$
7 $\frac{10}{3}x^{\frac{7}{3}} - 2x$ 8 $6 - 2x^{-2} + 2x^{-3}$ 9 $1 - 4x$ 11 (i) $3x^2 - 6x + 2$ (ii) $-21x^6 + 6x^5 + 18x^2 - 4x$

12 v and $\dfrac{dv}{dx}$ do not exist when $x = 0$.

Exercise 9.3   (p. 114)

1 $\dfrac{6}{(5 - x)^2}$ 2 $\dfrac{-x^2 - 6x - 4}{(x^2 - 4)^2}$ 3 $\dfrac{-2}{(2x + 3)^2}$ 4 $\dfrac{-8x}{(1 + x^2)^2}$ 5 $\dfrac{24x}{(3x^2 + 2)^2}$ 6 $\dfrac{4x^2 - 18x - 1}{(x^2 + x - 2)^2}$

7 $\dfrac{2 + 9x^2 - 4x^3}{(x^3 + 1)^2}$ 8 $\dfrac{-x^3 + 3x - 4}{x^3(x - 2)^2}$ 9 $5 + \dfrac{6}{x^4}$

Exercise 9.4   (p. 116)

1 $y = u^5$ where $u = x - 3$ 2 $y = u^8$ where $u = 2x + 1$ 3 $y = u^3$ where $u = x^2 + 4x - 5$
4 $y = \sqrt{u}$ where $u = x^3 + 5$ 5 $y = u^4$ where $u = 1 + 2\sqrt{x}$ 6 $5(x - 3)^4$ 7 $16(2x + 1)^7$

8 $3(x^2 + 4x - 5)^2\,(2x + 4)$ 9 $\dfrac{3x^2}{2\sqrt{x^3 + 5}}$ 10 $\dfrac{4(1 + 2\sqrt{x})^3}{\sqrt{x}}$ 11 $-12(1 - 4x)^2$ 12 $9(2x^2 + x)^8\,(4x + 1)$

13 $\dfrac{1 - x}{\sqrt{3 + 2x - x^2}}$ 14 $-10(5x + 1)^{-3}$ 15 $\dfrac{4}{3x^2}\left(7 - \dfrac{1}{x}\right)^{\frac{1}{3}}$

16 $y = (x^2 + x + 1)^3$, $\dfrac{dy}{dx} = 3(x^2 + x + 1)^2\,(2x + 1)$

$y = x^6 + x^3 + 1$, $\dfrac{dy}{dx} = 6x^5 + 3x^2$

17 $8x + 12$ 19 (i) $y = (4 + \sqrt{1 - 2x^2})^3$, $\dfrac{dy}{dx} = \dfrac{-6x(4 + \sqrt{1 - 2x^2})^2}{\sqrt{1 - 2x^2}}$

(ii) $\dfrac{6(4x - 1)^2}{\sqrt{(4x - 1)^3 + 1}}$ 20 $\dfrac{dy}{du} = \frac{1}{3}u^{-\frac{2}{3}} = \frac{1}{3}x^{-2}$, which does not exist when $x = 0$.

Exercise 9.5   (p. 118)

1 $(4x + 9)^2\,(16x + 9)$ 2 $2x^3(5 - 2x)^4\,(10 - 9x)$ 3 $6(3x - 4)^4\,(2 - x)^2\,(7 - 4x)$
4 $(x^2 + 1)^3\,(1 - x)^6\,(-15x^2 + 8x - 7)$
5 $\dfrac{5x^2 - 2x}{\sqrt{2x - 1}}, x > \frac{1}{2}$ 6 $\dfrac{1 - 2x^2}{\sqrt{1 - x^2}}, -1 < x < 1$ 7 $\dfrac{4(4x - 1)^{\frac{2}{3}}\,(9x^2 - x + 10)}{3(x^2 + 1)^{\frac{1}{3}}}$

8 $\dfrac{-2x(x+2)}{(x-1)^4}$, $x \neq 1$ **9** $\dfrac{(2x+5)^3 (59+6x)}{(3-x)^8}$, $x \neq 3$ **10** $\dfrac{x(3x+8)}{2\sqrt{(2+x)^3}}$, $x > -2$

11 $\dfrac{3x+2}{(x+1)^2 \sqrt{3x^2-2}}$, $x < -\sqrt{\dfrac{2}{3}}$ or $x > \sqrt{\dfrac{2}{3}}$, $x \neq -1$ **12** $\dfrac{x(x-4)}{(x-1)^2 (x+2)^2}$, $x \neq 1, -2$

13 $\dfrac{2(2x-5)}{(x-2)^2 (x-3)^2}$, $x \neq 2, 3$ **14** $\dfrac{-3}{\sqrt{(3x+1)(3x-1)^3}}$, $x < -\frac{1}{3}$ or $x > \frac{1}{3}$

15 $\dfrac{-2(x^2+1)}{3x^{\frac{1}{3}}(x^2-1)^{\frac{5}{3}}}$, $x \neq 0, \pm 1$ **16** $\dfrac{12(3x+1)^4 (3+54x-5x^3)}{\sqrt{(9-x^2)^3}}$, $-3 < x < 3$

Exercise 9.6   (p. 123)

1 $\frac{5}{3}x^3 - \frac{1}{4}x^8 + 3x + C$ **2** $-\dfrac{1}{x^2} + \dfrac{1}{3x} + C$ **3** $\frac{2}{7}x^{\frac{7}{2}} - 2x^{\frac{5}{2}} + 4x^{\frac{3}{2}} + C$ **4** $\frac{1}{2}x^4 - x^3 + C$ **5** $\dfrac{x^3+8}{2x} + C$

6 $\dfrac{9-2x-3x^2}{6x^4} + C$ **7** $\frac{1}{24}(4x+3)^6 + C$ **8** $\frac{1}{2}(2x-1)^5 + C$ **9** $\frac{1}{8}(x-2)^8 + C$ **10** $-\frac{1}{42}(3-7x)^6 + C$

11 $-\frac{1}{15}(1-3x)^5 + C$ **12** $\frac{1}{3}\sqrt{(2x+5)^3} + C$ **13** $-\frac{1}{2}\sqrt{5-4x} + C$ **14** $\frac{4}{3}(x+3)^{\frac{3}{4}} + C$ **15** $\dfrac{-1}{6(3x-1)^2} + C$

16 $-(3+\frac{1}{3}x)^{-3} + C$ **17** $-6(1-\frac{1}{2}x)^{\frac{1}{3}} + C$ **18** $-\frac{5}{32}(3-4x)^{\frac{8}{5}} + C$ **19** $\frac{1}{2}$ **20** $\frac{1}{5}$ **21** 2   **22** $\frac{38}{3}$

23 $\frac{1}{36}$ **24** 3   **25** (i) $\frac{4}{3}x^3 + 2x^2 + x + C$ (ii) $\frac{1}{6}(2x+1)^3 + C$; 57   **26** Impossible   **27** $\frac{3}{7}$ **28** Impossible

29 Impossible   **30** Impossible   **31** $\frac{45}{4}$ **32** (ii) $\frac{1}{7}x^7 + x^4 + 4x + C$

Exercise 9.7   (p. 126)

1 All except (iv), (vi), (viii) are graphs of functions.   **2** (i) $\frac{1}{2}x$ (ii) $\frac{1}{4}(3-x)$ (iii) $x-3$ (iv) $x^{\frac{1}{3}}$

(v) $4+\dfrac{1}{x}$ (vi) $\dfrac{3}{8-x}$ (vii) $\dfrac{3x-1}{4x+2}$ (viii) $x^{-\frac{1}{3}} - 4$

3 (i) $4-x$ (ii) $\dfrac{1}{x}$ (iii) $\dfrac{1-x}{1+x}$ **4** $\dfrac{dy}{dx} = -y^2$; -4; $f^{-1}(x) = \dfrac{1}{x-1}$

5 (i) $\frac{1}{9}$ (ii) $\frac{1}{27}$ **6** (i) $\frac{1}{6}$ (ii) $\frac{1}{81}$ **7** $x \leqslant 1$; $-\frac{1}{6}$ **8** $0 \leqslant x \leqslant 4$;   (i) $\frac{1}{12}$ (ii) $\frac{1}{9}$

Exercise 10.2   (p. 136)

2 $6 \cos 6x$ **3** $-2 \sin 2x$ **4** $6 \cos 2x + 15 \sin 3x$ **5** $-6 \cos \frac{3}{2}x$ **6** $3 \sin (\frac{1}{4}\pi - 3x)$

7 $2x \cos (x^2)$ **8** $3 \sin^2 x \cos x$ **9** $-8 \cos x \sin x$ **10** $6 \sin 3x \cos 3x$ **11** $24 \cos^2 4x \sin 4x$

12 $2 \sec^2 2x$ **13** $6 \sec 2x \tan 2x$ **14** $-2 \operatorname{cosec}^2 (2x+1)$ **15** $-6 \sec^2 (\frac{1}{3}\pi + 3x)$

16 $-3 \cot^2 x \operatorname{cosec}^2 x$ **17** $8 \operatorname{cosec}^4 x \cot x$ **18** $6 \sec^2 3x \tan 3x$ **19** $\cos^2 x - \sin^2 x$

20 $\sec^3 x + \sec x \tan^2 x$ **21** $3 \cos 3x \cos 5x - 5 \sin 3x \sin 5x$ **22** $\cos x - x \sin x$ **23** $2x \cot x - x^2 \operatorname{cosec}^2 x$

24 $x^2 \sin x$ **25** $\dfrac{x \sec^2 x - \tan x}{x^2}$ **26** $\dfrac{-2 \sin 3x \sin 2x - 3 \cos 2x \cos 3x}{\sin^2 3x}$ **27** $\dfrac{\cos x}{2\sqrt{1+\sin x}}$

28 $\dfrac{-1}{1-\cos x}$ **29** $4x \sin (x^2) \cos (x^2)$ **30** $2 \sin x \cos x$, $-2 \cos x \sin x$, 0   **31** $-\dfrac{\pi}{180} \sin (x°)$

32 (i) $\dfrac{\pi}{180} \sec^2 (x°)$ (ii) $-\dfrac{\pi}{90} \sin (2x°)$ (iii) $-\dfrac{\pi}{180} \cos (60° - x°)$

Exercise 10.3   (p. 139)

1 $\frac{1}{3} \sin 3x + C$ **2** $-\frac{1}{2} \cos (2x-3) + C$ **3** $-3 \cos x + 2 \sin x + C$ **4** $-2 \sin (\frac{1}{4}\pi - \frac{1}{2}x) + C$

5 $\frac{1}{2} \sec 2x + C$ **6** $-3 \operatorname{cosec} \frac{1}{3}x + C$ **7** $\frac{3}{2} \tan (\frac{1}{3}\pi + \frac{2}{3}x) + C$ **8** $-\frac{1}{5} \cot 5x + C$ **9** $-\cot x - x + C$

10 $\frac{1}{2} \tan 2x - x + C$ **11** $\frac{1}{2}x - \frac{1}{12} \sin 6x + C$ **12** $\frac{1}{2}x + \frac{1}{8} \sin 4x + C$ **13** $-\frac{1}{14}\cos 7x - \frac{1}{2}\cos x + C$

14 $\frac{1}{16}\sin 8x + \frac{1}{8}\sin 4x + C$ **15** $\frac{1}{6}\sin 3x - \frac{1}{10}\sin 5x + C$ **16** $\dfrac{\sqrt{3}}{2}$ **17** 1   **18** 1   **19** 1   **20** 1   **21** 4

22 $1 - \frac{1}{4}\pi$ **23** π **24** $\frac{1}{8}\pi$ **25** $\frac{1}{2}\pi$ **26** $-\frac{1}{2}$ **27** $\dfrac{2\sqrt{2}}{5}$ **28** $\frac{1}{12}\cos 3x - \frac{3}{4}\cos x + C,\ \frac{2}{3}$ **29** 1

Exercise 10.4   (p. 141)

2 $y = A\sin 2x + B\cos 2x$ **3** $y = A\sin\frac{2}{3}x + B\cos\frac{2}{3}x$ **4** $y = A\sin(\sqrt{3}x) + B\cos(\sqrt{3}x)$
5 $y = A\sin\frac{1}{4}x + B\cos\frac{1}{4}x$ **6** $y = 4\sin x + \cos x$ **7** $y = -5\sin\frac{1}{2}x$
8 $y = \frac{2}{3}\sin 3x + 4\cos 3x$ **9** $y = -3\sin\frac{1}{3}x$ **10** $y = -6\sin\frac{5}{2}x - \frac{8}{5}\cos\frac{5}{2}x$

Exercise 10.5   (p. 146)

1 (i) $\frac{1}{6}\pi, \frac{1}{2}\pi, 0, -\frac{1}{3}\pi, -\frac{1}{2}\pi$ (ii) $\frac{1}{3}\pi, \frac{5}{6}\pi, \pi$ (iii) $\frac{1}{4}\pi, 0, -\frac{1}{3}\pi$ (iv) $\frac{1}{3}\pi, -\frac{1}{2}\pi, \frac{1}{6}\pi$

6 $\dfrac{2}{\sqrt{1-4x^2}}$ **7** $\dfrac{-4}{\sqrt{1-16x^2}}$ **8** $\dfrac{2}{4+x^2}$ **9** $\dfrac{2x}{\sqrt{1-x^4}}$ **10** $\dfrac{1}{2x^2+2x+1}$ **11** $\dfrac{4x^3}{1+x^8}$ **12** $\dfrac{2\sin^{-1}x}{\sqrt{1-x^2}}$

13 $\dfrac{-1}{\sqrt{8+2x-x^2}}$ **14** $\dfrac{-x^2}{\sqrt{1-x^2}} + 2x\cos^{-1}x$ **15** $\dfrac{1}{x\sqrt{1-x^2}} - \dfrac{\sin^{-1}x}{x^2}$ **16** $\dfrac{2\cos x}{\sqrt{1-4\sin^2 x}}$

17 $\dfrac{1}{2\sqrt{x}\,(1+x)}$ **18** $\dfrac{1}{\sqrt{x^2(4x^2-1)}}$ **19** $\dfrac{-1}{\sqrt{x^2(x^2-1)}}$ **20** $\dfrac{-1}{1+x^2}$

Exercise 10.6   (p. 148)

1 $\frac{1}{4}\pi$ **2** $\frac{1}{2}\pi$ **3** $\frac{1}{4}\pi$ **4** $\frac{1}{3}\pi$ **5** $\frac{1}{12}\pi$ **6** $\frac{1}{2}\sin^{-1}2x + C$ **7** $\frac{1}{4}\tan^{-1}4x + C$
8 $\frac{1}{6}\tan^{-1}(\frac{2}{3}x) + C$ **9** $\frac{1}{30}\tan^{-1}(\frac{6}{5}x) + C$ **10** $\frac{1}{3}\sin^{-1}(\frac{3}{2}x) + C$ **11** $\frac{1}{4}\sin^{-1}(\frac{4}{5}x) + C$

Exercise 11.1   (p. 154)

1 (i) 22, 14, 8 m s^{-1}; -7 m s^{-2} (ii) 14 m s^{-1}; 1.1 s; 27, 10 m s^{-1}
2 (i) 10 s   (ii) 25 m s^{-1} (iii) $v = 50 - 10t$, $a = -10$; when $t = 2$, $y = 80$, $v = 30$, $a = -10$, moving up, speed decreasing; when $t = 7$, $y = 105$, $v = -20$, $a = -10$, moving down, speed increasing.
3 (i) 8 m to right of O; speed 2 m s^{-1} towards right; speed decreasing at 10 m s^{-2}.   (ii) 4 m to left of O; speed 1 m s^{-1} towards left; speed decreasing at 8 m s^{-2}.
8 (i) $x = 10, 5, -5, -10, 0, 10$; $v = 0, -27.2, -27.2, 0, 31.4, 0$ (ii) $a = -\pi^2 x$ (iii) $10\pi, 10\pi^2$
9 16 cm, -24 cm s^{-1} **10** (a) 10 m, -5 m s^{-2} (b) (i) 3 m   (ii) $\frac{1}{2}$ m s^{-1} (iii) -0.175 m s^{-2}

Exercise 11.2   (p. 160)

1 $a = 12t - 16$, $x = 2t^3 - 8t^2 + 6t$; $t = 0, 1, 3$ **2** 6 m s^{-1}, $4\frac{2}{3}$ m   **3** $33\frac{1}{3}$ s, 1440 m   **4** 10 cm s^{-2}, 3
6 $t = 3$ (i) $6\frac{3}{4}$ m   (ii) $2\frac{1}{4}$ m s^{-1} (iii) 3 m s^{-2} **7** $x = \frac{8}{3}t - \frac{2}{9}\sin 3t$; $x = 26.9$ m, $v = 2.56$ m s^{-1}
8 (i) 180 m   (ii) 29.1 m s^{-1} (iii) $\frac{5}{3}$ m s^{-2} (iv) 1.7 m s^{-2} **9** 1 m s^{-1} **10** 20.2 m

Exercise 11.3   (p. 164)

1 100 m, 60 km h^{-1}, $\frac{2}{3}$ m s^{-2}; 5 m s^{-2}, 4 s   **2** (i) $15t$ (ii) (a) t^2 (b) $24t - 144$ (iii) 16   (iv) 240 m
3 60 s   **4** (i) $4\frac{1}{2}$ min   (ii) $\frac{5}{18}$ m s^{-2} **5** (i) 1, 2 s   (iii) (a) $3\frac{1}{3}$ m   (b) 6 m   (iv) $3\frac{1}{9}$ m s^{-1}
6 (a) 15 m s^{-1} (b) 300 m   (c) 20 s   150 m   (d) $11\frac{1}{4}$ m s^{-1}

Exercise 11.4   (p. 169)

1 (ii) 4, 2π (iii) $\dfrac{d^2x}{dt^2} = -x$ (iv) 4   (v) 4, 0   **2** (ii) 1, π (iii) $\dfrac{d^2x}{dt^2} = -4(x-3)$ (iv) 2   (v) 3, 2

3 (ii) $8, \frac{1}{2}$ (iii) $\dfrac{d^2x}{dt^2} = -16\pi^2 x$ (iv) 32π (v) $4\sqrt{3}, 16\pi$

4 (ii) $5, \frac{1}{4}\pi$ (iii) $\dfrac{d^2x}{dt^2} = -64x$ (iv) 40 (v) $-4, 24$

5 $34.6\,\text{cm}$ from O, $20.9\,\text{cm}\,\text{s}^{-1}$, $-38.0\,\text{cm}\,\text{s}^{-2}$ **6** $4.52\,\text{m}\,\text{s}^{-1}$, $341\,\text{m}\,\text{s}^{-2}$

7 $2.05\,\text{s}$ **8** $5\pi\,\text{s}$; $1.61\,\text{s}$, $30\,\text{cm}$ **9** $6\,\text{s}$ **10** $\frac{8}{\pi}\,\text{m}\,\text{s}^{-1}$ **11** $10\,\text{cm}, \frac{2}{3}\pi\,\text{s}, 0.057\,\text{s}$

12 $x = 24 \sin 4t + 7 \cos 4t$; $\frac{1}{2}\pi, 25, t = 0.32$

Exercise 12.1   (p. 172)

1 For A: (i) $50\pi\,\text{cm}^2$ per cm   (ii) $200\pi\,\text{km}^2$ per km.   For C: (i) $2\pi\,\text{cm}$ per cm   (ii) $2\pi\,\text{km}$ per km.

2 $\dfrac{dS}{dr} = 8\pi r$, $\dfrac{dv}{dr} = 4\pi r^2$; $48\pi\,\text{m}^2$ per m, $144\pi\,\text{m}^3$ per m   **3** (i) $200\pi\,\text{cm}^3$ per cm   (ii) $25\pi\,\text{cm}^3$ per cm

4 $-5\,\text{N}$ per m   **5** (i) 905 (ii) 0 (iii) $-640\,\text{m}$ per radian   **6** $-\frac{1}{4}\,\text{cm}$ per cm

7 (i) h (ii) $S = \pi h^2$ (iii) $\frac{1}{3}\pi h^3$. $\dfrac{dv}{dh} = \pi h^2 = S$ **8** $\delta V \approx S\,\delta h$; $21\pi\,\text{cm}^2$

Exercise 12.2   (p. 176)

1 a^2 **2** $\frac{32}{81}\pi R^3$ **3** $\theta = \frac{1}{3}\pi$; $130\,\text{m}^2$ **4** $\sqrt{27}\,\text{cm}$ **5** £4; £1000   **6** $r = 2, h = 2\sqrt{2}$ **7** $4.5\,\text{m}^3$

8 $40 \sin\theta$, $20(\cos\theta + \sin\theta)$; $\theta = \frac{3}{8}\pi$; $200(1 + \sqrt{2})\,\text{cm}^2$ **9** $100\pi\,\text{cm}^2$ **10** (i) $r = 16, h = 16$

(ii) $r = 12, h = 24$ (iii) $r = 24, h = 0$

Exercise 12.3   (p. 180)

1 (i) Falling at $0.52\,\text{m}$ per h. (ii) Rising at $0.74\,\text{m}$ per h.   **2** 0.8 radians per s

3 $30 - t$, $12 + 2t$, $A = (30 - t)(12 + 2t)$; $24\,\text{cm}^2$ per s; $648\,\text{cm}^2$

4 $6\pi\,\text{m}$ per s   **5** $\dfrac{1}{10\pi}\,\text{m}$ per s; $0.1\,\text{m}^3$ per s   **6** $(\frac{1}{3}\pi h^3)\,\text{m}^3$; $2.12\,\text{cm}$ per s   **7** $11.2\,\text{cm}$ per s

8 305.7; $2.05\,\text{N}\,\text{m}^{-2}$ per s   **9** $\dfrac{1}{8\pi}\,\text{cm}$ per s   **10** $31.2\,\text{cm}\,\text{s}^{-1}$ towards the mirror.

11 $\frac{1}{15}\pi$ radians per s;   (i) 0.21 (ii) 0.84 (iii) $6.95\,\text{km}\,\text{s}^{-1}$

12 $10 \sin\theta$, $10 \cos\theta$;   (i) $\frac{1}{4}$ radians per s   (ii) $1\frac{1}{2}\,\text{m}\,\text{s}^{-1}$

Exercise 12.4   (p. 183)

1 (i) $66\,\text{cm}$ (ii) $3\frac{1}{3}\,\text{h}$ (iii) $98\,\text{cm}, 4\frac{2}{3}\,\text{h}$ **2** 184 cents

3 (i) $t = \frac{1}{4}x^2 - 4$ (ii) $10\,\text{m}$ **4** (i) 9 days   (ii) 12 days   **5** $133\frac{1}{3}\,\text{s}$

Exercise 12.5   (p. 187)

1 $\dfrac{dm}{dt} = km$ **2** $\dfrac{dI}{dt} = -\frac{1}{80}I$ **3** $\dfrac{d\theta}{dt} = -\dfrac{4}{45}(\theta - 15)$ **4** $\dfrac{dx}{dt} = \dfrac{12}{x}$ **5** $\dfrac{dv}{dt} = -k\sqrt{v}$ **6** $\dfrac{dr}{dt} = \dfrac{k}{8\pi r^2}$

7 $\dfrac{dx}{dt} = -\dfrac{1}{35}(1 + 2x)$ **8** $\dfrac{dM}{dt} = 0.12\,M$ **9** $\dfrac{dm}{dt} \neq km(20 - m)$ **10** $\dfrac{dr}{dt} = -\dfrac{1}{200\pi}$, $1000\pi\,\text{s}$

Exercise 12.6   (p. 190)

1 $300\,\text{s}$ **2** $31.6\,\text{s}$ **3** $3980\,\text{s}$ **4** $150.8\,\text{s}$ **5** $15\,\text{s}, 7.5\,\text{m}\,\text{s}^{-1}$ **6** $7.60\,\text{cm}$ **7** $7\frac{5}{7}\,\text{h}, \frac{7}{60}\,\text{cm}$ per h

8 $4.8\,\text{m}\,\text{s}^{-1}, 0.096\,\text{m}\,\text{s}^{-2}$ **9** 10 months   **10** $.26\,\text{g}, 0.8\,\text{g}$ per day   **11** $3.6\,\text{m}, 1.09\,\text{m}$ per h

12 $37.5\,\text{h}$ **13** $6\,\text{h}$ **14** $x = 40 - 8t$ **15** $277\frac{1}{3}\,\text{s}, 0.25\,\text{cm}$ per s

Exercise 12.7   (p. 193)

1 (i) $80\pi\,\text{cm}^3$ (ii) $16\pi\,\text{cm}^2$ **2** $\dfrac{25}{256\pi}\,\text{cm}$, $12.5\,\text{cm}^2$ **3** 5.5 days   **4** 0.013 cm, 2.51 cm³

5 0.0076 radians   **6** 3.99 cm   **7** 3.032 m   **8** $S = \dfrac{Ah^2}{(h-x)^2}$, $\delta S \approx \dfrac{2Ah^2}{(h-x)^3}\delta x$

Exercise 12.8   (p. 196)

1 (i) 1% (ii) 2% **2** $1\tfrac{1}{2}\%$ **4** 0.1% **5** (i) 4% increase   (ii) 2% decrease   **6** $2\tfrac{1}{2}\%$

7 $\dfrac{2+0.2p}{2+0.1p}z\%$ (i) 4.8% (ii) 6% **8** (i) $q = 93 - 10p$, $\eta = \dfrac{10p}{93-10p}$, $£4.65$ (ii) 1.50 decreased.

Exercise 12.9   (p. 197)

1 $1.32\,\text{m}^2$ **2** 0.8% **3** $V = 523.6 \pm 12.6\,\text{cm}^3$; $S = 314.2 \pm 5.0\,\text{cm}^2$ **4** 0.53% **5** 2.2%
8 $A = \pi(2ax - x^2)$; $0.66\,\text{cm}^2$

Revision paper C1   (p. 200)

1 (i) (a) $\tfrac{1}{4}$ (b) $-\tfrac{1}{8}$ (c) $\tfrac{1}{49}$ (d) 2.34   (ii) $x = -2$, $x = -\tfrac{2}{5}$ **2** (i) $\tfrac{1}{2}$ (ii) (a) $4 + \tfrac{8}{3}\sqrt{2} \approx 7.77$ (b) $3\tfrac{3}{4}$
3 (i) $\tfrac{2}{3}$ (ii) $\dfrac{dy}{dx} = x\cos x + \sin x$, 0.4   **4** $x = 4$, 1.75 min late   **5** 300 s, 4600 m   **6** 14.4 s, $\tfrac{5}{36}\,\text{cm s}^{-1}$, $-\tfrac{50\pi}{3}$

Revision paper C2   (p. 201)

1 (i) (a) $2(x+2)^{-2}$ (b) $2x(x^3+1)(4x^3+1)$ (ii) $\dfrac{d^2y}{dx^2} = -2\cos 2x$
2 (i) (a) 1.22   (b) 0   (c) 4.5   (ii) 6.25π **3** $0.96\pi l (\approx 3.02 l)$, 3.36 cm   **4** $t = 24$
5 (i) $\dfrac{30-t}{30}$ (ii) $15\,\text{m s}^{-1}$ (iii) 600 m   (iv) $10\,\text{m s}^{-1}$ **6** $16\tfrac{1}{2}\,\text{s}$, $a/27$

Miscellaneous revision questions: paper C3   (p. 202)

1 (i) $-10x(1-x^2)^4$; $-(1-x^2)^4(11x^2+10x-1)$; $x = \pm 1$, $\tfrac{1}{11}$ (ii) $(4-2x)(x^2+4)^{-2}$
2 (i) (a) $4x(1+4x^2)^{-\frac{1}{2}}$ (b) $10\sin 5x\cos 5x = 5\sin 10x$ (ii) $\tfrac{1}{108}$
3 (i) $-1, 2$ $k = -13$, 19 or -8 **4** ± 2, $60°$, $180°$
5 $v = \dfrac{2000\pi}{9\sqrt{3}}$ $(\approx 403.1\,\text{cm}^3)$ **6** 0, $-\tfrac{5}{2}$, 0; 0, -1.5, 0; 2.78%, 50.5%
7 $4y - x = 8$, 72π **8** $\tfrac{1}{8}(x^2+1)^4$, $76\tfrac{4}{5}\pi$
9 (i) $40\pi\,\text{cm}^3\,\text{s}^{-1}$, $0.04\,\text{cm s}^{-1}$ (ii) (a) 2% decrease   (b) 2% decrease   (iii) $\approx 7\tfrac{1}{14}$
10 $t > \tfrac{11}{4}\,\text{s}$; $t = \tfrac{1}{2}$, 5   **11** $x = 12$, $\dfrac{dx}{dt} = -20$, $t = 0.29$ **12** $24.425\,\text{ms}^{-1}$

Exercise 13.1   (p. 205)

1 $(-\tfrac{1}{2}, 3)$ min   **2** $(-1, -\tfrac{1}{2})$ min; $(7, \tfrac{1}{14})$ max   **3** $(2, \tfrac{2}{9})$ max   **4** $(0, 1)$ max; $(2, 9)$ min
5 $(-3, \tfrac{1}{2})$ min; $(0, -1)$ max   **6** $(-2, 1)$ min; $(-\tfrac{1}{2}, -2)$ max   **7** $(-1, 2)$ min; $(1, 2)$ min
8 $(-2, 0)$ min   **9** $(0, 0)$ inflexion; $(1, -1)$ min   **10** $(\tfrac{1}{4}\pi, \sqrt{2})$ max; $(\tfrac{3}{4}\pi, -\sqrt{2})$ min
11 $(\tfrac{7}{6}\pi, -3\sqrt{3})$ min; $(\tfrac{11}{6}\pi, 3\sqrt{3})$ max   **12** $(\tfrac{1}{4}\pi, 2\sqrt{2})$ min; $(\tfrac{5}{4}\pi - 2\sqrt{2})$ max

Exercise 13.2 (p. 208)

1 $(0, 0)$; $(2, -2)$ **2** $\left(-\dfrac{1}{\sqrt{3}}, \dfrac{3}{4}\right)$; $\left(\dfrac{1}{\sqrt{3}}, \dfrac{3}{4}\right)$ **3** $(-1, 1)$; $(0, 1)$ **4** $(0, 2)$ **5** $(-1, 0)$; $(1, 6)$

6 $(-\frac{1}{2}, 0)$ **7** $(0, 0)$; $(\frac{1}{3}\pi, \frac{3}{2}\sqrt{3})$; $(\pi, 0)$; $(\frac{5}{3}\pi, -\frac{3}{2}\sqrt{3})$; $(2\pi, 0)$ **8** $(0, 0)$; $(\pi, 0)$; $(2\pi, 0)$

Exercise 13.3 (p. 217)

1 $x = 0$ **2** $x = -1$ **3** None   **4** $x = 1$, $x = 3$ **5** $x = 0$; $x < -1$ **6** $x = 1$; $x < 1$
7 $x = 0$, $x = \pi$, $x = 2\pi$, ...   **8** $x = 0$, $x = \frac{1}{2}\pi$, $x = \pi$, $x = \frac{3}{2}\pi$, $x = 2\pi$, ...   **9** $x = \frac{1}{3}\pi$, $x = \frac{5}{3}\pi$, ...
10 No asymptotes; $x < -\frac{1}{5}$, $x > \frac{1}{5}$ **11** $y \to 0$ **12** $y \to +\infty$ **13** $y \to -1$ **14** $y \to +\infty$
15 $y \to +\infty$ as $x \to +\infty$; $y \to -\infty$ as $x \to -\infty$ **16** $y \to 0$
17 $y \to -\infty$ as $x \to +\infty$; $y \to +\infty$ as $x \to -\infty$ **18** $y \to 2$ **19** $y \to 0$ as $x \to +\infty$ **20** $y \to 0$

Exercise 13.4 (p. 221)

1 $-\dfrac{x}{3y}$ **2** $\dfrac{8x+2}{3y^2-3}$ **3** $\dfrac{3x+2y}{5y-2x}$ **4** $\dfrac{1-3x^2y-y^3}{1+3xy^2+x^3}$ **5** $\dfrac{\cos x}{2\sin y}$

6 $\frac{3}{4}$ **7** $7x + 11y = 32$ **8** $16x - 10y = 33$ **9** $(-7, 7)$ **10** $2, -\frac{56}{5}$

11 $\dfrac{dy}{dx} = \dfrac{x-2}{4-4y}$; $(2, 2)$ is a stationary point; tangent at $(0, 1)$ is parallel to the y-axis.
12 $(-1, 3)$ min; $(0, 0)$ max   **13** Maximum   **14** $(-4, 2)$ min; $(4, -2)$ max

Exercise 13.5 (p. 229)

19 $y = 5x - 11$ **20** $(5\frac{1}{2}, -3)$ **21** (i) 72π (ii) 144π **22** $\frac{8}{3}\pi$ **23** $\frac{2}{3}\pi$ **24** $\frac{16}{3}$ **25** $\frac{64}{3}$

Exercise 13.6 (p. 231)

1 6   **2** $y \to +\infty$ **3** 1   **4** $y \to -\infty$ as $x \to 0$ from below; $y \to +\infty$ as $x \to 0$ from above.
5 0   **6** 0   **7** 0.33   **8** -1 **9** 1.39   **10** 1

Exercise 13.7 (p. 235)

1 2   **2** 19   **3** 0   **4** $y \to +\infty$ **5** 1   **6** 1
7 $y \to -\infty$ as $x \to 0$ from below; $y \to +\infty$ as $x \to 0$ from above.   **8** 6   **9** $\frac{1}{2}$ **10** $-\frac{2}{3}$ **11** $\frac{1}{2}$ **12** 0

Exercise 13.8 (p. 238)

1 $-3, 0$ **2** $\pm\frac{1}{2}\pi, \pm\frac{3}{2}\pi, \ldots$ **3** $-2, 2, 0, \pm\pi, \pm2\pi, \ldots$ **4** 3   **5** 4
6 $f(x)$ continuous for all x; $f'(x)$ discontinuous at $x = 0$.
7 $f(x)$ continuous for all x; $f'(x)$ discontinuous at $x = 3$.   **8** $a = -1, b = 2, c = 3, d = 3$

Exercise 14.1 (p. 242)

7 $x = \frac{1}{2}, y = -\frac{1}{2}$

Exercise 14.2 (p. 245)

1 $\dfrac{1}{t}, -\dfrac{1}{6t^3}$ **2** $-\dfrac{1}{t^2}, \dfrac{1}{t^3}$ **3** $-\frac{3}{4}\cot t, -\frac{3}{16}\csc^3 t$ **4** $\frac{1}{3}\csc t, -\frac{1}{9}\cot^3 t$ **5** $\frac{2}{3}, -\frac{2}{9}$ **6** 1

7 $(-3, -9), (-\frac{20}{9}, -\frac{40}{27}); (-4, -8)$ **8** $(\frac{9}{5}, \frac{32}{5}), (-\frac{9}{5}, -\frac{32}{5})$

9 $(8, 8), (8, -8)$ **10** $\dfrac{3}{2t^2}, -\dfrac{3}{2t^3}, \dfrac{2t^2}{3}, \dfrac{4t^3}{9}$ **11** $(0, 0)$ min; $(2, -4)$ max   **12** $-\tan t$

Exercise 14.3   (p. 247)

1 $2x + 3y = 6$ **2** $8x + 6y = 11$ **3** $y = \sqrt{2}x - 1$ **4** $2y = 3t^2x - 4t^3 + 3t^2$ **5** $y = (2\tan t)x - 3\sin t$
6 $x - ty + at^2 = 0, tx + y = at^3 + 2at$ **7** $x\cos t - y\sin t = a\cos 2t$
8 $y = 6x - 8, (1, -2); y = 6x + 16, (-1, 10)$ **9** $y = 3x - 8, (2, -2)$
10 $x + t^2y - 6t = 0; x + y = 6, (3, 3); x + 49y + 42 = 0, (-21, -\frac{3}{7})$
11 $y + tx = 2t^3 + 4t; y = 0, 2x + y = 24, 2x - y = 24$ **12** $(-5, -1)$

Exercise 14.4   (p. 249)

1 84   **2** $27\frac{1}{5}$ **3** 1   **4** $\pi - \dfrac{3\sqrt{3}}{4}$ **5** $3\pi a^2$ **6** $t = -2, 0; 9\frac{3}{5}$ **7** $\frac{8}{3}$ **8** $\frac{8}{15}$

Exercise 14.5   (p. 252)

1 $\sqrt{13}, \sqrt{37}; \left(\dfrac{9}{16}, \dfrac{11}{16}\right), \dfrac{5}{\sqrt{2}}$
2 Max 4 at $(0, 3), (10, -3)$; min 3 at $(4, 0), (-4, 0)$.
3 $2\pi a$ **4** 74   **5** 24   **6** 32.577   **7** 8π

Exercise 15.1   (p. 258)

1 $\frac{1}{12}(2x - 7)^6 + C$ **2** $-\frac{1}{40}(3 - 4x)^{10} + C$ **3** $-\frac{2}{5}\sqrt{1 - 5x} + C$ **4** $\frac{1}{5}(x + 2)^5 - \frac{1}{2}(x + 2)^4 + C$
5 $\frac{1}{14}(2x + 3)^{\frac{7}{2}} - \frac{3}{10}(2x + 3)^{\frac{5}{2}} + C$ **6** $-\dfrac{1}{4(2x - 1)} - \dfrac{1}{8(2x - 1)^2} + C$ **7** $\frac{1}{8}(2 + x^2)^4 + C$ **8** $-\dfrac{1}{6(1 + 2x^3)} + C$
9 $\sqrt{x^2 - 1} + C$ **10** $\frac{1}{3}\sin^3 x + C$ **11** $\frac{1}{3}\tan^3 x + C$ **12** $2\sqrt{1 + \sin x} + C$ **13** $\frac{1}{4}\sin^{-1} 4x + C$
14 $\frac{1}{15}\tan^{-1}\dfrac{5x}{3} + C$ **15** $\sin^{-1}\left(\dfrac{x - 1}{2}\right) + C$ **16** $\frac{2}{5}(x + 1)^{\frac{5}{2}} - \frac{2}{3}(x + 1)^{\frac{3}{2}} + C$ **17** $\sin x - \frac{1}{3}\sin^3 x + C$
18 $\frac{1}{5}\cos^5 x - \frac{1}{3}\cos^3 x + C$ **19** $\frac{1}{4}\sin(x^4) + C$ **20** $\tan x + \frac{1}{3}\tan^3 x + C$
21 $\frac{1}{5}\sec^5 x - \frac{1}{3}\sec^3 x + C$ **22** $\frac{1}{6}\tan^{-1}(2x^3) + C$ **23** $\sin^{-1}(\frac{1}{2}\sin x) + C$
24 $\sec^{-1} x + C$ **25** (i) $\frac{1}{2}\sin^2 x + C$ (ii) $-\frac{1}{2}\cos^2 x + C$ (iii) $-\frac{1}{4}\cos 2x + C$

Exercise 15.2   (p. 263)

1 $23\frac{13}{15}$ **2** $3\frac{1}{3}$ **3** 1081.1   **4** 0.305   **5** $\frac{2}{5}$ **6** $\frac{2}{7}$ **7** $\frac{1}{4}$ **8** $\frac{1}{6}\pi$

9 $\dfrac{1}{24}\pi$ **10** $\dfrac{\pi}{12} - \dfrac{\sqrt{3}}{8}$ **11** $\frac{1}{2}\pi$ **12** $\dfrac{4}{35}$ **13** $\frac{3}{2}(1 - 2^{-\frac{2}{3}})$ **14** $\frac{1}{6}\pi, -\frac{1}{12}\pi$

Exercise 15.3   (p. 265)

1 $\frac{1}{10}(x^2 - 1)^5 + C$ **2** $-\frac{1}{5}(1 - x^2)^{\frac{5}{2}} + C$ **3** $-\frac{1}{9}\sqrt{(1 - 2x^3)^3}$ **4** $-\dfrac{1}{2(1 + x^2)} + C$

5 $\frac{2}{3}\sqrt{x^3 - 2} + C$ **6** $-\dfrac{1}{12(x^6 - 4)^2} + C$ **7** $-\frac{1}{2}\sqrt{1 - x^4} + C$ **8** $\sqrt{x^2 - 2x + 4} + C$

9 $\frac{2}{5}(1 + 2\sqrt{x})^{\frac{5}{2}} + C$ **10** $\frac{1}{5}\sin^5 x + C$ **11** $-\frac{1}{6}\cos^6 x + C$ **12** $\frac{1}{2}\tan^2 x + C$
13 $\frac{2}{3}\sqrt{(1 - \cos x)^3} + C$ **14** $\frac{2}{3}\sqrt{\tan^3 x} + C$ **15** $2\sqrt{\sin x + \cos x} + C$ **16** $\frac{1}{2}\sin(x^2) + C$

17 $-\frac{1}{3}\cos(x^3+1)+C$ **18** $-2\cos\sqrt{x}+C$ **19** $\frac{1}{2}(\sin^{-1}x)^2+C$ **20** $\frac{2}{5}(\tan^{-1}x)^{\frac{5}{2}}+C$

21 $1\frac{7}{8}$ **22** $\frac{1}{6}$ **23** $\frac{2}{3}$ **24** $2\sqrt{2}-2$ **25** $\dfrac{\pi^3}{648}$

Exercise 15.4   (p. 268)

1 $-x\cos x+\sin x+C$ **2** $-\frac{1}{3}x\cos 3x+\frac{1}{9}\sin 3x+C$ **3** $\frac{1}{4}x\sin 4x+\frac{1}{16}\cos 4x+C$
4 $-3x\cos\frac{1}{3}x+9\sin\frac{1}{3}x+C$ **5** $\frac{1}{3}(3-2x)\sin 3x-\frac{2}{9}\cos 3x+C$ **6** $x^2\sin x+2x\cos x-2\sin x+C$
7 $\frac{1}{27}(2+18x-9x^2)\cos 3x+\frac{2}{9}(x-1)\sin 3x+C$ **8** $x\cos^{-1}x-\sqrt{1-x^2}+C$ **9** $x\sin^{-1}3x+\frac{1}{3}\sqrt{1-9x^2}+C$
10 $\frac{1}{4}x^2-\frac{1}{4}x\sin 2x-\frac{1}{8}\cos 2x+C$ **11** π **12** -8 **13** $\dfrac{\pi^2}{8}-\frac{1}{2}$ **14** $\dfrac{\pi}{24}+\dfrac{\sqrt{3}}{4}-\frac{1}{2}$ **15** $-3\pi^2+12$

Exercise 15.5   (p. 275)

1 $2\sqrt{x^3}+2\sqrt{x}+C$ **2** $-\frac{1}{5}\cos^5 x+C$ **3** $\sqrt{x^2-1}+C$ **4** $\dfrac{1}{9(5-3x)^3}+C$

5 $\dfrac{5}{8(5-2x)^2}-\dfrac{1}{4(5-2x)}+C$ **6** $-\frac{1}{3}\cos(x^3)+C$ **7** $\frac{1}{2}\sin^{-1}\left(\dfrac{2x}{5}\right)+C$

8 $-\frac{1}{3}x\cos 3x+\frac{1}{9}\sin 3x+C$ **9** $\frac{1}{5}\tan^5 x+C$ **10** $\dfrac{1}{2(3-2\sin x)}+C$ **11** $\frac{1}{2}x+\frac{1}{20}\sin 10x+C$

12 $\frac{1}{9}\cos^9 x-\frac{1}{7}\cos^7 x+C$ **13** $\frac{1}{3}\tan 3x-x+C$ **14** $-\frac{1}{14}\cos 7x-\frac{1}{6}\cos 3x+C$ **15** $-\frac{1}{2}\tan(\frac{1}{6}\pi-2x)+C$
16 $\frac{2}{9}\sqrt{(2+x^3)^3}$ **17** $\frac{1}{3}\sqrt{(4-x^2)^3}-4\sqrt{4-x^2}+C$ **18** $\frac{3}{8}x-\frac{1}{4}\sin 2x+\frac{1}{32}\sin 4x+C$

19 $\frac{1}{2}\sqrt{1+x^4}+C$ **20** $\frac{1}{2}\tan^{-1}(x^2)+C$ **21** $\cos\left(\dfrac{1}{x}\right)+C$ **22** $\frac{2}{5}\sqrt{(x+2)^5}-\frac{8}{3}\sqrt{(x+2)^3}+8\sqrt{x+2}+C$

23 $\frac{1}{2}(\tan^{-1}x)^2+C$ **24** $\frac{1}{28}\tan^{-1}\left(\dfrac{4x}{7}\right)+C$ **25** $2\sin^{-1}\sqrt{x}+C$

26 $\frac{1}{2}x^2\sin 2x+\frac{1}{2}x\cos 2x-\frac{1}{4}\sin 2x+C$ **27** $\frac{1}{6}x^3-\frac{1}{4}x^2\sin 2x-\frac{1}{4}x\cos 2x+\frac{1}{8}\sin 2x+C$
28 $-\frac{1}{4}x\cos 2x+\frac{1}{8}\sin 2x+C$ **29** $-\frac{1}{4}(1-2x^2)\sin^{-1}x+\frac{1}{4}x\sqrt{1-x^2}+C$ **30** $2\tan^{-1}\sqrt{x}+C$

31 -2 **32** 98 **33** $\frac{10}{3}$ **34** $\frac{1}{12}\pi$ **35** $\frac{4}{3}$ **36** $\dfrac{\pi}{4}-\frac{1}{2}$ **37** $\frac{2}{3}$ **38** $\dfrac{1}{\sqrt{2}}$ **39** $\frac{1}{2}\pi,\frac{3}{8}\pi^2$ **40** $\dfrac{5\pi^2}{12}-4$

41 $\dfrac{2}{\pi},\dfrac{1}{\sqrt{2}}$ **42** $\dfrac{100}{\sqrt{30}}$ **43** $y=2\sqrt{x^2+3x+6}-5$ **44** $y=3\tan^{-1}x-\frac{3}{4}\pi$ **45** $y=3\sin(x+\frac{1}{6}\pi)$

Exercise 16.1   (p. 280)

1 1.52   **2** $-0.53, 0.65, 2.88$ **3** 0.618   **4** $(-1,-2), (3,-34)$; 5.05   **5** 1.90   **6** 4.43
7 1.35, 5.96   **8** 3; -0.42036

Exercise 16.2   (p. 283)

1 0.405   **2** 1.025   **3.** 0.259   **4** 2.285   **5** 1.317   **6** 21.100
7 2.979   **8** 60   **9** $\frac{2}{3}h^5$ **10** (i) π (ii) $\pi;\pi$

Exercise 16.3   (p. 288)

1 $f'(x)$: 1.97985,   1.97315,   1.88780,   1.72725
  $f''(x)$: 0.331, -0.465, -1.242, -1.969
2 $-0.301, -0.239$ **3** 0.4315, 0.5489, 0.6850   **4** 0.26466, 0.52976, 0.79545   **5** 3.02, 3.060404, 3.122053
6 4.1071, 5.3415; 6.0357, 6.7902   **7** 0.3129, 0.6025, 0.8423; 1.205, 1.059, 0.795

Exercise 16.4   (p. 292)

1 $0.95x^2 + 0.685x + 1.069$; 1.786   **2** $\frac{3}{2}x^3 - \frac{9}{2}x^2 + 2x + 1$ **3** $0.4613x^3 - 1.5204x^2 + 2.0591x$

Revision paper D1   (p. 293)

1 $\dfrac{2x}{1+x^4}$ **2** $1 < x < 4$, 2, $m = 1$ **3** $x = -4$, $y = 1\frac{1}{3}$ max   **4** (b) $\dfrac{dy}{dx} = \dfrac{y - x^2}{y^2 - x}$ **5** (ii) $\dfrac{9\pi}{8}$ **6** 0.41, 0.41

Revision paper D2   (p. 294)

1 $-\dfrac{1}{\sqrt{1-x^2}}$, $\dfrac{\pi}{2} - 1$ **2** (i) $\frac{2}{3}k^2$ (ii) $\frac{2}{15}\pi k^3$ (iii) 0.2 %, 0.3 %

3 $\dfrac{2(1+x^2)}{(1-x^2)^2}$; $x < -1$; $0 \leqslant x < 1$ **4** (b) $\dfrac{\pi}{\sqrt{5}}$ (c) 1.06   **5** 3.2

Miscellaneous revision questions: paper D3   (p. 295)

1 (a) $\dfrac{2}{(x+1)^2}$ (b) $-3\sin(6x+8)$ (c) $(\sin x)^{\frac{1}{2}} + \dfrac{x}{2}\cos x\,(\sin x)^{-\frac{1}{2}}$

2 (i) $11\frac{1}{3}$ (ii) 6   **3** $\frac{4}{27}\pi x^3$ (i) $1\frac{25}{27}$ s   (ii) $\frac{9}{8}$ cm s^{-1} **4** 36   **5** $b = \frac{1}{8}$, volume $= 8.4\pi$

6 $\frac{3}{2}$, $\frac{11}{6}$; -3.73, -0.27 **7** (i) (0, 2), (0, -2), (1, 0)   (ii) $\dfrac{\pi}{4} - \frac{1}{2}$

9 $(\frac{1}{2}x^2 - \frac{1}{4})\sin 2x + \frac{1}{2}x\cos 2x + C$; π; $\dfrac{\pi^3}{6} - \dfrac{\pi}{4}$; $\dfrac{\pi^3}{6} - \dfrac{31}{64}\pi$ **10** (1, 1)   **11** 1.95   **12** 0.524

Exercise 17.1   (p. 301)

1 $3 - x + 2x^2$ **2** $2 + 5x - \frac{3}{2}x^2 + x^3$ **3** $-3 + 2x + 2x^2 - \frac{1}{3}x^4$
4 1, 2, 6, 24; $y = 1 + 2x$, $y = 1 + 2x + 3x^2$, $y = 1 + 2x + 3x^2 + 4x^3$ **5** $y = x - \frac{1}{6}x^3$ **6** $f'(0)$ does not exist.

Exercise 17.2   (p. 303)

1 $9 - 4x - x^2 + x^3$ **2** -1, 1, 6; $9 - 11x + 3x^2$ **3** $4 - 6x + 4x^2 - x^3$
4 $-1 + \frac{3}{2}x + \frac{3}{16}x^2$ **5** $(1 - \frac{1}{2}\pi + \frac{1}{8}\pi^2) - (\pi - 2)x + 2x^2$

Exercise 17.3   (p. 307)

1 $1 - \frac{1}{2}x^2 + \frac{1}{24}x^4 - \dfrac{1}{720}x^6$; 0.955 34   **2** $x + \frac{1}{3}x^3$; 0.020 0027   **3** $1 + \frac{3}{2}x + \frac{3}{8}x^2 - \frac{1}{16}x^3 + \dfrac{3}{128}x^4$

4 $1 - \frac{2}{3}x + \frac{5}{9}x^2 - \dfrac{40}{81}x^3$ **5** $2 + \frac{1}{4}x - \frac{1}{64}x^2 + \dfrac{1}{512}x^3$; 2.0248   **6** $\frac{1}{9} + \frac{4}{27}x + \frac{4}{27}x^2 + \dfrac{32}{243}x^3$

7 $x^2 - \frac{1}{6}x^4$ **8** $1 + x^2$ **9** $x^2 - \frac{1}{3}x^4$; 0.0612   **10** $2x - \frac{4}{3}x^3 + \frac{4}{15}x^5$

Exercise 17.4   (p. 309)

1 $\dfrac{\sqrt{3}}{2} + \frac{1}{2}h - \dfrac{\sqrt{3}}{4}h^2 - \frac{1}{12}h^3$; 0.8829   **2** $\dfrac{1}{\sqrt{2}} - \dfrac{1}{\sqrt{2}}h - \dfrac{1}{2\sqrt{2}}h^2 + \dfrac{1}{6\sqrt{2}}h^3$; 0.7193   **3** $1 + 2h + 2h^2 + \frac{8}{3}h^3$

4 $1 - \frac{1}{2}h^2 + \frac{1}{24}h^4 - \dfrac{1}{720}h^6$ ($= \cos h$)   **5** $2 + 2\sqrt{3}h + 7h^2$ **6** $\frac{1}{6}\pi + \dfrac{2}{\sqrt{3}}h + \dfrac{2}{3\sqrt{3}}h^2$; 0.4893

Exercise 17.5   (p. 313)

1 $2x - \frac{4}{3}x^3 + \frac{4}{15}x^5$ **2** $1 - \frac{9}{2}x^2 + \frac{27}{8}x^4$ **3** $1 - \frac{1}{2}x^4 + \frac{1}{24}x^8$ **4** $x^2 - \frac{1}{3}x^4 + \frac{2}{45}x^6$ **5** $3x - \frac{57}{2}x^3 + \frac{2801}{40}x^5$

6 $1 - \frac{13}{2}x^2 + \frac{313}{24}x^4$ **7** $1 + 3x + 6x^2$ **8** $1 - \frac{1}{2}x^2 + \frac{3}{8}x^4$ **9** $4 - x^2 - \frac{1}{16}x^4$ **10** $x - \frac{1}{2}x^2 + \frac{5}{24}x^3$

11 $1 + x^2 + \frac{7}{6}x^4$ **13** $1 + \frac{1}{2}x^2 + \frac{3}{8}x^4 + \frac{5}{16}x^6$; $x + \frac{1}{6}x^3 + \frac{3}{40}x^5 + \frac{5}{112}x^7$; $\pi \approx 3.1412$

14 $0.46365, 0.32175$; $\pi \approx 3.1416$ **15** $1 + \frac{1}{2}x + \frac{1}{8}x^2 + \frac{5}{48}x^3 + \frac{17}{384}x^4$; 1.1059

Exercise 17.6   (p. 315)

1 $C + x - \frac{1}{2}x^2 + \frac{1}{3}x^3 - \frac{1}{4}x^4 + \ldots$; 0.0953 **2** $C + x + \frac{1}{5}x^5 + \frac{1}{9}x^9 + \ldots$ **3** $C + x + \frac{1}{9}x^3 - \frac{1}{45}x^5 + \ldots$
4 $C + \frac{1}{3}x^3 - \frac{1}{42}x^7 + \frac{1}{1320}x^{11} - \ldots$; 0.041481 **5** $C + x + \frac{1}{6}x^3 + \frac{1}{24}x^5$; 0.2013

Exercise 17.7   (p. 318)

1 $1 + 2x + 2x^2 + \frac{4}{3}x^3 + \frac{2}{3}x^4 + \ldots$; 1.2214 **2** $3x - \frac{3}{4}x^2 + \frac{1}{8}x^3 - \frac{1}{64}x^4 + \ldots$; -0.631
3 $f(1 + h) = 2 + 3h + \frac{5}{2}h^2 + \frac{7}{6}h^3 + \frac{7}{24}h^4 + \ldots$; 2.926 **4** $f(1 + h) = 1 + h + \frac{3}{2}h^2 + \frac{3}{2}h^3 + \ldots$; 1.117
5 $1 - x + \frac{3}{2}x^2 - \frac{4}{3}x^3 + \ldots$; 0.954

Exercise 17.8   (p. 319)

1 18 **2** -1 **3** $\frac{1}{4}$ **4** $\frac{1}{64}$ **5** 2 **6** $\frac{1}{5}$ **7** $\frac{1}{2}$ **8** $-\frac{1}{2\pi}$ **9** 2 **10** $\frac{1}{6}$

Exercise 18.1   (p. 328)

1 $k \approx 1.099$ **2** $k = \lim\limits_{\delta x \to 0} \left(\dfrac{3^{\delta x} - 1}{\delta x}\right) = 1.099$ **3** (i) 0.993 (ii) 1.030; $e = 2.71 \ldots$

Exercise 18.3   (p. 331)

1 $-e^{-x}$ **2** $6e^{2x} - 20e^{-5x}$ **3** $2xe^{x^2}$ **4** $\dfrac{1}{2\sqrt{x}}e^{\sqrt{x}}$ **5** $(x + 1)e^x$ **6** $-e^x \sin(e^x)$ **7** $\cos x\, e^{\sin x}$

8 $\dfrac{(x - 2)e^x}{x^3}$ **9** $e^x(\cos x + \sin x)$ **10** $2\sec^2 2x\, e^{\tan 2x}$ **11** $\dfrac{e^x}{1 + e^{2x}}$ **12** $(3x^2 - 2x^3)e^{-2x}$ **13** $\dfrac{(2x - 5)e^x}{(2x - 3)^2}$

14 $e^{-3x}(-2\sin 2x - 3\cos 2x)$ **15** $e^x e^{e^x}$ **16** $\dfrac{-e^{-\frac{1}{2}x}}{2\sqrt{1 - e^{-x}}}$ **17** $\dfrac{-6e^{2x}}{(1 + e^{2x})^4}$ **18** $-\dfrac{1}{x^2}e^{\frac{1}{x}}$

19 $e^{-x}(2x\cos x - x^2\cos x - x^2\sin x)$ **20** $\dfrac{e^x(1 + x - x^2)}{\sqrt{(1 - x^2)^3}}$ **21** $(1, e^{-1})$ max   **22** $(-\frac{2}{3}, \frac{4}{9}e^{-2})$ max, $(0, 0)$ min

23 $(1, \frac{1}{2}e)$ inflexion   **24** $\left(\frac{1}{4}\pi, \frac{1}{\sqrt{2}}e^{-\frac{1}{4}\pi}\right)$ max, $\left(\frac{5}{4}\pi, -\frac{1}{\sqrt{2}}e^{-\frac{5}{4}\pi}\right)$ min   **25** $(-2, -e^{-4})$

26 $(-1, e^{-\frac{1}{2}}), (1, e^{-\frac{1}{2}})$ **27** $(0, 1), (\pi, -e^{-\pi}), (2\pi, e^{-2\pi})$ **28** $y = e^{-2}(3 - 2x)$; $(\frac{3}{2}, 0)$ **29** $-\frac{1}{2}e^{x-2y}$; $y = 2x$
30 $e^{-2t}, -2e^{-3t}$ **31** $y = (t - 1)x + e^{-t}$ **32** $-1.37, 1.92$

Exercise 18.4   (p. 334)

1 $\frac{1}{5}e^{5x} + C$ **2** $-\frac{1}{2}e^{-2x} + C$ **3** $6e^{\frac{1}{2}x} + C$ **4** $-\frac{1}{3}e^{7 - 3x} + C$ **5** $\frac{1}{2}e^{2x} - \frac{2}{3}e^{3x} + \frac{1}{4}e^{4x} + C$ **6** $\frac{1}{2}e^{x^2} + C$

7 $-\frac{1}{2}e^{-x^4} + C$ **8** $e^{\sin x} + C$ **9** $-\frac{1}{2}e^{\cos 2x} + C$ **10** $-e^{-2\sqrt{x}} + C$ **11** $\frac{1}{3}e^{3\tan x} + C$ **12** $-\dfrac{1}{2(4 + e^x)^2} + C$

13 $-2\sqrt{1-e^x}+C$ **14** $\sin^{-1}(e^x)+C$ **15** $\tan^{-1}(e^x)+C$ **16** xe^x-e^x+C **17** $-\frac{1}{2}xe^{-4x}+\frac{5}{8}e^{-4x}+C$
18 $\frac{1}{3}x^2e^{3x}-\frac{2}{9}xe^{3x}+\frac{2}{27}e^{3x}+C$ **19** $\frac{1}{2}e^{-x}(\sin x-\cos x)+C$ **20** $\frac{1}{25}e^{3x}(3\sin 4x-4\cos 4x)+C$
21 $\frac{1}{3}(1-e^{-6})\approx 0.3325$ **22** $\frac{1}{2}(1-e^{-1})\approx 0.316$ **23** $\frac{3}{4}e^4+\frac{1}{4}\approx 41.20$ **24** $\frac{1}{3}(e-e^{-1})\approx 0.783$
25 $\frac{2}{5}(e^{-\frac{1}{2}\pi}+1)\approx 0.483$ **26** $y=6-(2x+3)e^{-x}$ **27** $\frac{1}{4}(e^6-e^2)\approx 99.0$
28 $2(1-e^{-2})\approx 1.73$; $2\pi(1-e^{-4})\approx 6.168$ **29** $e-2\approx 0.718$ **30** 0.8862

Exercise 18.5   (p. 336)

1 $y=Ae^{2x}$ **2** $y=Ae^{-5x}$ **3** $y=Ae^{\frac{1}{3}x}$ **4** $y'=Ae^{-\frac{7}{4}x}$
5 $y=3e^{-x}$ **6** $y=-e^{\frac{1}{2}x}$ **7** $y=5e^{3(x-2)}$ **8** $y=-4e^{-\frac{2}{3}(x-5)}$

Exercise 18.6   (p. 339)

1 (i) 3.320 117   (ii) 0.818 731   **2** $1-2x+2x^2-\frac{4}{3}x^3+\frac{2}{3}x^4$ **3** $x^2-x^3+\frac{1}{2}x^4-\frac{1}{6}x^5$
4 $1+x-\frac{1}{3}x^3-\frac{1}{6}x^4$ **5** $x-\frac{1}{2}x^2-\frac{1}{6}x^3+\frac{5}{24}x^4$ **6** $x^2-2x^3+\frac{5}{3}x^4-\frac{2}{3}x^5$
7 $1, 0, 1, -2, 9, -44; 1+\frac{1}{2}x^2-\frac{1}{3}x^3+\frac{3}{8}x^4-\frac{11}{30}x^5$ **8** $1-\frac{1}{2}x^2+\frac{1}{6}x^4$ **9** $1+x+\frac{1}{2}x^2+\frac{1}{2}x^3$
10 $0, 1; x+2x^2+\frac{11}{6}x^3+x^4+\frac{41}{120}x^5+\frac{11}{180}x^6$ **11** $\frac{1}{4}\pi+\frac{1}{2}x-\frac{1}{12}x^3; 0.8353$ **12** $1-\frac{1}{4}x+\frac{3}{32}x^2-\frac{7}{384}x^3$

Exercise 18.7   (p. 341)

7 (i) $\to +\infty$ (ii) $\to 0$ (iii) $\to 1$ **8** (i) $\to +\infty$ (ii) $\to 0$ **11** 3   **12** 1   **13** $\frac{1}{2}$ **14** 2   **15** 1

Exercise 19.1   (p. 345)

1 (i) $\dfrac{1}{x}$ (ii) x^2 (iii) $-\cos x$ **3** $\ln(364.5)$ **4** $\ln[x^2(x+3)]$ **5** $\ln\left(\dfrac{(2x+1)^3\sqrt{x}}{(4-x)^2}\right)$

6 $\frac{1}{2}\ln(2+x)-\frac{1}{2}\ln(2-x)$ **7** $3x+2\ln(1+x)-\frac{3}{2}\ln(6x-5)$ **8** $y=e^{x^2-1}$ **9** $y=\frac{1}{2}\ln\left(\dfrac{x+1}{x-1}\right)$

10 $y=\dfrac{x+3}{x^2}$ **11** $y=-\ln(2x+3)-4x$ **12** $y=\dfrac{2(\sqrt{x}+1)}{\sqrt{x}-1}$

Exercise 19.2   (p. 348)

1 $\dfrac{3}{3x+2}$ **2** $\dfrac{-7}{4-7x}$ **3** $\dfrac{4x-1}{2x^2-x+7}$ **4** $\dfrac{3}{x}$ **5** $\dfrac{x}{x^2+1}$ **6** $\dfrac{-26}{(2-5x)(3x+4)}$ **7** $\dfrac{1-2x^3}{x(1+x^3)}$

8 $\dfrac{1}{2x\sqrt{\ln x}}$ **9** $\cot x$ **10** $\sec x$ **11** $1+\ln x$ **12** $\dfrac{x(2\ln x-1)}{(\ln x)^2}$ **13** $e^{-x}\left(\dfrac{1}{x}-\ln x\right)$ **14** $\dfrac{1}{x\ln x}$

15 $\dfrac{2}{(2x+3)\ln 10}$ **16** $\dfrac{2x^2}{(x^2+1)\ln 10}+\log_{10}(x^2+1)$ **17** $\dfrac{x^5}{\sqrt{3x+5}}\left(\dfrac{5}{x}-\dfrac{3}{2(3x+5)}\right)$

18 $\dfrac{x^3(2x-1)^5}{(x+1)^2}\left(\dfrac{3}{x}+\dfrac{10}{2x-1}-\dfrac{2}{x+1}\right)$ **19** $e^{5x}x^4\cos 3x\left(5+\dfrac{4}{x}-3\tan 3x\right)$ **20** $10^x\ln 10$ **21** $2^{x^2+1}x\ln 2$

22 $-x^{-x}(1+\ln x)$ **23** $x^{\frac{1}{x}-2}(1-\ln x)$ **24** $x^{\sin x}\left(\dfrac{\sin x}{x}-\cos x\ln x\right)$ **25** $(e^{-\frac{1}{2}}, -\frac{1}{2}e^{-1})$ min

26 $\left(e, \dfrac{1}{e}\right)$ max   **27** (e^3, e^3) min   **28** $(1, 1)$ min   **29** $(0.368, 0.692)$ min   **30** $(e, e^{1/e})$ max

31 $(e^{1.5}, 1.5e^{-1.5})$ **32** $\dfrac{y(2x^2-1)}{x(1+2y^2)}$ **33** $\dfrac{t-1}{t+1}, \dfrac{2(t-1)}{(t+1)^2}; y=-x$ **34** $-\ln 4-1$ **35** 1.37   **36** 18.9

Exercise 19.3   (p. 355)

1 $\frac{1}{3}\ln|x|+C$ **2** $\ln|x+5|+C$ **3** $\frac{1}{2}\ln|2x-3|+C$ **4** $-\frac{1}{4}\ln|7-4x|+C$ **5** $\ln|x^2-4|+C$
6 $\frac{1}{4}\ln|x^4-1|+C$ **7** $\ln|\sin x+\cos x|+C$ **8** $\ln|\sin x|+C$ **9** $\frac{1}{8}\ln(3+4e^{2x})+C$

10 $2\ln(1+x^2)-3\tan^{-1}x+C$ **11** $\frac{1}{8}\ln(1+4x^2)-\frac{1}{2}\tan^{-1}2x+C$ **12** $x-2\ln|x+2|+C$

13 $\ln|x-4|-\dfrac{4}{x-4}+C$ **14** $x-2\ln|x+1|+\dfrac{1}{x+1}+C$ **15** $\ln\left|\dfrac{x-1}{x}\right|+C$ **16** $\ln\left|\dfrac{x-5}{x-2}\right|+C$

17 $\frac{1}{7}\ln\left|\dfrac{x-2}{3x+1}\right|+C$ **18** $\frac{1}{2}\ln\left|\dfrac{(x-2)^5}{(x+2)^3}\right|+C$ **19** $2\ln(\sqrt{x}+1)+C$ **20** $-\cos(\ln x)+C$

21 $\frac{1}{3}x^3\ln x-\frac{1}{9}x^3+C$ **22** $-\dfrac{\ln x}{3x^3}-\dfrac{1}{9x^3}+C$ **23** $\frac{1}{3}(\ln x)^3+C$ **24** $x[(\ln x)^2-2\ln x+2]+C$

25 $\ln|\ln x+1|+C$ **26** $\frac{1}{2}(\ln x)^2+C$ **27** $\frac{1}{2}\ln\frac{3}{11}$ **28** $\ln 2+\frac{3}{4}\pi\approx 3.05$ **29** $4+\ln 5\approx 5.61$

30 $\ln\dfrac{32}{27}\approx 0.170$ **31** $\dfrac{e^2-3}{4e^2}\approx 0.148$ **32** $\ln|x-3|$ is not defined when $x=3$ **33** $\frac{1}{2}\ln 5\approx 0.805$

34 area $=6-6\ln 2\approx 1.84$; volume $=\pi(18-24\ln 2)\approx 4.29$ **35** $\frac{1}{4}\ln 5$ **36** 0.264

Exercise 19.4   (p. 358)

1 $-0.051\,293$ **2** 1.61 **3** $\ln 2$ **4** $\frac{2}{3}\ln 2$ **5** $-2x-2x^2-\frac{8}{3}x^3$; $-\frac{1}{2}\leqslant x<\frac{1}{2}$
6 $x^2-\frac{1}{2}x^4+\frac{1}{3}x^6$; $-1\leqslant x\leqslant 1$ **7** $5x-\frac{5}{2}x^2+\frac{35}{3}x^3$; $-\frac{1}{3}<x\leqslant\frac{1}{3}$ **8** $\frac{3}{2}x+\frac{1}{4}x^2+\frac{1}{2}x^3$; $-1<x<1$
9 $\ln 4-\frac{5}{4}x-\frac{25}{32}x^2$; $-\frac{4}{5}\leqslant x<\frac{4}{5}$ **10** $-x-\frac{3}{2}x^2-\frac{4}{3}x^3$; $-1\leqslant x<1$ **11** $-3x^2-\frac{9}{2}x^3-\frac{17}{4}x^4$; $-\frac{1}{3}\leqslant x<\frac{1}{3}$
12 $x-\frac{1}{2}x^2+\frac{5}{6}x^3$; $-1<x<1$ **13** $-\frac{1}{2}x^2-\frac{1}{12}x^4$ **14** $\ln 2+\frac{1}{2}x+\frac{1}{8}x^2-\frac{1}{192}x^4$
15 $x-\frac{1}{2}x^2+\frac{2}{3}x^3$ **16** $\frac{1}{2}x^2-\frac{1}{12}x^4$

Exercise 19.5   (p. 361)

1 (i) $+\infty$ (ii) 0 (iii) 0 (iv) $+\infty$ (v) 0 (vi) $+\infty$ (vii) 0 (viii) $+\infty$ (ix) $+\infty$ (x) $+\infty$
2 (i) $-\infty$ (ii) $-\infty$ (iii) $-\infty$ (iv) 0 (v) $-\infty$ (vi) 0 (vii) $+\infty$ (viii) 0 (ix) $+\infty$ (x) 0
7 $\ln y\to 0$, $+\infty$; $y\to 1$, $+\infty$ **8** $\ln y\to -\infty$, 0; $y\to 0$, 1 **9** 1 **10** $\frac{1}{2}$ **11** 29.96 **12** 0.449

Exercise 20.1   (p. 365)

6 $y=x^2-\frac{1}{3}x^3+C$ **7** $x=-\dfrac{1}{3y}+C$ **8** $y=Ae^{-5x}$ **9** $y=A\sin 4x+B\cos 4x$ **10** $x=\frac{1}{2}\ln\left|\dfrac{y}{2-y}\right|+C$

11 $y=\frac{1}{2}x+\frac{1}{8}\sin 4x+3$ **12** $x=5-e^{-y}$ **13** $y=-3\sin 2x+2\cos 2x$ **14** $x=3-\cot y$

15 $y=1-\cos x$ **16** $y=-e^{-x}$ **17** $x\dfrac{dy}{dx}=y$ **18** $x\dfrac{dy}{dx}+2y=0$ **19** $\dfrac{dy}{dx}+y^2=0$ **20** $x\dfrac{dy}{dx}=y+2x^3$

Exercise 20.2   (p. 372)

1 $y=Ae^{2x}$ **2** $y=Ae^{-\frac{5}{3}x}$ **3** $y=Ae^{-x}+2x-7$ **4** $y=Ae^{2x}-3$ **5** $y=Ae^{-4x}+\frac{1}{4}x^2+\frac{5}{8}x-\frac{21}{32}$
6 $y=Ae^{3x}-\frac{1}{3}x^2-\frac{2}{9}x-\frac{2}{27}$ **7** $y=Ae^{-\frac{1}{2}x}+x-3$ **8** $y=Ae^x+\sin x-\cos x$
9 $y=Ae^{-2x}+\frac{2}{13}\sin 3x-\frac{3}{13}\cos 3x$ **10** $y=Ae^{-3x}+\frac{1}{4}e^x$ **11** $y=Ae^{\frac{3}{4}x}-\frac{2}{15}e^{-3x}$ **12** $y=Ae^x+2xe^x$
13 $y=2e^{\frac{3}{8}x}$ **14** $y=0$ **15** $y=e^{2-x}$ **16** $y=\frac{1}{2}e^{-2x}+\frac{5}{2}$ **17** $y=9e^x-2x-9$
18 $y=\frac{15}{13}e^{-3x}+\frac{3}{13}\sin 2x-\frac{2}{13}\cos 2x$ **19** $y=-e^{-x}$ **20** $y=2e^{-4x}+2xe^{-4x}$

Exercise 20.3   (p. 375)

1 $y=\dfrac{3}{x}+C$ **2** $y=\dfrac{4}{C-x}$ **3** $2x^2-y^2=C$ **4** $y=Ax$ **5** $x^2+y^2+2x-2y+C=0$

6 $y=\dfrac{A}{1-x}-1$ **7** $e^y=Ax^2y$ **8** $y=Ae^{\frac{1}{2}x^2}-2$ **9** $y=\dfrac{3}{C-x^3}$ **10** $y^2=\dfrac{Ax^2}{x^2+1}-1$

11 $3y^2+2y^3=6\ln x+3x^2+C$ **12** $\sqrt{y^2-1}=\sqrt{x^2-1}+C$ **13** $y=\dfrac{1}{C+\cos x}$ **14** $\cos y=A\cos x$

15 $y=Ae^{2\tan x}$ **16** $y=e^{x^2-1}$ **17** $x^2+y^2=25$ **18** $y=-\ln(1-4\tan^{-1}x)$ **19** $y=x-1$ **20** $y=0$

Exercise 20.4 (p. 379)

1 $y = Ae^{x^2} - 1$ **2** $y = \dfrac{C}{x} - \frac{1}{2}x$ **3** $y = Ae^{-x} + xe^{-x}$ **4** $y = \frac{1}{3}x^4 + Cx$ **5** $y = \dfrac{\sin^{-1}x + C}{\sqrt{1-x^2}}$

6 $y = \sin x + A\cos x$ **7** $y = x\sec x + A\sec x$ **8** $y = \dfrac{3x + x^3 + C}{(1+x^2)^2}$ **9** $y = \sin^2 x + A\sin x$

10 $y = \frac{1}{4}x^4 e^x + Ae^x$ **11** $y = \frac{1}{5}x^3 + \dfrac{9}{5x^2}$ **12** $y = \dfrac{4x^3 + 3x^4}{12(1+x^2)}$ **13** $y = 3\sin x - \cos x$

14 $y = \dfrac{(x+1)(e^x - e)}{x}$ **15** $y = Ae^{2x} - \frac{3}{2}$

Exercise 20.5 (p. 383)

1 $\tan^{-1}\dfrac{y}{x} = \ln(A\sqrt{x^2 + y^2})$ **2** $\dfrac{x}{x-y} + \ln(x-y) = C$ **3** $(x+2y)^2(x-y) = A$ **4** $x^2 - y^2 = Ax$

5 $xy(y-x) = A$ **6** $3xy^2 = x + 2y$ **7** $(2x-y)^2(x+y) = 16$ **8** $y = e^x(x+C)$

9 $y - \ln(x+y+1) = \frac{1}{2}x^2 + C$ **10** $y = \frac{1}{2}x + \dfrac{C}{x}$

Exercise 20.6 (p. 391)

1 $y = Ae^x + Be^{3x}$ **2** $y = Ae^x + Be^{-4x}$ **3** $y = Ae^{-2x} + Be^{-4x}$ **4** $y = Ae^{3x} + Be^{-3x}$

5 $y = Ae^{2x} + Be^{-\frac{2}{3}x}$ **6** $y = Ae^{\left(\frac{\sqrt{13}-1}{2}\right)x} + Be^{-\left(\frac{\sqrt{13}+1}{2}\right)x}$ **7** $y = A\sin x + B\cos x$

8 $y = A\sin 5x + B\cos 5x$ **9** $y = e^{2x}(A\sin 4x + B\cos 4x)$ **10** $y = e^{-3x}(A\sin x + B\cos x)$

11 $y = e^{\frac{1}{2}x}(A\sin x + B\cos x)$ **12** $y = e^{-\frac{3}{2}x}\left(A\sin\dfrac{\sqrt{3}}{2}x + B\cos\dfrac{\sqrt{3}}{2}x\right)$ **13** $y = (Ax + B)e^x$

14 $y = (Ax + B)e^{-3x}$ **15** $y = Ae^{2x} + B$ **16** $y = \frac{2}{3}e^{2x} - \frac{2}{3}e^{-x}$ **17** $y = 4e^x - e^{5x}$

18 $y = \frac{1}{2}e^{4x} + \frac{1}{2}e^{-4x}$ **19** $y = -\frac{1}{2}\sin 2x + 2\cos 2x$ **20** $y = e^{-x}\sin 2x$

Exercise 20.7 (p. 399)

1 $y = Ae^x + Be^{4x} + 2$ **2** $y = Ae^{2x} + Be^{-2x} - \frac{1}{2}x - \frac{5}{2}$ **3** $y = Ae^{3x} + Be^{-x} - \frac{1}{3}x^2 - \frac{8}{9}x + \frac{55}{27}$

4 $y = Ae^{-x} + Be^{-2x} + \frac{2}{5}\sin x - \frac{6}{5}\cos x$ **5** $y = Ae^x + Be^{-x} + \frac{1}{4}e^{3x}$ **6** $y = Ae^{2x} + Be^{4x} - \frac{3}{2}xe^{2x}$

7 $y = Ae^{5x} + Be^{-x} - \frac{2}{3}xe^{-x}$ **8** $y = A\sin 2x + B\cos 2x + \frac{3}{4}$ **9** $y = A\sin x + B\cos x + 3x - 1$

10 $y = A\sin 3x + B\cos 3x + \frac{1}{8}\sin x$ **11** $y = e^{-x}(A\sin x + B\cos x) + \frac{3}{2}e^{-2x}$ **12** $y = (Ax + B)e^{2x} + \frac{1}{4}x + \frac{1}{4}$

13 $y = A\sin x + B\cos x + \frac{1}{2}x\sin x$ **14** $y = (Ax + B)e^{-x} + \frac{3}{2}x^2 e^{-x}$ **15** $y = Ae^{-x} + B + x^2 - x$

16 $y = Ae^{4x} + B - \frac{1}{20}\sin 2x + \frac{1}{10}\cos 2x$ **17** $y = Ae^x + B + 4xe^x$ **18** $y = e^{3x}(\frac{1}{2}x^2 + Ax + B)$

19 $y = \frac{3}{2}e^x - \frac{11}{18}e^{3x} + \frac{4}{3}x + \frac{10}{9}$ **20** $y = -\frac{4}{15}e^{3x} + \frac{4}{15}e^{-3x} - \frac{4}{5}\sin x$ **21** $y = -\frac{1}{18}e^{\frac{1}{2}x} + \frac{1}{54}e^{-\frac{3}{2}x} + \frac{1}{27}e^{2x}$

22 $y = -\frac{1}{4}e^{-x} + \frac{1}{4}e^{-3x} + \frac{1}{2}xe^{-x}$ **23** $y = -2\sin x + 3$ **24** $y = 3$ **25** $y = -\frac{5}{4}e^{-2x} + \frac{5}{4} - \frac{5}{2}xe^{-2x}$

Exercise 20.8 (p. 400)

1 $y = 4e^{-3x}$ **2** $y = \dfrac{-1}{4x+6}$ **3** $y = -\ln(1 - \tan x)$ **4** $y = (\frac{9}{2} - xe^{-2x} - \frac{1}{2}e^{-2x})^{\frac{1}{2}}$

5 $y = 3e^{x^2 + x} - 1$ **6** $y = \frac{1}{12}e^{-6x} + \frac{1}{2}x - \frac{1}{12}$ **7** $y^2 = 4x^2 + 5$ **8** $\sin y = A\sin x - 1$

9 $y = 2e^{-\frac{3}{2}x} - e^{-2x}$ **10** $y = -\ln(2-x)$ **11** $y = (2x-y)^3$ **12** $y = x^2 e^{x^2}$

13 $x^2 y = x^2 e^x - 2xe^x + 2e^x + e$ **14** $\ln(1 - \sin y) = 1 - \sin x$ **15** $y = 2x^2 - 1$ **16** $y = \frac{1}{4}e^{-2x} + \frac{1}{2}x - \frac{1}{4}$

17 $x + y = A(5x + y)^3$ **18** $y = \sin x(\sin x + 1)$ **19** $y = x^4 - 1$ **20** $(y+1)e^{-y} = (1-x)e^x + C$

21 $y = x^2 + \dfrac{Ax}{x-1}$ **22** $\ln y = \frac{1}{2}x^2\ln x - \frac{1}{4}x^2 + \frac{1}{4}$ **23** $y = x\ln[\ln(e^c x)]$ **24** $\sin y = \frac{1}{2}\ln x$

25 $y = x^2 e^{-2x} + Cx^{-3}e^{-2x}$ **26** $y = 2e^{-3x} - 2$ **27** $y = \frac{1}{4}\sin 2x - \frac{1}{4}\cos 2x + \frac{1}{4}$

28 $y = 8\sin\frac{1}{2}x + 8\cos\frac{1}{2}x - 8$ **29** $y = \frac{9}{2}e^{\frac{1}{3}x} - \frac{15}{2}e^{-\frac{1}{3}x} + 1$ **30** $y = e^{3x} - 2e^{2x} + x + \frac{5}{6}$
31 $y = 3e^{2x} + e^{-4x} + 2e^{3x}$ **32** $y = 2e^{-x}(1 - \sin 2x - \cos 2x)$ **33** $y = -\frac{1}{4}e^x + \frac{1}{4}e^{-x} + \frac{1}{2}xe^x$
34 $y = e^{2x}(2 - \sin x)$ **35** $y = 4e^x - 4e^{-\frac{5}{3}x} + 2x + \frac{4}{5}$ **36** $y = e^x(\frac{2}{5}\sin x - \frac{4}{5}\cos x) + \frac{2}{5}\sin x + \frac{4}{5}\cos x$
37 $y = 2 - 4\sin 3x$; -2; $\frac{2}{3}n\pi + \frac{1}{18}\pi$, $\frac{2}{3}n\pi + \frac{5}{18}\pi$ **38** $y = e^{2x}(x^2 + Ax + B)$ **39** $y = 4x + 2x^4 - 4x^3$
40 $\left(\dfrac{dy}{dx}\right)^2 = y\dfrac{d^2y}{dx^2}$; ± 6

Exercise 21.1   (p. 408)

1 $ct - \dfrac{c^2}{g}(1 - e^{-gt/c})$ **2** (i) 2.77   (ii) $v = \dfrac{1}{1 + 4e^{-\frac{1}{2}t}}$ **3** $mv\dfrac{dv}{dx} + P + Qv^2 = 0$; (b) $\lambda = \dfrac{m}{2Q}$, $\mu = \dfrac{Q}{P}$; 512 m

4 1.44   **6** $x = \dfrac{Q}{\lambda^2}(1 - \cos \lambda t)$; SHM with centre at $x = \dfrac{Q}{\lambda^2}$ and period $\dfrac{2\pi}{\lambda}$.   **8** 92.1 newtons

Exercise 21.2   (p. 412)

1 $\dfrac{dT}{dt} = -k(T - T_0)$ **2** $y = 27 + 36e^{-\frac{1}{6}t}$; 11 min; 27 degrees   **3** 4.74 min

5 (i) $i = \dfrac{F}{\omega^2 - p^2}(\cos pt - \cos \omega t)$ (ii) $i = \dfrac{F}{2\omega}t \sin \omega t$ **6** $x = ae^{-2t}$; $y = \frac{4}{3}a(e^{-\frac{1}{2}t} - e^{-2t})$; $\frac{4}{3}\ln 2$

Exercise 21.3   (p. 415)

1 0.2   **3** 57 000   **4** $\dfrac{dN}{dt} = \frac{1}{2}N - \frac{1}{4}N_0$, $N = \frac{1}{2}N_0(1 + e^{\frac{1}{2}t})$; $t = 2\ln 3$ **5** (a) 0.805 min   (b) 679

6 $\dfrac{dM}{dt} = k(f(t) - \lambda M)$; 21.2 days   **7** $\dfrac{\ln 3}{V\beta}$, $m = \dfrac{V}{\alpha(1 + 3e^{-V\beta t})}$ **8** $x = \left(p - \dfrac{m}{k}\right)e^{kt} + \dfrac{m}{k}$; 231

Exercise 21.4   (p. 421)

1 2.8, 3.84, 5.168, 6.8416, 8.9299   **2** 0, 0.008, 0.0394   **3** 0.2718, 0.6009; 0.7211
4 2.9284; 4.1314, 5.7009, 7.6917, 10.2176; true values   2.9284, 4.1509, 5.7327, 7.7532, 10.3097
5 -0.5212; -0.5343, -0.5383, -0.5323 **6** $y = x + \dfrac{x^3}{6} + \dfrac{x^5}{120}$; 0.100167, 0.2010; 0.2013
7 $(2 + 3h)y_3 = 4y_2 - (2 - 3h)y_1 + 2h^2(8x_2^2 + \sin y_2)$; 0.3015, 0.3058, 0.3145   **8** (i) 0.5   (ii) 0.5357

Revision questions: paper E1   (p. 422)

1 (i) (a) $2^x(\ln 2 \tan x + \sec^2 2x)$ (b) $\sec^2 2x(1 + 4x \tan 2x)$ (ii) 0, 1, 0; $\sec^2 x \approx 1 + \dfrac{x^2}{2}$; 0.258

2 $-1, \frac{1}{2}, -\frac{5}{2}$ or 0   **3** $x = 0, y = 1$ **4** (i) (a) 1   (b) $\frac{1}{6}\tan^{-1}\left(\dfrac{3x - 2}{2}\right) + C$ (ii) e^2

5 1.5571   **6** 0.019   **7** $n = \dfrac{300}{2e^{-t} + 1}$, 1.4 days

Revision questions: paper E2   (p. 423)

1 $a = \pm 1$ **2** Max $\frac{1}{2}$, min $-\frac{1}{2}$, inflexions 0, $\pm\dfrac{\sqrt{3}}{2}$

4 (a) (i) $\dfrac{x^7}{7} + \dfrac{x^4}{4} + x + C$ (ii) $-\frac{1}{3}(1 - x^2)^{\frac{3}{2}} + C$ (iii) $\dfrac{x^2}{2}\ln x - \dfrac{x^2}{4} + C$ (b) $\ln 4$
5 5.357, 3.961, 2.996, 2.280, 1.702   **6** 0.1826, error $\approx 2.5 \times 10^{-4}$ **7** $V = gT\exp(x/gT^2)$

Miscellaneous revision questions: paper E3 (p. 424)

1 $\frac{1}{2} - \frac{5}{16}(x-1) + \frac{43}{256}(x-1)^2$, 0.488   2 (a) (i) $\sec^2 x\, e^{\tan x}$ (ii) $\frac{1}{15}$ (b) max 0, 2π min $\frac{4}{3}\pi$

3 $\pi[4h + 4e^h + \frac{1}{2}e^{2h} - \frac{9}{2}]$; 0.38 units s^{-1} 4 $0 \leqslant \theta \leqslant 0.615^C \left(= \tan^{-1}\dfrac{1}{\sqrt{2}} \right)$

5 (i) (a) $\dfrac{\frac{1}{2}\sec^2 x}{\sqrt{\tan x}}$ (b) $-x(1 + 2\ln x)$ (ii) $l = 100\,m$, $r = \dfrac{100}{\pi}m$ 6 $-1 < a < 1$ 7 $\pi a^3(\ln 2 - \frac{2}{3})$

9 $a = 4$, $n = 3$ (a) 4.826   (b) 4.8224   10 max -4 (at $x = 2$), min 0 (at $x = 4$);
$\ln 5 \approx 1.6222$ 11 (i) 2.178   (ii) 2.18

12 $x = Ae^{-3t} + Be^{2t} - \frac{1}{6}e^{-t}$; $x = \frac{19}{2}e^{-3t} - \frac{1}{6}e^{-t}$, $y = \frac{19}{12}e^{-3t} - \frac{7}{12}e^{-t}$

14 $a < 0$, $\pi - 2a$; $0 \leqslant a \leqslant \pi$, $\pi - 2\sin a$; $a > \pi$, $2a - \pi$ 15 $\dfrac{dv}{dt} > 0$, $v \to \sqrt{\dfrac{2k}{ma} + u^2}$

Exercise 22.1 (p. 435)

1 $\sqrt{2x+1} + C$ 2 $\frac{1}{2}x - \frac{1}{4}\ln|2x+1| + C$ 3 $\frac{1}{4}\ln(2x^2+1) + C$ 4 $\frac{2}{9}\sqrt{(3x+2)^3} - \frac{2}{9}\sqrt{3x+2} + C$

5 $\frac{1}{3}x^3 - 4x - \dfrac{4}{x} + C$ 6 $\frac{1}{6}\tan^{-1}\left(\dfrac{2x}{3}\right) + C$ 7 $\frac{1}{2}\ln\left|\dfrac{x}{x+2}\right| + C$ 8 $\ln|x+2| + \dfrac{2}{x+2} + C$

9 $\frac{1}{12}(1+x^2)^6 + C$ 10 $\ln\left|\dfrac{2x+1}{x+1}\right| + C$ 11 $\frac{1}{3}\tan^{-1}\left(\dfrac{2x-1}{3}\right) + C$ 12 $\frac{1}{2}\ln\left|\dfrac{x^2-1}{x^2}\right| + C$

13 $x + 4\ln\left|\dfrac{x+2}{x+1}\right| + C$ 14 $\ln\left|\dfrac{x+2}{2x-1}\right| - \dfrac{5}{2x-1} + C$ 15 $\frac{1}{2}\ln\left(\dfrac{1+x^2}{(1-x)^2}\right) + \tan^{-1}x + C$

16 $\frac{4}{15}$ 17 $\ln\frac{32}{27}$ 18 $\frac{1}{12}\pi + \frac{1}{2}\ln 2$ 19 $-\frac{1}{2}\ln 18$ 20 $\frac{1}{2}\ln 3 + \frac{2}{3} - \dfrac{\pi}{6\sqrt{3}}$

Exercise 22.2 (p. 439)

1 $\frac{1}{2}x - \frac{1}{12}\sin 6x + C$ 2 $-\frac{1}{10}\cos 5x + \frac{1}{2}\cos x + C$ 3 $\ln|\sin x| + C$ 4 $-\cos x + \frac{1}{3}\cos^3 x + C$

5 $\ln(1 + \sin x) + C$ 6 $x \sin x + \cos x + C$ 7 $-\frac{1}{2}xe^{-2x} - \frac{1}{4}e^{-2x} + C$ 8 $\frac{1}{8}x - \frac{1}{32}\sin 4x + C$

9 $-x^2\cos x + 2x\sin x + 2\cos x + C$ 10 $-\dfrac{\ln x}{2x^2} - \dfrac{1}{4x^2} + C$ 11 $\frac{1}{2}x^2\ln 2x - \frac{1}{4}x^2 + C$

12 $\frac{1}{3}\sin^3 x - \frac{1}{5}\sin^5 x + C$ 13 $2\tan x - 2\sec x - x + C$ 14 $\ln|1 + \tan\frac{1}{2}x| + C$ 15 $x \tan x + \ln|\cos x| + C$

16 $\dfrac{2}{e}$ 17 $\dfrac{\sqrt{3}}{2}$ 18 $\frac{1}{4}\pi - \frac{1}{2}\ln 2$ 19 $-\frac{8}{9}$ 20 $\frac{1}{4}\ln 2 + \frac{1}{8}\pi$

Exercise 22.3 (p. 442)

1 $\dfrac{1}{\sqrt{3}}\sin^{-1}\left(\dfrac{\sqrt{3}(x-1)}{2}\right) + C$ 2 $\sin^{-1}\left(\dfrac{x-1}{3}\right) + C$ 3 $\sin^{-1}(x-3) - \sqrt{-x^2 + 6x - 8} + C$

4 $-x\sqrt{1-x^2} + C$ 5 $\frac{3}{8}\sin^{-1}x - \frac{1}{2}x\sqrt{1-x^2} + \frac{1}{8}x(1 - 2x^2)\sqrt{1-x^2} + C$

6 $\frac{1}{8}\sin^{-1}x - \frac{1}{8}x(1 - 2x^2)\sqrt{1-x^2} + C$ 7 $\sqrt{x^2+4} - 2\ln(x + \sqrt{x^2+4}) + C$ 8 $\dfrac{1}{\sqrt{2}}$ 9 $\frac{1}{3}\sqrt{2}$ 10 $\dfrac{\pi}{\sqrt{3}}$

Exercise 22.4 (p. 442)

1 $-\sqrt{25 - x^2} + C$ 2 $\frac{1}{2}\sin x - \frac{1}{6}\sin 3x + C$ 3 $-\sqrt{1-2x} + C$ 4 $\ln(e^x + 1) + C$

5 $8 \ln|3x - 8| + 3x + C$ **6** $\frac{1}{5}\cos^5 x - \frac{1}{3}\cos^3 x + C$ **7** $-\frac{1}{2}x\cos 2x + \frac{1}{4}\sin 2x + C$ **8** $\frac{1}{3}\ln\left|\dfrac{(1+3x)^2}{(1-x)^3}\right| + C$

9 $\dfrac{1}{\sqrt{3}}\sin^{-1}\left(\dfrac{\sqrt{3}x}{2}\right) + C$ **10** $\tan\frac{1}{2}x + C$ **11** $x(\ln x)^2 - 2x\ln x + 2x + C$ **12** $-\frac{1}{10}\cos 5x - \frac{1}{2}\cos x + C$

13 $\frac{1}{2}x^2 - 2x + \ln(x^2 + 2x + 2) + 4\tan^{-1}(x+1) + C$ **14** $x\ln(x + \sqrt{x}) - x + \sqrt{x} - \ln(1 + \sqrt{x}) + C$

15 $\frac{1}{2}\ln\left(\dfrac{(x-1)^4}{x^2+1}\right) + C$ **16** $\frac{1}{2}xe^{2x} - \frac{1}{4}e^{2x} + C$ **17** $\ln\left|\dfrac{x+2}{x+3}\right| + C$ **18** $\frac{1}{2}e^{x^2} + C$

19 $\frac{1}{3}\tan^{-1}\left(\dfrac{x-1}{3}\right) + C$ **20** $-\dfrac{1}{2(1+e^{2x})} + C$ **21** $\frac{1}{4}\ln\left|\dfrac{2+e^x}{2-e^x}\right| + C$ **22** $x + \ln\left|\dfrac{(x-1)^2}{x-2}\right| + C$

23 $\frac{1}{3}x^3\ln x - \frac{1}{9}x^3 + C$ **24** $\frac{1}{2}\sec x\tan x + \frac{1}{2}\ln|\sec x + \tan x| + C$ **25** $\frac{1}{3}\ln 4$ **26** $\dfrac{\pi^2}{72}$ **27** $\frac{1}{3}$

28 $\frac{8}{15} - \frac{43}{120}\sqrt{2}$ **29** $\frac{2}{3}$ **30** $\frac{1}{4}\pi$ **31** $4\sqrt{3} - \frac{1}{3}\pi - 4$ **32** (a) $-\pi$ (b) 4π (c) 0   (d) 0

Exercise 22.5   (p. 447)

1 $2\frac{5}{6}$ **2** $11\frac{2}{3}$ **7** $\int_0^4 (2x+1)\,dx = \int_0^4 (9 - 2x)\,dx = 20$

Exercise 22.6   (p. 449)

1 0   **2** 0   **3** 66   **4** 2   **5** $\frac{1}{3}\pi$ **6** 0, 0   **7** $\frac{1}{4}$ **8** $\frac{1}{7}\pi$ **9** $\frac{1}{4}\pi$

Exercise 22.7   (p. 453)

1 $\frac{1}{2}$ **2** $\frac{1}{6}\pi$ **3** $\frac{1}{4}\pi$ **4** $\frac{1}{16}$ **5** $\frac{1}{4}$ **6** $\frac{1}{2}$ **7** 1   **8** 6   **9** 3   **10** 1   **11** π

12 $A = \ln a$, $V = \pi\left(1 - \dfrac{1}{a}\right)$; $V \to \pi$

Exercise 22.8   (p. 454)

1 $\dfrac{28}{15}$ **2** $\dfrac{24}{85}(e^\pi - 1), \dfrac{3}{13}(e^\pi + 1)$ **3** $\frac{1}{8}(e^2 + 3)$ **4** $-\frac{1}{3}(4 - x^2)^{\frac{3}{2}} + C$; 2π **5** $I_n = nI_{n-1} - \dfrac{1}{e}$; $24 - \dfrac{65}{e}$

6 $\dfrac{3}{32}\pi + \frac{1}{4}$ **7** $\frac{13}{15} - \frac{1}{4}\pi$

Exercise 22.9   (p. 458)

1 (i) $\dfrac{1}{1+x}$ (ii) $\sec x$ (iii) $\dfrac{x^3}{2+x^4}$ (iv) $\dfrac{2(1-2x)}{1+2x}$ (v) $3x^2\tan(x^3)$ (vi) $x\sin x\cos x$

(vii) $\dfrac{2\cos x}{\sqrt{1-x^2}}$ (viii) 0   **3** $\dfrac{2}{1-x^6}$ **4** $\sqrt{1+x^3}$, $F(x) + x\sqrt{1+x^3}$ **5** $f(x)$, $G(x) + xf(x)$ **6** $2x\sin^3(x^2)$

Exercise 22.10   (p. 461)

4 2   **5** $\frac{1}{5}$ **6** $\frac{1}{2}$ **7** $2\ln 2 - 1$ **8** $\frac{1}{6}\pi$

Exercise 23.1   (p. 462)

1 Odd, even, odd, odd.
2 (i) $\cosh 2x + \sinh 2x$ (ii) $(2\cosh 3x - 4\cosh 5x) + (2\sinh 3x + 4\sinh 5x)$
(iii) $\frac{1}{2}(\cosh 4x - \cosh 2x) + \frac{1}{2}(\sinh 4x - \sinh 2x)$

Exercise 23.2 (p. 466)

2 $\cosh 2\theta = \cosh^2\theta + \sinh^2\theta$ **3** $\cosh 2\theta = 2\cosh^2\theta - 1$ **4** $\cosh 2\theta = 1 + 2\sinh^2\theta$

5 $\cosh(A-B) = \cosh A \cosh B - \sinh A \sinh B$ **6** $\sinh(A+B) = \sinh A \cosh B + \cosh A \sinh B$

7 $\tanh(A+B) = \dfrac{\tanh A + \tanh B}{1 + \tanh A \tanh B}$ **8** $2\sinh A \cosh B = \sinh(A+B) + \sinh(A-B)$

9 $\sinh 3\theta = 3\sinh\theta + 4\sinh^3\theta$ **10** $\cosh C - \cosh D = 2\sinh\left(\dfrac{C+D}{2}\right)\sinh\left(\dfrac{C-D}{2}\right)$

12 $x = 0$ or $x = \ln 2$ **13** $\ln\frac{5}{3}$ **14** $-\frac{1}{2}\ln 3$ **15** $x = \ln 3,\ y = \ln 2$

Exercise 23.3 (p. 469)

1 (i) $4\sinh 4x$ (ii) $\frac{1}{2}\operatorname{sech}^2\frac{1}{2}x$ (iii) $-4\operatorname{sech} 2x \tanh 2x$ (iv) $4\cosh 2x \sinh 2x$ (v) $\tanh x$
(vi) $6\tanh x \operatorname{sech}^2 x$ (vii) $-4\operatorname{cosech}^2 2x \coth 2x$ (viii) $-3\operatorname{cosech}^2 3x$ (ix) $-\operatorname{cosech} x$
(x) $\cosh^2\frac{1}{3}x \sinh\frac{1}{3}x$ (xi) $3x^2\cosh(x^3)$ (xii) $2\operatorname{sech} 2x$
2 (i) $\frac{1}{2}\cosh 2x + C$ (ii) $2\sinh\frac{1}{2}x + C$ (iii) $\ln\cosh x + C$ (iv) $-\frac{1}{3}\coth 3x + C$ (v) $\frac{1}{2}x + \frac{1}{4}\sinh 2x + C$
(vi) $\frac{1}{12}\sinh 6x - \frac{1}{2}x + C$ (vii) $2\ln|\sinh\frac{1}{2}x| + C$ (viii) $2\tan^{-1}(e^x) + C$ (ix) $x - \tanh x + C$
(x) $\frac{1}{12}\cosh 6x - \frac{1}{4}\cosh 2x + C$ (xi) $\frac{1}{3}\cosh^3 x - \cosh x + C$ (xii) $\frac{1}{5}\cosh^5 x - \frac{1}{3}\cosh^3 x + C$
3 (a) (i) $x\sinh x - \cosh x + C$ (ii) $\frac{1}{3}x\cosh 3x - \frac{1}{9}\sinh 3x + C$ (iii) $(x^2+2)\cosh x - 2x\sinh x + C$
(b) $\frac{1}{4}e^{2x} - \frac{1}{2}x + C$ **4** 4   **5** $y - a\sinh\theta = \coth\theta\,(x - a\cosh\theta)$
6 $1.221,\ -1.221$ **7** $\sec^{-1}(\cosh y) + C,\ 0.436$ **10** $y = A\cosh 2x + B\sinh 2x - \frac{1}{3}\sinh x$

Exercise 23.4 (p. 474)

1 $-1 < x < 1$ **2** $\frac{1}{2}\ln\left(\dfrac{1+x}{1-x}\right)$

3 (i) $\ln(2+\sqrt{3}) \approx 1.316\,96$ (ii) $\ln(\frac{1}{2}\sqrt{5} - \frac{1}{2}) \approx -0.481\,212$ (iii) $\frac{1}{2}\ln 19 \approx 1.472\,22$

4 (i) $\dfrac{2}{\sqrt{4x^2-1}}$ (ii) $\dfrac{4}{\sqrt{16x^2+25}}$ (ii) $\dfrac{3}{1-9x^2}$ (iv) $\dfrac{1}{\sqrt{x^2-2x+2}}$ (v) $\dfrac{4x}{\sqrt{4x^4-1}}$ (vi) $\dfrac{-1}{2x^2+6x+4}$

5 $\cosh^{-1}\dfrac{x}{2} + C$ **6** $\sinh^{-1}\dfrac{x}{2} + C$ **7** $\frac{1}{3}\cosh^{-1}\left(\dfrac{3x}{2}\right) + C$ **8** $\frac{1}{2}\sinh^{-1}\left(\dfrac{2x}{\sqrt{3}}\right) + C$

9 $\dfrac{1}{\sqrt{2}}\sinh^{-1}\left(\dfrac{2x-1}{3}\right) + C$ **10** $\cosh^{-1}\left(\dfrac{x+4}{3}\right) + C$ **11** $\frac{1}{2}x\sqrt{x^2-25} - \dfrac{25}{2}\cosh^{-1}\dfrac{x}{5} + C$

12 $\frac{9}{4}\sinh^{-1}\left(\dfrac{2x}{3}\right) + \frac{1}{2}x\sqrt{4x^2+9} + C$ **13** $x\sinh^{-1}x - \sqrt{x^2+1} + C$ **14** $\cosh^{-1}4 - \cosh^{-1}2$

15 $\frac{1}{2}\sinh^{-1}2 + \sqrt{5}$ **16** $2\sinh^{-1}1$ **17** (i) $2\cosh^{-1}2 - \sqrt{3}$ (ii) $\frac{1}{3}\pi - \frac{1}{2}\cosh^{-1}2$
18 (i) $\cosh^{-1}2$ (iii) $\cosh^{-1}3 - \cosh^{-1}2$

Exercise 23.5 (p. 476)

2 $1 + 2x^2 + \frac{2}{3}x^4$ **3** $\frac{1}{2}x^2 + \frac{1}{48}x^6 + \frac{1}{3840}x^{10}$ **4** $x - \frac{1}{3}x^3 + \frac{2}{15}x^5$ **5** $1 - \frac{1}{2}x^2 + \frac{5}{24}x^4$
6 $x - \frac{1}{6}x^3 + \frac{3}{40}x^5 - \frac{3}{112}x^7$ **7** $x + \frac{1}{3}x^3 + \frac{1}{5}x^5 + \frac{1}{7}x^7$ **8** (i) $\cos x$ (ii) $j\sin x$ (iii) $j\tanh x$

Exercise 24.1 (p. 479)

3 $8a$ **5** $y = ae^{-s/a}$ **6** (b) $\frac{1}{2}a(\sinh^{-1}T + T\sqrt{1+T^2})$ **7** $(0,0)$

Exercise 24.2 (p. 481)

1 $\frac{1}{3}a^2\pi^3$ **2** $2ab$

Exercise 24.3 (p. 485)

1 $\dfrac{125\sqrt{3}c}{144}$ **2** $\dfrac{5\sqrt{5}a}{4}$; $(3a, \frac{9}{4}a)$ **3** $\rho = 4.762$, $s = 4.7$ **6** $\left(-\frac{1}{2}\ln 2, \dfrac{1}{\sqrt{2}}\right)$, $k = 0.385$, $(-1.846, 2.828)$

Exercise 24.4 (p. 487)

1 $\frac{8}{3}\pi a^2[(1+\alpha)^{\frac{3}{2}}-1]$ **3** $15 + 4\ln 2$; 384π **4** (a) $\frac{3}{2}a$ (b) $\frac{6}{5}\pi a^2$ **5** $\frac{64}{3}\pi a^2$

Exercise 24.5 (p. 490)

1 $(\frac{3}{2}, \frac{6}{5})$ **2** 0.067; $(2.48, 2.48)$ **3** $2\frac{1}{16}$ **4** $\frac{58}{35}a$ **5** $\frac{4}{3}a$; $(\frac{2}{5}a, \frac{11}{24}a)$
6 $\frac{11}{24}\pi a^3$, $\frac{5}{24}\pi a^3$; $\frac{21}{88}a$, $\frac{27}{40}a$ from base.   **9** $2\pi^2 ab^2$ **10** (a) $\frac{8}{15}$ (b) $(\frac{5}{16}, \frac{5}{84})$ (i) $\frac{4}{63}\pi$ (ii) $\frac{1}{3}\pi$

Exercise 24.6 (p. 493)

1 $\frac{9}{8}\pi$ **2** $\frac{1}{8}\pi a^2$ **3** 11π **4** $\dfrac{a^2\pi^3}{6}$ **5** $2 - \frac{1}{2}\pi$ **7** $\left(\dfrac{1}{\sqrt{2}}, \pm\dfrac{2\pi}{3}\right)$; $\sqrt{3} - \frac{1}{3}\pi$ **8** $\frac{3}{2}\pi a^2$

Exercise 25.1 (p. 498)

4 $2x$, $2y$ **5** $2x + 3y$, $-2y + 3x$ **6** $\dfrac{1}{y}$, $-\dfrac{x}{y^2}$ **7** $3x^2 - 3y$, $3y^2 - 3x + 2$

8 $-e^x \sin(x - 3y) + e^x \cos(x - 3y)$, $3e^x \sin(x - 3y)$ **9** $\dfrac{-y}{x^2 + y^2}$, $\dfrac{x}{x^2 + y^2}$ **10** $\dfrac{4(x+y)}{\cos z}$, $\dfrac{4(x-y)}{\cos z}$

Exercise 25.2 (p. 499)

4 108 **5** $(0, 0, 0)$ saddle, $(\frac{1}{6}, -\frac{1}{12}, -\frac{13}{432})$ min   **6** $(0, 0, 0)$ max, $(\frac{1}{3}, \frac{1}{3}, -\frac{1}{27})$ saddle
7 $(2, 1, 2)$ min, $(2, -1, 6)$ saddle, $(-2, 1, 34)$ saddle, $(-2, -1, 38)$ max
8 $(0, 0, 0)$ saddle, $(-\frac{1}{12}, \frac{1}{6}, -\frac{13}{432})$ min
9 Stationary points $\left(0, \pm\dfrac{2}{\sqrt{3}}, 0\right)$, $(+2, 0, +4)$, $(-2, 0, -4)$ $\left(0, \dfrac{2}{\sqrt{3}}, 0\right)$ saddle, $\left(0, -\dfrac{2}{\sqrt{3}}, 0\right)$ saddle

10 $\left(\dfrac{a}{\sqrt{3}}, 0, \dfrac{-2a^3}{3\sqrt{3}b^2}\right)$ min, $\left(-\dfrac{a}{\sqrt{3}}, 0, \dfrac{2a^3}{3\sqrt{3}b^2}\right)$ max

Exercise 25.3 (p. 502)

2 $\delta = \frac{1}{2}(b\gamma + c\beta)\sin A$; $a = 36.5 \pm 0.45$ mm, $\Delta = 436.5 \pm 14.3$ mm^2

Revision paper F1 (p. 503)

1 (a) $\tan\alpha = \frac{2}{3}$, $\dfrac{d^n y}{dx^n} = 13^{n/2}\sin(2x + n\alpha)$ (b) $\dfrac{\pi}{3\sqrt{3}}$, $\dfrac{\pi}{2}$, $\frac{1}{3}(8\ln 8 - 7)$

2 $A = 0.524$, $B = 0.470$, $c = 0.405$

3 (i) (a) $\tanh x$ (b) $2x\cosh(x^2)$ (c) $\dfrac{2}{\sqrt{1-4x^2}}$ (ii) (a) $\sinh^{-1}\frac{1}{2}$ (b) $\dfrac{\pi}{6}$

4 8π **5** (i) $a = 0, b = -\dfrac{1}{2n}, y = \dfrac{1}{n^2}\sin nx - \dfrac{1}{2n}x\cos nx$ (ii) $y = A - \dfrac{1}{x}$

7 $(t^2 + 1)(t^6 - 4t^5 + 5t^4 - 4t^2 + 4) - 4$

Revision paper F2   (p. 504)

1 (a)$I_p = \dfrac{1}{p+1} - \dfrac{2}{p+2} - \dfrac{1}{p+3}; \dfrac{1}{120}$ (b) $\dfrac{\sqrt{3}}{2} - 1 + \dfrac{\pi}{6}\ (= 0.195)$

2 (b) $\ln\dfrac{8}{3\sqrt{3}} = 0.432$ **3** $x - \frac{1}{6}x^3, 1$ **4** Max at $(0, -\frac{1}{4})$, asymptotes $x = \pm 2, y = 1; 1.537$

5 PS $\cos 3t + 3\sin 3t$; CF $e^{-2t}(A\cos 3t + B\sin 3t), x = (3 + e^{-2t})\sin 3t + \cos 3t$

7 $1 + 10x^3 + 45x^6 + 120x^9, 0.204$

Miscellaneous revision questions: paper F3   (p. 505)

1 (i) $\dfrac{\pi}{2} - 1$ (ii) $e - 1 + \dfrac{\pi}{2} - 2\tan^{-1}e$ (iii) $\dfrac{3\pi}{16}$ **2** (c) $\dfrac{2\pi}{3} - \sqrt{3}(= 0.3623)$

3 Area $\dfrac{\pi}{2} - 1$ **5** $\dfrac{1}{2\sqrt{2}}\sin^{-1}\frac{1}{4}$ **6** $y^{(n)} = 0$ n even; $y^{(n)} = (n-1)^2 (n-3)^2 \ldots 2^2$, n odd.

8 $\pi\left(\dfrac{\pi}{2} - 1\right)$ **9** (i) $y = \frac{1}{2}\sin 2x$ (ii) $k = 3, y = 3e^{-2x}(1 - \cos 5x)$

10 $12a, 18\pi a^2$ **11** (i) $2x^{\frac{1}{2}} - 3x^{\frac{1}{3}} + 6x^{\frac{1}{6}} - \ln(1 + x^{\frac{1}{6}}) + C$

(ii) $\frac{1}{2}e^{-x}(-\sin x - \cos x); \dfrac{1}{2}\left(\dfrac{1 + e^{-\pi}}{1 - e^{-\pi}}\right)(1 - e^{-(n+1)\pi})$ **12** (a) $Cxe^x - x^3 - 2x^2 - 2x, C = 5$

(b) $y = \frac{1}{2}(x\sinh x + e^x - e^{-x}) = \sinh x(1 + \frac{1}{2}x)$ **13** $\dfrac{1}{k^2}(1 - \cos kt)$ **15** 4.122 04, 4.04.

Index

Acceleration   150
amplitude   165
angular velocity   179
arbitrary constant   67
area between curves   78
area function   57
area of sector   479, 491
area under curve   51, 76, 94, 248
astroid   250
asymptote   109, 208, 225, 231, 236, 240
average value   82, 274

Binomial series   307

Centre of curvature   481, 483
centroid   488, 492
chain rule   116
circle   224
circle of curvature   481
complementary function   368, 391
composite function   114
conic   222, 241
continuity   236
contour lines   496
curvature   481, 492
curve sketching   36, 103, 108, 129, 208, 226, 234, 240,
    329, 340, 344, 360
cycloid   251

Decreasing function   36
definite integral   63, 72, 259, 443, 448
dependent variable   170
derivative   21
derived function   14
differential coefficient   21
differential equations   35, 69, 139, 181, 285, 287, 315,
    335, 362, 403
differentiation
    from first principles   27
    of algebraic functions   117
    of circular functions   132
    of composite functions   115
    of exponential functions   330
    of hyperbolic functions   467
    of implicit relations   219
    of inverse circular functions   144
    of logarithmic functions   345

of parametric equations   242
of polynomials   15
of powers   19
of products   111
of quotients   113
of x^n 127
discontinuity   32, 62, 74, 122, 236
displacement   149
distance travelled   157

e   328
elasticity of demand   195
ellipse   223, 248, 480
errors   197, 500
Euler's method   418
even function   84
exponential function   328

First principles   27, 460
function   2
function of a function   114
fundamental theorem   59, 456

Gradient function   21, 23
gradient
    of a curve   6, 22, 26, 90
    of a line   4
    of inverse graph   125

Higher derivatives   34
homogeneous differential equation   381
hyperbola   225, 470
hyperbolic functions   462

Implicit relations   218
improper integral   451
increasing function   36
indefinite integral   67
independent variable   170
individual solution   368
inequalities   205, 445, 459
infinity (behaviour at)   209
integral   66
integrating factor   375
integration
    by inspection   264, 271
    by parts   265, 438

by substitution   253, 271
from first principles   460
of algebraic functions   119, 430, 440
of circular functions   132, 435
of exponential functions   332
of hyperbolic functions   467
of logarithmic functions   353
using inverse circular functions   146
using logarithmic functions   349
interpolation   289
inverse circular functions   142
inverse function   124
inverse graph   123
inverse hyperbolic functions   470
iterative process   278, 324

Lagrange interpolating formula   290
length of curve   101, 250, 477, 492
L'Hôpital's rule   318
limit   230, 318, 339, 358
limits of integration   73
line (equation of)   5
linear differential equation   366, 384
linear function rule   119
logarithmic differentiation   347
logarithmic functions   343
lower bound   52, 94

Maclaurin series   305, 336, 356, 475
maxima and minima   43, 173
maximum point   38
minimum point   38

Nested multiplication   3
Newton's method   278, 324
normal   47, 220, 245
numerical differentiation   284, 286, 320
numerical solution of differential equations   285, 287, 417

Odd function   83
Osborn's rule   463
oscillations   165

Parabola   222
parametric equations   239
partial derivatives   497
partial fractions   353, 431, 434

particular solution   368, 369
percentage change   193
percentage error   197
period of oscillation   165
perpendicular lines   46
phase angle   166
pi (π)   99, 102, 312
point of inflexion   38, 206
polar coordinates   491
polynomial   298
polynomial approximations   299, 301
powers   16
power series   297
power series for integrals   314
product rule   112
proportionate change   193

Quotient rule   113

Radius of curvature   481
rate of change   42, 170, 176
rectangular hyperbola   228
reduction formula   453
relative error   197

Second derivative   34
second order differential equations   384, 419
separable variables   372
series   460
simple harmonic motion   165
Simpson's rule   281, 323
small changes   191, 497
solid of revolution   85, 489
standard integrals   269, 428, 473
standard series   311
stationary points   38, 204, 220, 243, 498
surface   494
surface of revolution   485, 489

Tangent   45, 220, 245
Taylor series   308
trapezium rule   96, 322
turning point   38

Upper bound   53, 95

Velocity   150
volume of revolution   85, 100